PERGAMON INTERNATIONAL LIBRARY
of Science, Technology, Engineering and Social Studies

*The 1000-volume original paperback library in aid of education,
industrial training and the enjoyment of leisure*

Publisher: Robert Maxwell, M.C.

BIOLOGICAL OCEANOGRAPHIC PROCESSES

THE PERGAMON TEXTBOOK
INSPECTION COPY SERVICE

An inspection copy of any book published in the Pergamon International Library will
gladly be sent to academic staff without obligation for their consideration for course adoption or recommendation. Copies may be retained for a period of 60 days from receipt
and returned if not suitable. When a particular title is adopted or recommended for adoption
for class use and the recommendation results in a sale of 12 or more copies, the inspection
copy may be retained with our compliments. The Publishers will be pleased to receive
suggestions for revised editions and new titles to be published in this important International
Library.

OTHER TITLES OF INTEREST

BIOLOGICAL OCEANOGRAPHIC PROCESSES

SECOND EDITION

TIMOTHY R. PARSONS

and

MASAYUKI TAKAHASHI

Institute of Oceanography
University of British Columbia

and

BARRY HARGRAVE

Marine Ecology Laboratory
Bedford Institute

PERGAMON PRESS

OXFORD · NEW YORK · TORONTO · SYDNEY
PARIS · FRANKFURT

U.K.	Pergamon Press Ltd., Headington Hill Hall, Oxford OX3 0BW, England
U.S.A.	Pergamon Press Inc., Maxwell House, Fairview Park, Elmsford, New York 10523, U.S.A.
CANADA	Pergamon of Canada, Suite 104, 150 Consumers Road, Willowdale, Ontario, M2J 1P9, Canada
AUSTRALIA	Pergamon Press (Aust.) Pty. Ltd., P.O. Box 544, Potts Point, N.S.W. 2011, Australia
FRANCE	Pergamon Press SARL, 24 rue des Ecoles, 75240 Paris, Cedex 05, France
FEDERAL REPUBLIC OF GERMANY	Pergamon Press GmbH, 6242 Kronberg-Taunus, Pferdstrasse 1, Federal Republic of Germany

First edition 1973

Second edition 1977

Reprinted (with corrections) 1979

Library of Congress Cataloging in Publication Data

Parsons, Timothy Richard, 1932-
Biological oceanographic processes.

(Pergamon international library of science, technology, engineering, and social studies)
Bibliography: p.
Includes index.
1. Marine ecology. 2. Marine plankton.
3. Primary productivity (Biology) I. Takahashi, Masayuki, joint author. II. Hargrave, B., joint author. III. Title.
QH541.5.S3P37 1977 574.5'2636 76-58832
ISBN 0-08-021502-5 hard cover
ISBN 0-08-021501-7 flexicover

Printed in Great Britain by A. Wheaton & Co. Ltd., Exeter

CONTENTS

ACKNOWLEDGEMENTS

ACKNOWLEDGEMENTS for the use of copyright material are given as follows: Cambridge University Press, Hutchison Publishing Group Ltd. and the Fisheries Research Board of Canada for parts of Fig. 3; Botanical Society of America for Fig. 26; Academic Press Inc. Ltd. for Figs. 40 and 56 and Table 28; Botanical Society of Japan for Fig. 32; American Society of Limnology and Oceanography for Figs. 8, 13, 16, 19, 29, 38, 41, 47, 75, and Table 11; Conseil International pour l'Exploration de la Mer for Figs. 7, 17, 36, 39, and 77; International Commission for the Northwest Atlantic Fisheries for Fig. 56; National Academy of Sciences for Fig. 12; Fisheries Research Board of Canada for Figs. 21, 22, 53, 81, and Tables 4, 8 and 27; International Association of Geochemistry and Cosmochemistry for Fig. 33; Microforms International Marketing Corp. for Figs. 9, 10, and 14A; Oceanographic Society of Japan for Fig. 4; Hokkaido University for Figs. 45, 50 and Table 32; *Journal of Phycology* for Fig. 37; Springer-Verlag for Tables 6, 7, 15, 18, and Figs. 49 and 76; Marine Biological Association for Table 9; Scottish Marine Biological Association for Figs. 1 and 9; Oliver & Boyd, and Otto Koeltz Antiquariat for Fig. 58 and Tables 25 and 31; *Journal of Marine Research* for Table 30; *Fishery Bulletin* for Table 19. Table 14, Figs. 2 and 14B are copyright 1970, 1963 and 1969, respectively, by the American Association for the Advancement of Science. Figure 6 and parts of Fig. 3 were originally published by the University of California Press; reprinted by permission of the Regents of the University of California. Tables 1 and 5 are published with permission from George Allen & Unwin Ltd; Fig. 15 is with permission from the Koninklijke Nederlandse Akademie van Wetenschappen; Fig. 54 is with permission from the North-Holland Publishing Co. Figure 52 is with permission from Biologische Anstalt Helgoland; Figs. 67 and 71 are with permission from the Geological Society of America; Figs. 68, 74 and part of Fig. 70 are with permission from Ophelia; part of Fig. 69 is with permission from The English Universities Press. Other parts of Fig. 69 are with permission from McGraw-Hill Book Co. and the Linnaean Society; Fig. 72 is with permission from the Freshwater Biological Association; Figs. 78 and 80 are with permission from Oikos; Fig. 82 is with permission of Harvard University Press; Fig. 73 is with permission from Blackwell Scientific Publications.

INTRODUCTION

BIOLOGICAL oceanography may be defined as a study of the biology of the oceans, including the pelagic and benthic communities. Having adopted this broad definition we have found it necessary to limit our discussion of the subject in several different ways. In the earlier edition of this text we confined ourselves to the pelagic environment; by inviting Dr. Hargrave to contribute a chapter on benthic processes we now feel that one criticism of the first edition has been met. Other readers have offered several comments, in particular Dr. Karl Banse of the University of Washington has been most helpful in his criticisms of certain points in our earlier edition. However, the original purpose of the book, as an introduction to the field of quantitative biological oceanography, remains the same. The title of the book has been chosen to reflect this fact and the book should not be regarded as a literature review. We have selected reference material illustrative of certain types of biological oceanographic processes which can generally be explained either in terms of an empirical equation or through the use of definite biological or chemical descriptions. By definition these empirical relationships are advanced as being among the most acceptable at the time of writing the text but it is to be expected that researchers will improve or disprove many of the processes discussed, in the light of further scientific advancement. Such is the nature of science. As an introduction to the subject, however, we feel that students, physical oceanographers, engineers, hydrologists, fisheries experts, and scientists in a number of other professions may require some quantitative expressions of biological ocean-ographic phenomena. In some cases we have drawn on examples from the freshwater environment; in practically all examples, however, we have referred to processes in the near-surface pelagic environment and the benthic community; littoral, coral reef, bathypelagic, and a number of other special marine environments are largely excluded from discussion. The book does not cover descriptive oceanography or biological oceanographic models; rather, it is intended to bridge the gap between these two subjects which are covered by other texts referred to at the end of this section.

In the first two chapters we have attempted to describe the plankton community in terms of its composition and distribution of organisms. We felt that these descriptions were necessary for the reader to gain an introduction to the processes described in Chapters 3 to 5. Chapters 1 and 2 give some indication, therefore, of the complexity of both plankton distributions and the chemistry of plankton. Added to this complexity, however, are the results obtained by different scientists using different techniques in widely separated areas of the world. Consequently a synthesis of results reported in the literature is difficult. Instead we have tended to present data which offer the reader examples of biological oceanographic variability rather than trying to convince people of the acceptability of any one author's results in an absolute sense.

Chapter 3 deals with the primary formation of particulate material which is the beginning of the food chain in the pelagic environment. Feeding processes and the kinetics of food exchange in the pelagic food web are discussed in Chapter 4. In Chapter 5 an attempt is made to relate various processes into cycles which emphasize the interdependence of all processes in the sea. Chapter 6 is intended to serve as an introduction to biological processes in the benthic community.

In the final chapter we have given a number of examples of problems in the marine environment which we feel require the particular attention of the biological oceanographer. These have been chosen as representing areas in which there are also interdisciplinary interests between biological oceanographers and other professionals working in the marine habitat, including physical oceanographers and fisheries experts. The chapter is not intended to solve any problems but to suggest, by example, where there is a basis for the solution of such problems, using some of the information contained in the first five chapters.

Symbols used in equations and figures have presented us with a problem since aquatic biologists have tended to use the same symbol for several different entities. Thus R is commonly used to represent 'ration' and 'respiration'; P is used to represent the element phosphorus, photosynthesis and production. Where possible we have kept the common usage of the symbols and defined each one in the immediate context of its use. Where two processes are defined by the same letter in the same equation we have differentiated between the symbols used. In referring back to the literature we felt that the preservation of the popular symbols would be less confusing than introducing a large number of new symbols. Biological oceanographers have also tended to use different units for measuring the same parameters. Added to these differences is the fact that there is a lack of uniformity in the use of abbreviations for the same units. For example, light has been reported both in energy units and units of illumination, while the distance of one micron is sometimes abbreviated as $1\,\mu$ or $1\,\mu m$. It is not our purpose to endorse any uniform use of units or abbreviations but it is our purpose to clarify the literature by showing how units can be converted and where any similarity in abbreviations may exist.

For more detailed coverage of specific subjects in biological oceanography, the reader will find a number of recent reviews in multi-author texts or in journals; also several books can be particularly recommended for coverage of the literature up to the date of their publication. The latter include *The Chemistry and Fertility of Sea Waters* by H. W. Harvey (Cambridge University Press, 1957); *Measuring the Production of Marine Phytoplankton* by J. D. H. Strickland (Queens Printer, Ottawa, 1960), *Plankton and Productivity in the Oceans* by J. E. G. Raymont (Pergamon Press, Oxford, 1963) and *The Structure of Marine Ecosystems* by J. H. Steele (Harvard University Press, Cambridge, Mass., 1974).

Finally the authors would like to acknowledge the advice of Dr. G. L. Pickard, Director of the Institute of Oceanography, University of British Columbia, and other members of the Institute who read the text prior to its publication. Many thanks are also extended to Dr. R. W. Eppley, Scripps Institution of Oceanography, and Dr. G. H. Geen, Simon Fraser University, who were invited by the authors to criticize a draft edition of the text. In acknowledging the assistance of these scientists we do not hold them responsible for any errors or misconceptions which may have been included in the text. This work was in part supported by the National Research Council of Canada and by the Fisheries Research Board of Canada. Part of the new edition was prepared when one of us (T. R. P.) was a guest at the Institut für Meereskunde, Kiel. Chapter 6 on benthic communities was reviewed by Dr. Eric Mills, Dalhousie University, and we are grateful to him for his helpful advice and suggestions.

CHAPTER 1

DISTRIBUTIONS OF PLANKTON AND NUTRIENTS

1.1 TAXONOMIC, ENVIRONMENTAL
AND SIZE SPECIFIC GROUPS OF PLANKTON

Organisms which are unable to maintain their distribution against the movement of water masses are referred to as 'plankton'. Included in this group are bacterioplankton (bacteria), phytoplankton (plants) and zooplankton (animals). Generally all plankton are very small and in many cases, microscopic. However, relatively large animals, such as the jellyfish, are also included in the definition of plankton. Some plankters, including both plants and animals, are motile but their motility is weak in comparison with the prevailing movement of the water. Animals, such as fishes, which can maintain their position and move against local currents are known as 'nekton'. However, the division between plankton and nekton is not precise and some small fish, especially fish larvae, may be a part of the plankton community.

The biomass or weight of plankton or nekton per unit volume or area of water is referred to as the 'standing stock'; typical units used for standing-stock measurements are $\mu g/l$, mg/m^3, g/m^2, $kg/hectare$, etc., where the weight should be specified as referring to wet weight, dry weight, or carbon. The productivity of organisms is defined in terms of 'primary productivity', 'secondary productivity', and 'tertiary productivity'; units are the same as in standing-stock measurements when expressed per unit time (e.g. per hour, day, or year). Ideally, primary productivity represents the autotrophic fixation of carbon dioxide by photosynthesis; secondary productivity represents the production of herbivorous animals and tertiary productivity represents the production of carnivorous animals feeding off the herbivore population. However, these definitions are not precise since some plants may utilize growth factors, such as vitamins (auxotrophic growth), and others are capable of taking up organic substrates as a source of energy (heterotrophic growth). Thus the particulate material grazed by secondary producers may be derived from a variety of processes and include phytoplankton and bacterioplankton. Similarly many filter-feeding zooplankton which might be nominally classed as herbivores may at times feed upon other small animals, such as Protozoa; thus the boundaries between components in the aquatic biosphere are difficult to define with the same precision as is used in chemistry of physics. Biological associations are better considered *in toto* as an ecosystem in which various components react with each other to a greater or lesser degree. Components of an ecosystem can be defined in

1

terms of their taxonomy or chemistry; interactions between components can then be expressed quantitatively by empirical equations. Thus a phytoplankton standing stock may be described as consisting of 10^6 cells per litre of a species, *Skeletonema costatum*, or as being represented by a chlorophyll *a* concentration of $1\ mg/m^3$, or in a trophic sense as a ration for zooplankton, such as is represented in eqn. (90), Section 4.2.1. Attempts to synthesize the components of biological production into simulated models of various ecosystems are currently being made, but the results of this work are at present largely experimental.

The bacterioplankton of the oceans have been discussed in some detail by a number of authors including ZoBell (1946), Kriss (1963) and Wood (1965). According to Wood (1965) the principal genera of bacteria represented in the oceans are the *Micrococcus*, *Sarcina*, *Vibrio*, *Bacillus*, *Bacterium*, *Pseudomonas*, *Corynebacterium*, *Spirillum*, *Mycoplana*, *Nocardia* and *Streptomyces*. Among these genera are various morphological differences including coccoid, rod and spiral forms. A large number of the most common bacteria are motile and gram negative. The bacterioplankton do not usually contribute significantly to the total biomass of particulate organic matter but in association with detritus (see Section 2.4) they may form an appreciable organic reserve during times of low phytoplankton density. Their role in the oceans is more important in recycling elements and organic material back into the food chain (see Fig. 41).

MacLeod (1965) has discussed the specific identity of marine bacteria as compared with bacteria from a terrestrial origin. His findings showed that marine bacteria have special requirements for inorganic ions which include a highly specific need for Na^+ and a partial need for halide ions which could be satisfied by either bromine or chlorine ions. Mg^{2+} and Ca^{2+} were also required, usually at concentrations higher than are normally needed for terrestrial bacteria.

Marine bacteria living at the sea surface are usually recognized as a specific community (e.g. Sieburth, 1971; Tsyban, 1971). In samples from the eastern Pacific Ocean (between 10 and 30°N), Sieburth (1971) found the bacterial flora at the surface (bacterioneuston) were predominantly atypical pseudomonads having marked lipolytic and proteolytic activity. Tsyban (1971) describes the bacterioneuston from the Black Sea and northeast Pacific as having specific biochemical properties and consisting of brightly pigmented strains. The number of organisms at the sea surface increases during periods of wind and wave action and this has given rise to the question of whether bacterioneuston are accumulated largely through physical forces or through propagation (Marumo *et al.*, 1971).

Phytoplankton taxonomy has undergone frequent revision but for the purposes of our discussions, the "Check-list of British Marine Algae—2nd Revision" by Mary Parke and Peter Dixon (1968) has been followed in reporting on various taxonomic groups and species. The major classes of algae which contain representatives of the phytoplankton are as follows:

Taxonomic class	*Common name*
Cyanophyceae	(Blue–green algae)
Rhodophyceae	(Red algae)
Cryptophyceae	—
Dinophyceae	(Dinoflagellates)
Haptophyceae	(including the coccolithophorids)
Chrysophyceae	(Yellow–brown algae)

Xanthophyceae	(Yellow algae)
Prasinophyceae	—
Chlorophyceae	(Green algae)
Bacillariophyceae	(Diatoms)

Among these classes, the Bacillariophyceae, Dinophyceae, and Haptophyceae are generally considered to be the most important in the sea; under some conditions the Cryptophyceae, Chlorophyceae, Cyanophyceae, and Xanthophyceae may be very abundant. The principal exception to the classification given above is that the Dinophyceae are sometimes included in the animal kingdom among the phylum, Protozoa.

The zooplankton include practically every major taxonomic group of animals, either for their entire life cycle or, for short periods, as in the case of the larval stages of some fish and molluscs. The major phylla represented in the following list have been taken in part from a much more detailed summary provided by Newell and Newell (1963).

Animal phylum	*Some representative among the zooplankton*
Protozoa	Oligotrich and Tintinnid ciliates, Radiolaria and Foraminifera
Coelenterata	Hydrozoa, Scyphozoa (Jellyfish)
Ctenophora	(Ctenophores)
Chaetognatha	(Arrow worms)
Annelida	(Polychaete worms)
Arthropoda (class) Crustacea	(Copepods, cladocerans, mysids, euphausiids, ostracods, cumaceans, amphipods, isopods)
(Sub-Phylum) Urochordata	(Salps and appendicularians)
Mollusca	(Heteropods)

In addition to the above, there are many large invertebrates having larval planktonic stages (e.g. polychaetes, crustaceans, gastropods, lamellibranchs, and echinoderms). Among vertebrates, fish eggs and larvae both occur as members of the plankton. Important commercial species which have a planktonic state include the herring, anchovy, tuna, and bottom feeders, such as cod and plaice. From among the types of zooplankton listed above, by far the most important group are the Crustacea, and of these, the copepods are the most predominant.

Biogeographical distributions of plankton have been based on the very early recognition that specific environmental factors, such as light, temperature, salinity, and nutrient requirements, to some extent determined the occurrence and succession of species. Smayda (1958 and 1963) has reviewed a number of the terms used to describe plankton from similar environments. Plankton with a tolerance to a wide range of temperatures is described as 'eurythermal', while a narrow range of temperature tolerance is described as 'stenothermal'; similarly salinity (euryhaline and stenohaline), pH (euryionic and stenoionic), and light (euryphotic and stenophotic). A further classification of light response has been used to obtain a vertical separation of plankton communities into those inhabiting the 'euphotic' zone (where the net rate of photosynthesis is positive) as opposed to the 'aphotic' zone. Unfortunately, these words usually lack precise quantitative description but some knowledge of their meaning may be useful in reading other publications.

The terms 'oceanic' and 'neritic' have been used quite extensively in describing plankton associated with the oceans and with coastal waters, respectively. The classification may be particularly useful in reporting taxonomic data collected from commercial vessels, such as with a Hardy recorder as illustrated in Fig. 1. From this figure it is easy to see that certain 'indicator species' belong to each region. Over large areas of ocean there may be several oceanic groups. Bary (1959 and 1963) defined such plankton distri-

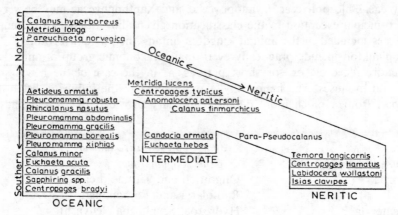

FIG. 1. Distribution series of copepods in the North Sea and the north-eastern Atlantic arranged in such a way that the distribution of each organism is most similar to those of the neighbouring organisms in the list (redrawn from Colebrook *et al.*, 1961).

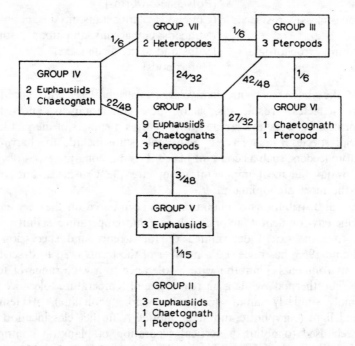

FIG. 2. Composition of zooplankton groups in the north Pacific. Fractions are the ratios of the number of observed species–pair connections between groups to the maximum number of possible connections; for example, there are six possible intergroup species pairs between group IV and VII but only one of these showed affinity at the 'significance' level used (redrawn from Fager and McGowan, 1963).

butions in terms of their temperature and salinity tolerances (called T–S–P diagrams). He emphasized that the importance of such diagrams was in showing the distribution of plankton in certain water bodies rather than the exact geographical location of the samples. Fager and McGowan (1963) used an elaborate mathematical technique to show probable affinities between zooplankton species in the north Pacific. They classified the zooplankton into six groups and showed interrelationships between groups as illustrated in Fig. 2. Perhaps the most important aspect of all such groupings is that the occurrence of species in another group (or changes in the between group associations) may indicate changes in the ocean and coastal environments, such as might occur from a shift in the direction of a current. However, Fager and McGowan (1963) found that many of the usually measured properties of water (e.g. temperature, thermocline depth, etc) were not closely correlated with differences in zooplankton abundance. From this it was concluded that the organisms reacted to a more complex interaction of known properties or to some environmental factors yet to be elucidated.

Phytoplankton species which produce resting spores or have a sedentary phase are known as 'meroplankton' as opposed to 'holoplankton' (Smayda, 1958). However, these terms were originally used to refer to zooplankton and in this sense 'meroplankton' refers to organisms which are only temporarily members of the plankton community (e.g. some bivalve larvae) while 'holoplankton' refers to a permanent member of the plankton community (e.g. most copepods). A general classification of plankton abundance based on availability of nutrients is used in describing waters as 'eutrophic', 'mesotrophic', and 'oligotrophic', in decreasing order of plankton abundance (see Hutchinson, 1969, for a further discussion of these terms). Plankton may be grouped by the depth zone in which they are found in the 'pelagic' or open-sea environment. These zones have received a number of different classifications but the simplest approximate

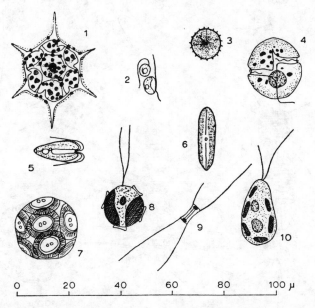

Fig. 3A. Examples of nanoplankton: flagellates [*Distephanus* (1), *Thalassomonas* (2), *Gymnodinium* (4), *Tetraselmis* (5), *Coccolithus* (7), *Pontosphaera* (8), *Cryptochrysis* (10)], diatoms [centrate (3), pennate (6), *Chaetoceros* (9)]. (Redrawn from Wailes, 1939, Cupp, 1943, Fritsch, 1956, and Newell and Newell, 1963.)

definitions appear to be 'epipelagic' (0 to 150 m), 'mesopelagic' (150 to 1000 m), 'bathypelagic' (1000 to 4000 m), and 'abyssopelagic' (4000 to 6000 m) (see Hedgpeth, 1957, for further definitions). Plankton (or other particulate matter) produced within a designated ecosystem is referred to as 'autochthonous' while 'allochthonous' material is imported into the ecosystem. A large number of other groupings have been employed by systematists in describing plankton communities. In some cases these terms lack universal usage because of the specificity of their original definition; others have acquired common scientific usage while lacking a precise definition (e.g. see Smayda, 1958). Some attempt has been made to adopt universal definitions of certain terms, and the Report of the Committee on Terms and Equivalents (1958) has been followed where possible in this text.

From the point of view of food chain studies one of the most useful groupings for plankton and larger organisms is to consider all particulate material on a single size

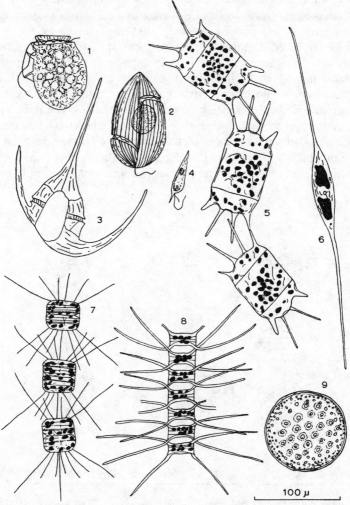

100 µ

FIG. 3B. Examples of microphytoplankton: dinoflagellates [*Dinophysis* (1), *Gyrodinium* (2), *Ceratium* (3), *Prorocentrum* (4)], diatoms [*Biddulphia* (5), *Nitzschia* (6), *Thalassiosira* (7), *Chaetoceros* (8), *Coscinodiscus* (9)]. (Redrawn from Wailes, 1939, and Cupp, 1943).

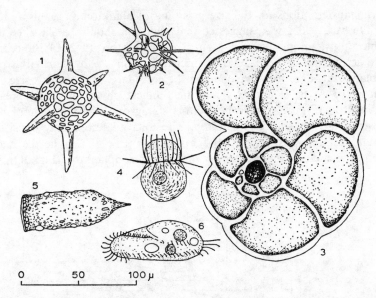

FIG. 3C. Examples of microzooplankton: radiolarians [*Hexastylus* (1), *Plectacantha* (2)], foramini-feran [*Pulvinulina* (3)], ciliates [*Mesodinium* (4), *Tintinnopsis* (5), *Amphisia* (6)]. (Redrawn from Wailes, 1937 and 1943, and Cushman, 1931).

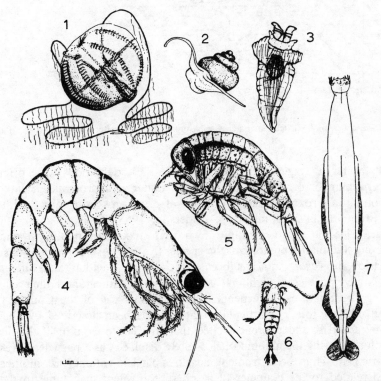

FIG. 3D. Examples of macro- and mega-zooplankton: ctenophore [*Pleurobrachia* (1)], mollusc ptero-pods [*Limacina* (2), *Clione* (3)], euphausiid [*Thysanoessa* (4)], amphipod [*Parathemisto* (5)], copepod [*Calanus* (6)], chaetognath [*Sagitta* (7)]. (Redrawn from LeBrasseur and Fulton, 1967).

scale. The original definitions for the size grouping of planktonic organisms have been discussed by Dussart (1965), who suggested that plankton should be classified according to the scheme, 'ultrananoplankton' ($< 2\,\mu$); 'nanoplankton' (2–20 μ); 'microplankton' (20–200 μ); 'macroplankton' (200–2000 μ); 'megaplankton' ($> 2000\,\mu$)*. Illustrations of the types of planktonic organisms which might be considered in Dussart's size categories are shown in Fig. 3 A, B, C, and D. Thus the smallest phytoplankton represent most of the nanoplankton (Fig. 3A) while below this size group, the ultrananoplankton are chiefly represented by the bacterioplankton. The microplankton may consist both of large phytoplankton (Fig. 3B) and small zooplankton (Fig. 3C), while the macro- and megaplankton are generally represented by the largest zooplankton (Fig. 3D), including

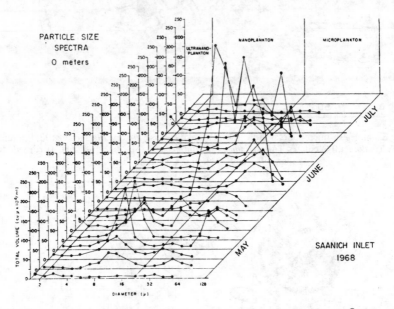

FIG. 4A. Size spectra of particulate material measured with a Coulter Counter® (from Parsons, 1969).

jellyfish and the larval stages of some fish. While these definitions were originally intended for grouping of organisms as measured under a microscope, it later became possible to obtain approximate size groupings of organisms using filters with different pore sizes. From the latter studies (e.g. Anderson, 1965; Teixeira *et al.*, 1967; Saijo and Takesue, 1965) it was generally shown that maximum photosynthetic activity occurred among the smaller phytoplankton species (*ca.* 5–50 μ). The method of separation lacked precision, however, and has been criticized by Sheldon and Sutcliffe (1969) because the stated pore sizes of filters are not a good indicator of their effectiveness for separating size fractions of suspended particles. The use of the Coulter Counter® for analysing size fractions of particulate material has been described by Sheldon and Parsons (1967 a and b, and references cited therein). The operation of this instrument depends upon measuring the amount of electrolyte displaced as a particle moves through a small electric field. Thus the instrument measures particle volume; size in linear dimensions is then related to the diameter of a sphere equivalent in volume to the original

*1 μ, or micrometer (μm) $\equiv 10^{-3}\,$mm $\equiv 10^{-6}\,$m.

particle. Size spectra of particulate material measured with a Coulter Counter® are shown in Fig. 4A. The figure illustrates differences in the size of particles (1 to 100 μ) in the surface waters of a fiord on the coast of British Columbia. Figure 4B shows an idealized representation of data that can be obtained by expanding the scale to include zooplankton. Unfortunately, at present there is no instrument of similar accuracy to a Coulter Counter® for measuring large particles. The size scale used in this spectrum is based on measuring the number of particles in size groups which progress by doubling

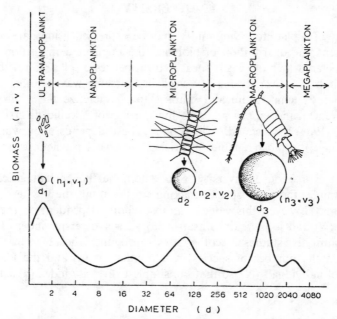

Fig. 4B. Particle spectrum representing biomass ($n \times v$) of material in different size categories determined by the diameter (d) of a sphere equivalent in volume (v) to the original particle, times the number of particles (n).

the volume in each particle size category. The equivalent diameters obtained from these volumes give a grade scale (based on $2^{1/3}$) in which the size categories suggested by Dussart (1965) appear at equal intervals along the abscissa (e.g. Fig. 4B). The total biomass of material measured with the Coulter Counter® is statistically related to such parameters as the weight of particles, chlorophyll a concentration and particulate carbon (e.g. Zeitzschel, 1970). The advantages of using a continuous size spectrum for particle distributions in the sea are (i) the size group of plankters contributing most to the total standing stock of plankton can be readily identified as peaks in the spectrum, (ii) biomass diversity indices (Wilhm, 1968) can be calculated from the spectrum, and (iii) the growth increment, or grazing loss, of different size categories can be determined independently of the total biomass of phytoplankton, zooplankton, and other particles (Parsons, 1969; Parsons and LeBrasseur, 1970). However, particle size spectra *per se* do not relate to taxonomic groups and microscopic identification of the principal components in a plankton crop is recommended when using this technique. Also the results include all detrital particulate material, and special methods are sometimes necessary to differentiate between detritus and growing cellular material (Cushing and Nicholson, 1966).

Attempts to measure zooplankton biomass in different size categories have been suggested by Chislenko (1968) who has produced nomographs giving the different weights from the length measurements of a large number of different zooplankton shapes. More recently an automated apparatus for measuring the length of zooplankton has been described by Fulton (1972).

1.2 DIVERSITY

The diversity of a plankton community may be expressed using data on the number of species present, the distribution of biomass, the pigment composition, or a number of other parameters which are easily measured properties of plankton. From a heuristic approach, an index of diversity may be used in the same way as other environmental parameters, such as temperature and salinity, to characterize the environment. There are both theoretical and empirical bases for the use of specific diversity indices, but as Lloyd *et al.* (1968) have stated, "which one is 'best' depends upon which one proves in practice to give the most reliable, surprising ecological predictions and the greatest insight".

The simplest expression of diversity is to determine the percentage composition of species in a sample; the more species making up the total, the greater is the diversity of the organisms. However, this value is almost wholly dependent on the total number of individuals (N) and is therefore unsatisfactory as a diversity index. From some of the earliest quantitative studies in ecology it was recognized, however, that a relationship existed between the number of species in a population (S) and the logarithm of the total number of individuals (N), so that the simplest diversity index (d) can be expressed as:

$$d = \frac{S}{\log_{10} N}.$$

(1)

This value will be very small under conditions of a plankton bloom and generally high in tropical plankton communities. A better expression which reduces to 0 when all the individuals are from the same population was given by Margalef (1951):

$$d = \frac{S - 1}{\ln N}.$$

(2)

Margalef (1957) introduced the idea that the 'information content' could be used as a measure of diversity in a plankton sample. Thus the diversity of a collection containing a total of N individuals and $n_1, n_2, \ldots n_i$ individuals of each species can be written as:

$$\frac{N!}{n_1! n_2! \ldots n_i!},$$

(3)

and information (H in 'bits')* per individual as:

$$H = \frac{1}{N} \log_2 \frac{N!}{n_1! n_2! \ldots n_i!}.$$

(4)

---

* Information is commonly expressed in 'bits' when using $\log_2$, or 'nats' when using $\log_e$. The following conversion may be useful: $\log_{10} = \log_e \times 2\cdot303$ and $\log_2 = \log_e \times 1\cdot443$.

The information content as expressed above, eqn. (4), can be interpreted as the degree of uncertainty involved in predicting the species identity of a randomly selected individual. If N is large and none of the n_i fraction are too small, information content per individual (H' in bits) can be approximated from the expression:

$$H' = -\Sigma p_i \log_2 p_i, \tag{5}$$

where $p_i = n_i/N$ and is the proportion of the collection belonging to the ith species (Shannon and Weaver, 1963). Under some circumstances eqn. (5) will be more easily determined than eqn. (4) but Lloyd et al. (1968) have provided examples and tables for the solution of both equations for values of N from 1 to 1050. Margalef (1961) made a statistical comparison between the diversity of plankton samples calculated from the diversity index [eqn. (2)] and the theoretical diversity [eqn. (4)]. There was a highly significant correlation between the two although there was a difference in the regression line depending on whether diversity was determined for a diatom or dinoflagellate population. Using eqn. (5) Lloyd and Ghelardi (1964) related diversity to the maximum possible value for a given number of species if they were all equally abundant. This term was called the 'equitability' (ε) and was expressed as the ratio of H' to a theoretical maximum (M) for the same number of species, where $n_1 = n_2 = \dots n_i$. A table of M values for 1 to 1000 species is given by the authors. The value of ε in describing a collection may be more useful if units of biomass rather than number are employed to determine diversity (e.g. Wilhm, 1968).

'Equitability', as defined by Lloyd and Ghelardi (1964) is the opposite of 'dominance' which expresses the most abundant species in a population. Hulburt et al. (1960) expressed the dominance of a plankton community as the ratio of the concentration of the most abundant species to the total cell concentration. Also if the presence of one species in a population is nearly always accompanied by another species, the amount of information gained is small. This can be expressed as the 'redundancy' (R) in terms of the equation (Patten, 1962a):

$$R = \frac{H_{max} - H}{H_{max} - H_{min}}, \tag{6}$$

where H_{max} is the diversity when the species are equally distributed and H_{min} is the diversity when all the individuals belong to one species. The value R, varies between 0 and 1, and is also partially an index of 'dominance'.

The general use of diversity of the type shown in eqns. (4) and (5) above has been discussed by Pielou (1966) who points out that the diversity of a sample should not be regarded as the diversity of a larger population from which it was obtained; the sample itself may, however, be treated as a population and defined. Patten (1959) discusses the absolute diversity of an aquatic community in terms of its information content. As an approximation he calculated that in a Florida lake, the community could be described in terms of 3×10^{24} bits/cm^2/year. If the average information content of a printed page is 10^4 bits, it is apparent that the amount of information required annually to describe the Florida lake community is many orders of magnitude larger than the information contained in the largest libraries! Thus diversity, as discussed in this section, is a property of the entity from which the data are collected and not of the whole environment.

The species diversity indices discussed above [e.g. eqn. (2)] are dependent to some extent on the size of the sample, especially for small numbers ($N < 100$). In a method

TABLE 1. THE RAREFACTION METHOD OF CALCULATING DIVERSITY—A HYPOTHE-
TICAL PLANKTON SAMPLE CONSISTING OF 100 INDIVIDUALS AND 10 SPECIES
(A LARGER EXAMPLE IS GIVEN BY SANDERS, 1968)

Rank of species by abundance	Number of individuals	% of sample	Cumulative sample %
1	45	45	45
2	31	31	76
3	8	8	84
4	5	5	89
5	4	4	93
6	3	3	96
7	1 each	1 each	100
Total	100	100	

used by Sanders (1968), however, samples of benthic fauna from the same environment, ranging in size from 35 to 2514 individuals, showed no tendency for smaller samples to be less diverse. The technique is described as the 'rarefaction' method and it depends on determining the shape of the species abundance curve rather than obtaining an expression for the absolute number of species per sample. The method is demonstrated as follows using a hypothetical plankton sample shown in Table 1. The problem is to generate a species abundance curve for different sample sizes (Fig. 5). If a sample size of twenty individuals is selected, this will represent 100% of the individuals and each individual will represent 5% of the sample. In the original sample, four species represented 5% or more of the total or 89% of the sample (Table 1). Therefore it is assumed that each of these species will be in the reduced sample which leaves 11% of the sample for the remaining six species; since none of them form more than 5% of the total, none is likely to be represented by more than one individual. Thus the total number of species present in a sample of 20 is $4 + 11/5 = 6\cdot2$. Similarly the number of species in a sample of ten individuals is 5 and so on; the shape of the species abundance curve generated by Table 1 is shown in Fig. 5. Also, two curves are shown with lower diversity (bottom) and higher diversity (top). The author emphasizes that in order to use this method, samples must be taken by the same sampling technique, from similar environments (i.e. it would be inappropriate to compare the diversity of attached algae and phytoplankton) and curves may not be extrapolated. The curves (Fig. 5) cannot have confidence limits applied to them but curves from samples with similar diversities emerge as a family of curves which strongly indicate that the diversity differences are real.

Changes in the diversity index of samples from a plankton community are shown in Fig. 6. From these data Margalef (1958) recognized three stages of succession. Stage 1 was typical of turbulent waters in which a few species survived and in which there was an occasional bloom of diatoms; stage 3 was characteristic of highly stratified waters in which there was a mature phytoplankton crop and a high diversity following nutrient depletion. Stage 2 was characteristic of inflowing waters which may have transported allochthonous species into the area of study, thus increasing the diversity of organisms present in any sample. Hulburt et al. (1960) studied changes in species diversity in the Sargasso Sea and recognized a succession of three species groups. These consisted of a sparse population with a normal distribution of abundant and rare species, a winter period in which a single species was dominant over all other species, and

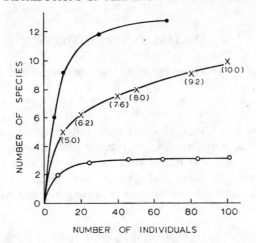

FIG. 5. Plot of number of species at different population levels using the rarefaction diversity method for different plankton samples. (× — × curve generated from Table 1; ○—○ low diversity sample; ●—● high diversity sample).

a period of thermal stratification in which dominance was shared by several species. The most extensive field tests of various diversity indices have been carried out by Travers (1971) in the Mediterranean. From this study the author concluded that the degree of maturation and of organization of an ecosystem can be appreciated by means of several diversity indices. From the use of different indices he concluded that a diversity index based on plankton pigments (Margalef, 1965) was a poor method, especially where plankton levels were low; diversity indices based on information theory were considered the best measure of structure although less laborious calculations of diversity can often be used [e.g. eqn. (2)]. Heip and Engels (1974) have recently compared diversity indices from the point of view of the statistical significance of observed differences or similarities. They conclude by recommending use of the Shannon–Weaver function [eqn. (5)] together with a new index of 'evenness' or 'equitability'.

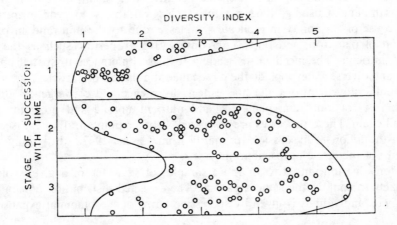

FIG. 6. Stage of succession dependence of diversity index [eqn. (2)] in a number of samples from the surface waters at the bay of Vigo (redrawn from Margalef, 1958).

1.3 SPATIAL DISTRIBUTIONS

1.3.1 STATISTICAL CONSIDERATIONS

In carrying out a series of replicate analyses on a single, well-mixed sample of sea water, small differences in the values obtained may be attributed to analytical technique. Such errors are caused by a lack of instrument reliability, sub-sampling, and slight variations in the way an individual analyst repeats each analysis. It may be assumed that these errors are randomly distributed and that if a large number of replicate analyses are made on one sample, the mean and standard deviation (s) of the analysis can be determined. The precision of the method can then be expressed with a 95% confidence limit for n determinations, as

$$m \pm 2s/\sqrt{n}, \tag{7}$$

where m is the mean of the replicate samples. The principal exception to the use of these statistics for analytical techniques is in the counting of plankton from a settled volume of sea water. In this case the distribution of plankton cells may not be random and special methods may have to be used in order to determine the degree of contagion (Holmes and Widrig, 1956).

The collection of samples of seawater, or plankton from the ocean, introduces much larger differences between replicate samples than can be ascribed to analytical errors alone. Thus Cushing (1962) has summarized a number of reports and showed that in calm weather, the % variability (expressed as the coefficient of variation $s/m \times 100$) of the number of three species of plankton in individual hauls varied from 15 to 70%; under conditions of rough weather this variability was increased up to 300%. Cassie (1963) estimated that the coefficient of variation for large samples is most often in the range of 22 to 44%, with obvious exceptions being made for rough weather, or highly stratified environments. Wiebe and Holland (1968) have summarized data on the 95% confidence limits for *single* observations of zooplankton abundance; the range for most data was between *ca.* 40 and 250%.

While the use of sampling gear itself may contribute a small amount of variability to ocean sampling (Cushing, 1962, assigns *ca.* 5% variability to gear operation), the principal cause of variability in replicate samples is due to the non-random or patchy distribution of plankton, and other non-conservative properties, such as the concentration of nutrients. The mechanisms leading to these differences in spatial abundance are many and diverse. They include the physical accumulation of particles by the vertical and horizontal movement of water masses (e.g. divergence and convergence), differences in growth rates of individual plankters, and nutrient uptake and predation patterns of the food chain. These processes are discussed at other points in the text (see Section 1.3.5) and the following discussion (primarily from Cassie, 1962a) deals only with the extent and not the cause of distributions.

In random distributions, two or more samples of seawater of a given volume are equally likely to contain the same organism. The expected distribution of samples with n_1, n_2, n_3, etc., individuals is given by successive terms of the binomial expansion

$$(q + p)^k,$$

where k is the maximum number of individuals a sample could contain, p is the probability of an organism's occurrence and $q = 1 - p$. The population mean (μ) and variance (σ^2) of a binomial distribution are

$$\mu = kp \text{ and } \sigma^2 = kpq$$

from which

$$\sigma^2 = \mu - \mu^2/k. \tag{8}$$

Since for plankton in a seawater sample $k \to \infty$, the variance become equal to the mean,

$$\sigma^2 = \mu. \tag{9}$$

The above relationship expresses a special case of the binomial distribution, in which the probability of an organism occurring ($p = \mu/k$) is small; this is known as the Poisson distribution and the experimental value of the variance (s^2) and mean (m) for replicate plankton collections can be expressed as s^2/m and used to determine if the plankton are randomly or 'over-dispersed'. Theoretically if the value σ^2/μ is greater than 1, the population will be over-dispersed. However, in practice if s^2/m is calculated for a series of samples their distribution can be represented as $\chi^2/N-1$; thus with twenty samples (19 degrees of freedom), s^2/m should be less than $30{\cdot}14/19 = 1{\cdot}6$ for a 95% probability that the organisms are distributed randomly (Holmes and Widrig, 1956). However, in most cases involving the collection of plankton over an area, the value s^2/m will be significantly greater than 1; thus the ratio can be used as a dispersion coefficient (Ricker, 1937) which along with other parameters (e.g. diversity indices) may be useful in characterizing a body of water.

When populations are over-dispersed the presence of one organism in a sample increases the probability of additional organisms of the same species occurring in the same sample. This is the opposite of a binomial distribution and it can be expressed theoretically as the negative binomial distribution which is given by an expansion of the expression

$$(q - p)^{-k}$$

where $q = 1 + p$. The variance (σ^2) is

$$\sigma^2 = \mu + \mu^2/k \tag{10}$$

As Cassie (1962b) has pointed out, since p and k are negative, they cannot have the same meaning as they did in the binomial distribution. In particular, however, k appears as a useful parameter for expressing the degree of patchiness, or contagion, in a population. Cassie (1962a) has given an estimate of $1/k$ as $\hat{c}$, where

$$\hat{c} = \frac{s^2 - m}{m^2}, \tag{11}$$

s^2 and m being the sample variance and mean, respectively. The expression [eqn. (11)] was used by Cassie (1959) as a coefficient of dispersion and he concluded that $\hat{c}$ was better than s^2/m, since $\hat{c}$ was not strongly correlated with the mean. This allows for a comparison of dispersion to be made between samples with different means.

In practice it may be found easier to establish an empirical relationship than to fit raw data to a theoretical distribution. Barnes (1952) and Cassie (1962a) have discussed transformations which may be suitable for marine biological data. The most convenient transformation is to convert the raw data to logarithms and the transformed frequencies may then have a log-normal distribution; an example of transformed plankton data is shown in Fig. 7. The mean (m') of the transformed data is the geometric mean and after taking antilogarithms, the value m' will be less than the arithmetic mean, m. Variability about the mean can be expressed as the 'logarithmic coefficient of variation' (Winsor and Clarke, 1940);

$$V' \text{ per cent} = 100(10^{s'} - 1), \tag{12}$$

where s' is the standard deviation calculated from the logarithms of the raw data and V' is the logarithmic coefficient of variation. The advantage of using V' has been illustrated by Cassie (1968) who considered two samples with means of 100 and coefficients

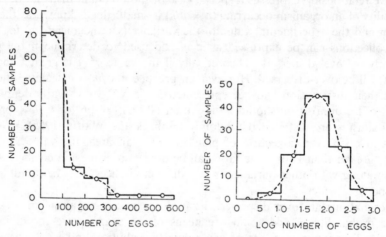

FIG. 7. Frequency distribution of pilchard egg counts in arithmetic and logarithmic forms (redrawn from Barnes, 1952).

of variation V (normal distribution) and V' (log-normal distribution) both of 110%. The mean with one standard deviation for the normal distribution was,

$$100 \pm (1 \cdot 10 \times 100) = -10 \text{ to } 210$$

and for log-normal

$$100 \overset{\times}{\div} 2 \cdot 1 = 48 \text{ to } 210.$$

The negative lower limit in the arithmetic range above is meaningless in the context of a normal distribution.

In a detailed study of spatial heterogeneity, Platt *et al.* (1970) separated the sources of variation in a single analysis as follows:

$$\sigma_T^2 = \sigma_0^2 + \sigma_1^2 + \sigma_2^2, \tag{13}$$

where σ_T^2 was the total variance, σ_2^2 was the variance due to real differences between stations, σ_1^2 was the variance between repeated samples taken at the same station and σ_0^2 included sub-sampling and other analytical errors. The authors found that σ_0^2 and σ_1^2 were about the same size and accounted for *ca.* 10% of the variance each. Real differences between stations were generally much larger, both in time and space. Thus the log-coefficient of variation for a single chlorophyll *a* observation in a near-shore community increased rapidly from 14% at 0·625 sq. mile, to 70% at 1 sq. mile, thereafter remained relatively constant out to 4 sq. miles. During a study of temporal variations over a period of 5 weeks, the log-coefficient of variation varied from 11 to 111% (mean 42%) at nine stations covering 13 sq. miles in the same location. During this period phosphate varied from 10 to 60%. Rapid temporal variations in V' sometimes occurred over a few days and ranged from 21 to 45% in the case of chlorophyll, and from 32 to 64% for phosphate. Since this study was conducted under relatively ideal weather conditions, the values quoted may be considered to be representative of maximum σ_2^2 values, but minimal σ_0^2 and σ_1^2 values. Thus Cassie (1962b) showed that patchiness was greatest in calm seas and least at times of turbulent mixing, such as during storms; conversely the operation of gear over the side at a single point, as well as laboratory analyses on board vessels, becomes much more difficult during rough weather and consequently the terms σ_0^2 and σ_1^2 may be expected to increase under such conditions.

Most of the discussion above pertains to overall descriptions of the patchiness of plankton communities in terms of plankton numbers, chlorophyll a or nutrients. Quite a different distributional problem arises when a description is required of the total number of species that might define the fauna of a particular area. The question is then a matter of how large a sample must be filtered in order to include all the species. In experiments conducted while following a current drogue, McGowan (1971) showed that for three groups of plankters (fish larvae, molluscs, and euphausiids) the amount of water to be filtered varied with the group. In the case of euphausiids there was no increase in the number of species with the volume of water filtered over the range 128 to 1510 m³, but for fish larvae and molluscs the amount of water to be filtered in order to include all species was in excess of 10,000 m³. However, the relationship between the number of species and the volume filtered was logarithmic; for example, ca. 80% of the fish larvae species were found in ca. 1000 m³ of water. The amount of water filtered by a plankton net can be increased either by increasing the size of the net or the length of the tow. Wiebe (1971) studied the precision of replicate tows with different nets and towing distances and his conclusion was that the length of the towing distance was considerably more important in determining precision than the size of the net.

1.3.2 AREAL DISTRIBUTIONS

A large amount of data has been collected on the areal distribution of plankton and nutrients. These data are generally the result of observations carried out at points along the cruise track of a research vessel; consequently while the data are useful for surveys of very large areas (e.g. seas, oceans, and the hydrosphere) they are of very little use in trophic studies. The latter subject is discussed later in the text, but for the present it must be apparent that plankton distributions mapped from samples collected miles apart are probably not representative of the food supply for a larval or juvenile fish, which may travel less than 100 m in a day. Thus small-scale plankton distributions are particularly important in assessing the food supply and hence, in part, the survival of very young fish.

It has been recognized from the time of the earliest explorers of the hydrosphere that plankton may sometimes occur in dense swarms or blooms (see Bainbridge, 1957, for historical references). Scientific observations (e.g. Barnes, 1949; Barnes and Marshall,

1951) showed that in general planktonic organisms were more often clumped or aggregated than randomly distributed. For example, Cassie (1959) showed in a study on the occurrence of plankton over a distance of 1 m that the distribution of the diatom (*Coscinodiscus gigas*) was non-random. The problem of collecting detailed samples over appreciable distances was solved in 1936 for larger plankters with the invention of apparatus which could be towed behind ships and continuously collect plankton on

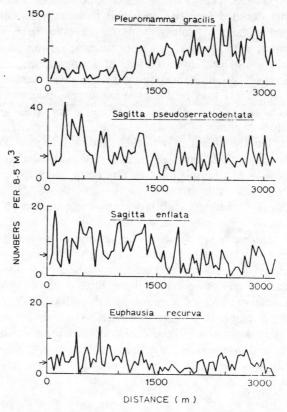

FIG. 8. Plots of abundance versus distance illustrating the presence of large-scale patchiness on which is superimposed smaller-scale patchiness (redrawn from Wiebe, 1970).

a slowly moving fine-mesh belt (Hardy, 1936). An adaptation of this apparatus (Longhurst *et al.*, 1966) for studying micro-distributions of plankton has been particularly important for trophodynamic studies. Using this apparatus Wiebe (1970) was able to show areal patchiness in the distribution of zooplankton species over distances of less than 20 m. Some of the results obtained by Wiebe are shown in Fig. 8. From these data it is apparent that there is both a small-scale and a larger-scale patchiness in the distribution of the species reported. Due to the mechanical ability of the apparatus, the minimum distance over which the plankton patches could be detected with Longhurst–Hardy recorder was *ca.* 14 m.

The most extensive descriptions of large-scale plankton species distributions are contained in reports from the Oceanographic Laboratory, Edinburgh (published in Bulletins of Marine Ecology). Zooplankton data are obtained from samples collected with Hardy plankton recorders towed behind commercial vessels (Hardy, 1936); an example of the descriptive data is given in Fig. 9. The numbers of zooplankton are reported as averages by rectangular sub-divisions for the North Sea and the Atlantic approaches to the British Isles. Similar data are collected for the larger phytoplankton species which are reported as a percentage incidence for each species.

Some data on regional differences in the plankton on an oceanic scale have been reported on for all of the world's oceans. For example, Omori (1965) has defined three oceanic regions in the north Pacific based on the distribution of three species-groups

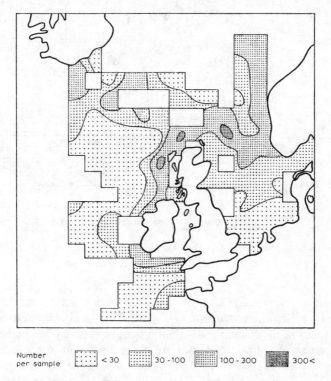

Number per sample ⬚ < 30   ⬚ 30 -100   ⬚ 100 - 300   ⬛ 300<

FIG. 9. The distribution of *Calanus finmarchicus*, stages V and VI, from data obtained with a Hardy plankton recorder. Data show the number of animals per sample; blank rectangles indicate insufficient data (redrawn from Colebrook *et al.*, 1961).

of copepods. These are (1) a cold offshore water region characterized by *Calanus plum-chrus–C. cristatus*, (2) a warm offshore water region associated with *Calanus pacificus* and, (3) a neritic water mass region represented by *Pseudocalanus minutus–Acartia longir-emis*. The latter region is oceanographically very complex and large differences in plank-ton concentrations are encountered on oceanic approaches to neritic environments. Ad-ditional information on the oceanic distribution of certain plankton species can be obtained from a study of sediments. For example, in the case of coccolithophores, the calcium carbonate coccoliths are often preserved both in the surface sediments, and in fossil remains. McIntyre and Bé (1967) have used this technique to describe species–specific zones of coccolithophore production in the Atlantic Ocean and similar maps have been drawn to show the distribution of diatoms and planktonic foraminifera.

Various attempts have been made to summarize productivity data on a global scale. Koblentz–Mishke *et al.* (1970) have reported primary productivity data for the world's

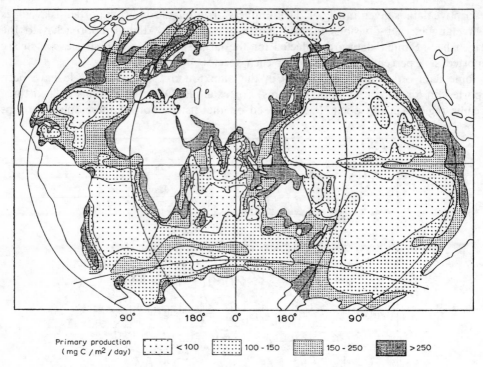

Primary production (mg C / m² / day)   < 100   100 - 150   150 - 250   > 250

FIG. 10. Distribution of primary production in the World Ocean (redrawn from Koblentz–Mishke *et al.*, 1970).

oceans, based on a review of a large number of reports and a modified version of their original figure has been redrawn in Fig. 10. From these results it is apparent that over large areas of the Pacific and Atlantic oceans, primary production is relatively low but that higher primary productivities are generally found in the proximity of land masses. There are some exceptions to this, such as where the South Equatorial current in the Pacific Ocean causes a band of relatively high primary productivity to occur along the equator.

Differences in biological production associated with near-shore processes will generally give rise to greater patchiness of plankton in neritic compared with oceanic environments. This is true both for large-scale difference in productivity, such as the influence of an estuary on biological production (e.g. Ketchum, 1967) and for small-scale patchiness (e.g. Venrick, 1972). In the latter reference it is shown, for example, that over a 10-mile distance, a lack of aggregation was observed in some species of oceanic phytoplankton; this is interpreted to mean that the randomizing of phytoplankton by turbulent processes is generally greater, relative to biological factors causing differences in productivity, in oceanic compared with neritic environments. Differences in coastal productivity may be associated with several factors-including tides (e.g. Kamykowski, 1973), local morphogeography (e.g. LaFond and LaFond, 1971), and offshore physical processes (e.g. Platt *et al.*, 1972). A special case of higher biological production associated with islands is sometimes referred to as the 'island mass effect'. Gilmartin and Revelante (1974, and references cited therein) generally attribute increased oceanic production in the vicinity of islands to nutrient enrichment caused by local turbulent upwelling in the island passages as well as to the effects of possible nutrient additions from local runoff.

The development of automated analysers has greatly assisted in the description of nutrient distributions in the ocean. An illustration of nitrate distribution obtained with an Auto-analyser® during eight transects of an approximate 10 sq. mile area is shown in Fig. 11. The illustration has been chosen to show changes in nutrient concentration in an area where nutrient depletion was general but in which there was some upwelling.

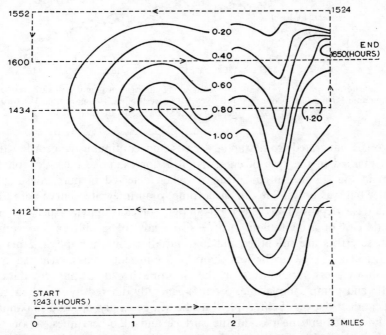

FIG. 11. Nitrate (μg at/l) at the surface off Punta Colnett Baja California (29°57′N, 116°20′W) 9 July 1965. —nitrate concentration; ---cruise track (redrawn from Armstrong et al., 1967).

Over larger areas of ocean, trends in nutrient concentration can be seen which are generally larger than the small-scale differences shown in Fig. 11. This is illustrated in Fig. 12 for nutrient observations carried out with an Autoanalyser® on two cruise tracks across the Pacific Ocean. The greater variability in the results in the western Pacific compared with the Gulf of Alaska, reflects the mixture of the two very different water masses (the Oyashio and the Kuroshio Currents) off the coast of Asia.

1.3.3 VERTICAL DISTRIBUTIONS

Until quite recently, most vertical profiles of biological parameters were made either with water bottles, which collected samples from discrete depths, or with plankton nets designed to open over some depth interval. The use of automated sampling gear, as well as recent advances in echo-sounding equipment, have greatly improved the data which are now being collected on vertical distributions. Strickland (1968) made direct comparisons between nutrients and chlorophyll *a* as measured in samples pumped from 0 to 75 m and the same data represented by a standard hydrographic cast. Differences

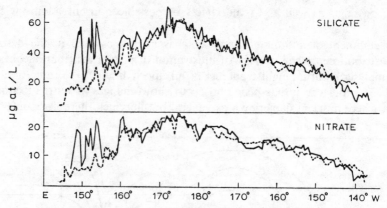

FIG. 12. Silicate and nitrate at 3 m as measured along an eastward and westward cruise track in the subarctic Pacific Ocean during April 1969 (—— westbound, ———eastbound, Victoria, B.C. to Tokyo (redrawn from Stephens, 1970).

between profiles integrated from standard casts and continuously recorded data were particularly marked in the case of chlorophyll a; for example, the chlorophyll a peak at *ca.* 20 m in Fig. 13 measured 2.9 mg/m^3 when detected in pumped samples using a fluorometer but was only 1.3 mg/m^3 according to an integrated curve based on bottle casts at standard depths. Thus the total amount of chlorophyll a per m^2 integrated from a bottle cast or a continuous profile also tends to be different; to some extent, however, these errors are smoothed out and variations in chlorophyll a per m^2 were found by Strickland (1968) to be less than 25%. Nutrient analyses carried out at the same time showed less variability than the chlorophyll a data; nutrient data for integrated bottle and pump samples (i.e. per m^2) generally differed by less than 10%.

The vertical distribution of chlorophyll in the sea generally shows a maximum which may sometimes be found near or at the surface and at other times, at or below the apparent euphotic depth (Steele and Yentsch, 1960). A deep chlorophyll maximum appears to be a seasonal feature of summer vertical profiles as far north as 45 to 50° in both the Atlantic and Pacific Oceans. Anderson (1969) found the chlorophyll

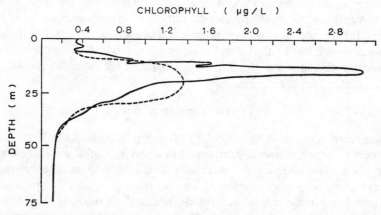

FIG. 13. Chlorophyll profiles integrated from standard bottle casts (———) and continuously recorded (—) data (redrawn from Strickland, 1968).

maximum off the Oregon coast at *ca.* 60 m was formed by photosynthetically active cells which were apparently adapted to very low light intensity. South of 40°N Venrick *et al.* (1973) have described a deep chlorophyll maximum at 100–150 m. This appears to be a more or less permanent feature of oceanic latitudes as far south as the region of tropical upwelling at 10°, north and south of the equator. In the southern hemisphere the deep chlorophyll maximum starts again south of 10°S and is sometimes found below 200 m. The depth of the chlorophyll maximum shows some seasonal variation but is generally well below the 1% light level. The maximum is believed to be associated with the nutricline, where the absorption of nutrients by phytoplankton causes a decrease in the sinking velocity of phytoplankton cells. Hobson and Lorenzen (1972) showed that chlorophyll maxima were associated with pycnoclines which occurred at various depths in the Atlantic Ocean and Gulf of Mexico. Further it was apparent that increased concentrations of microzooplankton were associated with the chlorophyll maxima.

Large differences in the concentration of zooplankton at specific depths have been encountered using the Longhurst–Hardy recorder (Longhurst *et al.*, 1966), which collects samples continually by integrating catches over very short invervals of *ca.* 10 m. An example of the zooplankton concentrations down to 400 m as measured with this apparatus is shown in Fig. 14A. The approximate 6-fold increase in zooplankton biomass at 300 m is a unique feature of this profile which would have been difficult to observe using data collected with a conventional plankton net. A similar result can be obtained with high-frequency echo sounders as is illustrated in Fig. 14B from Barraclough *et al.* (1969). In this example the presence of echo-sounding material is indicated as a discrete band on the echogram. When the layer was sampled with conventional nets, however, the discrete accumulations of zooplankton appear more as a smoothed maximum in concentration. This is due to the generally unavoidable collection of zooplankton at intermediate depths while nets are being lowered and brought up from specific

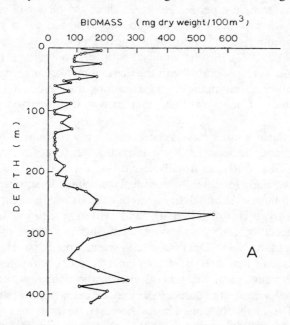

FIG. 14A. Vertical distribution of zooplankton biomass in the eastern Pacific Ocean; data obtained using a Longhurst–Hardy recorder (redrawn from Longhurst *et al.*, 1966).

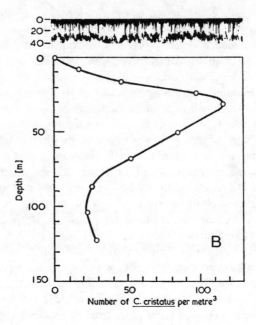

Fɪɢ. 14B. Depth profile of *Calanus cristatus* showing the actual echogram obtained with a 200-kHz recorder (top) and depth samples (O) using Miller nets (bottom) (redrawn from Barraclough *et al.*, 1969).

sampling depths. One solution to this problem is the use of specially designed nets (e.g. the Clarke–Bumpus sampler) which can be made to open and close at the beginning and end of a specific sampling period. However, these samplers still represent integrated concentrations for the distance over which the net is towed when open. Alternatively discrete *in situ* location of zooplankton now appears possible using a towed electronic particle counter (Boyd, 1973). Results obtained with this apparatus showed that aggregations of zooplankton were associated with the thermal microstructure in a 250-m vertical profile. However, not all zooplankton aggregations are associated with temperature structure (e.g. Fasham *et al.*, 1974) and other causes are discussed at the end of this chapter.

A number of detailed studies have been carried out on sound scattering layers (for review, see Hersey and Backus, 1962); in particular there are many reports on deep scattering layers (DSL) which are usually found between 100 and 500 m in the oceans. Animals which have been found in scattering layers include squids, euphausiids, fish, and certain siphonophores. Although the types of animals which occur at discrete depths in the ocean may vary, it is probable that sound scattering with low frequency sounders (12-kHz) is only caused by certain specific animals, including large fish and particularly animals containing gas bubbles. Data from echograms, Longhurst–Hardy recorders, and closing nets all indicate that species of zooplankton and nekton generally occur over quite limited depth ranges and the general classifications of organisms by depth zone (e.g. epipelagic, mesopelagic, etc., Section 1.1) may be employed in referring to the vertical distribution of animals. In some cases, however, animals may migrate vertically over distances of up to 1000 m in one day; in these cases specific depth location has little meaning.

The distributions of other biologically important materials generally show less patchiness with depth than occurs with the animals discussed above. Holm–Hansen *et al.* (1966) made detailed studies of a depth profile of nutrients (including nitrates, phosphates, ammonia, dissolved organic nitrogen, and Vitamin B_{12}) and particulate material (including particulate carbon, nitrogen, and phosphorus). Analyses were performed down to 1300 m off the coast of southern California. High concentrations of particulate materials were generally confined to the top 50-m euphotic zone; below 50 m the concentration of particulate carbon, nitrogen, and phosphorus was about one-fifth of the surface values and there was some indication of a further gradual decrease in particulate material to a tenth of surface values below 1000 m. One exception to these results was a high concentration of particulate carbon and phosphorus at 600 m in the profile. Nutrients (nitrates, phosphates, and Vit. B_{12}) showed the opposite distribution, being high in water below 50 m and low at the surface. These results are rather similar to a great many less-detailed descriptions of vertical profiles reported in the literature. From such descriptive data alone it is difficult to determine whether differences on a vertical axis are due to properties of the water column, such as vertical migrations, or if vertical differences are caused by lateral (Section 1.3.2) or temporal (Section 1.3.4) changes. Vertical differences in a water mass over a considerable lateral distance are illustrated in Fig. 15 for the distribution of particulate carbon along a transect in the Gulf of Aden. In this profile there is an apparent high concentration of particulate material between 400 and 600 m and another increase in the deep trench at 2000 to 3000 m. Similar

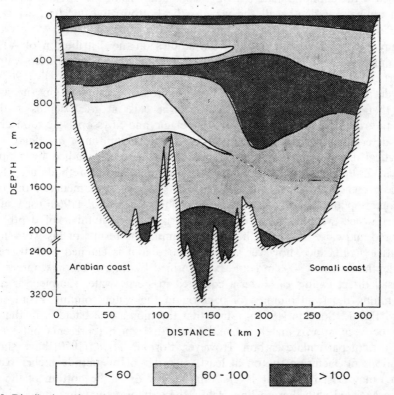

Fig. 15. Distribution of particulate carbon (μg/l) in the Gulf of Aden (redrawn from Szekielda, 1967).

differences in depth distribution of organic material were found by Menzel (1964) in the western Indian Ocean. On the other hand, Menzel (1967), reporting on the depth distribution of both dissolved and particulate organic carbon, showed that there was a relative constancy in the particulate (15 ± 5 μg/l) and dissolved (0.6 ± 0.1 mg/l) carbon in the tropical Atlantic, with slightly lower but also constant values for the tropical Pacific Ocean, below 200 m. Some differences in technique, both in the collection of samples and analyses (see Section 1.3.1), may account for these differences in observations; alternatively they represent real differences in vertical distributions in different oceans which will require further examination and explanation.

The decrease in particulate organic carbon with depth from 0 to 100 m in the North Pacific was shown by Nakajima and Nishizawa (1972) to follow an exponential function such that

$$C = C_0 e^{-k(Z - Z_0)} \tag{14}$$

where C is the concentration of carbon at depth Z and C_0 is the maximum concentration at depth Z_0. The depth of Z_0 was found to vary from 0 to 30 m. The decay constant, k, was attributed to zooplankton grazing and bacterial decomposition with the highest values of k being associated with the highest concentrations of C_0. Equation (14) might also be applied to the lateral transport of organic material from a plankton bloom over a distance (d_0 to d) or to the time course decay (t_0 to t) of a plankton bloom.

A number of investigators have attempted to determine how much of the particulate material below the euphotic zone is living. Thus apart from the obvious accumulations of certain animals at depths all the way to the bottom, the question arises whether there are appreciable quantities of unicellular organisms on which deep-water filter feeders may exist without migrating. From studies on the distribution of ATP in the ocean, Holm–Hansen (1969a) concluded that the living component of deep-water microscopic particulate material was only about 0.5 μg C/l, or 3% of Menzel's (1967) average value for deep-water particulate carbon. From microscopic examinations, it has become apparent that pigmented cells, as well as bacteria, exist at great depths in the ocean although the role of these cells in the bathypelagic and abyssopelagic food chains has yet to be determined (Bernard, 1963; Fournier, 1966; Hamilton et al., 1968).

The vertical distribution of the bacterioplankton has been studied by a number of authors (e.g. ZoBell, 1946; Kriss et al., 1960; Sorokin, 1964b). Methods used for determining the presence of bacteria have led to very different assessments of the total biomass of bacterioplankton in the water column. Thus Kriss et al. (1960) used submerged glass slides which were liable to collect bacteria from a number of depths as they were lowered and raised through the water column. The results of his work may have some relative significance, however, and in particular it is claimed that discrete water types at depths down to 4000 m could be identified by their bacterial content. Sorokin (1964b) used direct counts of bacteria collected in sterile water samplers as a method for determining the total biomass of bacteria in the water column. From studies in the central Pacific, Sorokin (1964b) found that the biomass of bacteria in the euphotic zone was between approximately 10 and 50 mg/m³ which represents only a few per cent of the total particulate carbon. However, Sorokin et al. (1970) have shown that concentrations of bacterioplankton at least an order of magnitude higher (i.e. 300 to 700 mg/m³) may exist in discrete layers, often below or at the bottom of the euphotic zone and associated with thermocline. Below the euphotic zone the number of bacteria decreased sharply to 2000 or 3000 per ml, while below 1000 m the number of bacteria were generally less than 1000 per ml or a total biomass of $<0.2\,mg/m^3$. The general form of these results agrees with biomass estimates of total living material made by Holm–Hansen (1969a) using ATP analyses as an indication of the amount of living organic carbon; however, the total amount of living carbon determined from ATP analyses was considerably greater than that found from direct bacterial counts alone. This indicates that bacterioplankton, both in the euphotic zone and in deeper water, are generally a small fraction of the total biomass of living material and only a few per cent of the total particulate organic carbon in the water column. Relative changes in the biomass of bacterioplankton are generally more important, therefore, as an indication of recycling processes than as a direct source of food. However, the methodology of bacterial counting still leaves a lot to be desired and it is probably correct to assume that many of the detrital particles in the ocean have bacteria accumulated on their surfaces and that this material is not included in direct counts of bacterial cells and clumps.

The presence of bacteria in the deep oceans (below 3800 m) has been reviewed by ZoBell (1968) who concluded that bacteria were widely but unevenly distributed at all depths. Numbers ranged from nil to $10^6/ml$, the most dense populations having been found in materials from the sea floor. In near-shore environments, sedimentary material may be very rich in bacteria and it is believed that bacteria on the sedimented particles are the chief source of food for some benthic animals (Newell, 1965; Seki et al., 1968).

Pelagic populations of organisms in the uppermost surface of the sea are referred to under the general term of neuston. Zaitsev (1961) first drew attention to the importance of neuston in the marine environment and a review of the subject has recently been given by Hempel and Weikert (1972). These authors subdivide the depth distribution of the neuston community into the 'euneuston'—organisms with maximum abundance in the surface where they stay night and day; 'facultative neuston'—organisms which concentrate at the surface only during certain hours, mostly during darkness; 'pseudoneuston'—the maximum concentrations of these organisms do not lie at the surface but at deeper layers; however, the range of their vertical distribution reaches the surface layer at least during certain hours. The largest change in the population of neuston organisms occurs during the evening and at night when many organisms among the facultative neuston and pseudoneuston join the comparatively few euneuston species. Hempel and Weikert emphasize that while the population density of neuston organisms may be very high, the total biomass of neuston in the water column is small since only a very thin layer is occupied by this community. Organisms which are permanently fixed to the sea surface by their own buoyancy and subject to wind drift are referred to as 'pleuston'. This group would include such organisms as the seaweed *Sargassum natans* in the Sargasso Sea and the coelenterates *Physalia* and *Velella*. Some other organisms, such as the blue green alga *Trichodesmium* which forms blooms at depth in tropical seas (e.g. Marumo and Asaoka, 1974), may float to the surface and form dense mats of senescent organisms. As such they can be temporarily described as pleuston. Banse (1975) has drawn attention to some inconsistencies in the use of the words pleuston and neuston. In order to present a clear definition of the difference between these two words Banse (1975) has proposed that pleuston be defined as organisms specialized to live on or below, but close to the surface and neuston as organisms actually attached to, above (epi-) or below (hypo-) the surface film. In this sense pleuston becomes the general word for a surface living organism (i.e. living in the surface zone of 'pleustal') and neuston becomes a subordinate zone connected only with the surface film.

1.3.4 TEMPORAL CHANGES IN PLANKTON COMMUNITIES

The most rapid temporal changes in a plankton community can be observed by continuous monitoring at a fixed point; these changes are not due to changes within the plankton community *per se* but are caused by internal waves. Armstrong and LaFond (1966) studied changes in nutrients, transparency, and temperature at a fixed point in a highly stratified near-shore environment; continuous 3-hr records showed correlated fluctuations as illustrated in Fig. 16. The changes shown in Fig. 16 correspond to internal waves up to 5 m high with periods of around 10 min. Since the water intake used to collect these data was located at 9 m near the principal thermocline, the data reflect maximum changes due to an abrupt vertical gradient (e.g. from 0 to $12 \mu g$ at NO_3^-/l between 6 and 12 m). Superimposed on the short time scale changes in Fig. 16 are other oscillations caused by tide and alternating wind speed and direction.

Temporal changes within a plankton community itself are largely determined by the growth, mortality, sinking, and migration rates of the individual plankters and their predators. The most rapid growth or reproduction rates for phytoplankton are of the order of several hours but whole populations generally require at least a day or more to double in size. Bacterioplankton may generate within a matter of hours, depending on temperature and substrate concentration. Zooplankton growth rates vary enormously from less than a week for some protozoa to 2 years for antarctic euphausiids. Large populations of temperate copepods having one generation per year normally grow from egg to adult in 2 or 3 months, depending on food supply and temperature. Average

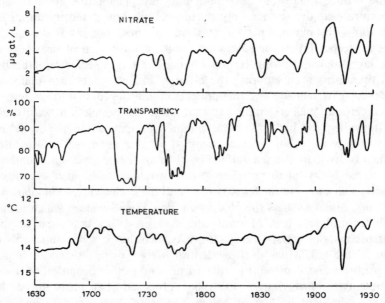

FIG. 16. Continuous 3-hour records of nitrate, transparency, and temperature at a fixed point (9·2 m from the sea floor) in a highly stratified environment off S. California (redrawn from Armstrong and LaFond, 1966).

mortality rates are generally less than growth rates or species would become readily extinct; however, over short periods, predation by planktivorous animals may cause mortality to exceed growth in a population. Sinking rates are discussed in Section 1.3.5; these, together with the rate of animal migrations, may range from less than a meter to several thousand meters per day. Consequently observed temporal changes in a plankton community may vary depending on the frequency of an investigator's observations.

Changes which occur in a plankton community regularly every 24 hr are referred to as 'diel' changes; daily and nightly occurrences are called 'diurnal' and 'nocturnal', respectively. Included among diel changes are animal migrations, changes in photosynthetic potential, and, inshore, changes in plankton communities associated with the tidal cycle.

A detailed account of changes in an animal community caused by diel migrations is given by Bary (1967) as an example of a number of studies in this field. Using a 12-kHz sounder, Bary (1967) has described four stages in the ascent of organisms from a deep scattering layer (DSL) in a coastal environment. These stages started with a gradual vertical spreading of the DSL, 1 or 2 hr before sunset, followed by a period of slow ascent which started just before and ended after sunset. A period of rapid ascent followed about 1 hr after sunset; during this period the migrating organisms came up at speeds ranging from *ca.* 1 to 8 m/min. The final stage in the ascent migration was characterized by a reduced rate of ascent as the animals approached the surface, followed by a period when the animals gradually dispersed themselves in the surface layers. The descent followed the reverse process and commenced 1 or 2 hr before sunrise.

Diel variations in the rate of photosynthesis are quite apparent in that photosynthetic organisms require light for autotrophic growth; however, less obvious diurnal changes occur in the physiological response of phytoplankters to light. Shimada (1958) showed, for example, that photosynthesis of a plankton community reached a maximum during the early morning and declined during the rest of the day. This was apparently caused by a decrease in the amount of chlorophyll *a* during the latter part of the day. Steemann Nielsen and Jørgensen (1962) attributed this daily chlorophyll *a* rhythm to the fact that under laboratory conditions, no chlorophyll *a* was synthesized during the latter part of the day and consequently if herbivore grazing remained constant, a net decrease in chlorophyll *a* would be observed in nature. However, Sournia (1967) and others have shown that the photosynthetic index (mg C assimilated per mg Chl *a*) of phytoplankton is higher before noon than after noon. More recent data reported by Malone (1971) has indicated that changes in the photosynthetic index are complex and related to at least three factors including the time of day, the size of cells, and the availability of nutrients. Thus the photosynthetic index of microphytoplankton in tropical waters was highest after noon while the same index for nanoplankton was highest before noon. However, in eutrophic waters nanoplankton showed their highest photosynthetic index after noon. These differences were not attributable to changes in cellular chlorophyll *a*. Thus apart from grazing effects, there is an obvious physiological change in photosynthesis which may effectively slow down any potential increase in phytoplankton standing stock during different parts of the day (see also Section 3.1.5).

Tidal changes in near-shore communities cause very marked fluctuations in the relative abundance of plankton and nekton. This is particularly apparent in estuarine communities where there is a large change in the type of water at a fixed point over a single tidal cycle. Welch and Isaac (1967) showed that chlorophyll *a* values could vary up to 800% in an estuary during 24 hr.

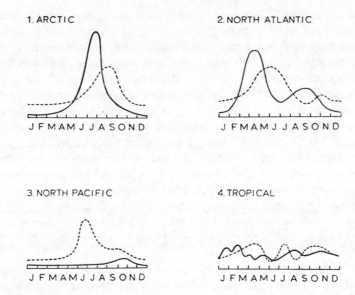

FIG. 17. Summary of seasonal cycles in plankton communities (——changes in phytoplankton biomass;———changes in zooplankton biomass; modified from Heinrich, 1962).

Large-scale temporal variations are associated with seasonal cycles in oceanic and neritic environments. A summary of seasonal cycles in plankton communities has been prepared by Heinrich (1962) and is represented in Fig. 17 with some modifications as suggested by the work of Sournia (1969). Four different seasonal cycles may generally be recognized in oceanic environments based on changes in standing stock of phytoplankton and zooplankton. The first of these (Fig. 17,1) is characteristic of arctic or antarctic waters where the amount of light is only sufficient for a single plankton bloom during the summer. The second seasonal cycle is characteristic of North Atlantic temperate waters (Fig. 17,2) where breeding of zooplankton cannot start until an increase in primary productivity has occurred in the spring. Thus as a result of the increased primary production, there is a temporary increase in the standing stock of phytoplankton followed by a decrease, as the grazing pressure from an increased standing stock of zooplankton becomes effective. In these areas there are generally two maxima in phytoplankton and zooplankton standing stocks, a large one occurring in the spring and a smaller one in the autumn. Theoretical discussions of factors causing an increase in plankton during the spring are given in Section 3.1.5; the occurrence of a second plankton maximum in the autumn is associated with similar factors including the amount of nutrient mixed into the water column after summer stabilization, the presence of sufficient light to cause photosynthesis and the occurrence of zooplankton species capable of taking advantage of an autumn increase in the standing stock of phytoplankton.

A third type of seasonal increase in plankton (Fig. 17,3) is found in the north Pacific Ocean. In this environment neither the beginning of the zooplankton breeding nor the size of the zooplankton standing stock is dependent on the presence or abundance of phytoplankton in the early spring. Nauplii of the zooplankton species, *Calanus plumchrus* and *Calanus cristatus*, which predominate in this area are hatched from adults which have wintered in deep water (>200 m) and laid their eggs without feeding in the spring. Consequently the young stages of these animals can take immediate advantage of any increase in primary productivity. In such environments it is difficult to observe any change in the phytoplankton standing stock (Fig. 17,3) except possibly during the autumn when a relaxation in zooplankton grazing allows a small increase in phytoplankton standing stock before winter.

A fourth type of seasonal cycle (Fig. 17,4) is found in tropical oceans. According to Sournia (1969) and Blackburn et al. (1970) there is very little evidence for predominant maxima and minima associated with seasonal events. A succession of small increases and decreases in phytoplankton and zooplankton standing stocks may occur throughout the year and these are largely determined by local weather conditions and the movement of water masses. Thus where a condition of upwelling occurs, such as in the Pacific approximately at the equator, a relatively high standing stock of plankton can be observed which is largely determined by the exact location of the equatorial counter current (Blackburn et al., 1970). However, Owen and Zeitzschel (1970) observed a statistically significant seasonal change in primary production in the eastern tropical Pacific; apparently this change was not reflected in changes in the plankton standing stock as discussed above. While the seasonal cycles in Fig. 17 indicate clear differences between different oceanic environments, it is also apparent that there are large zones of transition in the oceans. In sub-tropical waters at 32°N, for example, Menzel and Ryther (1960) observed a definite seasonal cycle in primary production in spite of an almost constant euphotic zone depth of 100 m.

While the illustrations in Fig. 17 indicate a smooth transition from periods of low to periods of high biomass, actual data often show considerable scatter and it would be more correct to describe the onset of seasonal conditions as occurring in a series of pulses. In addition the biomass changes shown do not indicate any seasonal change in species composition. For phytoplankton such a seasonal progression might involve a change in predominant species from diatoms, to dinoflagellates, to blue-green algae during the period spring to autumn (e.g. Bodungen et al., 1975).

Annual differences in the quantity of plankton measured at a fixed location may also be due to the strength of currents. This has been illustrated by Wickett (1967), who showed that annual variations in the concentration of zooplankton in the surface layers off California varied directly with the southerly transport of water during the previous year. Quantitative changes in zooplankton biomass and growth over a 25-year period are illustrated in Fig. 18 from Glover et al. (1974). These results show that both the total number of copepods and the zooplankton biomass have significantly decreased. These decreases have either been paralleled by, or may in part be tied to, a delay in the advent of the spring phytoplankton bloom from March to April and a consequent shortening of the zooplankton growing season by about 1 month. The results in Fig. 18 are for the North Sea but similar results were obtained for the northeast Atlantic. The longest time series changes in a plankton community has been documented by the Plymouth laboratory (Russel et al., 1971; Southward, 1974). Their data from the English Channel show an approximate 40-year cycle of events. Starting in the 1920s, the biological community was characterized by a high plankton abundance, a herring fishery and the indicator organism, Sagitta elegans; these all virtually disappeared during a 40-year period through to the late 1960s when they returned. During the 40-year period they were replaced by low plankton abundance, a pilchard fishery and the indicator organism, Sagitta setosa. While a number of theories have been put forward for these changes, it is generally believed that they are related to climatic changes. Warming

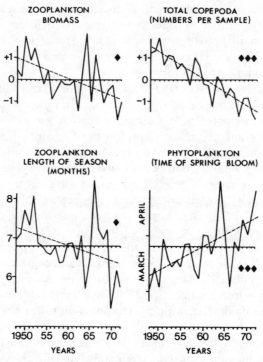

Fig. 18. Fluctuations in the plankton abundance in the North Sea showing (top) mean abundance plus or minus one standard deviation and (bottom) monthly variation from 25-year mean value. Calculated trend lines drawn and the significance of the fit shown by one, two or three * corresponding to $P = <5\%$, $<1\%$, and $<0.1\%$ respectively (redrawn from Glover *et al.*, 1974).

of the Arctic would affect circulation in the North Sea by allowing Atlantic water to extend farther north; cooling would have the reverse effect. At the same time, population centers of more northerly and southerly plankton populations and fisheries would shift their position (Russel *et al.*, 1971).

Conditions governing seasonal changes in neritic environments are much more complex than in oceanic environments due to the added effects of local geography, river discharge, and tides. Barnes (1956) described the seasonal occurrence of phytoplankton blooms and development of meroplanktonic barnacle larvae in the Firth of Clyde. From a study extending over 10 years the author concluded that the normal spring increase in phytoplankton could be heavily suppressed in some years by strong winds which resulted in a sequential catastrophic reduction in the development of barnacle larvae. In the vicinity of large rivers the neritic oceanographic climate may be generally modified to provide nutrient entrainment and a more stable water column resulting in a higher standing stock of plankton. This is apparent, for example, in data presented by Anderson (1964) on seasonal changes in chlorophyll *a* off the Washington and Oregon coasts. In this study it is shown that chlorophyll *a* was higher in an area influenced by fresh water from the Columbia River (the Columbia River plume) than in either adjacent oceanic or neritic areas. Ice cover, and the effect of melting ice on the stability of the water column, may also cause earlier seasonal changes in plankton abundance in polar regions than in temperate oceanic waters located at lower latitudes (Marshall, 1958; Bunt and Lee, 1970).

Oscillatory variations in permanently stratified seas removed from the immediate influence of land have been reported and are more difficult to explain in terms of known physical processes. Steven and Glombitza (1972) showed that *Trichodesmium* blooms occurred in the tropical Atlantic with a frequency of about 120 days, measured over a 3-year period. This they suggest could either be explained by some unknown variation in deep-water circulation, or more simply, by assuming that 120 days is the minimum time for biological factors controlling productivity to produce a bloom of *Trichodesmium*.

Studies on the temporal variation in bacterial numbers and species have been reviewed by Sieburth (1968) using specific examples from Narragansett Bay, Rhode Island. The author showed that there were apparent effects of phytoplankton species, solar radiation and temperature on the number and type of bacteria present. For example, two different colonies of flavobacteria were present during the year, a yellow-pigmented colony being present during periods of low radiation and an orange-pigmented colony being present in larger numbers during periods of high solar radiation. Also inverse relationships between genera were found to occur independently of season. Thus pseudomonads appeared to be dependent on phytoplankton blooms and a percentage increase in the isolates of pseudomonads was accompanied by a percentage decrease in the isolates of arthrobacters. While these relationships appear as real temporal variations, it should be emphasized that their cause may be more complex than indicated by seasonal factors or by the presence of other organisms.

1.3.5 SOME PROCESSES GOVERNING PATCHINESS OF PLANKTON DISTRIBUTIONS

Stavn (1971) has summarized the principal factors determining the non-random or patchy distributions of planktonic organisms as follows:
1. Physical/chemical boundary conditions including light, temperature, and salinity gradients.
2. Advective effects as in wind or water transport, including small-scale variations due to turbulence.
3. Reproduction rates within the population.
4. Social behaviour with populations of the same species.
5. Coactive factors determined by competition between species.

While these factors may not account for all forms of patchiness, the environmental parameters included in the first and second items, together with the resultant physiological responses of planktonic organisms, are probably the principal causative agents in both small- and large-scale plankton patchiness. However, as Platt *et al.* (1975) have commented, the reported patchiness of planktonic organisms is always difficult to completely dissociate from the sampling method used. Data collected over a period of time, or distance, may be examined by a mathematical technique known as spectral analysis. Properly applied, the technique reveals where there are dominant and subdominant cycles in the occurrence of plankton; the purpose of the analysis is to be able to fit the frequencies of these cycles to known physical and biological processes (Platt and Denman, in press). From a study of spatial variations in chlorophyll Platt *et al.* (1975) have suggested that length scale variations of less than 100 m are due to physical turbulence. The effect of turbulence on small-scale patchiness of marine populations is further not confined to chlorophyll distributions; for example, Frontier (1973) found similar effects on zooplankton. For spatial scales greater than 100 m, patchiness of organisms could occur due to non-linear interactions between adjacent trophic levels resulting in a breakdown of steady-state conditions (e.g. Steele, 1974b). The idea that patches of plankton can be generated by differences in the coefficients of biological processes has led to the concept of 'contemporaneous disequilibrium' (Richerson et al., 1970). This concept views the aquatic environment as a series of contiguous microhabitats in which the biology of one microhabitat develops different growth and feeding characteristics to that of a neighbouring habitat. Obviously the size of such microhabitats has to be sufficiently large for the biological processes which tend towards diversity to be greater than physical processes (including turbulent diffusion) which tend towards uniformity. In experiments in which four plankton populations were isolated in large (ca. 60-ton) plastic water columns at a distance of ca. 30 m from each other, Takahashi et al. (1976) found practically no biological divergence in the four communities over a period of 30 days. On the other hand, Platt and Filion (1973) found persistent differences over several months in phytoplankton P/B ratios at six stations separated by ca. 2 km intervals in a small bay. Thus, the predominance of physical over biological processes, or vice versa, appears to be in part a difference of the spatial scale on which the observations are based.

The question of why so many species of phytoplankton occur in the apparently homogeneous environment of any water body is not unrelated to the problem of patchiness. A number of papers deal with this problem among lake phytoplankton communities. Peterson (1975) suggests that a diversity of coexisting species is possible in oligotrophic environments where there are many nutrients which may be limiting at one time, each species being limited by the availability of a different nutrient. In contrast, under eutrophic conditions, there is a decrease in species diversity because there are fewer rate-limiting nutrients.

From studies on light, temperature, and salinity gradients, it is apparent that planktonic organisms react strongly to various boundary conditions. For example, in experimental studies on salinity gradients Lance (1962) showed that the migration of the copepod Acartia bifilosa was inhibited when the bottom salinity (34‰) was diluted at the top to more than 20 to 24‰. More sensitive changes were found by Harder (1968) among a wider group of planktonic organisms, some of which responded to salinity changes of less than 1‰. The light responses of plankton in the deep scattering layer has already been described (Section 1.3.3) but experimental studies, particularly with the freshwater cladoceran, Daphnia, have yielded some additional properties of the zooplankton/light response. Smith and Baylor (1953) showed that Daphnia was sensitive to different coloured light and that light of over 500 nm caused upward swimming while light of less than 500 nm caused downward swimming. McNaught and Hasler (1964) found an approximate linear relationship between the rate of vertical movement and the rate of change in the logarithm of the light intensity. Stavn (1971) showed that there was a minimum light intensity of 70 ergs/cm^2/s below which Daphnia did not respond to water movement; the author concluded that orientation in moving water was affected through visual detection of currents.

The distribution of plankton under various conditions of water movement is in part a passive response. Thus particles which tend to float will gather at a boundary where there is a convergence resulting in downwelling; conversely particles which tend to sink will accumulate at a divergence where there is upwelling (Stommel, 1949). Both convergences and divergences can be induced by wind and currents. In the case of a layer of warm water being pushed over a body of colder water, a convergence will result at the boundary of the two water masses. Conversely, a current which divides on meeting a headland may cause a divergence or upwelling. Plankton patchiness due to small-scale effects of wind was first observed by Nees (1949), who discovered that plankton samples collected parallel with the wind direction were more variable than samples collected at right angles to the wind. The reason for this is that under conditions of light wind, plankton accumulate in rows parallel to the wind (a property which can be observed visually in the Sargasso Sea, where *Sargassum* weed floats on the surface forming familiar wind rows). Thus if plankton are collected across the direction of the wind, the samples result in an integrated collection but if the samples are collected parallel to the wind, the resulting samples may come from a column of water in which plankton are concentrated, or from the relatively barren water in between the wind rows. The exact explanation for plankton patchiness under conditions of light wind may include several factors but in general it appears to be related to Langmuir circulation (Sutcliffe *et al.*, 1963; Stavn, 1971). The principal property of this circulation is that the water tends to move in vortices which result in 'micro-zones' of upwelling and downwelling water (Fig. 19 and Faller, 1971, for a discussion of the physics of

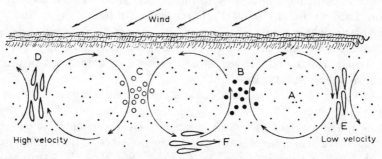

FIG. 19. Langmuir vortices and plankton distributions redrawn from Stavn (1971). A—neutrally buoyant particles randomly distributed. B—particles tending to sink, aggregated in upwellings. C—particles tending to float, aggregated in downwellings. D—organisms aggregated in high velocity upwelling, swimming down. E—organisms aggregated in low-velocity downwelling, swimming up. F—organisms aggregated between downwellings and upwellings where there is less relative current velocity than within the vortices.

Langmuir circulation). Conditions for particle accumulations A, B, and C are based only on the buoyancy of particles as suggested by Stommel (1949); these conditions may apply particularly to phytoplankton and detrital accumulations or to all planktonic organisms in the absence of light. Under conditions, D, E, and F, however, experimental results obtained by Stavn (1971) are included; these take into account the relative current velocities of the water and the swimming speed of the animals together with their light reaction which (in the case of *Daphnia*) may be used to sense the currents. Thus during the day planktonic animals may swim into a current (low velocity) since this is the most stable swimming position. In this case the animals will be aggregated in downwelling water and swimming up. At intermediate current velocities the plankton will be clumped in between the spirals and swimming horizontally against the current. At high current velocities animals may be trapped in a high-velocity upwelling where they are attempting to swim down in order to avoid surface light which is generally too bright for many zooplankters (negative phototaxis).

Reproduction as a factor in causing plankton patchiness can be seen from examples in both the animal and plant communities. Figure 35 shows the development of a phytoplankton bloom which results in an uneven depth distribution of phytoplankton due to differences in the growth rate with light intensity. Uneven zooplankton distributions can result from the tendency of some zooplankters to release clumps of eggs; this contagious distribution of eggs may be carried through in the developing stages of copepods. Social behaviour among planktonic animals is probably only of minor significance in causing patchiness but of much greater significance among the nekton and in benthic communities. Coactive factors determined by competition between species may be caused by subtle effects, such as the release of substances toxic to other species, or more directly by predation patterns caused by different trophic levels in the plankton community.

Large-scale patchiness, such as is depicted in Fig. 10 for phytoplankton production in the hydrosphere, can be mostly accounted for in terms of ocean circulation. LaFond and LaFond (1971) have discussed the patterns of water movement which lead to areas of high and low plankton productivity. The general mechanism involves some process which carries nutrient-rich deep water to the surface where there is adequate light for photosynthesis. Of particular importance in this respect is the influence of a land mass on circulation. LaFond and LaFond (1971) recognized three types of land influences on water movement which result in an upwelling of deep water; islands which create large eddy currents on the leeward side of the principal current flow, land promontories which cause similar eddies, and changes in underwater topography which cause turbulence in the flow of near surface currents.

In addition to these physical land mass effects, boundary conditions between currents moving in opposite directions may lead to large-scale upwelling. This is particularly apparent along the equator in the Pacific Ocean where the equatorial current in the northern and southern hemispheres tends to move in a northerly and southerly direction, respectively (due to the Coriolis force, see Pickard, 1964, Fig. 27). This results in a divergence near the equator and an upwelling of deep water. Similarly a commonly occurring seasonal upwelling may be produced by winds blowing parallel to a coastline which can result in water being displaced offshore (an upwelling, or divergence) or onshore (a downwelling, or convergence). These observations are in accordance with physical oceanographic theory and are illustrated schematically in Fig. 20. According to Wooster and Reid (1963) the principal areas in the hydrosphere where an exchange occurs between near-surface and deeper waters are (1) in high latitudes, (2) along the equator, and (3) in coastal regions, particularly on the eastern sides of oceans. This is apparent in the higher biological productivity of waters in such areas as the California current, the Peru current, and the Benguala current.

The vertical component of plankton patchiness may be partially influenced by factors discussed above, but in particular the processes of passive sinking and active vertical migration will influence plankton concentration at any depth.

Sinking rates of phytoplankton have been the subject of an extensive review by Smayda (1970a). From a theoretical approach, the rate of sinking (v) can be expressed as:

$$v = \frac{2gr^2}{9} \frac{(\varrho' - \varrho)}{\eta \cdot \phi_r}, \tag{15}$$

where r equals the radius of a particle, g is the acceleration due to gravity, ϱ' and ϱ represent

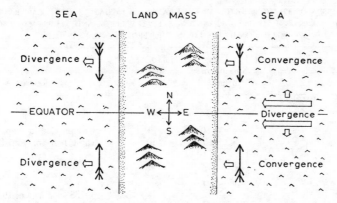

FIG. 20 Schematic effect of wind ($\ggg$) and water ($\Rightarrow$) movement in producing areas of upwelling (divergence) and downwelling (convergence). Wind direction shown from the south toward the equator in the southern hemisphere and from the north toward the equator in the northern hemisphere. Water movement caused by this wind pattern results in a divergence off the west coast and a convergence off the east coast. Water movement is reversed with a reversal in wind direction. Currents along the equator in the northern and southern hemispheres similarly tend to move water towards or away from their boundary area, depending on the principal direction of water flow.

the densities of the organism and the liquid, respectively, η is the viscosity of the liquid and ϕ_r is the 'coefficient of form resistance'. The latter term is given by

$$\phi_r = \frac{V_s}{V_a},$$
(16)

where V_s is the terminal velocity of a sphere of volume and density equivalent to a non-spheroid body having a terminal velocity of V_a. From the above equations it is apparent that sinking rate will increase with diameter and density, but decrease with departures from a spherical shape (i.e. increase in ϕ_r); the subject is discussed by Munk and Riley (1952). For particles of *ca.* 5 μm, the latter term is lowest for organisms having shapes approximating a plate, cylinder and sphere in increasing order; at about 50-μm plates and cylinders sink at about the same rate, but slower than spheres of the same volume, while for particles greater than 500 μm spheres sink faster than cylinders but slower than plates. The sinking rates of plankton and other particles have been studied by a number of authors and a summary of their results has been made by Smayda and is represented in Table 2. From these results it is apparent that dead phytoplankton sink much faster than living phytoplankton and that larger-sized organisms sink faster than smaller, as predicted by the above equation. However, it is difficult to generalize on the data in Table 2 since within the groups reported there are large differences in size and ash content (e.g. silica frustules among the phytoplankton) which cause a wide range in the sinking rates for any one group.

Mechanisms which tend to modify the sinking rate of phytoplankton include apparent changes in cell buoyancy and, among flagellates, the ability to swim in response to stimuli (e.g. light). Since some small flagellates can be observed under the microscope to swim a distance of their body length in less than a second, maximal swimming speeds for some species may be around 100 to 1000 μm per second, or in the approximate range of 1 to 10 m per day. Loeblich (1966) measured the swimming speed of *Gonyaulax polyedra* and from his observations, as well as others which he quotes, it appears that

TABLE 2. A COMPARISON OF PHYTOPLANKTON SINKING RATES (AS m DAY^{-1}) WITH SOME RATES OBTAINED FOR DEAD MARINE PROTOZOANS, FISH EGGS, ZOOPLANKTON, FAECAL PELLETS, AND VIABLE MACROSCOPIC ALGAE, AS REPORTED ON IN THE LITERATURE FROM SMAYDA (1970a)

Group	Sinking rate (m day^{-1})	No. of species used	Reference
Phytoplankton			
Living	0–30	~25	
Palmelloid stage	~5–6150	1	As compiled in Appendix
Dead, intact	<1–510	10	Table I in Smayda, 1970a
Fragments	1·5–26 × 10^3	?	
Protozoans			
Foraminifera	30–4800	—	
Radiolarians	~350	—	p. 253 in Kuenen, 1950
Zooplankton			
Amphipoda	~875	1	Apstein, 1910
Chaetognatha	~435	1	Apstein, 1910
Cladocera	~120–160	2	Apstein, 1910
Copepoda	36–720	14	Apstein, 1910; Gardiner, 1933;
Ostracoda	400	1	Apstein, 1910
Pteropoda	760–2270	5	Apstein, 1910
Salpa	165–253	2	Apstein, 1910
Siphonophora	240	1	Apstein, 1910
'Animal plankton'	~225–500	?	Seiwell and Seiwell, 1938
Faecal pellets	36–376	—	Osterberg et al., 1963; Smayda, 1969
Fish eggs	215–400	2	Apstein, 1910

large dinoflagellates may swim at speeds from *ca.* 2 *to* 20 m per day, in other words, 25 to 50 times the cell length per second (Throndsen, 1973).

Changes in the buoyancy of phytoplankton cells may be accomplished through a change in the cellular constituents. Thus an increase in lipids or a change in the ionic content of the cell will alter the cell density. Of these two mechanisms, Smayda (1970a) considers that the latter mechanism is the more important, especially in cells having large vacuoles. The author reports on the work of Beklemishev, Petzikova, and Semina who analysed cell sap in the diatom, *Ethmodiscus rex*, and found that the total ion content of living cells was 23·9 mg/ml compared with 33·3 mg/ml in dead cells and 33·8 mg/ml in the sea water medium. The selective exclusion of heavy ions from cell sap as a means of buoyancy was first proposed by Gross and Zeuthen (1948) but it is apparent from observations made by Eppley *et al.* (1967) and theoretical calculations made by Smayda (1970a) that the theory does not entirely account for the range of density changes observed among the phytoplankton. Other physiological mechanisms yet to be fully explored are changes in buoyancy due to nutrient enrichment (Steele and Yentsch, 1960) and the effect of light (Smayda and Boleyn, 1966).

Mechanisms which would tend to increase sinking rates compared with the data given in Table 2 have been discussed by Smayda (1971). In this reference Smayda found it necessary to account for persistent reports of chlorophyll-containing phytoplankton at great depths. The author assumes that the cells could not have reached depths of several thousand meters in such a healthy state by sinking at rates generally reported in Table 2. In order to account for this accelerated sinking of plankton Smayda suggests four mechanisms. These are density inversion currents, wind-induced downwelling, aggregate formation of phytoplankton cells [increase in r, eqn. (15)] and accelerated sinking due to fecal pellet formation in which live phytoplankton cells can sometimes be found to have survived copepod digestion.

The vertical migration of zooplankton and their aggregation at specific depths has been reviewed by Banse (1964). In general vertical movement appears to be related either to the life history of a species or to its feeding habits. The former generally cause seasonal differences in the vertical distribution of plankton while the latter cause diel changes in vertical abundance. Vinogradov (1955) postulated in the latter case that food material was transported to great depths by a series of overlapping animal migrations. Thus it was considered that some bathypelagic animals obtained their food by migrating into the mesopelagic zone while some mesopelagic animals migrated into the epipelagic zone; the exact depth ranges being dependent on the species.

Several theories have been offered as explanations for the vertical migration of zooplankton populations. One suggestion (McLaren, 1963) is that an animal can conserve energy by reducing its metabolism during part of the day (or life cycle) if it migrates into cold water below the seasonal thermocline. Several theories on the optimization of phytoplankton resources have been advanced (e.g. Petipa and Makarova, 1969; McAllister, 1970; Kerfoot, 1970). These theories, while not precisely the same, have in common a separation of phytoplankton from zooplankton during maximum light thus allowing the phytoplankton a period in which they can increase exponentially in abundance instead of being continually grazed by zooplankton. In another theory (McLaren, 1974) it is suggested that for populations in which size and fecundity are negative functions of temperature, it is advantageous to migrate vertically in thermally stratified waters.

CHAPTER 2

CHEMICAL COMPOSITION

2.1 SEA WATER

The following section introduces a few important facts concerning the chemistry of sea water in relation to biological productivity. Since the chemistry of sea water has been treated in a number of comprehensive texts (e.g. Harvey, 1957; Riley and Chester, 1971) only a brief summary has been made here to cover some essential features of sea-water chemistry as they pertain to biological oceanographic processes. In addition, methods for determining some biologically important chemical constituents of sea water have already been given in previous texts (e.g. Barnes, 1959; Strickland and Parsons, 1968).

TABLE 3. CHEMICAL COMPOSITION OF THE MAJOR CATIONS AND ANIONS IN SEA WATER OF 35‰ SALINITY (Cl = 19·374‰) BASED ON 1967 ATOMIC WEIGHTS (AFTER MORCOS, 1973)

Cations	g/kg	Anions	g/kg
Na^+	10·76	Cl^-	19·35
Mg^{2+}	1·30	SO_4^{2-}	2·71
Ca^{2+}	0·41	Br^-	0·07
K^+	0·40	HCO_3^-	0·14
Sr^{2+}	0·01	H_3BO_3	0·03

Oceanic water contains about 35 g of salt per kilogram of sea water; the major chemical components of the salt are shown in Table 3. In addition to these ions, sea water contains many minor elements and radicals, some of which have great biological importance in spite of their relatively low concentration. Elements which do not significantly change in concentration due to biological or geochemical activity (e.g. see Table 3) are referred to as being among the 'conservative' properties of sea water, while the 'non-conservative' substances include all materials whose concentrations are affected by biological or geochemical events. This distinction can be applied to short time periods but is not wholly correct for geological periods of time since the conservative compounds are in fact maintained at their apparent concentrations by an equilibrium which removes an amount equivalent to that being added by river runoff and other processes. A more

40

exact distinction can be made in defining a 'conservative' constituent as one whose chemical activity is so low relative to the rate of physical oceanographic processes (such as mixing and advection) that its distribution is essentially controlled by physical oceanographic processes. Thus a non-conservative element is one whose distribution reflects the effects of short-term biological or geochemical activity as well as the effects of physical oceanographic processes. Among the latter group are the nutrient salts, particularly nitrate and phosphate which may be present in deep ocean waters at concentrations of *ca.* 30 and 2·5 μg at/l, respectively. The concentrations of these salts vary with the water mass, season, and depth. In many tropical and subtropical environments as well as in temperate waters following a phytoplankton bloom, the concentration of nitrates and phosphates may be below the limit of detection by conventional methods; in such cases it may be possible to demonstrate that the principal source of nitrogen is present in low concentrations (<5 μg at/l) of ammonia, urea, or some organic nitrogen compounds. Thus changes in the nutrients of sea water are usually reflected in a redistribution of biologically important elements; this is illustrated in Fig. 21, which shows changes in the concentration of three forms of phosphorus in a coastal environment over a period of 1 year. Biologically important constituents of sea water also include silicate, iron, manganese, cobalt, molybdenum, and other trace elements used in cellular metabolism. Carbon dioxide is present in sea water, mostly as bicarbonate in concentrations of *ca.* 25 mg C/l; this level is a large excess over the amount generally required for plant growth and consequently the pH of sea water, which is largely determined by the bicarbonate/borate concentration, is usually in the range 7·5 to 8·5. Changes in pH within this approximate range are due to photosynthesis and respiration of

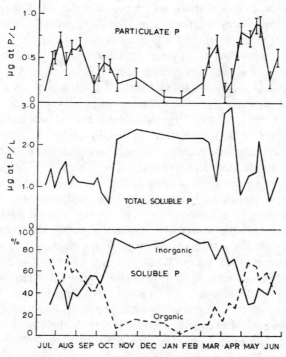

FIG. 21. Annual variations in the presence of three forms of phosphorus in a coastal environment, Departure Bay, British Columbia (redrawn from Strickland and Austin, 1960).

marine organisms and low pH values are reflected in high concentrations of total carbon dioxide. Since carbon dioxide and oxygen are the products of biologically opposite processes, it is also usual to find high concentrations of oxygen (*ca.* 4 to 8 ml/l) and higher pH values in areas of high photosynthesis, and lower oxygen concentrations (<4 ml/l) where respiratory processes predominate.

Dissolved organic material is present in open ocean sea water at a concentration of *ca.* 0·4 to 2 mg C/l while particulate organic material, including both plankton and detritus (see Section 2.4), is present in concentrations ranging from *ca.* 10 μg C/l in deep water to surface values of *ca.* 100 to 500 μg C/l. The total concentration of organic material may not be as biologically important as the presence of certain organic constituents. Vitamin B_{12}, for example, may occur in sea water at concentrations of $\mu\mu$g/ml* and still be utilized by a number of organisms which require this vitamin for growth.

The most biologically important physical mechanism for altering the chemistry of sea water is the process of mixing. Near the sea surface, waters tend to become stratified, either by thermal heating or by differences in total salt content. In the tropical oceans, for example, heating causes stratification which effectively prevents deep nutrient-rich sea water from reaching the surface throughout the year. A similar situation can occur during the summer in temperate waters and nutrient exhaustion through plant growth in the surface layers results in low productivity in the absence of some mechanism for breaking down the stratified water column. In temperate latitudes, winter storms cause mixing of the water column so that deep-water nutrients are seasonally restored to the surface layers. In other circumstances, wind and current-generated upwelling (such as off the coast of Peru) may cause an area of high productivity to occur during most of the year. Rivers entering the sea also cause mixing to occur and river water will entrain many times its own volume of deep salt water as it enters the sea; this, together with a supply of organic nutrients from the land, generally alters the chemistry of sea water in estuarine environments so as to cause local areas of very high productivity. Elementary discussions of physical processes governing mixing and stability of water masses have been given by Sverdrup *et al.* (1946) and Pickard (1964).

The nature of chemical changes caused by the admixture of river water with sea water to some extent depends upon the specific chemistry of local river waters. In general, however, both inorganic and organic constituents of sea water will be altered in the immediate vicinity of a river. Silicate is usually higher in river water than sea water and certain trace elements, such as thorium and cerium, may be several orders of magnitude more concentrated in river water (Goldberg, 1971). The natural heavy metal concentration of river waters has been difficult to determine in view of the widespread industrial use of metals such as copper, lead, and mercury. A sharp gradient in heavy metal concentration appears to exist in coastal waters, particularly in areas where there is some accountable source of heavy metal pollution (Abdullah *et al.*, 1972). The total dissolved organic constituents of large rivers entering the sea are generally higher than the surrounding sea water. While the types of organic compounds contributed by rivers are diverse and poorly investigated, particular biological interest has been placed on the vitamin content (e.g. Vitamin B_{12}; Burkholder and Burkholder, 1956) and the presence of chelating compounds. The latter may be closely associated with the humic acid content of river water; aside from their chelating properties, Prakash (1971) has indicated that humic acids may also enhance plant growth through certain specific physiological and biochemical reactions.

* 1 $\mu\mu$g $\equiv$ 1 picog $\equiv$ 10^{-3} nanog $\equiv$ 10^{-6} microg $\equiv$ 10^{-9} mg $\equiv$ 10^{-12} g.

2.2 PHYTOPLANKTON

Table 4 shows the percentages of protein, carbohydrate, and fat found in different species of phytoplankton harvested during the exponential phase of growth. Identification of major metabolites for the species reported was $100 \pm 10\%$ of the total organic matter. Considering the diversity of size and taxonomic groups involved, the cellular composition of these species appears to be remarkably similar and shows, in contrast

TABLE 4. PHYTOPLANKTON COMPOSITION OF MAJOR METABOLITES (FROM PARSONS et al., 1961)

| Species | Approx. cell volume (μ^3) | Percentage composition, ash free dry wt. | | |
		Protein[1]	Carbo-hydrate[2]	Fat[3]
Prasinophyceae				
Tetraselmis maculata	310	68	20	4
Chlorophycae				
Dunaliella salina	400	58	32	7
Bacillariophyceae				
Skeletonema costatum	1390	58	33	7
Chrysophycae				
Monochrysis lutheri	28	53	34	13
Dinophyceae				
Amphidinium carteri	740	36	39	23
Exuviaella sp.	780	35	42	17
Myxophyceae				
Agmenellum quadruplicatum	1·5	44	38	16

[1]Nitrogen × 6·25.　　[2]Anthrone reaction.　　[3]Saponifiable fraction only.

to terrestrial plants, a general predominance of protein over other constituents. However, among the two representatives of the Dinophyceae and the one Myxophyceae, the ratio of protein to carbohydrate, or lipid, is lower than among the diatoms. Haug and Myklestrad (1973) attributed the low protein to carbohydrate ratio of dinoflagellates to carbohydrate material associated with cell walls. In their results the lipid fraction of both diatoms and dinoflagellates was less than 10% and the protein to carbohydrate ratio was suggested as a measure of the physiological state of the diatoms, as indicated by data in Table 11. Myklestrad (1974) has also shown that while different species of phytoplankton have similar composition during the logarithmic phase of growth as shown in Table 4, the composition changes appreciably between species during the stationary growth phase (i.e. after exhaustion of nutrients). Thus the ability of cells to make glucan was shown to range from 25 to 60% of the dry weight among six different species of diatom in the stationary growth phase.

The relationship between cell size and total organic content of phytoplankton has been studied extensively by Strathmann (1967 and references cited therein), who gives two regression equations for the carbon content of different cell volumes. The first of these for phytoplankton, other than diatoms, is given as:

$$\log C = 0{\cdot}866 \log V - 0{\cdot}460. \tag{17}$$

Since diatoms contain a vacuole, the quantity of carbon per unit volume is less than for other phytoplankton species and is given as:

$$\log C = 0.758 \log V - 0.422, \tag{18}$$

where C is the carbon per cell in picograms and V is the cell volume in cubic microns.

For energy budgets it is necessary to convert the carbon content of phytoplankton to their calorific value. Platt and Irwin (1973) determined the calorific value for a mixed blooms of phytoplankton having different proportions of the major organic components and found a simple regression equation ($r^2 = 0.91$) such that

$$\text{cal/mg dry wt.} = 0.632 + 0.086\,(\%C). \tag{19}$$

From acid hydrolysates of different phytoplankton species, Cowey and Corner (1966) showed that most of the organic nitrogen could be accounted for in terms of amino acid nitrogen and that in confirmation of earlier work, the amino acid spectrum of different algae was very similar, as illustrated in Table 5. By comparison with protein of known nutritional value (casein), algal proteins show some differences in the proportion of amino acids but in general there is a well-balanced distribution which includes all the essential amino acids. Feeding experiments with rats, using the dinoflagellate, *Gonyaulax polyedra*, have further confirmed the dietary adequacy of phytoplankton protein (Patton *et al.*, 1967). In a very extensive examination of the amino acid spectra in thirty-one species of marine phytoplankton, Chau *et al.* (1967) and Chuecas and Riley (1969) agreed with the general distribution of amino acids found by earlier workers but noted also a number of additional characteristics. These included the occurrence

TABLE 5. AMINO ACID COMPOSITION OF UNICELLULAR ALGAE; WHOLE CELL HYDROLYSATES EXPRESSED AS g AMINO ACID N/100 g TOTAL N (FROM COWEY AND CORNER, 1966)

	Organism					
Amino acid	*Skeletonema costatum*	*Phaeo-dactylum tricornutum*	*Monochrysis lutheri*	*Cricosphaera elongata*	*Chlorella ellipsoida*	Whole[1] casein
Aspartic acid	7·92	7·05	6·13	6·52	5·7	7·1
Threonine	4·22	3·71	3·59	3·23	3·6	4·9
Serine	5·63	4·55	4·29	3·23	4·4	6·3
Glutamic acid	7·15	7·99	6·32	5·75	7·9	22·4
Proline	4·05	4·71	3·18	3·23	3·3	10·6
Glycine	8·89	6·69	6·72	5·61	7·9	2·0
Alanine	7·03	7·08	8·25	5·74	6·2	3·2
Valine	5·25	5·75	4·96	3·99	3·8	7·2
Methionine	0·71	1·29	1·59	2·27	0·8	2·8
Iso-leucine	4·14	3·33	2·91	2·04	2·4	6·1
Leucine	5·94	5·70	6·67	5·47	5·1	9·2
Tyrosine	1·51	1·80	2·19	2·55	2·1	6·3
Phenylalanine	3·22	3·03	2·79	2·95	3·0	5·0
Lysine	7·93	8·32	8·16	5·75	7·1	8·2
Histidine	3·08	3·28	3·55	3·29	3·2	3·1
Arginine	10·67	10·79	11·34	10·62	14·4	4·1
Tryptophan	—	—	1·74	0·98	2·1	1·7
Cystine/2	—	1·12	1·07	—	0·8	0·3
Total	87·34	86·19	91·94	78·53	89·9	110·5

[1]From Gordon and Whittier, 1966: (—) not reported.

TABLE 6. LIPID COMPOSITION OF DIATOMS FROM LEE *et al.* (1971)

	Lauderia borealis	*Skeletonema costatum*	*Chaetoceros curvisetus*
	(% recovery by weight)		
Hydrocarbon	11	16	2
Wax ester	0	0	0
Triglyceride	16	14	12
Sterol	17	15	10
Free fatty acid	5	12	16
Phospholipid	51	57	50
Total % dry wt.	13·2	8·6	9·1

of amino acid derivatives of butyric and adipic acids, the frequent occurrence of serine as the principal amino acid in diatoms, and differences between diatoms and other algae in the relative proportions of a number of other amino acids.

The principal lipids found in three species of marine diatoms are shown in Table 6 (Lee *et al.*, 1971). The particular characteristics of these data are the high concentration of phospholipid and the absence of wax esters. The latter are often the principal lipids in copepods which may feed upon diatoms. Fatty acids are included in the triglyceride, free fatty acid, and phospholipid fractions in Table 6 and their distribution in 8 classes of phyto-plankton is shown in Fig. 22 (from Ackman *et al.*, 1968). Early data on the

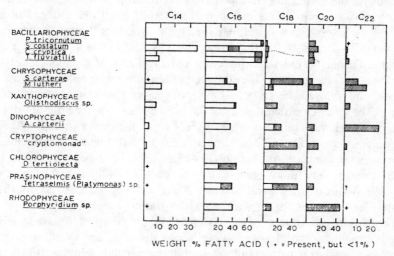

FIG. 22. Fatty acid spectrum of twelve marine phytoplankters. ☐ saturated or one double bond; ▨ polyunsaturated acids (redrawn from Ackman *et al.*, 1968).

fatty acid composition of plankton are difficult to interpret since most of the long-chain fatty acids were missed in analyses. However, in general marine phytoplankton fatty acids contain between 10 and 30% palmitic acid (C_{16}) as well as appreciable amounts of C_{18}, C_{20}, and C_{22} polyunsaturated acids*. Among the Bacillariophyceae the major saturated acids are C_{14} and C_{16} and the unsaturated acids are 16:1, 16:3 and 20:5. As an exception to other classes of algae, C_{18} acids appear to be very minor constituents

* C_{18} indicates a chain length of 18 C atoms; 16:1 indicates a chain length of 16 C atoms and 1 double bond; 20:5 ω 3 indicates 3 carbon atoms between the terminal methyl group and the middle of the double bond nearest the terminal methyl group.

of the Bacillariophyceae. The Chrysophyceae differ from the Bacillariophyceae in containing more polyunsaturated acids, particularly C_{18} and C_{22}. The Dinophyceae appear to be characterized by 16:0 and polyunsaturated C_{20} and C_{22} acids. However, Chuecas and Riley (1969) showed that two species of Dinophyceae contained large amounts of 16:0 fatty acid, while in a third the principal fatty acid was 16:1. Four members of the Cryptophyceae examined by Chuecas and Riley (1969) were similar and the principal acids were 16:0, 16:1, 18:3, 18:4, 20:1 and 20:5. These results differ from the results found by Ackman *et al.* (1968), who reported a virtual absence of 16:1 and 20:1 acids while Chuecas and Riley (1969) described a relatively high concentration of the latter as a characteristic of the Cryptophyceae. The representatives of the Chlorophyceae and Prasinophyceae shown in Fig. 22 have very similar compositions; the principal components were 16:0 and polyunsaturated C_{16} and C_{18} acids. In *Porphyridium* sp. the principal fatty acids were palmitic acid and C_{20} polyunsaturated acids including the unusual occurrence of over 20% of 20:4. The taxonomic position of the Xanthophyceae, *Olisthodiscus* sp. is uncertain, but its fatty acids composition was similar to the two Chrysophyceae. Parker *et al.* (1967) investigated the fatty acid composition of eleven representatives of the Cyanophyceae and found a predominance of 16:0, 16:1, 18:1 and 18:2 acids; however, there were obvious differences between species including one species in which half of the fatty acids were C_{10}. Branch-chained fatty acids were absent from the Cyanophyceae but were a major component in marine bacteria.

The carbohydrates of marine phytoplankton occur either as storage products of cellular metabolism or as constituents of cell wall material; small quantities of other carbohydrates may be associated with cellular metabolic processes. Using *Skeletonema costatum*, Handa (1969) demonstrated that approximately half of the algal carbohydrate was water soluble (Table 7) and that this consisted primarily of a glucose polymer, β-1, 3-glucan; the water-insoluble residue consisted of mannose with lesser amounts of rhamnose, fucose, and xylose. The carbohydrate storage product of diatoms is very similar to laminarin, the reserve polysaccharide of the brown seaweeds. The diatom polysaccharide is known as chrysolaminarin which differs from laminarin in that although it is a glucose polymer it contains no mannitol which is a constituent of laminarin. Members of the Chrysophyceae may also contain chrysolaminarin (Beattie *et al.*, 1961, and references cited therein). The dinoflagellate, *Thecadinium inclinatum*, has been reported to contain an α-1, 4-glucan (Vogel and Meeuse, 1968); water-soluble extracts of the Rhodophyceae may also contain an α-glucan, floridean starch.

The carbohydrate constituents of cell wall material are complex but are usually known to include homopolysaccharides, such as cellulose, mannan, and xylan, or some heteropolysaccharides, such as glucuromannan found in the Bacillariophyceae (Ford and Percival, 1965). From these observations Handa and Tominaga (1969) concluded that the cell wall carbohydrate, represented by the residue in Table 7, was not a single compound

TABLE 7. CARBOHYDRATE FRACTIONS OBTAINED FROM *Skeletonema costatum* (FROM HANDA, 1969)

Carbohydrate fraction	Yield (%)	Principal sugar
Water—extractable	45·6	91% glucose
Soluble in ethanol–acetone	(10·8)	Glucose and oligosaccharides
Insoluble in ethanol–acetone	(34·8)	β-1, 3-glucan
Residue	54·4	55% mannose

but that the diverse monosaccharide constituents found in acid hydrolysates represented sugars from several polysaccharides.

Monosaccharides obtained from acid hydrolysates of whole cells are shown in Table 8 for algal species harvested during the exponential phase of growth. The amount of 'crude fibre' found in different species may be interpreted as indicating the proportion of the total carbohydrate employed as cell wall material. Glucose, galactose, and ribose were found in all species analysed; glucose was always the predominant sugar which agrees with more recent results for three species of Bacillariophyceae analysed by Handa and Yanagi (1969). However, these authors found approximately 30% of the monosaccharides to be mannose which is a considerably larger fraction than is shown for the Bacillariophyceae in Table 8. Further differences between the results in Table 8 and other authors may also occur. For example, Lewin *et al.* (1958) reported the presence of fucose in *Phaeodactylum tricornutum*. A note of caution is required in accepting data presented in Table 8 since these analyses were performed using paper chromatography and better methods are now available. Haug *et al.* (1973) used gas/liquid chromatography to examine the monosaccharides in extracts from natural phytoplankton. They found that an acid extractable fraction contained β-1, 3-glucan but that an alkali extractable fraction contained a complex group of monosaccharides, similar to data given in Table 8.

A soluble polysaccharide excreted by the diatom, *Chaetaceros affinis*, has been shown by Myklestad *et al.* (1972) to consist of a sulphate ester; monosaccharides tentatively identified in an acid hydrolysate consisted primarily of rhamnose and fucose with lesser

TABLE 8. MONOSACCHARIDE COMPOSITION OF WHOLE HYDROLYSATES OF UNICELLULAR ALGAE (FROM PARSONS *et al.*, 1961)

	Crude fibre (percentage of total carbo-hydrate)	Principal sugars (percentage dry weight of cells)										
		Glucose	Galactose	Mannose	Ribose	Xylose	Arabinose	Rhamnose	Fucose	Fructose	Hexosamine	Hexuronic acids
CHLOROPHYCEAE												
Dunaliella salina	9·8	17·2	11·8	–	1·7	–	–	–	–	–	–	+
PRASINOPHYCEAE												
Tetraselmis maculata	12·6	11·9	2·3	–	0·95	–	–	–	–	–	–	+
CHRYSOPHYCEAE												
Monochrysis lutheri	3·6	22·1	4·4	–	1·3	3·5	–	–	–	–	–	+
HAPTOPHYCEAE												
Syracosphaera carterae	1·7	9·2	7·1	–	1·5	0·8	1·9	–	–	–	–	+
BACILLARIOPHYCEAE												
Chaetoceros sp.	22·8	3·3	1·5	0·79	0·71	0·4	–	2·8	+	–	–	+
Skeletonema costatum	9·6	16·4	1·8	0·87	1·2	–	–	1·0	0·9	–	–	+
Coscinodiscus wailsii	29·0	2·1	0·4	0·41	+	–	–	0·7	0·5	–	–	+
Phaeodactylum tricornutum	2·5	10·7	2·7	3·7	0·72	0·7	–	1·5	–	–	–	+
DINOPHYCEAE												
Amphidinium carteri	2·0	19·0	8·4	–	0·9	–	–	+	–	–	–	–
Exuviaella sp.	37·0	26·8	8·3	–	+	+	+	+	–	–	–	–
MYXOPHYCEAE												
Agmenellum quadruplicatum	17·4	17·4	3·2	–	1·5	–	–	–	–	3·5	0·3	+

[+]Sugars detected but not estimated.   [-]Sugars not detected.

amounts of arabinose and galactose. Another extracellular polysaccharide from the diatom, *Nitzschia frustulum*, was also shown to be a sulphate ester but to contain different
sugars, including large amounts of rhamnose and mannose with lesser amounts of six
other sugars (Allan *et al.*, 1972).

Major metabolites and some minor cellular constituents in five species of unicellular
algae have been reported by Ricketts (1966a). Among minor cellular constituents, nucleic
acids ranged from 1 to 7%; DNA 0·31 to 0·86%; RNA 0·7 to 6·65%; phospholipids
1·0 to 4·0%; acid soluble phosphorus 0·13 to 0·69%; phosphoprotein phosphorus 0·02
to 0·14%; total phosphorus 0·49 to 1·19% of the dry weight. Total phosphorus has
been analysed by a number of authors and may range from about 0·5 to 3% of the
dry weight depending to some extent on the supply of phosphate in the surrounding
water. Holm–Hansen (1969b) measured the DNA content of ten species of marine phytoplankton and found that the amount ranged from 0·01 to 200 pg per cell. This range
was strongly correlated with the amount of carbon per cell and the ratio of DNA
to cell carbon was 1:100.

The ash content of some phytoplankton species may be very high due mainly to
cell wall constituents, such as calcium carbonate in coccolithophores and silica in diatoms. Silicon is widely distributed in small amounts throughout all plants and is generally present in phytoplankton at levels between 0·1 and 1% dry weight; among diatoms
the silica frustule may represent up to *ca.* 40% of the dry weight of the cells but this
can vary by a factor of at least 5 depending on the availability of silicon in the surrounding water (Lewin, 1957; Vinogradov, 1953; Parsons *et al.*, 1961). In a series of papers
Coombs *et al.* (1967 a, b, and c) have shown that silicon uptake and deposition in
the cell wall of the diatom *Navicula pelliculosa* required ATP and that metabolic changes
accompanied silicon starvation. These included a decrease in the net synthesis of proteins, carbohydrates, chlorophyll, and fucoxanthin, and an increase in the net synthesis
of diadinoxanthin and lipid. Among coccolithophores, the calcium carbonate coccoliths
originate from within the cell (Paasche, 1962) and may form a dense cover over the
cell wall or at other times they may be totally absent (e.g. Braarud, 1963). The distribution of eighteen trace metals in fifteen species of phytoplankton has been studied by
Riley and Roth (1971) and compared with earlier work by Vinogradova and Koval'skiy
(1962); some of the former analyses are reported in Table 9. The principal conclusion
reached from analysing species from the same class was that trace element distribution
was not correlated with taxonomy and further that the concentration of individual
elements may vary considerably between species. From further experiments Riley and
Roth (1971) were able to show that in general the trace metal content of algae could
be increased by increasing the concentration of metals in the medium in which the
organisms were grown.

The photosynthetic pigments of unicellular algae are probably their most exhaustively
analysed components. Early work has been summarized by Strain (1951) and Goodwin
(1955); later contributions include reports by Goodwin (1957), Strain (1958), Haxo and
Fork (1959), 'OhEocha and Raftery (1959), Allen *et al.* (1960 and 1964), Dales (1960),
Jeffrey (1961), Parsons (1961), Jeffrey and Allen (1964), Ricketts (1967 a and b, 1970),
Riley and Wilson (1967), and Riley and Segar (1969). Some of the more recent reports
have included extensive quantitative data. However, apart from the use of quantitative
data on chlorophyll *a* and total carotenoids in ecological studies, the chief importance
of pigment data lies in the obvious differences in pigment composition of different taxonomic groups. A summary of these data is presented in Table 10, which shows that

TABLE 9. DISTRIBUTION OF TRACE ELEMENTS IN PHYTOPLANKTON (FROM RILEY AND ROTH, 1971)

	Chlorella salina[1]	Asterionella japonica[1]	Phaeo-dactylum tricornutum[1]	Sea plankton[2]
% ash, dry wt.	9·5	24·1	7·9	45·8
Element (ppm)				
Mn	48	54	73	118
Zn	301	115	325	282
Cu	25	105	110	36
Ag	4·6	10	6·6	3·3
Pb	<10	<20	46·3	900
Sn	<9	35	101	34
Ni	3·1	<12	6·2	48
Be	2·7	<4	3·8	<6
V	3·7	<5	<2	3·1
Al	118	1750	490	<5000
Ti	13·5	85	16	940
Ba	70·5	75	95	248
Cr	<3	5·5	4·4	7·5
Sr	31·4	<7	7·6	70

[1]Grown on synthetic medium.   [2]Sample from the Irish Sea.

all marine algae contain chlorophyll a and some accessory pigments, including chlorophylls and carotenoids. Secondary chlorophylls appear to be absent in the Myxophyceae and possibly in the Xanthophyceae. The tentative presence of chlorophyll c in the latter is based on its presence in Olisthodiscus, which may have been wrongly placed in the Xanthophyceae (Riley and Wilson, 1967). Chlorophyll c occurs in the Bacillariophyceae, Dinophyceae, Chrysophyceae, Cryptophyceae, and Haptophyceae; chlorophyll b occurs in the Chlorophyceae and the Prasinophyceae. Beta carotene is the predominant carotenoid in all the unicellular algae except in the Cryptophyceae where α-carotene predominates. The Prasinophyceae appear to be unique in containing several carotenes and possibly lycopene (Ricketts, 1970). Fucoxanthin is usually the predominant xanthophyll in the Bacillariophyceae, Chrysophyceae, Xanthophyceae, and Haptophyceae while peridinin characterizes the Dinophyceae. Other xanthophylls are more or less characteristic of the taxomonic classes. However, the Prasinophyceae contain a large number of xanthophylls and many of these have not been properly identified (e.g. see Ricketts, 1970). In addition, inconsistencies occur; for example, Riley and Wilson (1967) have reported on the occurrence of fucoxanthin as the principal xanthophyll in Gymnodinium veneficum—peridinin was absent from this species; astaxanthin, which is usually a characteristic pigment in some marine animals, was found by Jeffrey (1961) to occur in a Chlorophyceae; antheraxanthin was found by Parsons (1961) to be the principal xanthophyll in a Myxophyceae in which both myxoxanthin and myxoxanthophyll were apparently absent. The reporting of chlorophyll c as a single compound in Table 10 may be open to question since Jeffrey (1969) has identified two spectrally different components in chlorophyll c preparations. In general, however, it remains the conviction of pigment chemists (e.g. Strain, 1966) that there is a relationship between the occurrence of particular pigment systems and the taxonomic classification of an organism. In this respect it has been suggested (Ricketts, 1966b) that the very diverse spectrum of pigments found in the Prasinophyceae may reflect their possible position at a branching point in protistan evolution.

TABLE 10. PIGMENTS OF MARINE PHYTOPLANKTON

Pigment		Bacillario-phyceae	Dino-phyceae	Chryso-phyceae	Chloro-phyceae	Myxo-phyceae	Xantho-phyceae	Crypto-phyceae	Prasino-phyceae	Hapto-phyceae
Chlorophyll	a	+++	+++	+++	+++	+++	+++	+++	+++	+++
	b				++				++	
	c	++	++	+			(+)	++		++
Carotene	α							+++	+	
	b	+++	+++	+++	+++	+++	+++		+++	+++
	g								+	
Xanthophylls										
Fucoxanthin		+++	(+)	+++			+++			+++
Neofucoxanthin		++		++						++
Diadinoxanthin		++	++	++			++			++
Diatoxanthin		+					+			+
Dinoxanthin			+							
Peridinin			+++							
Neoperidinin			+							
Lutein					+++				++	
Zeaxanthin					+				+	
Flavoxanthin					+					
Violaxanthin				+	+				++	
Neoxanthin					+					
Alloxanthin (1 + 2)								+++		
Monodoxanthin								+		
Crocoxanthin								++		
Myxoxanthin						++				
Myxoxanthophyll						++				
Anthraxanthin						(+)				
Siphonaxanthin					+*				+	
Number of unidentified pigments		2		1			1	2	8	1
Phycobilins						++		++		

( ) Presence uncertain, + + + principal pigment, + + generally reported as present, + sometimes reported as present.
*Only in the siphonous (non-planktonic) members.

From a consideration of photosynthetic mechanisms the diversity of pigments discussed above can be divided into three major pigment systems. These are known as

1. the chlorophyll *a* and *b* system,
2. the chlorophyll *a*, *c* and carotenoid system,
3. the chlorophyll *a* and phycobilin system.

Photosynthetic processes in the Chlorophyceae, Bacillariophyceae, and Myxophyceae are characteristic of the three systems, respectively.

The chemical composition of unicellular algae can be greatly affected by changes in environmental conditions; one of the earliest illustrations of this was with *Chlorella* (Spoehr and Milner, 1949). Subsequent investigations have been carried out by a number of authors and some differences between the results of different analysts may be attributed to a lack of clear definitions of the environmental conditions under which the algae were grown. Environmental conditions which may affect composition include nutrient levels, light, temperature, and salinity. An illustration of changes in the composition of a phytoplankton bloom is given in Table 11. The results are taken from Antia et al. (1963). In the experiments reported, a large volume of sea water was enclosed in situ within a plastic sphere and changes in the sea water chemistry and phytoplankton were followed over a period of 30 days. The principal phytoplankton species which grew during the experiment were a mixture of diatoms (Thalassiosira rotula, T. aestivalis, Skeletonema costatum, Stephanopyxis turris, Chaetoceros pelagicus, Navicula sp.) and a dinoflagellate (Gyrodinium spirale). From the results of the experiment it may be seen that the nitrogen content of the phytoplankton decreased when nitrate became exhausted; carbohydrate and lipid increased as organic nitrogen decreased. A summary of the overall changes in cell components, together with a summary of values obtained from the literature (Strickland, 1960) are presented in the bottom half of Table 11. These ratios are useful in ecological studies involving trophic relationships where it is often impossible to make separate measurements of plant and detrital particulate material in situ. More recently, Banse (1974 a and b, in press) has re-examined some aspects of the literature on the ratios of the principal elements and compounds in phytoplankton. From his studies it is clear that some ecological data have led to incorrect interpretations and one of these has been corrected in Table 11. However, a major point made by Banse (in press) is that all particulate material ratios are subject to the time frame of sample collection. Thus, for example, an actively growing phytoplankton population in recently upwelled water may have a carbon:chlorophyll ratio of 30 but as the water ages the increase in detrital and microzooplankton particulate carbon may lead to an observed increase in the carbon:chlorophyll ratio before phytoplankton themselves have departed from a ratio of ca. 30, due to nutrient exhaustion. From these observations it is apparent that while the cellular composition of phytoplankton is heavily influenced by environmental conditions, the accurate collection of data on such changes may be difficult under field conditions.

Studies carried out by Handa (1969) on chemical changes in a pure culture of Skeletonema costatum showed that the transfer of healthy cells from light to dark resulted in a 57% decrease in carbohydrate followed by a 28% decrease in protein and a 44% decrease in lipid. The principal glucose constituent utilized was the water-soluble fraction (β-1, 3-glucan) with little change occurring in the quantity of cell wall constituents. Ackman et al. (1968) studied time-dependent changes in the fatty acid content of three species of phytoplankton. Indefinite results were obtained with one species but Dunaliella tertiolecta showed a decrease in C_{16} unsaturated acids and an increase in C_{18} unsaturated acids over a 2-week period. The effect of temperature on fatty acid synthesis in Monochrysis lutheri was also investigated; total polyunsaturated acids were found in twice the quantity at 10°C compared with 20°C cultures, and 18:1ω7 was reduced at lower temperatures from 2·5 to 0·8%. There is an apparent parallel in these results and results which have been obtained with terrestrial plants (Ackman et al., 1968). Degens (1970) studied changes in protein and amino acids of three species of unicellular algae grown at different temperatures in nutrient-enriched water and following respiration in the dark for up to 19 days. The author concluded that there was a gain in protein content with increasing water temperature but that the total protein content decreased with the length of respiration. The conclusion reached by Degens (1970) that in the initial stages of respiration, protein is lost more rapidly than carbohydrate, is not borne out by Handa's results with respect to the rapid utilization of the water-soluble carbohydrate fraction during respiration.

TABLE 11. CHANGES IN THE CHEMISTRY OF SEA WATER AND PHYTOPLANKTON DURING AN ALGAL BLOOM* (FROM ANTIA et al., 1963)

Day	Nitrate (µg at/l)	Phosphate (µg at/l)	Silicate (µg at/l)	$\dfrac{\text{Carbohydrate}}{\text{Carbon}}$	$\dfrac{\text{Lipid}}{\text{Carbon}}$	Phytoplankton ratios					
						$\dfrac{\text{C}†}{\text{N}}$	$\dfrac{\text{C}}{\text{P}}$	$\dfrac{\text{Si}}{\text{C}}$	$\dfrac{\text{C}}{\text{Chl } a}$	$\dfrac{\text{Carotenoids}}{\text{Chl } a}$	$\dfrac{\text{N}}{\text{P}}$ (atoms)
12	17	1·5	48	0·2	0·1	3	17	0·5	37	1·0	13
14	9	0·9	42	0·3	0·2	3	22	0·6	23	0·9	16
16	0	0·2	15	0·6	0·3	15	27	1·0	25	0·9	18
18	0·5	0·2	10	1·1	0·3	15	31	0·9	49	1·0	19
20	0·5	0·3	2	1·0	0·4	15	37	0·9	49	1·0	19
24	1·0	0·4	1	1·2	0·3	15	33	—	52	1·1	16
27	0·5	0·3	2	1·1	0·4	15	32	—	66	1·2	15
30	0	0·5	4	0·9	0·3	15	35	—	79	1·2	17
Vigorously growing phytoplankton with excess nitrate in the water				0·2	0·15	3	20	0·6	25	0·95	12–15
Unhealthy phytoplankton in nitrate-depleted water				1·1	0·35	15	33	1·0	60	1·15	15–18
Values suggested by Strickland (1960)				—	—	6 ±2	40 ±15	0·8	30	0·9	14·5

*Predominantly diatom.   †Ratios suggested by Banse (1974a).

The effect of light quality on the chemical composition of phytoplankton has been studied extensively by Wallen and Geen, both in the laboratory (1971 a and b) and under natural conditions (1971c). From laboratory studies it was apparent that the colour of light influenced the pathway of $^{14}CO_2$ metabolism; for example, blue or green light favoured protein synthesis while white light favoured carbohydrate synthesis. Minor constituents, such as DNA, RNA, and pigment concentrations were also influenced by the colour of light. These results were largely borne out by field studies. For example, the relative activity of ^{14}C in the phytoplankton carbohydrate fraction decreased with depth while the ^{14}C incorporated into the protein fraction increased with depth. From these results the authors concluded that depth differences in light quality were independent of changes in light intensity in determining the chemical composition of phytoplankton *in situ*.

It has been suggested (Blumer *et al.*, 1971, and references cited therein) that differences in the hydrocarbon content of different species of marine phytoplankton are sufficiently great to be used as a means of identifying algal classes and possibly species. This suggestion may be particularly useful in determining the composition of mixed phytoplankton populations. In addition, since hydrocarbons are not readily destroyed in passing up the food chain, it appears that a useful technique may be available for diagnosing food preferences of zooplankton and higher organisms, *in situ*. The predominant hydrocarbon found in the Bacillariophyceae, Dinophyceae, Cryptophyceae, Haptophyceae, and Euglenophyceae was the unsaturated n-21:6 heneicosahexaene (abbr. HEH). This compound was absent in the Cyanophyceae, Rhodophyceae, Xanthophyceae, and Chlorophyceae, which were generally characterized by n-14 to n-17 hydrocarbons and in particular by either n-pentadecane or n-heptadecane. The hydrocarbons of phytoplankton appear to be very different from the hydrocarbons of zooplankton (mostly C_{19} and C_{20} isoprenoid alkanes and alkenes) and of mineral oils, in which olefins are absent but which contain branched chain, alicyclic, and aromatic compounds in relatively high proportions.

2.3 ZOOPLANKTON

Major chemical constituents of the zooplankton from three areas are shown in Table 12. While some disagreement between similar groups may be due to differences in analytical technique (especially drying), it is apparent that the data show general similarities. Thus the dry to wet weight ratio of copepods and euphausiids appears to be in the range 10–20% with possibly higher values found among amphipods; high dry weights in pteropods may be associated with inorganic shell material, which is generally reflected in an ash content of greater than 20%. On a dry weight basis, the ash content of 70% encountered among the ctenophores and tunicates is due to inorganic salts contained in the gelatinous bodies of these animals. The carbon content of a wide variety of zooplankton ranges from *ca.* 30 to 40% of the dry weight, except when a large amount of ash is present; similarly nitrogen and phosphorus values generally lie in the range 5–10 and 0·5–1·0%, respectively. The C/N ratio for more than 80% of the samples analysed by Omori (1969) was in the range 3–8.

Platt *et al.* (1969) found a high degree of correlation ($r = 0.94$) between the carbon content and caloric equivalent of marine zooplankton. Using their own data and data from other reports, the best fitting equation was given as

$$y = -3370 + 136x - 0.514x^2, \tag{20}$$

TABLE 12. MAJOR CHEMICAL CONSTITUENTS OF ZOOPLANKTON

Organism or group	Dry wt. as % wet wt.	Carbon	Nitrogen	Hydrogen	Phosphorus	Ash	Comment
			(all constituents expressed as a % of dry wt.)				
Copepods*	11·6–16·3	35–48	8·2–11·2		0·7–0·8		Sargasso Sea
Euphausiids &							plankton from
mysids*	14·5–18·0	35–43	9·4–10·5		1·4–1·6		Beers (1966)—
Chaetognaths*	6·0–7·4	22–34	6·3–9·4		0·5–0·7		range for each group
Fish/fish larvae*	11·9–16·0	33–42	8·3–10·7		0·9–1·8		
Polychaetes*	5·7–27·0	16–44	4·4–11·2		0·4–1·8		
Siphonophores*	0·3–6·1	3–16	1·0–4·4		<0·1–0·2		
Hydromedusae*	0·3–10·1	5–10	1·4–6·2		0·1–0·4		
Pteropods*	22–32	21–25	2·7–4·2		0·2–0·4		
Copepods+	10·2–15·8	32–42	4·7–7·1		0·4–0·8	18–23	Continental
Euphausiids+	19·0–20·0	33–37	5·2–7·1		0·9–1·2	19–22	shelf off New
Ctenophores+	4·7–5·0	(6·4)	0·2–1·1		0·1–0·2	70–75	York, from
Pteropods+	3·5–19	26–28	2·2–5·0		0·3–0·6	24–64	Curl (1962)—
Tunicates+	4·0–4·1	7–11	0·3–1·5		0·1–0·3	71–77	range for each group
Copepods*	9·2–33·9	39–66	5·1–13·1	6·7–10·3		2–6	North Pacific
Amphipods*	18·4–36·6	26–48	4·4–8·2	4·4–7·6		10–37	from Omori
Euphausiids*	20·2–21·3	39–47	10·0–10·7	6·7–7·6		8–9	(1969)—range
Chaetognaths*	11·6–14·1	44–48	10·7–11·1	7·2–7·6		4–5	for each group
Pteropods*	25·0–36·4	17–29	1·5–6·0	1·1–3·8		29–43	

*Dried at 60°C.   +Dried at 105°C.

where y is in cal/g dry wt. and x is the % organic matter in the dry plankton. If data on ash content were included a higher correlation ($r = 0.98$) was obtained. For the authors' area of study this was given as

$$cal/g\,dry\,wt. = 1351 + 106(\%\,carbon) - 21.2\,(\%\,ash). \qquad (21)$$

Raymont et al. (1969a and references cited therein) have analysed a large number of zooplankton species (decapods, mysids, and euphausiids) for their biochemical constituents in terms of total protein, carbohydrate, and lipid. The authors concluded that in all their zooplankton analyses, protein was high (53–64% dry weight), carbohydrate was extremely low (1–3%), and lipid content was variable. Some areal and vertical differences were encountered; neritic mysids had a high protein (70–72%) and low lipid (13–14%) content, while some deep-sea and offshore species had a lipid content of greater than 20%. Ash content was highest in deep-sea decapods (15–24%) and lowest in neritic mysids (7–8%). The chitin content of all species of crustacea ranged from 3 to 8%. Differences reported above may in part be due to seasonal changes in composition for a single species. Thus Raymont et al. (1969 a, b) showed that for the euphausiid, *Meganyctiphanes norvegica*, seasonal changes in lipid content varied from 10 to 30%, and in protein content from 50 to 60%; the seasonal relationship between protein and ash content was the reciprocal of the lipid content. With *Euphausia superba* it was shown that the lipid content of animals increased seasonally from spring to summer and with body weight. An accompanying decrease in the body protein gave large animals a protein to lipid ratio of nearly 1 (Ferguson and Raymont, 1974).

The amino acid constituents of two species of euphausiid (from the south and north Pacific) and one species of copepod from the English Channel are compared with the amino acids from casein in Table 13. According to Cowey and Corner (1963) the proportions of amino acids found in *Calanus helgolandicus* are very similar to those found in phytoplankton and particulate matter on which the animals feed. There are some differences between the proportion of amino acids in the two euphausiids compared with *Calanus helgolandicus*, particularly with reference to the amounts of glycine and alanine and the sulfur amino acids, cystine and methionine. However, in general the amino acid spectrum of the three crustaceans appears well balanced and comparable to a nutritionally reliable animal protein, casein. Raymont *et al.* (1973) analysed the

TABLE 13. AMINO ACID COMPOSITION OF ZOOPLANKTON

Amino acid	Species (g amino acid/100 g protein)			
	Euphausia pacifica[1]	*Euphausia superba*[1]	*Calanus helgolandicus*[2]	Whole casein[3]
Alanine	5·61	5·46	8·1	3·2
Glycine	5·35	4·67	8·9	2·0
Valine	5·19	5·90	6·6	7·2
Leucine	7·83	7·70	7·8	9·2
Isoleucine	5·16	5·10	4·8	6·1
Proline	3·47	4·21	4·1	10·6
Phenylalanine	6·50	6·47	4·1	5·0
Tyrosine	4·15	4·06	1·5	6·3
Tryptophan	1·57	1·50	—	1·7
Serine	4·82	4·95	4·1	6·3
Threonine	4·83	4·70	4·0	4·9
Cystine/2	1·35	1·45	0·7	0·3
Methionine	3·25	3·03	1·2	2·8
Arginine	5·95	6·22	7·8	4·1
Histidine	2·22	2·30	1·8	3·1
Lysine	7·84	8·58	8·1	8·2
Aspartic acid	13·7	12·2	9·4	7·1
Glutamic acid	14·7	14·6	11·8	22·4
Glucosamine	2·04	3·45	—	—
Amine N	1·40	1·37	—	—
Taurine	—	—	2·1	—

[1]From Suyama *et al.* (1965).   [2]From Cowey and Corner (1963), and converted from % amino N to % amino acid.   [3]From Gordon and Whittier (1966).

amino acid composition of carnivorous zooplankton including *Sagitta setosa* and *Pleurobrachia pileus*. They found that except for minor differences in the predominance of one or two amino acids, the spectrum from protein hydrolysates of whole animals was remarkably similar to other planktonic organisms assayed.

Jeffries (1969) has studied the free amino acid pool in a temperate zooplankton community. From these studies it was shown that free amino acids were generally higher in summer than in winter and that major changes occurred during periods of community stress, such as those accompanying environmental changes during spring and early winter. Differences were reflected in quantitative changes in the concentration of free amino acids per unit weight of tissue, while relative amino acid composition was the same both in summer and winter zooplankton. The most abundant free amino acids were taurine, proline, glycine, alanine, and arginine. The changes reported are consistent with metabolic changes in other animals where the free amino acid pattern is reported

TABLE 14. COPEPOD LIPIDS (FROM LEE et al., 1970)

Fraction	Calanus helgolandicus[1]	Gaussia princeps[1]
	(% recovery by weight)	
Hydrocarbons	3	Trace
Wax esters	30	73
Triglycerides	4	9
Polar lipds[2]	17	}17
Phospholipids[3]	45	
Total lipid as % dry weight	15	28·9

[1]Collected off La Jolla, California.
[2]Free acids, cholesterol, mono- and diglycerides.
[3]Lecethin and phosphatidyl ethanolamine.

to represent a picture of an organism's metabolic activities; the author suggests that these changes could be used to monitor subtle environmental changes.

The major lipid fractions in two copepods are shown in Table 14. The most characteristic feature of this table is the presence of large amounts of wax esters, which are not generally found among the phytoplankton. Lee et al. (1971) found that the lipid content, and spectrum of fatty acids and alcohols, was largely dependent on the copepods diet. Thus there was a linear correlation between the amount of food fed and total lipid content of the zooplankton; further, the composition of wax esters was changed with increased diet. The most obvious feature of this change was an increase in the amount of C_{30} ester from 12 to 37% in well-fed animals to ca. 50% or greater in animals on a minimal diet. The chain length and degree of saturation of fatty acids and alcohols of the wax esters are shown in Table 15, together with the fatty acid components of the phospholipids. The authors found that the triglyceride and free fatty acids generally resembled the dietary fatty acids but that the long-chain alcohols did not correspond either in chain length or degree of saturation with the dietary fatty

TABLE 15. Calanus helgolandicus LONG-CHAIN ALCOHOL AND FATTY ACIDS (FROM LEE et al., 1971)—% WEIGHT AS METHYL ESTERS

Component	Wax esters		Phospholipid
	Alcohols	Fatty acids	
14:0	1·5	11·8	1·5
16:0	13·0	29·0	39·8
16:1	1·0	13·2	tr
16:2	—	—	0·2
16:3	—	tr	tr
17:1	1·2	0·6	—
18:0	0·8	4·3	4·1
18:1	3·8	12·1	2·7
18:2	0·2	3·6	0·4
20:1	18·4	7·3	0·2
20:2	tr	5·1	0·3
20:3	—	0·6	—
20:4	40·9	7·6	0·3
20:5	—	2·0	13·3
22:3	8·4	—	—
22:6	8·1	2·1	36·5
24:3 to 7	1·7	—	—

acids. From this it was concluded that the alcohols were derived from dietary fats by a variety of different metabolic pathways. The fatty acids of the phospholipid fraction (Table 15) were dissimilar to the fatty acids of the wax esters as indicated primarily by the large amounts of 20:5 and 22:6 fatty acids found in the former. These fatty acids were not present in large quantities in the diet fed to the copepods and it is assumed that they must have been synthesized by the animals. The observation from feeding experiments that the phospholipid fatty acids were not affected by changes in the amount or type of food ingested is compatible with their structural function in animal metabolism; in contrast the wax esters appear to be entirely storage products of copepod metabolism. While these results have dealt almost exclusively with one species of copepod found off California, similar results regarding the high wax ester content of copepods has been found by Yamada and Ota (1970) with respect to *Calanus plumchrus* in the western subarctic Pacific Ocean. Two North Atlantic tunicate filter feeders, *Pyrosoma* and *Salpa cylindrica*, have been shown (Culkin and Morris, 1970) to contain appreciable amounts of 14:0 (myristic), 16:0 (palmitic), 16:1 (palmitoleic), and 18:1 (oleic), as well as the polyunsaturated 20:5 and 22:6 acids which were also found in the phospholipid fraction of *Calanus helgolandicus*. Lewis (1969) has reported that minor fatty acids in the amphipod, *Apherusa glacialis*, included branched-chain acids and positional isomers; small amounts of branched-chain fatty acids were also found in the two tunicates by Culkin and Morris (1970), who suggested that they may be synthesized within the zooplankton. In a study of fatty acids in phytoplankton, zooplankton, and fish, Williams (1965) showed that the proportional transfer of fatty acids from phytoplankton to zooplankton, discussed above, was also maintained between zooplankton and fish. Thus throughout the three levels of production in the pelagic food chain there is a ubiquitous predominance of palmitic acid with generally large proportions of myristic, palmitoleic, and oleic acids in many organisms.

The presence of wax esters in marine crustaceans has been found to increase in animals from deep water and in near-surface animals taken from subarctic and arctic environments (Lee *et al.*, 1971; Morris, 1972; Gatten and Sargent, 1973). The reason for the elaboration of large amounts of wax esters is believed to be associated with efficient energy storage among animals living in nutritionally sparse environments (e.g. deep water or the arctic). In these areas it is suggested that a brief input of food has to serve for long periods of food shortage. However, in this sense it is questionable why animals should store waxes instead of fats. According to Benson and Lee (1975) waxes and fats are similar in their physical properties, both having similar densities, caloric value per unit volume and compressibility. They appear to differ, however, in their coefficient of thermal expansion. Thus wax expands more than fat when warmed and this may have some effect in increasing buoyancy during diel vertical migrations.

Jeffries (1970) has discussed the effect of seasonal changes in the environment on the fatty acid composition of zooplankton in Narragansett Bay, Rhode Island. The ratio of palmitoleic to palmitic declined during thermal warming from 2·0 to *ca.* 0·3, which reflected a dietary change from diatoms to dinoflagellates. The two acids were found to vary reciprocally throughout the year, as were oleic and steraidonic acids. The author suggests that such variations in chemical patterns could be useful in studying biological organization at the community level.

Mayzaud and Martin (1975) have measured the mineral composition of marine zooplankton and compared their results with the mineral content of phytoplankton. They found that zooplankton had less of an ability to concentrate metals than phytoplankton and that metal concentrations in zooplankton varied with species and metal. Iron, manganese, copper, and zinc were all concentrated in zooplankton compared with sea water but the concentration factor was approximately 10 times less for zooplankton compared with phytoplankton. Since concentrations of these elements are dependent on their local concentration in sea water, Mayzaud and Martin (1975) suggested that the basis for a comparison of results between species as measured by different authors should be made in terms of a 'discrimination factor'. This term was defined as the ratio of the concentration factor of one element to the concentration factor of another element. From a comparison of discrimination factors it was shown that the ratios were different for each category of plankton covering three trophic levels from phyto-plankton, herbivore (*Calanus*) to carnivore (*Sagitta*). The authors believed that these differences indicated that the elements were being chemically adsorbed in various amounts rather than being incorporated into tissue.

2.4 PARTICULATE AND DISSOLVED ORGANIC MATERIALS

In this section particulate and dissolved organic carbon (POC and DOC, respectively) will be considered as being part of the non-living components of sea water. The actual division between what is alive and what is dead among the microscopic particles in sea water is difficult to make, either by microscopic or chemical analyses. Obviously part of the POC in sea water is composed of phytoplankton and bacteria; however, from microscopic examination the presence of small organisms (especially bacteria) adhering to fragments of organic or inorganic 'detritus' is difficult to detect. A chemical differentiation is sometimes attempted based either on the chlorophyll *a* concentration or the amount of Adenosine triphosphate (ATP). The former, times a factor which may range from *ca.* 25 to 250 depending on environmental conditions (e.g. see Table 11), is taken to represent the amount of phytoplankton carbon; the latter times a factor (*ca.* 285, Holm–Hansen, 1970) is taken to represent the amount of living organic carbon, since ATP is destroyed rapidly in dead organisms.

The non-living particulate debris of the sea is sometimes referred to as detritus and has been the subject of a number of reviews (e.g. Nishizawa, 1969; Riley, 1970). Micro-scopic observations of particulate material from sea water collected on a membrane filter, or in a settling cylinder, always show the presence of a large number of particles of irregular shapes and sizes. These particles are collectively referred to as detritus although Odum and de la Cruz (1963) have pointed out that scientists use a wide variety of terms to describe particulate material in aquatic environments. Thus the terms 'organic debris', 'suspended matter', 'particulate organic and inorganic material', 'lepto-pel', and (in lakes) 'tripton' may be assumed to refer to detritus unless specifically defined in another sense (e.g. the world leptopel in geology is sometimes used to describe a fine mud or clay). The term 'seston' should be used in referring to all particulate material including living organisms (plankton) and detritus. Detritus may be further qualified as being inorganic detritus and organic detritus, or bio-detritus; the latter terms imply that the detritus has originated from dead organisms. However, detritus has micro-organisms associated with it and might be more appropriately thought of as microcosm consisting of a particulate substrate in which bacteria and other micro-organisms may be embedded. From this concept the term 'organic aggregate' has often been used as a better description of detritus seen under the microscope. Kane (1967) differentiated between two types of aggregates found in the Ligurian Sea; a 'typical' aggregate was composed of a substrate to which various recognizable particles, such as bacteria and phytoplankton, adhered. These aggregates were usually brownish-yellow and were generally between 30 and 50 μm in their longest dimension. 'Granular' aggregates were composed of small inorganic grey-black granules and were considered to be a possible early stage in the development of 'typical' aggregates. Nemoto and Ishikawa (1969) used various stains to identify the nature of detrital aggregates in the East China Sea. Acid fuchsin, which is a general stain for cytoplasm, was found to stain many particles at all depths; however, the ratio of particles stained with Millon's reagent (protein stain) to particles stained with acid fuchsin increased with depth. A few particles stained with α-naphthol (carbohydrates) and with Sudan black (fats). From similar studies carried out in the north Atlantic, Gordon (1970a) differentiated between detrital particles which appeared as aggregates, flakes and fragments; judging from reactions to histochemical stains he concluded that the aggregates were chiefly carbohydrate, the flakes were chiefly protein and the fragments were entirely carbohydrate.

There is a lack of agreement among analysts on the ratio of carbon to nitrogen in detrital material from below the euphotic zone. The C:N ratio for compounds such as urea is less than one while for most proteins it is about 4·5; the latter value may also characterize some bacteria (Porter, 1946). In phytoplankton the ratio of C:N is 3 to 6 (See Table 11) while sediments generally have a C:N ratio of 10 or more (e.g. Seki *et al.*, 1968; Degens, 1970). It would be logical to expect, therefore, that deep-water detritus would have a C:N ratio between that of phytoplankton and sediments. Some authors (e.g. Holm–Hansen *et al.*, 1966; Handa, 1968; Gordon, 1970b) have obtained C:N ratios of deep-water detritus of greater than 10; others (e.g. Parsons and Strickland, 1962a; Menzel and Ryther, 1964; Dal Pont and Newell, 1963) have obtained values of less than 5. Differences also exist in the C:N ratio of soluble organic material. For example, Duursma (1960) found values in deep water of *ca.* 3 while Holm–Hansen *et al.* (1966) have reported values of *ca.* 10. While these values may be real, it must also be considered that differences in analytical procedures may have led to different results. For example, if the C:N ratio of deep-water soluble organic material is assumed to be low (*ca.* 3) and if this fraction is partly included with the particulate material (e.g. through adsorption on $MgCO_3$-coated membrane filters, such as were used by Parsons and Strickland, 1962a) it would tend to decrease the apparent C:N ratio of the particulate material. Alternatively if bacteria form an appreciable fraction of the detrital biomass they will tend to cause a lower C:N ratio than if the detritus is derived entirely from phytoplankton. In experimental studies on the decomposition of phytoplankton in the absence of zooplankton, Otsuki and Hanya (1968) showed that the C:N ratio of dissolved material released by phytoplankton gradually decreased over a period of 200 days from *ca.* 10 to <3. Decomposition of organic nitrogen in phytodetritus was rapid in the first 30 days; release of carbon and nitrogen during this period amounted to 65–70% of the total. Approximately 10% was converted to dissolved substances and the rest remineralized as CO_2 and NH_4^+.

Williams and Gordon (1970) studied the $^{13}C/^{12}C$ ratios (expressed relative to a standard) in dissolved and particulate organic material down to 4000 m in the Gulf of Mexico and off the coast of southern California. For deep water they found similar ratios ($-21\cdot2$ to $-24\cdot4$) regardless of location or season; further, the ratio for dissolved organic material was very similar to the ratio for particulate organic material ($-22\cdot0$ to $-24\cdot3$). It was also shown that these ratios corresponded most closely to the cellulose and 'lignin' fraction (residue after all other extractions) of phytoplankton ($-22\cdot4$ and $-23\cdot1$, respectively). These ratios were also similar to the ratio for organic material from sediments ($-20\cdot8$ to $-22\cdot3$) but were quite different from organic material derived from the Amazon River ($-28\cdot5$ to $-29\cdot5$). From these detailed studies the authors concluded that deep-water organic detritus is derived primarily from marine plankton and that the soluble and particulate fractions are similar in chemical composition. In another report Williams *et al.* (1969) determined the age of deep-water soluble organic materials being 3400 years old; this value agrees within an order of magnitude with values derived by Skopintsev (1966) from a theoretical approximation based on input and the rate of decomposition.

Various methods have been used to determine how much of the detrital particulate matter is biologically utilizable, particularly in the euphotic zone where it is associated with large populations of filter-feeding animals. The principal difficulty in this respect has been to separate or differentiate between the detrital particulate material and the living particulate material (mostly phytoplankton). In a study on the biological oxidation of organic detritus in which a correction was made for the amount of phytoplankton, Menzel and Goering (1966) found that in samples from 1 m in the north Atlantic, between 16 and 52% of the detritus was biodegradable; applying the same technique to samples taken from below the euphotic zone (200 to 1000 m) the authors could not detect any oxidation of organic material and concluded that the material was essentially all refractory. From feeding experiments, Paffenhöfer and Strickland (1970) found that *Calanus helgolandicus* would not feed directly off detritus. This is similar to the result obtained by Seki *et al.* (1968), who showed that *Artemia* would not feed directly off sedimented plant material but that they could feed off bacterial aggregates which were grown using the detritus as a substrate.

Gordon (1970b), using proteases, showed that approximately 20 to 25% of deep organic detritus was hydrolysable. Holm–Hansen and Booth (1966) determined the amount of ATP in particulate material at various depths off the coast of California. In deep-water samples, 500 to 1000 m, the authors concluded that 3% or less of the particulate material was alive; above 100 m, 14 to 79% was living while at intermediate depths approximately 6% was living with one exception of 27%. Thus it appears that the deep-water hydrolysable material found by Gordon (1970b) represented dead biodegradable detritus; this is in contrast with the results found by Menzel and Goering (1966) using a different technique.

Direct chemical analyses were made on particulate material taken from 400 m in the north Pacific (Parsons and Strickland, 1962a). From the discussion presented above, it may be concluded that most of the organic material in this fraction was dead. The analyses showed that the material was composed of protein and carbohydrate; the principal amino acids were glycine and alanine with glutamic acid, aspartic acid, lysine, arginine, serine, and proline also being detected. Degens (1970) has reported in greater detail on the amino acid composition of deep-water detritus in the Atlantic. Below 200 m there was an apparent increase in the proportions of serine, glycine, lysine, and arginine, and a decrease in alanine with depth down to 2500 m.

The carbohydrate fraction of deep-water detritus analysed by Parsons and Strickland (1962a) was 70% insoluble to treatment with weak acid and alkali (in contrast with 'crude fibre' values for healthy phytoplankton in Table 8). The principal sugars following total acid hydrolysis were glucose, galactose, mannose, arabinose, and xylose. The quantity of fat present was less than 1%. Glucosamine and hexuronic acids were not detected, indicating the lack of appreciable amounts of chitin from crustaceans or hexuronides from marine plants, respectively. However, Wheeler (1967) has reported on the presence of copepod carcasses between 2000 and 4000 m in the north Atlantic; as chitinous material these would contribute to the total organic detritus, but on the basis of their concentration per m^3, they would only account for between *ca.* 0·5 and 5% of the total organic carbon in deep water. Handa and Tominaga (1969) and Handa and Yanagi (1969) have carried out detailed carbohydrate analyses of particulate materials down to 700 m in the northwest Pacific. Their results show that the water-soluble carbohydrate fraction of phytoplankton disappeared between 50 and 300 m and that between 300 and 1000 m, only water-insoluble carbohydrates remained. Detrital material from these depths contained 50% less glucose than phytoplankton, but correspondingly higher proportions of galactose, mannose, xylose and, in contrast to Parsons and Strickland (1962a), appreciable quantities of glucuronic acid.

The inorganic fraction of particulate detritus may be a variable fraction of the total dry weight but generally it amounts to at least 70% (Wangersky, 1965). The exact chemical nature of the inorganic material has not been defined but calcium carbonate particles are known to occur in the open ocean (Wangersky and Gordon, 1965) and these may be intimately associated with the organic material in sea water (Chave, 1965 and 1970). In coastal areas inorganic particles are associated with clays and other minerals derived from the land; these may include appreciable amounts of silicon, iron, aluminium, and calcium (Armstrong and Atkins, 1950). More detailed data on the inorganic fraction of detritus can be derived from literature on marine sediments (e.g. Griffin *et al.*, 1968).

The quantity of chlorophyll *a* in particulate material decreases with depth so that below the euphotic zone most of the chlorophyll *a* has either disappeared or been converted to phaeophytin or phaeophorbide (Lorenzen, 1965). Saijo (1969) reported on chlorophyll pigments down to 4000 m in the northwest Pacific. The total concentration of chlorophyll *a* in waters below 400 m was from <0·001 to 0·003 µg/l; phaeo-pigments were often present, however, at 10 times the concentration of chlorophyll *a*. Currie (1962) showed earlier that digestion of phytoplankton by zooplankton resulted in the almost total conversion of chlorophyll *a* to phaeo-pigments; this has been used by some scientists as a measure of zooplankton grazing (e.g. Levi and Wyatt, 1971; Malone, 1971b). From studies in the western Indian Ocean, Yentsch (1965) showed that the ratio of chlorophyll *a* to phaeopigments decreased with light intensity. This was concluded to be a reversible reaction since the chlorophyll to phaeophytin ratio of dark adapted phytoplankton could be increased when they were restored to the light. Carotenoids appear to be much more resistant to decomposition both within the water column (Yentsch and Ryther, 1959) or as a result of the digestive processes of animals. In the latter case, Fox *et al.* (1944) found an average content of 24 mg of carotenoid per 100 g dry weight of feces from mussels, but negligible quantities of chlorophyll. However, these authors did not measure chlorophyll *c* and while chlorophyll *a* is biologically and chemically unstable, chlorophyll *c* has been found (Jeffrey, 1974) to persist in detrital material and scenescent cells.

Considerable biological importance has recently been attached to the surface film of organic material in the sea (Garrett, 1965; Harvey, 1966). Using a specially constructed surface skimmer, Harvey (1966) showed that the surface film contained large amounts of living nanoplankton, structural components of disintegrated organisms, surface active substances, chlorophyll, and carotenoid pigments. Garrett (1964, 1967) showed that major chemical components of this layer were fatty acids, fatty acid esters, alcohols and hydrocarbons. Taguchi and Nakajima (1971) measured the ratio of particulate carbon in the surface layer (150 μm thick) compared with subsurface (10–15 cm deep) samples. The ratio of surface to subsurface particulate carbon was generally 2 to 5 with a maximum concentration factor of 17·6 in coastal environments. Similar high ratios were found by Nishizawa (1971) in the open ocean waters of the equatorial Pacific. Both Sieburth (1971), and Taguchi and Nakajima (1971) have drawn attention to the concentrations of living organisms (neuston) associated with the detrital surface film. Bacterial counts were generally one or two orders of magnitude higher in the surface film than just below the surface; the species of phytoplankton and zooplankton were also quantitatively greater and to some extent qualitatively different, in the surface layers compared with subsurface samples. These observations have lead to the conclusion that a neuston community exists at the sea surface which constitutes a micro-environment rich in a specific microflora and fauna.

The presence of this neuston community may in part be due to an accumulation of material floating on the sea surface. However, it may also be formed by the peculiar properties of the air–water interface. Sutcliffe et al. (1963) showed that when air is bubbled through filtered sea water, organic particles are formed. It was further shown that these particles could support the growth of brine shrimp (Baylor and Sutcliffe, 1963). Carlucci and Williams (1965) showed that the action of bubbling sea water tended to concentrate bacteria at the surface. However, Menzel (1966) and Barber (1966) have questioned some of the experimental evidence for these findings. In the latter reference it is reported that neither bacteria nor bubbling alone cause a significant increase in the amount of particulate material in sea water containing organic materials with a molecular weight of less than 100,000.

While most observations on detritus have been made on samples collected in water bottles it is apparent that the form of detritus as observed under the microscope may bear little relation to its appearance in situ. From some of the earliest in situ observations made under the sea it was noted that much of the detritus was aggregated into clumps or long streaks which were easily disintegrated by the action of sampling bottles. Suzuki and Kato (1953) described its appearance in situ as 'marine snow'. Aggregation of particulate materials has been studied by a number of authors (e.g. Sheldon et al., 1967; Parsons and Seki, 1970) and it appears that bacteria are probably involved in this process, either through certain species which tend to form clumps or through the release of organic polymers which tend to attach to other particles. The effect of aggregation is to increase the size and therefore the sinking rate of the particles. However, as Riley (1970) and Kajihara (1971) have observed, the sinking of detrital aggregates does not obey Stoke's law [eqn. (15)] since there is a decrease in the specific gravity of particles with increased size due to aggregation.

The division between particulate (POC) and dissolved (DOC) organic carbon is usually made on the arbitrary basis of filter pore size. Filters most frequently used to separate particulate from dissolved substances in sea water have a pore size of 1 μm, plus or minus 0·5 μm depending on individual choice. However, Sheldon and Sutcliffe (1967) showed that there was a considerable difference in the size of particles retained by commercially available filters compared with the advertised pore size. Sharp (1973) has shown that the arbitrary division of particulate and soluble organic matter in sea water is actually unjustified. The author showed that a continuous size distribution exists of microscopic, submicroscopic and colloidal particles over the size range 10^{-3} to 10^3 μm. In spite of this, it is still convenient to refer to POC and DOC as a functional division or organic matter in sea water. This is because from the point of view of microscopic examination of sea water, as well as for many filter feeding organisms, there is a definite distinction between what can and cannot either be seen or filtered. For analytical purposes this distinction is generally made by considering everything larger than *ca.* 1 μm diameter to be particulate.

From a summary of values in the literature, Williams (1975) concludes that near surface (<100 m) dissolved organic carbon values range from 0·6 to 2·0 mgC/l while deep-water values range from 0·4 to 1·5 mgC/l depending to some extent both on geographic location and the method of analysis used. The composition of the dissolved fraction in sea water can be divided into two principal categories of compounds on the basis of their biological activity. The first group of compounds are generally present at extremely low, or 'threshold', concentrations because at higher concentrations they are rapidly metabolized by heterotrophic activity. Compounds that belong to this group include monosaccharides, such as glucose, amino acids, acetate, and other simple compounds such as those found in the tricarboxylic acid cycle. Jannasch (1970) found that experimentally determined threshold concentrations of lactate, glycerol, and glucose for single species of bacteria were very high (*ca.* 1 mg/l). In the sea the total carbohydrate is generally found to be less than 1 mg/l (e.g. Handa, 1970) so that individual sugars must be present at much lower concentrations. This discrepancy is explained by Jannasch (1970) by assuming that in the sea, the many species of bacteria present with different uptake efficiencies will selectively reduce the concentration of any one readily metabolizable substrate to very low concentrations. Such concentrations of specific compounds have to be analysed by enzymatic methods; for glucose the concentration in sea water is generally found to be less than 10 μg glucose C/l with a few higher near surface values of *ca.* 50 μg glucose C/l (e.g. Vaccaro *et al.*, 1968; Andrews and Williams, 1971).

Oruga (1972a) has discussed the rate of decomposition of organic material from dead phytoplankton. He found that microbial decomposition rate of dead *Scenedesmus* was 0·02 to 0·05 day^{-1} during the first 15 days and 0·0004 day^{-1} for the next 100 days. Comparable decay rates for recently dead phytoplankton are found for dissolved organic carbon in surface waters. Decay rates for deep organic carbon are more comparable to the second value above (i.e. for phytoplankton after an initial brief period of rapid decomposition).

The second group of compounds found in the dissolved fraction are refractory materials which probably make up the bulk of the total organic carbon dissolved in sea water (e.g. Ogura, 1972b). From the ageing of DOC in deep water (*ca.* 3400 years, see Section 5.2) it is apparent that these compounds are not readily metabolized. While their exact nature has not been thoroughly investigated it is probable that these compounds include high molecular weight polymers which may include such compounds as lignins, humic acids, and proteins (e.g. see review by Williams, 1975).

Skopintsev (1971) has attempted to give an organic carbon budget for the oceans based on the average concentration of organic matter in the water column and the average rate of primary production in the hydrosphere. Assuming the average phytoplankton production to be 120 gC/m^2/yr (neglecting products of exudation), then the total production was calculated as $3·84 \times 10^{16}$ gC (assuming a surface area of $3·2 \times 10^{14}$ m^2). From experimental data and oxidation rates in the sea it was assumed that 92% of this production was respired back by higher trophic levels, 5% settled out on the bottom and 3% was contributed to the more refractory pool of organic carbon in the ocean. The latter figure amounts to 11.52×10^{14} gC to which Skopintsev added 1.8×10^{14} gC from terrigenous origin. The total input of organic carbon was then 13.3×10^{14} g which represents approximately 0.1% of the total organic carbon in sea water (i.e. assuming a concentration of 1 gC/m^3 and a total ocean volume of 1.3×10^{18} m^3). Hence the residence time of refractory organic matter in the ocean is in the order of 10^3 years, as calculated earlier by Skopintsev (1966).

CHAPTER 3

THE PRIMARY FORMATION
OF PARTICULATE MATERIALS

3.1 AUTOTROPHIC PROCESSES

In the ocean there are algae and some bacteria which can synthesize high-energy organic compounds from low-energy inorganic compounds such as water and carbon dioxide. The source of energy for these organisms is either light, or chemical energy derived from the oxidation of inorganic compounds; such organisms do not require organic materials as a source of energy. This life style is called 'autotrophy' and the organisms are called 'autotrophs'. When one considers the cycling of organic material in the oceans, autotrophic organisms are referred to as 'primary producers' because it is these organisms which are the only producers of original autochthonous organic material in the sea. The organic material produced by the primary producers is referred to as 'primary production' and primary production per unit time in a unit volume of water (or under a unit of area) is called 'primary productivity'. On the basis of differences in energy source for organic matter synthesis, autotrophy is divided into two different categories known as 'photosynthesis' (light energy) and 'chemosynthesis' (chemical energy).

3.1.1 BASIC PHOTOSYNTHETIC REACTIONS

The fundamental relationship governing the photosynthetic process can be summarized in the following equation:

$$nCO_2 + 2nH_2A \xrightarrow{\text{light}} n(CH_2O) + 2nA + nH_2O \qquad (22)$$

where reduced compounds, such as H_2O, H_2, H_2S, $H_2S_2O_3$, and some organic compounds may be used as the H-donor in H_2A but only light is used as the energy source.

The whole photosynthetic process is not a single reaction expressed by eqn. (22). For example, the photosynthetic process can further be described by three different steps: (1) capturing light energy and transferring the energy into chemical forms, (2) further changing the chemical forms into another suitable chemical form for bio-chemical

65

reactions (ATP and NADPH, see p. 67 for their complete names), and (3) fixing CO_2 using ATP and NADPH produced by the former steps. The first two steps are distinctive only for the photosynthetic organisms, but the third step is observed, widely in all autotrophic organisms including chemolithotrophs (see Section 3.1.7).

Photo-autotrophs in the ocean include representatives of the algae as well as photosynthetic bacteria; both types of these organisms are usually widely distributed in the ocean. However, quantitatively the algae are the most important photo-autotrophs in the ocean, with a few exceptions to this generalization to be found in neritic regions.

Photosynthetic algae require H_2O as the H-donor, and eqn. (22) can be modified for algal photosynthesis as follows:

$$nCO_2 + 2nH_2O \xrightarrow{\text{light}} n(CH_2O) + nO_2 + nH_2O. \tag{23}$$

This process requires energy of *ca.* 120 kcal per mole of carbohydrate formed. The energy is derived through the absorption of light by photo-synthetic pigments, which absorb the light mainly in the visible region from 400 to 720 nm. Each photosynthetic

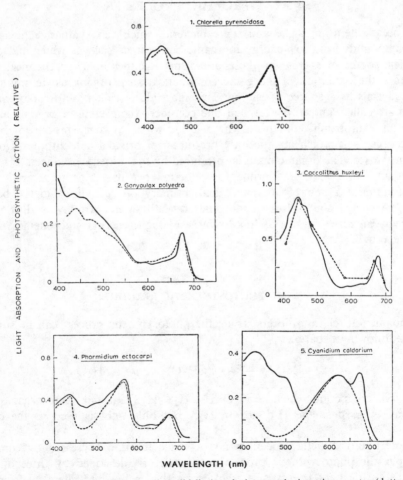

FIG. 23. Light absorption of intact cells (solid line) and photosynthetic action spectra (dotted line). 1, 2, 4, and 5 redrawn from Haxo (1960); 3, after Paasche (1966).

pigment has a distinctive light (precisely, photon or quanta) absorption characteristic depending on their molecular structure. Each group of organisms contain chlorophyll *a* and several accessory* pigments in the thylakoid membranes in the chloroplasts in the cells (in procaryotic algae, such as the blue green, there are no obvious intracellular organella and thylakoid membranes are suspended directly in the cells). Thus the light-absorption patterns are different in each algal group depending on their pigment systems (see Table 10). Fig. 23 shows light-absorption spectra of some intact marine algal cells. Light of wavelengths shorter than 600 nm is mainly absorbed by chlorophyll *a* and accessory pigments. Above 600 nm light for photosynthetic processes is only absorbed by chlorophyll. The latter absorption peak is normally observed at 680 nm; however, by improved techniques, such as derivative spectrophotometry and spectrofluorometry at very low temperatures, the existence of two other peaks, one at 670 and the other at 695 nm, were recognized (Kok and Hoch, 1961). These multiple peaks of chlorophyll *a* are distinctive for intact cells and are not observed in chlorophyll *a* extracted in organic solvents. From these findings it became clear that chlorophyll *a* has a complex stereo structure within the thylakoid membranes. The function of these absorption peaks is not fully understood but two separate photo-chemical reactions are now recognized based on light absorption at the shorter wavelength (called chlorophyll *a* 670) and the longer wavelength (called chlorophyll *a* 680).

Energy absorbed at the longer wavelength (chlorophyll *a* 680) is used directly for photochemical reactions or emitted as fluorescence (fluorescence at 730 nm; Fl 730), but energy absorbed at the shorter wavelengths is transferred by the accessory pigments to chlorophyll *a* 670 before being used or emitted as fluorescence (fluorescence at 684 nm and 695 nm; Fl 684 and Fl 695, respectively). The energy accepted by both types of chlorophyll *a* is used for photochemical reactions in two photosynthetic systems, I and II (Fig. 24). These two photosynthetic systems are conjugated by a series of electron transfers, involving quinone and cytochrome. System I, which is primarily mediated through energy derived from chlorophyll *a* 680, is mainly involved in electron transfers. The photochemical reaction catalysed by pigment System II liberates oxygen from water and transfers electrons to plastoquinone (PQ in Fig. 24). Energy transferred from the two photochemical reactions is used for (1) the reduction of nicotinamide adenine di-nucleotide phosphate (NADP) and (2) photophosphorylation of adenosine diphosphate (ADP) into the high-energy compound, adenosine triphosphate (ATP). This series of reactions is carried out in the light and they are collectively referred to as the 'light reaction'.

The reducing power of NADPH and the energy of ATP promote the reduction of CO_2 and produce carbohydrate as well as synthesizing proteins and fats. These reactions are carried out in the dark and are referred to collectively as the 'dark reaction'. Together the light and dark reactions are generally included in the term photosynthesis and this whole process takes place within the chloroplasts. The metabolic processes of the dark reaction were first demonstrated by Calvin and his colleagues, and thus it is also known as the 'Calvin/Benson cycle'. The details of the Calvin/Benson cycle can be found in the treatise of Calvin and Baasham (1962). The first product of CO_2 assimilation by the Calvin–Benson cycle is the three-carbon compound 3-phosphoglyceric acid (PGA) catalysed by the RuDP carboxylases. Recently another pathway was found for the dark reaction, the Hatch–Slack pathway (Hatch and Slack, 1970), in which the first products of CO_2 assimilation are not the three- but the four-carbon compound, oxaloacetic acid,

*'Accessory' may not be a suitable word because these pigments are the main light energy acceptors for the photochemical system II in the basic photosynthetic light reaction.

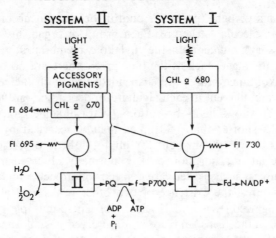

F<small>IG</small>. 24. Schematic presentation of photosynthetic system in algae, represented by two photochemical systems (I and II); PQ, plastiquinone; f, cytochrome; Fd, ferredoxin; NADP reductase; ADP and ATP, adenosine di- and tri-phosphates; P_i, inorganic phosphate (redrawn from Fujita, 1970).

catalysed by the PEP carboxylase. The carbon of the oxaloacetic acid is then decarboxylated. The CO_2 produced is fixed through by the Calvin–Benson cycle, and the three-carbon compound, phosphoglyceric acid, produced as a side product is then used again as the carbon carrier after phosphorylation. The plants which have the Hatch–Slack pathway are called the 'C4-plants' based on the carbon numbers of their first products of CO_2 assimilation. On the other hand, the plants which only have the Calvin–Benson cycle are called the 'C3-plants'. C4 plants are widely known in many species of terrestrial plants in which there is no obvious phylogenical relation recognized; even plants belonging to the same genus have different first products of photosynthesis. In algae, however, there is still only obscure information on which algal species belong to the C4 plants.

The basic photochemical reaction of photosynthesis is carried out by 'photons'; four photons (quantum energy) are required to produce one mole of ATP and NADPH. Since the quantum energy (E) is a function of wavelength,* the shorter wavelength photon has the greater energy. Considering a reaction requiring four photons (quantum energy), actual energy requirement will be much greater in the reaction with short-wavelength than with long-wavelength light. The efficiency of quantum energy transfer from pigments to photosynthetic systems is not always the same. The transfer efficiency can be estimated from the 'quantum yield' which expresses how many moles of CO_2 are fixed (or O_2 are produced) by one photon of light absorbed by pigments ($\{CO_2\}/h\nu$). Some examples of quantum yields are shown in Fig. 25 for different marine algae. The transfer efficiency of light to chlorophyll a is highest because the energy is transferred directly to the photosynthetic system; light is transferred via the accessory pigments with variable efficiencies. For example, the accessory pigments of diatoms, dinoflagellates, and coccolithophorids transfer light energy with an efficiency similar to that of chlorophyll a, but the accessory pigments of green and blue-green algae have relatively low transfer efficiencies. From a consideration of the photosynthetic efficiency of pigments, Fujita (1970) classified marine algal groups into (1) chlorophyll a and b type

---

*$E = h\nu = hC/\lambda$. h, Planck's constant, $6 \cdot 63 \times 10^{-34}$ joule/sec; ν, frequency; C, the speed of light in vacuum, 3×10^8 m sc^{-1}; λ, wavelength.

for green and euglenoid algae, (2) chlorophyll *a*, *c*, and carotenoid type for diatoms, dinoflagellates, and brown algae, and (3) chlorophyll *a* and phycobilin type for red and blue-green algae. The actual light utilization spectra of algae can be obtained by combining the light absorption of intact cells with the quantum yield; such a curve is called the 'action spectrum' (see Fig. 23). The action spectrum is considered to show the photosynthetic light utilization efficiency of a cell and it is one of the important

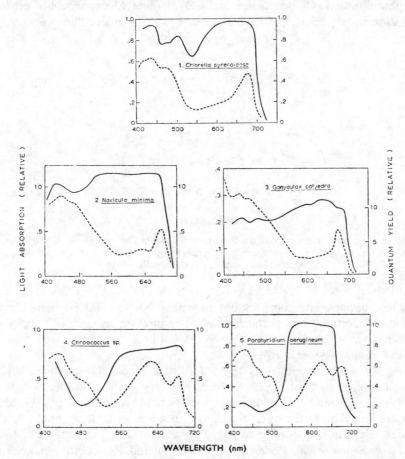

FIG. 25. Light absorption of intact algal cells (dotted line) and quantum yield (solid line). 1, 3, and 5 (light-absorption spectrum) redrawn from Haxo, 1960; 2 redrawn from Tanada, 1951 (quantum yield is absolute); 4 redrawn from Emerson and Lewis, 1942.

characteristics of a species since it determines the ability of phytoplankton to adapt to different light regimes in the ocean. The action spectrum is not stable, however, but sometimes changes depending on growth conditions such as differences in illumination, wavelength of light and nutrient concentration. Extreme examples of change are observed in the blue-green algae (Halldal, 1958; Fujita, 1970).

The final product of photosynthesis shown in eqns. (22) and (23) is carbohydrate, $n(CH_2O)$. This means that the photosynthetic quotient expressed as the ratio of evolved O_2 to absorbed CO_2 (PQ) is close to unity. The photosynthetic quotient becomes higher if other organic compounds are produced (e.g. *ca.* 1·25 for proteins and 1·43 for lipids). The photosynthetic quotient is not a stable property of a cell but it changes depending on the past history of the species and environmental conditions (e.g. Myers, 1953, observed changes from 1·04 to 2·50 in *Chlorella pyrenoidosa*).

Photosynthetic products are partly consumed by basic respiration occurring in mitochondria; this is generally the reverse reaction of photosynthesis. Respiration takes place both in the light and in the dark; however, it can usually only be detected experimentally as O_2 consumption or CO_2 production in the absence of light (i.e. with no disturbance either by photosynthetic O_2 production and CO_2 uptake). Measurement of phytoplankton respiration in the ocean, particularly open ocean, is practically impossible, firstly

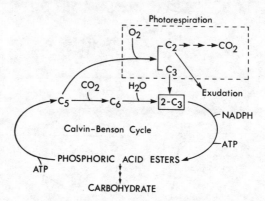

FIG. 26. Photosynthetic carbon fixation pathway of C_3 plant. The number of 'C' indicates the number of carbon atoms in the compound; C_2, phosphoglycolic acid; C_3, phosphoglyceric acid; C_5, ribulose diphosphate, etc. (for details, see text).

because of its very low activity and secondly because of the respiration of other microorganisms (e.g. bacteria and zooplankton). Steemann-Nielsen and Hansen (1959b) proposed an indirect estimation approach of algal respiration using $^{14}CO_2$ uptake technique, which gives reasonable estimate in coastal waters but a very high estimate for open ocean samples (Ryther, 1954).

According to many investigators (e.g. McAllister *et al.*, 1964; Humphrey, 1975), the basic dark respiration of algae obtained from many different species and growth conditions will be around 10% of maximum gross photosynthesis (P_{max}). When phytoflagellates are abundant in the water the ratio of respiration to P_{max} should be changed because high respiration rates of 35 to 60% P_{max} were observed in cultures of *Exuviaella cordata*, *Gymnodinium wulffii*, *Peridinium trochoideum*, and *Prorocentrum micans* (Moshkina, 1961). These high respiration rates were attributed to the motility of the flagellates.

Whether or not algal respiration is the same in the light as in the dark has been a big question for a long time. By applying the non-radioactive mass spectrometric technique using $^{18}O_2$, it has become possible to have direct measurement of algal respiration in the light. From these experiments it has been found that respiration increased in the light due both to additional basic respiration (mitochondrial respiration) and photorespiration; the latter is defined as a light-dependent O_2 uptake and CO_2 release that occurs in photosynthetic cells. At the time of writing, available information on photorespiration have been mainly collected from cultured (freshwater) green and blue-green algae. Unfortunately none of these becomes dominant in the sea, except for a few special cases. However, it may be anticipated that photorespiration will occur commonly in many algae, including marine dominant groups (cf. Tolbert, 1974). The following description of the process is given even though the information was obtained entirely from experiments using green or blue-green algae. As shown schematically in Fig. 26, photorespiration is divided into two steps: (1) to produce glycolate (C_2 compound) from ribulose diphosphate (C_5 compound) which is one of the photosynthetic products, and (2) to oxidize glycolate into CO_2. The whole process is sometimes called the 'glycolate pathway': it occurs in chloroplasts, does not conserve energy as ATP, and does not utilize substrates involved in the ordinary respiratory processes (i.e. tricarboxylic acid cycle). Some glycolate produced is also lost from cells by exudation. In most cases glycolate is the major photosynthetic product exuded by algae, but by no means the only compounds (Fogg, 1966, also see Section 3.2.1). Glycolate produced is also expected to be the major source for protein synthesis in many species of algae. The magnitude of photorespiration during photosynthesis in algae can vary, depending upon many parameters, from a few percent to nearly 100% of total CO_2 fixation (Baasham and Kirk, 1962). High O_2, low CO_2, high light intensity, high temperature and high pH of the medium all favour increased photorespiration. Among these factors, all except for pH and light intensity (particularly near the surface) are rather unfavourable for photorespiration in the marine environment compared with the terrestrial environment. The light quality in the water column seems also unfavourable for photorespiration because photorespiration is sensitive to red and white light but insensitive to blue light which is the most predominant in the marine environment (cf. Lord et al., 1970). Furthermore, algae do not lose much CO_2 during photorespiration because they refix the CO_2 by photosynthesis (Tolbert, 1974). However, all the factors mentioned above only reduce the possible magnitude of photorespiration among marine phytoplankton but it is not possible to eliminate its potential importance. It seems still too early to have quantitative verification of actual photorespiration of phytoplankton in natural aquatic environments.

3.1.2 LIGHT ENVIRONMENT IN THE SEA

The basic aspects of light having biological importance are quantity and quality. Both of these characters of light in the sea fluctuate, sometimes in great magnitude, depending on time (i.e. daily, seasonally, and annually), space (different location on the earth, and depth), weather condition, angular distribution (including direction of maximum flux and degree of diffusion and polarization). The control of many of these aspects originates above or at the surface (such as the change in flux due to rising and setting of the sun), but for other aspects it originates within the water (such as changes in diffusion due to suspended matter and spectral changes due to selective absorption).

Light is qualitatively described by its spectral distribution depending on the difference in wavelength (units:* $m\mu m$, nm, and Ångstrom). Light of wavelength longer than 760 nm is, roughly speaking, considered to be infrared (IR), and that of wavelength shorter than about 300 nm is designated as ultraviolet (UV). The wavelength range between UV and IR is called visible (VS), which is the most important fraction for biological aspects such as photosynthesis and visual sense of organisms. Some organisms are known to be able to sense UV and IR light as well.

From the photosynthetic action spectra shown in Fig. 22, it is obvious that the light energy required by algal photosynthesis is restricted to the wavelengths between 400

*1 $m\mu m$ = 1 nm = 10 Ångstrom (Å) = 10^{-9} m.

and 720 nm. This wavelength range is called 'photosynthetically active radiation' (PAR or PhAR).

Quantitative assessment of light is done by intensity measurement, which is generally taken in three different ways (or in different units) depending on its use and also the instruments available; (1) illumination, (2) energy, and (3) quanta.

Illumination is the indicator of brightness which only includes the visible portion of the energy spectrum (400 ~ 750 nm) and is defined as flux per unit area. The basic unit for luminous intensity is the international candle (cd), which is the intensity of a standard light detected by a standard observer who has a distinctive sensitivity for different wavelength luminosity factors; highest sensitivity is at 555 nm and becomes less sensitive towards the longer and the shorter wavelengths, over the total spectrum, 400 to 750 nm. The luminous flux per unit area is defined for illumination as follows:

> 1 international candle on 1 m^2 = 1 lux (lx),
> 1 international candle on 1 ft^2 = 1 ft cd.
> Then 1 ft cd $\simeq$ 10·76 lx.

Considering that most biological events occur in the visible light range, it is to be expected that the illumination unit is a convenient measure for biological problems in the sea. However, one should always remember that the basic luminous intensity (cd) is detected through a distinctive selective light filter (multiplied by a luminosity factor) as mentioned before, and therefore *precise* comparison of illumination units is only valid for the same type of light source used. If the spectral distributions in the visible range are known for the different light sources which are used (e.g. energy distribution actually measured or calculated based on the colour temperature of the light source), a suitable correction factor for the comparison of the illumination units can be estimated using the curve of the visual sensitivity characteristics.

Energy units for light-intensity measurement, on the other hand, apply for the whole energy spectrum including UV, VS, and IR light, and are expressed by the units such as watts·sec, erg, and gram-calories. All these energy units are interconvertible as follows:

$$1 \text{ g cal} = 4·185 \times 10^7 \text{ ergs * } = 4·185 \text{ watt·sec.}$$

The energy flux with the dimension of units time and unit area is the most suitable energy expression for the biological processes in the sea (e.g. g cal/cm^2/min or g cal/cm^2/day, etc.). The g cal/m^2 (i.e. without the time dimension) is defined as langley (ly):

$$1 \text{ g cal/cm}^2 = 1 \text{ ly.}$$

The measurement of light energy is entirely dependent on the quality of light (i.e. spectral differences). Thus when an energy unit is used for biological events, the light source (especially the energy spectrum distribution in the entire wavelength) should always be specified as well as additional information on experimental conditions, such as the thickness of water for incubation.

Only a rough conversion from an illumination unit to an energy unit (or vice versa) can be made; thus 1 lx = approx. 6×10^{-6} ly/min for the sunlight at the sea surface (Strickland, 1958), approx. 86×10^{-6} ly/min for a tungsten lamp (Hill and Whittingham, 1955), approx. 5×10^{-6} ly/min for white fluorescent lamps (Westlake, 1965).

---

*10^7 ergs = 1 joule.

The measurement of light in quanta measures the number of photons of a particular energy which is related to the wavelength:

$$1 \text{ g cal} = 2 \cdot 11 \times 10^{15} \times \text{Å quanta}$$
$$= 3 \cdot 50 \times 10^{-9} \text{ Å einsteins}$$

Then

$$1 \text{ einstein} = 6 \cdot 02 \times 10^{23} \text{ quanta} = \frac{2 \cdot 86 \times 10^8}{\text{Å}} \text{ g cal}$$

where Å is the wavelength in Ångstrom (10^{-10} m). This equation indicates that the quantum energy decreases with increasing wavelength. In the range of visible light (4000–7000 Å), the average energy of each photon will be estimated as follows (employing the average wavelength of visible light 5500 Å):
For visible light:

$$1 \text{ einstein} = \frac{2 \cdot 86 \times 10^8}{5500} \text{ g cal} \simeq 52 \times 10^3 \text{ g cal.}$$

Considering that the basic photosynthetic reaction is carried out by photons, the quanta unit may be the most ideal measurement for photosynthetic events.

In the natural environment, solar radiation is the most important light source, but it can show order of magnitude changes temporally, spatially, and geographically. The subject has already been reviewed by a number of authors including Strickland (1958), Clarke (1965), and Jerlov (1968).

The intensity of solar radiation reaching outside the earth's atmosphere lies mostly between 1·90 and 1·94 ly/min; on penetrating the atmosphere, energy is lost due to scattering and absorption, such as from water vapor, carbon dioxide, ozone, and dust. These substances are not equally distributed in the atmosphere (e.g. changes with cloud coverage). The angle of incoming solar light also effects the light penetrating into the sea surface. The lowering of the altitude of the sun makes the angle of incidence smaller and the path of light through the atmosphere longer with a corresponding reduction in intensity. Such changes in the sun's altitude result from differences in latitude as well as from changes in the season and in the time of day.

Theoretical solar radiation values, depending on latitude, season, and a clear sky have been determined (e.g. Haltiner and Martin, 1957, Fig. 7–3); actual solar radiation data are collected by a number of countries using instruments known as pyranometers (formerly, pyrheliometers). In Canada radiation is measured at thirty-four different stations and published monthly by the Department of the Environment. A comparison of these values and theoretical values shows, for example, that at 50°N in June and December the maximum theoretical radiation is 769 and 131 ly/day, respectively, while actual measurements at Nanaimo, B.C. (*ca.* 50°N) during 1969 were maximum in July (622 ly/day) and minimum in December (44 ly/day).

The duration of the period of irradiation is also an important factor (latitudinal differences of irradiation period and incoming sun angle are tabulated in the Smithonian Meteorological Tables, List 1958). On the equator the day is always 12 hr long, but in the temperate region the day grows longer as spring progresses. This effect is accelerated at higher latitudes, and the day becomes 24 hr long during the summer in the polar region. Up to moderately high latitudes the increase in length of day during summer has more effect on the total amount of light received per day than the reduction in solar radiation due to the greater angle of irradiance. This indicates that a considerable amount of solar radiation energy reaches the sea surface even at high latitudes during a limited time of the year. This is quite important in understanding seasonal differences in biological parameters (see Section 1.3.4).

It is known that the energy spectrum of solar radiation will also change depending on cloud coverage in the sky (Fritz, 1957) and the height of the sun. On a cloudless day, approximately 50% of the total solar radiation reaching the sea surface is the photosynthetically active radiation (Strickland, 1958). On the other hand, the ratio of visible light to total radiation increases with cloud (Vollenweider, 1969).

Some solar energy is lost by true reflection, and by scattering from particles (including foam) at the sea-surface. The actual value for surface varies considerably with conditions of the sea surface and sun angle; tables given by von Arx (1962) should be consulted for detailed values. On a fine day in summer, with a sun angle to the horizon of over 30°, the surface loss would be only a few percent under conditions of complete calm; this value increases to 5–17% with light winds and to over 30% for moderate to strong winds. As the sun angle decreases to less than 10°, reflection increases rapidly to over 30%. For field work, a mean value of 15% for total surface losses may be used as an approximation for conditions under which it is usually possible to carry out photosynthetic measurements.

Light penetrating into the water is reduced by selective absorption and scattering due to the seawater itself, and dissolved and suspended matter in the water. The reduction of light in the water column can be expressed in terms of the (vertical) extinction coefficient* (k, also called attenuation coefficient, generally defined as m^{-1}):

$$I_d = I_0\, e^{-kd} \tag{24}$$

where I_0 is the incoming light intensity, I_d the light intensity travelling a distance of (d). The value (k, m^{-1}) varies with the wavelength of light, being large for ultraviolet and infrared light; 0·033 at 425 nm, 0·018 at 475 nm, and 0·288 at 650 nm for the extinction coefficient of pure water (Jerlov, 1968). In clear oceanic waters blue-green light (maximum around 480 nm) can only penetrate to any appreciable depth. However, under turbid conditions due to particulate material in the water, blue light is selectively scattered and the spectral peak of transmitted light is moved up towards the red (maximum at ca. 550 nm). A discussion of these changes in ocean and coastal waters is given by Jerlov (1968). For the purpose of the most biological events, the average extinction coefficient (i.e. in the interval 400 to 720 nm) rather than the value at particular wavelengths is probably the most useful.

The average extinction coefficient in the water column (k') can be defined as follows:

$$k' = k'_w + k'_p + k'_s \tag{25}$$

where k'_w, k'_p, and k'_s are possible diffusion and scattering of light energy due to water (w), suspended particles (p), and dissolved matter (s), respectively. The suspended particles include many different forms such as clay particles, organic detritus, and organisms varying in size from less than 1 μm to a few mm. According to Clarke and James (1939), the suspended particles reduce the light transmission by less than 7% (actual transmittance, m^{-1}) in the continental shelf water, but they reduce 30 to 70% in the coastal water. They also observed the same trend in the effects for the fraction of dissolved matter, although the actual reducing effects were much smaller than for the

*Distinctive from the light-absorption coefficient.

particulates. Then it can be expected that the optical properties of a given water column are significantly controlled by the quantity and quality of the total suspended particles in the sea.

A summary of the conditions of light in the sea and its critical limiting effects is given in Fig. 27. Here the maximum depths are indicated for the growth of phytoplankton. Since the mean illumination necessary for the vision of aquatic animals is so very much smaller, the depth limitation is at a much greater level. Animals can respond to the difference between day and night at somewhat greater depths. Below this level no perceptible light from the surface penetrates and the water is completely dark except for light provided by luminescent organisms. Terms used in deep bodies of water for zones based on the light factor are as follows:

Euphotic zone: sufficient light for photosynthesis.

Disphotic zone: insufficient light for photosynthesis but sufficient light for animal responses.

Aphotic zones: no light of biological significance from the surface.

The actual depth limits of these zones differ widely according to transparency as indicated in Fig. 27 for the penetration of light in coastal waters.

For the measurement of solar radiation reaching the sea surface, one can use different types of sensors depending on what kind of radiation one wants to measure. For the total energy measurement, a pyranometer (almost no sensitivity difference at different wavelengths) is suitable. A photometer (detection range 400–700 nm with maximum sensitivity at 600 nm when a selenium cell is used) and quanta meter are suitable for illumination and quanta measurements, respectively. (Note: since there is no instrument available in order to detect quanta directly, a phototube with suitable light-selection filter is usually used.)

There are two approaches (one direct and one indirect) to estimate the (average) extinction coefficient (k') in a given water column. The direct approach involves the actual measurement of light intensity at different depths using a suitable sensor (mentioned above, but embedded in a water-tight box); the extinction coefficient can then

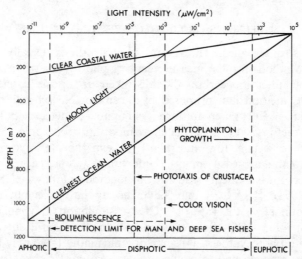

FIG. 27. Schematic diagram to show the penetration of sunlight into the clearest ocean water $(k = 0.033)$ and into clear coastal water $(k = 0.15)$ in relation to minimum intensity values for some biological light receptors (redrawn from Clarke and Denton, 1962).

be calculated using eqn. (24). Since all energy meters are relatively insensitive, a photo-meter (responding to as low as 10^{-3} of sunlight using photovoltaic cells, and 10^{-12} of sunlight using photomultiplier tubes) or a quantum meter, is generally used for the measurement of underwater irradiance.

The indirect estimation can be done from Secchi disc reading (D_s, in meter), using an empirical relation which was originally proposed by Poole and Atkins (1929) as follows:

$$k' = \frac{1 \cdot 7}{D_s}. \tag{26}$$

According to Idso and Gilbert (1974), the constant (1·7) always gave fairly close extinc-tion coefficients when compared to optically measured coefficients in a wide range of water visibility (turbid to clear ocean water) covering D_s between 1·9 and 35 m.

Morel and Smith (1974) found that the ratio of (total quanta)/(total solar energy) in the range of 400 to 700 nm was fairly constant both at the sea surface ($2 \cdot 77 \times 10^{18}$ quanta s^{-1} watt^{-1}) and in the water column ($\{2 \cdot 5 \pm 0 \cdot 25\} \times 10^{18}$ quanta s^{-1} watt^{-1}) under various weather conditions with sun altitudes above 22°. This result supports the fact that one can indirectly estimate total photosynthetic quanta (400 ~ 700 nm) from the measurement of total energy at those wavelengths. This is quite convenient when one considers the possible experimental difficulty of building and accurately cali-brating a quantum meter compared with an energy meter.

The solar beams penetrating the water column are expected to keep their incoming angle (refracted slightly at the air/sea water interface) at least near the surface. Since most organisms have three-dimensional shape, the actual light absorption by organisms might be closer to the absorption pattern given by 4π collectors (sphere, three dimen-sions) rather than 2π collectors (flat plate, two dimensions). From this reason 4π collec-tors as opposed to 2π collectors may be advisable for biological events in the aquatic environments.

3.1.3 EFFECTS OF LIGHT ON PHOTOSYNTHESIS

Light intensity strongly affects the rate of photosynthesis (usually expressed as mg C/mg Chl a/hr). Methods for the measurement of photosynthetic rate usually involve a measurement of either the carbon dioxide taken up or the oxygen produced per unit time. The ^{14}C-method first proposed by Steemann Nielsen (1952) is usually used for the measurement of CO_2 taken up since it is possible to detect very low photosynthe-tic rates with this method. This is particularly important in oceanic areas where photo-synthetic rates are especially low and where an experimenter may only have a few hours in which to make measurements. Unfortunately some doubt may exist as to the interpretation of measurements made by this method as follows (cf. Steemann Nielsen and Willemoës, 1971): (1) $^{14}CO_2$ may be taken up into cells but not incorporated in organic compounds, (2) the rate of $^{14}CO_2$ assimilation may not be the same as that of $^{12}CO_2$, (3) some $^{14}CO_2$ fixed may be lost due to respiration which takes place simultaneously with photosynthesis, (4) some organic matter (^{14}C) may be lost by exuda-tion during the experiment, and (5) some organic matter (^{14}C) may also be lost when the algae are filtered off from the experimental medium. Although the importance of none of these may be particularly pronounced, it is appropriate to introduce small corrections (cf. Steemann Nielsen, 1958). It is generally assumed, however, that in open waters, the ^{14}C-method measures the rate of increase in particulate carbon (e.g. Antia et al., 1963). If a separate measurement is made of the amount of carbon lost through exudation, then the sum of the increase in particulate carbon, plus the loss of dissolved organic carbon, would be a measure of net photosynthesis. If the loss of dissolved organic carbon is small, then the ^{14}C-method will approximate net photosynthesis. The true result may be higher or lower according to the species of phytoplankton and environmental conditions. Production estimates based on oxygen evolution can be made using either an oxygen electrode or the Winkler titration technique, but both of these methods are usually an order of magnitude less sensitive than the ^{14}C-technique.

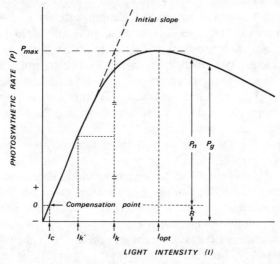

FIG. 28. Diagrammatic photosynthesis–light relationship. (P_{max}, photosynthetic maximum; I_c, light intensity at the compensation point; R, respiration; P_n, net photosynthesis; P_g, gross photosynthesis; I_{opt}, light intensity at P_{max}; I_k, see text).

Thus the oxygen technique is not particularly suitable for use in oceanic waters, but it may be used in some coastal areas, or in high-latitude oceanic waters having a high density of algae. The merits and actual procedures for both methods are described in detail in the following two manuals: Strickland and Parsons (1968) and Vollenwieder (1969).

The photosynthesis/light curve (or P vs. I curve) shown in Fig. 28 is a convenient reflection of environmental effects on photosynthesis and one which can be used to diagnose certain properties of algal species, or natural samples of phytoplankton. Rabinowitch (1951), Steemann Nielsen and Jørgensen (1968), and Yentsch and Lee (1966) are recommended as references on the general interpretations of P vs. I curves. From Fig. 28 it is apparent that photosynthesis increases with increasing light intensity up to some asymptotic value, P_{max} where the system becomes light saturated. The two most important properties of the curve are the slope ($\Delta P/\Delta I$) and P_{max}, in which the latter is also called the 'assimilation number' (or index). The initial slope is a function of the light reaction (see Section 3.1.1) and is not usually affected by other factors. In plant physiology, the initial slope of the P vs. I curve has been defined as 'the quantum yield' in which light intensity is expressed in the quantum unit. The quantum yield can then be expressed as the number of moles of oxygen evolved (or of carbon incorporated) per unit light intensity (in einsteins). According to the review by Kok (1960), the maximum quantum yield is about 0·10 mole oxygen evolved per einstein absorbed. The yield for carbon is likely to be slightly less than that for oxygen, as a result of nitrate reduction and the formation of some carbon compounds more reduced than carbohydrates [photo-synthetic quotients ($O_2/—CO_2$) for natural populations suggested by Ryther (1965b) are 1·1 to 1·3]. The quantum yield mentioned is the approximate upper limit when the wavelength of illumination and the physiological state of the organisms are both optimized. In natural environments, several factors can be anticipated to lead to a lower value for the quantum yield. First, quantum yield and action spectra indicate that some phytoplankton accessory pigments sensitize photosynthesis less efficiently than does chlorophyll (Figs. 22 and 24); the average spectral yield is generally 10 to 20% less than that at wavelengths absorbed solely by chlorophyll. Secondly, the quantum yield changes depending upon the physiological state and growth stage of organisms, and their growth condition. Bannister (1974) has suggested a yield of 0·06 mol $CO_2 \cdot$ einstein^{-1} absorbed for natural (healthy) populations. P_{max}, on the other hand, is a function of the dark reaction (see Section 3.1.1) provided no environmental factors are causing photosynthetic inhibition. If other environment factors are operative, such as low temperatures or nutrient limitation, P_{max} becomes a function of the environmental inhibitor. As a combination of the initial slope and P_{max}, Talling (1975b) proposed I_k which is the light intensity at the intersection of an extension from the initial slope and P_{max} (Fig. 28).

Recently, Tyler (1975) presented an approach in which he estimates the *in situ* quantum efficiency of natural phytoplankton populations based on the quantity of CO_2^- fixation measured *in situ* and the quanta actually absorbed by the phytoplankton populations. The efficiency obtained may include all possible effects mentioned above, and is also time averaged; this is because efficiency changes with light intensity which shows a marked daily variation. The efficiencies in Tyler's observations varied from 0 to 0·08 molecules CO_2 fixed $\cdot$ quanta^{-1} (= moles CO_2 fixed $\cdot$ einstein^{-1}) absorbed depending on the light intensity; the quantum efficiencies increased linearly with the logarithm of decreasing quanta absorbed in the range of 10^{21} and 10^{24} quanta/m$^2 \cdot$ half-day. The maximum quantum efficiency at very low light conditions is analogous, but not the same, to the maximum quantum yield.

Figure 28 also shows the difference between total photosynthesis (or gross photosynthesis, P_g) and net photosynthesis (P_n), which is the fraction of P_g minus respiration (R). An approximate P_n can be estimated directly by the oxygen method because oxygen is consumed by respiration at the same time as it is produced through photosynthesis in the light. However, care should be taken in interpreting the oxygen method because photorespiration (see Section 3.1.1) sometimes causes an over-estimation of respiration in the light. When P_g equals R, P_n is zero and the photosynthetic system is at the 'compensation point'. The light intensity at the compensation point is called the 'compensation light intensity', I_c, and photosynthetic micro-organisms held at the compensation light intensity should theoretically show no growth. In nature, phytoplankton are subjected to continually varied light conditions and, except during the summer in extreme latitudes, to no light at all during the night. Thus at noon the amount of light reaching a cell may be above the compensation light intensity; however, over a period of 24 hr there may still be a mean light level below which the algae will decrease in weight. The compensation point is therefore best expressed on a 24-hr basis and is usually determined as the 24-hr mean light intensity in ly/day or ly/min (Strick-

land, 1958). From studies on natural phytoplankton populations, not excessively contaminated with other microflora or fauna, an average 24-hr compensation light intensity
appears to be in the range 0·002 to 0·009 ly/min or some 3 to 13 ly/day in temperate
seas (e.g. Strickland, 1958; McAllister *et al.*, 1964; Hobson, 1966). The value is dependent
on the rate of respiration which has been found to be approximately 10% of P_{max}
(see Section 3.1.1). However, the basic assumption that respiration is constant in the
light and dark is difficult to check and may not be valid for cells which spend a long
time below the euphotic zone. For example, adaptation to photosynthesis at very low
light intensities (*ca.* 0·00014 ly/min) has been reported for phytoplankton growing under
ice (Wright, 1964). Since respiration is temperature dependent, the value for I_c may
be expected to be larger at higher temperatures. In field work the compensation depth
can be approximated from the depth of 1% of the surface radiation; this value can
either be determined using a bathyphotometer or may be approximated as three times
the depth of the Secchi disc visibility (Strickland, 1958; see also Section 3.1.4). Compensation depths may vary from a few meters in turbid coastal waters to over 150 m in
some tropical seas.

Respiration rate of phytoplankton has sometimes been expressed in terms of biomass
units (e.g. O_2/Chl a/time) but these values usually vary over a great range depending
on species, past growth history, and physiological status of a cell (e.g. Ryther and Guillard, 1962). Another way for expressing algal respiration rate has been made on the
simple assumption that respiration rate is proportional to the maximum photosynthetic
rate, P_{max}, as follows:

$$R = rP_{max} \tag{27}$$

where r is the proportionality constant (dimensionless), called the 'loss factor', which
characterizes the respiration economy of species or populations. Since P_{max} depends on
environmental conditions, growth history and physiological status of algae, no correction is generally needed in order to apply this relation, providing P_{max} is measured
under the same conditions for estimating respiration. In most cases eqn. (27) is used
with an assumption that respiration rate changes independently of light. According to
literature (see Section 3.1.1), the loss rate (r) for phytoplankton is ~0·1.

However, the assumption mentioned above may not be entirely true because it has
been known that respiration rate increases in the light, in proportion to the amount
of light. Tooming (1970) has extended eqn. (27) to account for additional respiration
loss occurring under light by assuming that the additional carbon loss in the light
was proportional to gross primary productivity which changes dependently of light
intensity:

$$R = r_1 P_{max} + r_2 P_g \tag{28}$$

where both r_1 and r_2 are (dimensionless) proportionality constants. Basic respiration
constant, r_1, is the factor for respiration loss due to maintaining the photosynthetic
and non-photosynthetic functions of algae, and it is the same as the 'loss factor' in
eqn. (27). Light respiration constant, r_2, on the other hand, is the factor for respiration
loss due to organic synthesis and growth which generally occurs under the light.
Respiration loss due to 'photorespiration' will also be included in the factor of 'r_2'.
Unfortunately at the time of writing, no datum is available for estimating 'r_2'.

If micro-organisms are exposed to a strong light above the point at which they are
light saturated, the P vs. I curve may show a depression in photosynthetic rate. This

phenomenon is named '(high) light inhibition' or 'photoinhibition'. Photoinhibition is not generally observed over short periods of time (e.g. 10 min) but may result from longer exposures and may also increase in magnitude with time (Takahashi *et al.*, 1971).

It appears from early work (e.g. Ryther, 1956) that there may exist a broad division between taxonomic groups with respect to their *P* vs. *I* curves. Ryther (1956) reported *P* vs. *I* curves for fifteen species of marine algae representing three taxonomic groups (green algae, diatoms, and dinoflagellates). Measurements were made under sunlight

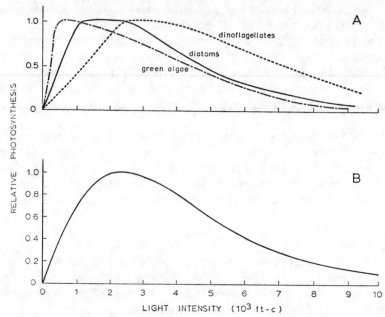

FIG. 29A. Relative photosynthesis–light curves in some marine phytoplankton. Green algae: *Dunaliella euchlora, Chlamydomonas* sp., *Platymonas* sp., *Carteria* sp., *Mischococcus* sp., *Stichococcus* sp., and *Nannochloris* sp. Diatoms: *Skeletonema costatum, Nitzschia closterium, Navicula* sp. and *Coscinodiscus excentricus*. Dinoflagellates; *Gymnodinium splendens, Gyrodinium* sp., *Exuviaella* sp., and *Amphidinium klebsi* (redrawn from Ryther, 1956).
FIG. 29B. Mean curve from FIG. 29A.

on a clear day at 20°C and the results showed a remarkable similarity in the photosynthetic behaviour of organisms within each taxonomic group, but a rather striking difference between those of different groups. A summary of Ryther's results is shown in Fig. 29; from these results it may be seen that the saturation light intensities for green algae are between 500 and 750 ft-c (or 3·3 to 4·9 × 10^{-2} ly/min), between 2500 and 3000 ft-c (or 16 to 20 × 10^{-2} ly/min) for the dinoflagellates, and at intermediate intensities for the diatoms. Photo-inhibition is apparent in all three algal groups within about 1000 ft-c (0·066 ly/min) of saturation. At intensities of 8000 to 10,000 ft-c (52 to 66 × 10^{-2} ly/min), which is comparable to full sunlight, the photosynthetic rate in green algae and diatoms is only 5 to 10% of that at saturation, while the photosynthetic rate for dinoflagellates is still 20 to 30% of P_{max}. The I_k of each curve in Fig. 29 is 400 ft-c (2·6 × 10^{-2} ly/min) for green algae, 1000 ft-c (6·6 × 10^{-2} ly/min) for diatoms and 2400 ft-c (16 × 10^{-2} ly/min) for dinoflagellates. In the sea, high I_k values are observed during the summer and in shallow algal communities, and low I_k values are observed during the winter and in deep-water communities (Steemann Nielsen and Hansen, 1959a and 1961; Ichimura *et al.*, 1962).

I_k gives a measure of the radiant energy or illumination at light saturation but it does not express the photosynthetic efficiency; consequently plants or phytoplankton communities may have the same I_k values but differ appreciably in the rate of photosynthesis at I_k. In terrestrial communities plants are divided into 'sun-' and 'shade-types' (Boysen Jensen, 1932) and a similar division is employed in algal communities. Thus sun-type algal communities are those who can utilize high light intensities with high

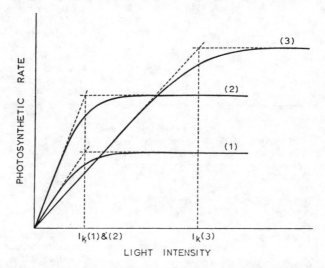

FIG. 30. Three types of *P* vs. *I* curves. (1) and (2) shade type algae showing similar I_k values but with higher photosynthetic efficiency in (2) than (1). Sun-type community (3) showing lower photosynthetic efficiency than (1) or (2) at lower intensity.

efficiencies while photosynthesis in shade-type communities is generally depressed by high light intensities. However, the absolute photosynthetic rates of shade-type communities are usually higher than those of the sun-type communities at low light intensities. These differences in I_k values and photosynthetic efficiencies are illustrated in Fig. 30.

Table 16 shows some accumulated values for the initial slopes of *P* vs. *I* curves measured under incandescent and natural light. Algal cultures of different species show values between 0·1 and 0·65 (average 0·38) mg C/mg Chl *a*/hr per klx.* Those of natural samples range between 0·05 and 0·8 (average 0·4) mg C/mg Chl *a*/hr per klx. The table shows that certain species or populations are adapted to low light intensities and others to high light intensities. Differences which are sometimes found between species, or within the same community over a period of time, may be due to inactivation of cells or changes in photosynthetic mechanisms (e.g. Ichimura *et al.*, 1968).

There are many reviews on the evaluation of P_{max} (Steemann Nielsen and Hansen, 1959a; Ichimura and Aruga, 1964; Yentsch and Lee, 1966). P_{max} is primarily influenced by environmental conditions (see Section 3.1.3) but under conditions of optimum temperature and sufficient nutrient P_{max} can be employed as the 'potential P_{max}' or 'potential photosynthetic rate'. The P_{max} of phytoplankton populations is usually measured at

*Kilolux (K.lux used in the first edition).

TABLE 16. INITIAL SLOPES OF P VS. I CURVES OF ALGAL CULTURES AND NATURAL POPU-
LATIONS***

Algal cultures

Species	Exp. temp. (°C)	Initial slope (mg C/mg Chl a/hr per klx)	Author
Chlorella vulgaris	20	0·50*	Steemann Nielsen, 1961
C. pyrenoidosa		0·48	Steemann Nielsen & Jørgensen, 1968
C. ellipsoidea	5–30	0·50**	Aruga, 1965b
Scenedesmus sp.	10–40	0·65**	Aruga, 1965b
Skeletonema costatum		0·52	Steemann Nielsen & Jørgensen, 1968
Skeletonema costatum	20	0·44**	Nakanishi & Monji, 1965
Chaetoceros sp.	20	0·13**	Nakanishi & Monji, 1965
Coccolithus huxleyi		0·13*	Jeffrey & Allen, 1964
Hymenomonas sp.		0·10*	Jeffrey & Allen, 1964

Natural populations

Situation	Dominant species	Exp. temp. (°C)	Initial slope (mg C/mg Chl a/hr per klx)	Author
Tokyo Bay	Skeletonema costatum	20	0·56**	Aruga, 1965b
Pond	Synedra sp.	10–30	0·65**	Aruga, 1965b
Pond	Anabaena cylindrica	10–30	0·25**	Aruga, 1965b
Lake	Cryptomonas sp.		ca. 0·8	Ichimura et al., 1968
Arctic		4·5–6	0·36	Steemann Nielsen & Hansen, 1959a
Tropical, temperate (summer, winter), Northern & Arctic (general)			0·35	Steemann Nielsen & Hansen, 1959a
Kuroshio (general)			0·05	Ichimura et al., 1962
Oyashio (general)			0·35	Ichimura et al., 1962
Mixed regions of Kuroshio & Oyashio (general)			0·17	Ichimura et al., 1962
Antarctic			0·42	Burkholder & Mandelli, 1965

*mg C/mg Chl $(a + b)$/hr per klx.
**mg C/mg Chl/hr per klx.
***The initial slope and P_{max}, as calculated from chlorophyll and photosynthetic data in the early literature, might contain some over-estimations for chlorophyll(s) (about 25%: Banse and Anderson, 1967) and photosynthesis (up to 30%; Steemann Nielsen, 1965) because of the methods used. The former will increase the initial slope (or P_{max}) and the latter decreases it. Consequently, it is fortuitous that the results quoted here appear to be approximately correct.

in situ temperature and nutrient concentrations; the results of some of these measurements, together with values for cultures, are shown in Table 17. The values range from between 1·1 to 6·2 mg C/mg Chl a/hr for cultures and from 0·1 to 6·0 mg C/mg Chl a/hr in natural samples. The data show some geographical differences in P_{max}; thus the values for polar waters are generally low, temperate waters are quite variable, and tropical waters have the highest values. Some exceptions to the general range of values have been reported; for example, *Skeletonema costatum* has sometimes been reported to give values during a bloom of 9·0 to 16·9 mg C/mg Chl a/hr (Hogetsu et al., 1959).

From the photosynthetic action spectra shown in Fig. 23, it is to be expected that light energy available for algal photosynthesis is restricted to wavelengths between 400 and 720 nm. Within these wavelengths the light absorbed by the phytoplankton pigments can be divided into two parts: (i) the light of >600 nm which is mainly absorbed by chlorophyll a and (ii) the light of <600 nm which is mainly absorbed by accessory pigments. Except in some blue–green and red algae (e.g. curve 5 in Fig. 23), the action

TABLE 17. P_{max} OF ALGAL CULTURES AND NATURAL POPULATIONS***

Algal cultures

Species	Culture conditions temp. (°C)	Culture conditions light int. (klx)	Exp. temp. (°C)	light int. (klx)	P_{max} (mg C/mg Chl a/hr)	Author
Chlorella ellipsoidea	20		30	20	5·5**	Aruga, 1965b
C. vulgaris		3		6	1·1*	Steemann Nielsen, 1961
		30		30	3·8	
Scenedesmus sp.	20		30	10	5·0**	Aruga, 1965b
Skeletonema costatum	20	8	20	20	6·2**	Nakanishi & Monji, 1965
	20		20	13	3·6**	Aruga, 1965a
Synedra sp.	20		20	15	2·0**	Aruga, 1965b
Cyclotella		3			2·1–3·4	Jørgensen, 1964
meneghiniana		30			3·6–4·4	Jørgensen, 1964
Chaetoceros sp.	20	8	20	20	1·5**	Nakanishi & Monji, 1965
Anabaena cylindrica	20		30	10	1·7**	Aruga, 1965b
Coccolithus huxleyi	14	38	>38		>2·2*	Jeffrey & Allen, 1964
Hymenomonas sp.	14	38	>38		>1·9*	Jeffrey & Allen, 1964

Natural populations

Situation	Depth (m)	Exp. temp. (°C)	P_{max} (mg C/mg Chl a/hr)	Author
Arctic (summer)	0		1·0–1·5	Steemann Nielsen & Hansen, 1959a
Antarctic (phytoplankton)			2·3	Burkholder & Mandelli, 1965
Antarctic (ice flora)		−1·6	2·6	Burkholder & Mandelli, 1965
Antarctic (ice flora)		−1·5	0·4	Bunt, 1964b
Northern (general) no vertical stability			2·9–3·4**	Steemann Nielsen & Hansen, 1959a
Western north Pacific (summer)	10	8–20	1·4–2·0	Takahashi et al., 1972
Temperate (summer)	0		4·0–4·2**	Steemann Nielsen & Hansen, 1959a
Temperate (winter)	0		1·5**	Steemann Nielsen & Hansen, 1959a
Oyashio (general, summer)			3–6**	Ichimura & Aruga, 1964
Kuroshio (general, summer)			0·3–0·7**	Ichimura & Aruga, 1964
Bays and coastal waters near Japan (general)			2–6**	Ichimura & Aruga, 1964
Tokyo Bay (Skeletonema bloom, summer)		20–25	9·0–16·9**	Hogetsu et al., 1959
Tropical (general)			8·0**	Steemann Nielsen & Hansen, 1959a
Tropical Pacific (autumn)		23–27	1·1–5·2	Takahashi et al., 1972
Tropical Pacific (spring, general) nitrogen-poor water			3·15	Thomas, 1970a
Tropical Pacific (spring, general) nitrogen-rich water			4·95	Thomas, 1970a

TABLE 17. (cont.)

Situation	Depth (m)	Exp. temp. (°C)	P_{max} (mg C/mg Chl a/hr)	Author
Tropical Atlantic	10	20	3·0–4·0**	Yentsch & Lee, 1966
Lake (general)				Ichimura & Aruga, 1964
eutrophic			2–6**	
mesotrophic			1–2**	
oligotrophic			0·1–1**	

*mg C/mg Chl $(a + b)$/hr.
**mg C/mg Chl/hr.
***The initial slope and P_{max}, as calculated from chlorophyll and photosynthetic data in the early literature, might contain some over-estimations for chlorophyll(s) (about 25%; Banse and Anderson, 1967) and photosynthesis (up to 30%; Steemann Nielsen, 1965) because of the methods used. The former will increase the initial slope (or P_{max}) and the latter decreases it. Consequently, it is fortuitous that the results quoted here appear to be approximately correct.

spectra at > 600 nm of all algae are similar, while those at < 600 nm are quite different and depend mainly on the light absorbed by accessory pigments. Since most of the light penetrating to depth is in the region 400 to 600 nm, it is quite apparent that the accessory pigments are most important as light absorbers in the ocean environment.

Another effect of light on photosynthesis is a diel rhythm. Although this subject, as recently reviewed by Sournia (1974), has not been fully understood yet, tentative generalizations only can be made. Diel oscillations are known for rates of photosynthesis (Doty and Oguri, 1957; Ichimura, 1960), of chlorophyll synthesis (Yentsch and Ryther, 1957), and rates of nutrient uptake (Goering et al., 1964). The diel maximum in photosynthetic rate is recorded for early morning in the ocean near the equator (Doty and Oguri, 1957), for later in the day in lake (Lorenzen, 1963), and in inshore marine environments (Newhouse et al., 1967). The amplitude of the diel oscillations generally decreases with increase in latitude (Doty, 1959) and is less pronounced on cloudy days and during the winter (Saijo and Ichimura, 1962). The main question is how are these rhythms evolved? Two hypotheses may be considered for possible explanations of photosynthetic rhythms. The first is that the photosynthetic potential oscillates in response to intrinsic organization of the cell with a light–dark cycle. In the second hypothesis, photosynthetic rate oscillates in response to a forcing from the external concentration of limiting nutrients while photosynthetic potential of the cell remains constant. These two hypotheses were tested in natural populations using elegant models by Stross et al. (1973). They have concluded that photosynthetic rhythms could result from both an intrinsic and a nutrient (forcing) oscillation.

P_{max} may change with the physiological state of algae themselves (e.g. with age) even under constant environmental conditions (Jørgensen, 1966). From experiments with synchronous cultures of Skeletonema costatum, it was shown that P_{max} increased in young cells (just after cell division) and reached a maximum value in full grown cells, just prior to cell division. Cell division started 6 hr after illumination and continued 4 hr after cells were placed in the dark.

From these observations it is apparent that adaptive changes in photosynthesis occur in algal cells in response to surrounding light conditions. This phenomenon is known as 'light adaptation', i.e. a physiological adjustment to surrounding conditions which has been observed to involve at least one of the following morphological or biochemical changes within the cell: (1) change in total photosynthetic pigment content, (2) change in the ratio of photosynthetic pigments, (3) change in the morphology of the chloroplast, (4) change in the arrangement of the chloroplasts, and (5) change in the availability of enzymes for the dark reaction. Specific examples of these changes have been demonstrated, for example, Fujita (1970) showed that in blue–green algae, red light induced phycocyanin synthesis and blue light increased the amount of phycoerythrin. In diatoms, the shrinking of chloroplasts and their aggregation under the influence of strong light has been observed as a reaction which is reversed under weak light (Brown and Richardson, 1968).

The time dependence of light adaptation of algae in the ocean is important in determining the day-to-day effects of light variation. According to Steemann Nielsen *et al.* (1962) *Chlorella* required about 40 hr to adapt to a change in light intensity from 3 to 30 klx at 21°C. Algal populations taken from the surface of strong vertically mixed water masses off Friday Harbor, Washington, U.S.A., took 3 days to adapt to low light conditions of about 5% of the surface illumination (Steemann Nielsen and Park, 1964). If algae are kept in the dark, there is a gradual loss in photosynthetic activity which in the case of *Nannochloris* amounted to 50% of the photosynthetic activity in about 40 hr at 20°C (Yentsch and Lee, 1966).

3.1.4 NUTRIENT AND TEMPERATURE EFFECTS ON PHOTOSYNTHESIS AND THE GROWTH REQUIREMENTS OF PHYTOPLANKTON

Photosynthesis of algae is also controlled by factors other than light, such as nutrients and temperature, and to a lesser extent a variety of factors, such as pH and salinity. Liebig (1840) postulated a simple rule for the effect of various factors on yield (i.e. net photosynthesis, P_n). His statement was that "growth of a plant is dependent on the minimum amount of foodstuff presented" and this has come to be known as "Liebig's law of the minimum". Sixty years after Liebig's work, Blackman (1905) suggested that a generalized form of this law could be applied to photosynthesis in order to explain field and experimental observations. He took from Liebig the idea that the rate of a biological process (in this case, photosynthesis) is determined, under given conditions, by a single limiting factor. However, in addition to the supply of material ingredients (the only kind of factors with which Liebig was concerned), Blackman considered also light intensity and temperature as limiting factors. He suggested that the rate of photosynthesis increased with an increase in the value of any one of these factors (F_1), as long as the particular factor was rate limiting, and that it ceased to be dependent on F_1 when one of the other factors (F_2, F_3 ...) became limiting. In other words, the plot of photosynthesis, P_n, versus a variable F_1 (at constant values of all other kinetic factors) was postulated by Blackman to have the shape 1, 2, 3 in Fig. 31. In actual fact the curve approaches the maximum asymptotically, without a sudden break as indicated by the point 2. It should be noted also that an excessive amount of the factor, F_1, eventually causes a depression in P_n. Assuming F_1 to be light intensity, the P_n vs. F_1 relation is the same as the P vs. I curve shown before in Section 3.1.3. In order that other factors, F_2, F_3 ... should conform to the same rule, it is apparent that for different values of a second parameter (F_2), photosynthesis (P_n) as a function of F_1 will be represented by a sequence of solid lines, such as 1.2.3, 1.4.5, 1.6.7, in Fig. 31. These coincide at low F_1 values (part 1.2 in the figure) but are distinguished at higher F_1 values by the position at which the ascending part of the curve becomes horizontal. This value is determined by the second factor (F_2), which causes an increase

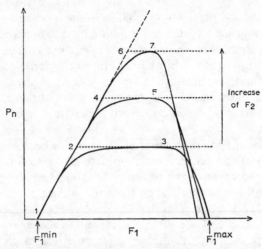

FIG. 31. Changes in photosynthetic rates, P_n, with a change in two environmental factors, F_1 and F_2, varied independently.

in P_n as represented by levels 3, 5, and 7. The horizontal plateau in Fig. 31 does not extend indefinitely, however, and P_n declines when F_1 causes an inhibition. The initial slope, which is dependent on F_1, tends to be similar for the sequence of curves but the decline may be more rapid with an increase in F_2. The maximum and minimum values F^{max} and F^{min}, in Fig. 31 are called the 'upper lethal limit' and the 'lower lethal limit' respectively, and these can be generally recognized in applying any factor to a growth process. This gives a numerical evaluation to descriptive terms which have been used to express a relative degree of tolerance. Thus the series of expressions which have come into general use in ecology (see Section 1.1) and which utilize the prefixes 'steno-' meaning narrow and 'eury-' meaning wide, with reference to the tolerance of a process, or an organism, can be given definite meaning in terms of Fig. 31. Furthermore, the ascending and descending slopes, together with the width and height of the plateau, are also important physiological characters to consider in the ecological adaptation of certain algal populations to a given aquatic environment.

An example of the relationship between the seasonal photosynthesis of mixed algal populations, and different phosphate concentrations and temperatures, is given by Ichimura (1967) for phytoplankton in Tokyo Bay (Fig. 32A). From these results it is apparent that phosphate regulation on P_{max} is governed by temperature; the relationship is quite similar to the Blackman-type limitation shown in Fig. 31, although the higher concentrations of phosphate were not sufficient to cause a break in the curve. Figure 32B from Aruga (1965b) serves as a second example of the general relationship in Fig. 31. In this case the influence of light and temperature on photosynthesis gives a more complete Blackman-type response including depression of P_{max} at high light intensities. While these results were obtained for local populations which had been reconditioned to a specific environment, it is also known that populations can adapt to some extent to different environments. Steemann Nielsen and Jørgensen (1968) showed, for example, that a phytoplankton population could become adapted to a new temperature regime with a few days. For some species, however, temperature adaptation has to be made in a series of small steps (e.g. 5°C changes) in order to exclude harmful effects caused by a sudden change in temperature. Experimental data on the effect of temperature

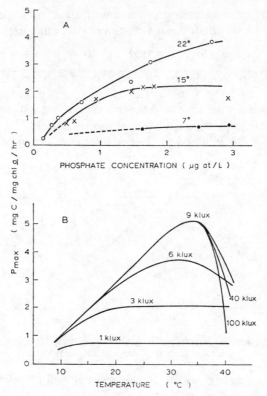

FIG. 32. Photosynthesis regulation by phosphate (A) and temperature (B). (A) Natural populations taken from Tokyo Bay (redrawn from Ichimura, 1967); (B) cultured *Scenedesmus* sp. (redrawn from Aruga, 1965a).

on photosynthesis are generally scarce but some studies have been made in the Antarctic (Bunt, 1974 a, b) and in the Pacific (Aruga *et al.*, 1968; Ichimura *et al.*, 1962); the latter results indicated that temperate Pacific phytoplankton had their highest P_{max} at about 20°C in spite of *in situ* temperatures which varied between −0·9 and 17·9°C.

P_{max} measured at *in situ* temperatures is a good indicator of what kinds of factors (other than light) are limiting photosynthesis. As an example, seasonal variations in P_{max} from Tokyo Bay are shown in Fig. 33. From the results it is apparent that from October to June, temperature regulates photosynthesis; during July to September P_{max} is depressed by the lack of nutrients which increase again due to autumn mixing in October. On the other hand, the potential photosynthetic rate, which is measured under optimum temperature and nutrient conditions, is constant throughout the year except in February and August. Low potential photosynthetic rates in these months have been attributed to a changing phytoplankton flora during the early spring and late summer. In high latitudes where there are low temperatures and high nutrient concentrations, P_{max} will be largely regulated by temperature. Phytoplankton communities near river mouths may also be regulated by temperature throughout most of the year (Ichimura, 1967). On the other hand, P_{max} in tropical and sub-tropical communities is more likely to be limited by nutrients.

Eppley (1972) has summarized the effect of temperature on algal growth rate. By plotting growth rate data, obtained by different people, vs. temperature, he found an

empirical relation for the maximum growth rate of algae over the temperature range
between 0 and 40°C, under conditions of continuous illumination:

$$\log_{10} \mu = 0.0275\,t - 0.070 \tag{29}$$

where μ is the maximum possible growth rate in doublings per day (see Section 3.1.5)
and t is temperature in degrees Celsius. The equation was deduced from data on algal
cultures which included a wide variety of taxonomic groups, cells with different comple-
ments of photosynthetic pigments, and diverse morphologies. The growth rate of each
algal species (or clone) can be fitted to the equation over a limited temperature range;

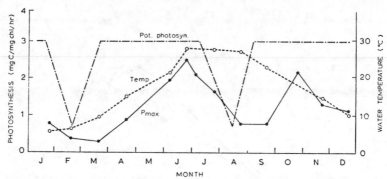

FIG. 33. Seasonal variations in temperature, P_{max} and potential photosynthetic rate in Tokyo Bay
(redrawn from Ichimura and Aruga, 1964).

growth ceases at temperatures above a supraoptimal point, which is a characteristic
of the species or clone (Eppley, 1972).

Algae require certain elements for their growth. Some of these elements, such as
C, H, O, N, Si, P, Mg, K, and Ca are needed in relatively large amounts and are
often known as 'macro-nutrients'. Other elements are required in very small amounts
and are referred to as 'micro-nutrients' or 'trace elements'; Hewitt (1957) has listed
ten trace elements required by green plants as follows: Fe, Mn, Cu, Zn, B, Na, Mo,
Cl, V, and Co. Most of these elements are contained in sufficient abundance for algal
growth in sea water; however, nitrogen and phosphorus may often limit plant growth
during the summer or in tropical and sub-tropical latitudes throughout the year. Thomas
(1968) made comprehensive tables on nutrient requirements and utilization of many
different species of algae, in which he defined requirements for vitamins, organic carbons
(sugars, alcohols, fatty acids, and other organic acids), amino acids, and other nitrogenous substances, including different inorganic forms of nitrogen compounds. His summary indicates which species of algae can grow on what kind of nutrients.

As an inorganic carbon source, algae can use free carbon dioxide, bicarbonate, and
carbonate. These three are measured as 'total carbon dioxide' and the amount is expected
to be ca. 90 mg CO_2/l in offshore pelagic waters. At this concentration phytoplankton
photosynthesis is not limited by the total amount of carbon dioxide. For example,
Talling (1960) did not find any stimulation or reduction in photosynthesis following
carbonate addition to a culture of Chaetoceros affinis in natural sea water. When phytoplankton grows vigorously, such as in a 'bloom', the total carbon dioxide content of
sea water (determined as the partial pressure) may show a negative correlation with
chlorophyll concentration in temperate waters (Gordon et al., 1971).

Nitrogen and phosphorus are of major importance as metabolites, and their concentration should always be considered first in determining possible limitations in primary production. The ratio in which nitrogen and phosphorus are taken up from sea water was studied by Redfield (1934) and Cooper (1937), who concluded that both in natural populations and cultures, the atomic ratio of cellular N:P was about 16:1 (see also Table 11). However, under some circumstances, depending on species and nutrient availability, different N:P ratios are encountered. For example, Ketchum's (1939b) N:P ratios for some representatives of the Chlorophyceae were about 7:1.

In the oceans, nitrogen exists mainly as molecular nitrogen and as inorganic salts, such as nitrate, nitrite, and ammonia, and some organic nitrogen compounds of amino acids and urea. The usual range of concentration for these compounds in sea water is 0·01 to 50 μg at/l for nitrate, 0·01 to 5 μg at/l for nitrite and 0·1 to 5 μg at/l for ammonia, 0·2 to 2 μg at N/l for amino acids (Clark *et al.*, 1972; Riley and Segar, 1970), and 0·1 to 5·0 μg at N/l for urea (Remsen, 1971). Molecular nitrogen is fixed by certain blue–green algae [e.g. *Calothrix* sp. (Allen, 1963) and *Trichodesmium* spp. (Dugdale *et al.*, 1964; Goering *et al.*, 1966; Dugdale and Goering, 1967)], yeasts [e.g. *Rhodotorula* sp. (Allen, 1963)] and bacteria [e.g. *Azotobacter* and *Closteridium* (Pshenin, 1963)]. However, compared with freshwater microorganisms, nitrogen fixers in the marine environment have not been clearly defined. *Trichodesmium* can also actively utilize both ammonia and nitrate but since the tropical habitat of *Trichodesmium* is usually relatively poor in nitrate, nitrogen fixation, together with ammonia utilization, are probably the most important sources of algal nitrogen in such environments (Goering *et al.*, 1966). Most other algae have no ability to fix molecular nitrogen and must utilize inorganic nitrogen salts and organic forms of nitrogen compounds. Algae generally show a preferential utilization for nitrate, nitrite, and ammonia (e.g. Guillard, 1963); some exceptions are found among the green algae (Braarud and Føyn, 1931) and a few flagellates (e.g. Strickland, 1965). Of those different forms of nitrogen compounds, ammonium is generally utilized in preference to nitrate (Eppley *et al.*, 1969a) and urea (McCarthy and Eppley, 1972) in different species of phytoplankton cultures and in natural phytoplankton populations. Ammonia is used directly for amino acid synthesis through transamination, but nitrate and nitrite must be reduced before being utilized by the cell. Reduction of nitrite appears to be carried out in the light (Hattori, 1962a; Grant, 1967; Eppley and Coatsworth, 1968) using photosynthetically reduced ferredoxin to reduce nitrite to ammonia (Hattori and Myers, 1966). Nitrate is reduced by the action of the enzyme, nitrate reductase, which is not dependent on light; the occurrence of this enzyme is induced by the presence of nitrate (Hattori, 1962 a and b), but not nitrite which is reduced by nitrite reductase (Eppley *et al.*, 1969a). Nitrate reductase would not be required for nitrite assimilation and neither enzymes are required for ammonium assimilation. Cellular content of each enzyme changes depending on the availability of nitrite or nitrate in the water. In some areas the principal source of nitrogen for phytoplankton may differ with depth in the same water column. For example, in the eastern subtropical Pacific Ocean off the continental shelf of Mexico, a pronounced discontinuity layer often exists at between 40 and 60 m within the euphotic zone, and phytoplankton in this layer use nitrate as their major nitrogen source; those in the overlying nitrate-impoverished water 0–40 m use ammonia which is derived largely from nitrogen recycled through zooplankton (Goering *et al.*, 1970). Natural living phytoplankton populations are known to utilize amino acids as a N-source even at low concentrations of natural levels (North, 1975). However, it seems that most phytoplankton have some preference to use inorganic forms of nitrogen compounds and urea, and do not start using amino acids effectively until other forms of nitrogen are depleted or reduced down to a certain level.

Phosphorus occurs in sea water in three principal phases: dissolved inorganic phosphorus, dissolved organic phosphorus, and particulate phosphorus. However, the existence of these forms is quite complex and special analytical procedures are needed to show the different forms of phosphorus in sea water (e.g. Strickland and Parsons, 1968). Phytoplankton normally satisfy their requirement for this element by direct assimilation of dissolved inorganic phosphorus (orthophosphate ion) and sometimes by utilizing dissolved organic phosphorus (Provasoli and McLaughlin, 1963; McLaughlin and Zahl, 1966). In polluted water, polyphosphate, which is inorganic, and organic soluble phosphorus detected as orthophosphate after acid hydrolysis, may be present in appreciable amounts. Some coastal algae, such as *Skeletonema costatum* and *Amphidinium carteri*, can use polyphosphate as a phosphorus source in the presence of excess nitrate. These species appear to be able to hydrolyse an external supply of polyphosphate more rapidly than it is utilized by the cells with the result that orthophosphate may accumulate in the surrounding water (Solōrzano and Strickland, 1968). Phosphorus absorbed into a cell becomes part of the structural component of a cell (e.g. in poly-P-RNA) and is in part continually turned over in the energetic processes of organisms (e.g. as adenosine di- and tri-phosphate). In this sense the role of phosphorus as a metabolite is quite different from nitrogen because nitrogen is used primarily as a structural component of cells and not directly in the energy cycle of the cell.

Diatoms and silicoflagellates take up a great amount of dissolved silicon and deposit it as hydrated silica to form their elaborately patterned valves. The concentrations of dissolved silicon are generally high in coastal and deep pelagic waters and low in surface waters away from the influence of estuaries. Diatoms grown in a medium low in silicate become silicon deficient but may remain viable for several weeks in the dark. However, on exposure to bright light they photosynthesize for a limited period and soon die (Riley and Chester, 1971). Diatom blooms deplete the surface waters of silicate and Menzel *et al.* (1963) have suggested that this is the chief mechanism leading to a species succession of non-silicious flagellates following the spring diatom bloom in sub-tropical waters.

Ferric iron as a hemin complex known as cytochrome plays an important role in cellular respiratory processes; in addition, the iron complex called ferredoxin is an essential part of the light reaction in photosynthesis (Section 3.1.1). Natural sea water contains from 1 to $60 \mu g/l$ of iron but only some of the iron detected analytically seems to be available for algal metabolism. A shortage of iron for phytoplankton growth has been demonstrated in the pelagic waters of the Sargasso Sea (Menzel and Ryther, 1961; Menzel *et al.*, 1963) and off the coast of Australia (Tranter and Newell, 1963). High concentrations of iron in sea water are generally associated with river runoff (see, e.g., Williams and Chan, 1966). Ryther and Kramer (1961) compared the iron requirements of a number of coastal and oceanic species of phytoplankton and found that coastal species could not grow at the low iron concentrations at which oceanic species would grow. In a reversal of this situation Strickland (1965) has suggested that the growth of some oceanic species of phytoplankton in coastal waters may be inhibited by relatively high trace-element concentrations. In some instances differences in the micro-nutrient concentration may be governed more by their availability through chelating agents than through their absolute concentration. The ability of sea water samples to support the growth of phytoplankton has been observed by a number of authors to be quite variable

TABLE 18. SYNTHETIC SEA WATER MEDIUM FOR THE GROWTH OF
PHYTOPLANKTON (ASP 2 FROM PROVASOLI et al., 1957)

Compound	wt./100 ml	Compound	wt./100 ml
NaCl	1·8 g	TRIS†	0·1 g
$MgSO_4 \cdot 7H_2O$	0·5 g	B_{12}	0·2 μg
		Vitamins*	1·0 ml
KCl	0·06 g	Na_2EDTA‡	3·0 mg
Ca(as Cl^-)	10 mg	Fe (as Cl^-)	0·08 mg
		Zn (as Cl^-)	15·0 μg
$NaNO_3$	5 mg	Mn (as Cl^-)	0·12 mg
		Co (as Cl^-)	0·3 μg
K_2HPO_4	0·5 mg	Cu (as Cl^-)	0·12 μg
$Na_2SiO_3 \cdot 9H_2O$	15 mg	B (as H_3BO_3)	0·6 mg

*Vitamins: 1 ml contains 0·05 mg thiamine HCl, 0·01 mg nicotinic acid, 0·01 mg pantothenate, 1·0 μg p-aminobenzoic acid, 1·0 mg biotin, 0·5 mg inositol, 0·2 μg folic acid, 0·1 μg thymine.
†TRIS buffer (2-amino-2-hydroxymethyl-propane-1,3 diol).
‡EDTA chelating agent (ethylenediaminetetra-acetic acid di-sodium salt).

(e.g. Johnston, 1963; Smayda, 1964 and 1970b; Barber et al., 1971). These studies have led to a general conclusion that subtle differences in the micro-nutrient composition of sea water, including the presence of organic substances, may play an important role in determining the total productivity of the water and the diversity of species present. The chemical identity of these substances remains obscure although some work on the origin of soluble organic substances (e.g. Prakash and Rashid, 1968; Pratt, 1966) has shown that humic acids, and the exudates from other phytoplankton, may inhibit or enhance the growth of algae.

The total spectrum of inorganic nutrients needed for phytoplankton growth can be illustrated from the work of Provasoli and others (e.g. Provasoli et al., 1957), who developed several types of artificial sea-water media for the growth of phytoplankton; one of these media which has been used extensively in phytoplankton culture experiments is given in Table 18. The ingredients illustrate the need for trace elements and organic compounds, such as the vitamins and a chelating agent (Na_2EDTA). The vitamin requirement of some organisms should be considered as a special case of autrophy; the name 'auxotroph' is used to describe organisms which have a physiological growth requirement for one or more organic compounds but which derive their carbon from CO_2 and energy from light. The need for such growth factors should not be confused with the use of organic materials as an energy source, which is concerned in the next section under 'heterotrophy'. Provasoli (1958) reviewed some of the growth factor requirements of the algae and showed that among the commonest groups of phytoplankton, the Dinophyceae exhibited the most extensive auxotrophic requirements which often included a need for thiamine, biotin, and vitamin B_{12}. Many phytoplankton do not require any of the organic growth supplements shown in Table 18. In studies on truly autotrophic organisms it should be necessary to add the inorganic compounds shown in Table 18, together with a buffer (Tris) and the chelating agent (Na_2EDTA), neither of which are used directly by the cells. Both substances may be necessary in cultures, however, since the small volumes of culture vessels are unrepresentative of the natural sea-water environment, and the pH and chelating properties in culture vessels can change much more drastically than in the sea. In addition, for large-standing stocks of phytoplankton it may be necessary to bubble CO_2 enriched air through a culture medium.

3.1.5 PHOTOSYNTHESIS AND GROWTH OF PHYTOPLANKTON IN THE SEA

Under natural conditions, photosynthesis of phytoplankton is regulated spatially and temporally by several environmental factors. Since the general effect of these environmental factors on photosynthesis is known, *in situ* photosynthesis can either be measured directly, or indirectly by a mathematical combination of data on environmental changes and physiological responses of the phytoplankton.

The direct measurement of photosynthesis is used extensively. In this method, samples are collected from various depths in the euphotic zone and each sample is used to fill at least one clear glass bottle and one opaque glass bottle, replicates at each depth being filled as desired. The bottles are innoculated with radioactive bicarbonate (Steemann Nielsen, 1952) and a set of light and dark bottles are then returned to the same depth from which the sample was taken. The vertical spacing of the samples to be collected and exposed depends mainly on the depth to which the water column is illuminated. For example, the sampling depths may be chosen as follows: 100, 50, 25, 10, 5, and 1% of the surface illuminaton. These depths can either be determined by using a known value for the extinction coefficient [eqn. (24) in Section 3.1.2], or approximated from the Secchi disc depth using the relationship, $k_e = 1 \cdot 7/T$ where k_e is the extinction coefficient (m^{-1}) and T is the depth of the Secchi disc in metres.

The light and dark bottles remain suspended from a buoy during the period of the incubation and the carbon dioxide taken up in the light bottle minus the same value for the dark bottle is considered to be the amount of particulate organic carbon produced by photosynthesis. This value is probably somewhere between a measure of net and gross photosynthesis, but it is often referred to as net photosynthesis if factors, such as the exudation of soluble organic carbon, are ignored.

The suspension of bottles in the water column is conveniently carried out during half a day (i.e. from dawn to mid-day, or mid-day to dusk). The fraction of the daily radiation occurring during the period of the incubation can either be measured with a pyranometer or calculated using a formula, such as that proposed by Vollenweider (1965) or Ikushima (1967). The transformation of short incubation periods into daily rates (Platt, 1971) does not, however, take into account that photosynthesis may decline during incubation. This may be due to a diel photosynthetic rhythm (Vollenweider and Nauwerck, 1961), or to qualitative and quantitative changes in the enclosed population (Ichimura and Saijo, 1958). Physiological reasons for the latter have not been studied extensively, but losses due to damage during manipulations could well be a contributing factor (e.g. to sensitive flagellates). Changes in the daily photosynthetic rhythm have been studied more extensively and it appears (Lorenzen, 1963) that maximum photosynthetic rate (per unit of chlorophyll *a*) usually occurs before noon and about 70% of the total daily photosynthesis at the surface is carried out in the morning (i.e. between dawn and mid-day). However, at different depths in the water column the amount of light will be a function of sun angle; consequently, the period for which light is available for photosynthesis is partially a function of depth and this tends to reduce the diurnal rhythm in the water column as compared with observed changes in surface samples. For the whole water column, Vollenweider and Nauwerck (1961) found that photosynthesis in the morning was 54 to 62% of the daily photosynthesis per unit area.

Another disadvantage to short exposures is that it ignores the influence of temporary changes in weather conditions as well as diurnal changes in illumination. Thus in spite of the physiological difficulties of long exposures mentioned above, half or full day incubations are usually performed, and correction factors are employed for day-to-day and seasonal differences in illumination and day length.

In many cases it is inconvenient to keep bottles suspended *in situ* for half- or one-day periods. Such situations arise due to the limited availability of ship time at one location or in some near-shore environments which are not readily accessible. In order to avoid this difficulty, extensive use has been made of various simulated *in situ* methods involving an experimental approach and a mathematical interpretation of the results. In this technique, instead of suspending bottles in the sea, each light and dark bottle is put under a suitable light filter adjusted to the light transmission at the depth at which the sample was taken. The sample is exposed to sunlight in a temperature-controlled incubator (Jitts, 1963; Vollenweider, 1969; Kiefer and Strickland, 1970).

The indirect method involves measuring the photosynthetic response of phytoplankton to different environmental factors; all results are then combined and integrated mathematically (Ichimura, 1956b; Talling, 1957b), or graphically (Ryther, 1956) in order to determine the total amount of carbon dioxide fixed in the water column. Since light intensity is the most changeable parameter under field conditions, spatial and temporal changes in radiation, and the response of phytoplankton, are the most important experimental results. For this technique a water sample is usually taken from just under the sea surface (or from specific light depths if time permits) and then photosynthesis is measured at the *in situ* temperature and under a series of sunlight illuminations (e.g. 100, 50, 25, 10, 5, and 1% of the surface illumination). Exposure to sunlight is made for a few hours (e.g. 3 hr) around local apparent noon. Other light sources, such as an incandescent lamp showing high colour temperature (e.g. 3200 K), or a daylight-type fluorescent tube, may be used conveniently.

Photosynthetic response of phytoplankton to light intensities (e.g. P vs. I curves) has already been discussed in Section 3.1.2 (Fig. 28). Basic parameters describing a P vs. I curve, being generally understood by ecologists and physiologists, are (1) maximum rate of photosynthesis and (2) initial slope of the P vs. I curve. Based on these two parameters, the P vs. I curve can be expressed in several different mathematical forms. The simplest form is a hyperbolic function as follows (Tamiya *et al.*, 1953; Tominaga and Ichimura, 1966):

$$P_g = P_{max} \cdot \frac{I/I_k}{1 + I/I_k} \text{ or } P_g = \frac{P_{max}I}{I_k + I} \tag{30}$$

where P_g is gross photosynthesis (usually expressed as mg C/mg Chl a/hr), P_{max} is the maximum photosynthesis (the same units as P_g), I is the light intensity in suitable energy or illumination units, I_k is the light intensity at the junction of initial slope and P_{max} (see Fig. 28); this is the point where P_g is $0.7 \ P_{max}$. I_k is $2I'_k$, in which I'_k is the light intensity when $P_g = P_{max}/2$; it is similar to the half-saturation constant in the Michaelis–Menten equation. The second function in eqn. (30) is in a form similar to the Michaelis–Menten expression for enzyme reactions (see Section 3.1.6). The initial slope of eqn. (30) ($\Delta P/\Delta I = P_{max}/I_k$), however, falls off too fast compared to actual curves measured. Correspondence with data can be improved by rewriting eqn. (30) as follows:

$$P_g = P_{max} \cdot \frac{I/I_k}{\{1 + (I/I_k)^2\}^{1/2}}. \tag{31}$$

This formulation, first suggested for photosynthesis by Smith (1936), was applied to both culture and natural populations by Talling (1957 a, b).

Under high light intensities photosynthesis is inhibited by photoinhibition (see Figs. 28 and 29) and this is not apparent in either eqns. (30) or (31). The mechanism of photoinhibition, however, has not been fully understood yet; empirical approximation is therefore the only way to formulate a P vs. I curve with photoinhibition. Several authors have proposed solutions to this difficulty (Steele, 1962; Vollenweider, 1965; Parker, 1974). In the solution proposed by Steele (1962), P vs. I curve was expressed as an exponential function:

$$P_g = P_{max} \cdot I/I_{opt} \cdot \exp(1 - I/I_{opt}) \tag{32}$$

where P_g,* P_{max}, and I are the same specified above, and I_{opt} is the light intensity maximizing P_g. The initial slope of this equation is,

$$\frac{\Delta P}{\Delta I} = P_{max} \cdot e/I_{opt}.$$

In practice it is difficult to fit this equation to actual data obtained experimentally, because the shape of the curve is essentially fixed, particularly at high light intensity. However, this handicap may be overcome to some extent by introducing eI_k for the first I_{opt} in eqn. (32) (Vollenweider, 1965).

Vollenweider (1965) and Parker (1974) have made further modification on eqns. (31) and (32), respectively, by introducing more parameters, which give more flexibility in fitting experimental data. In Parker's modification, he raised the entire right-hand side of eqn. (32) to a power. The altered function can then be written as:

$$P_g = P_{max}\{(I/I_{opt} \cdot \exp 1 - I/I_{opt})\}^{\beta} \tag{33}$$

where β is a dimensionless constant, and others are the same as specified previously. The initial slope ($\Delta P/\Delta I$) of this curve depends on β; $\Delta P/\Delta I = (P_{max}/I_{opt}) \exp(\beta)$. On the other hand, Vollenweider (1965) made his modification by introducing two more parameters into eqn. (31):

$$P_g = P_{max}\frac{I/I_k}{\{1 + (I/I_k)^2\}^{1/2}} \cdot \frac{1}{\{1 + (bI/I_k)^2\}^{n/2}} \tag{34}$$

where P_{max} is the maximum rate of photosynthesis when $b = 0$ or $n = 0$, b is a constant with dimension of (light intensity)$^{-1}$, and n is a dimensionless constant. In practice the parameters P_{max} and I_k are unmeasurable in eqn. (34), when there is any kind of inhibition. Actual methods required to evaluate these parameters, which requires a complicated treatment, have been discussed by Fee (1969). Both eqns. (33) and (34) are capable of fitting a large variety of P vs. I curves with and without photoinhibition, and are quite successful in describing experimental data. This is not surprising in view of the number of free parameters (three or four) which are available for fitting. However, there are some drawbacks accompanying this flexibility; these equations are cumbersome to integrate over depth and time, and there are some difficulties involved in evaluating the parameters as mentioned above. It should also be mentioned that there is one more approach, which is rather different from those mentioned above, but which is used to approximate any shape of P vs. I curve; this is the use of polynomical equation

---

*Steele (1962) expressed P_g in terms of actual growth rate, gC/g plant C/hr instead of gC/gChl a/hr.

which has the advantages of simpler parameter estimation and analytic integrability
as well as a probable low loss of fidelity in describing experimental data (cf. Takahashi
and Nash, 1973).

The derivative for each equation when it is evaluated for $I = 0$, which derives the
initial slope of the curve, always includes P_{max} as one of the parameters. In other words,
the derivative can also be rewritten for P_{max}. In the case of eqn. (30), for example,
the initial slope (ϕ will be used instead of $\Delta P/\Delta I$ for simplicity) can be expressed as,

$$\frac{\Delta P}{\Delta I} = \phi = P_{max}/I_k, \tag{35}$$

and then the whole equation can be rearranged for P_{max} as follows:

$$P_{max} = \phi I_k. \tag{36}$$

Generally, P_{max} varies widely depending on changes in environmental conditions and
the physiological state of algae; however, the initial slope is a more nearly constant.
Considering these fundamental characters for P_{max} and ϕ, eqn. (36) indicates that I_k
is entirely proportional to P_{max} because ϕ is constant. By replacing P_{max} in eqn. (30)
with ϕI_k, the equation becomes simple without losing any of the capabilities of the
original equation:

$$P_g = \phi I_k \cdot \frac{I/I_k}{1 + I/I_k} = \phi \frac{I}{1 + I/I_k} \tag{37}$$

where ϕ must have the same light intensity unit as I and I_k. Actual values for ϕ can
be obtained either from direct measurement or data already reported (cf. Table 16).
This advancement for empirical approximation using ϕ was first suggested by Bannister
(1974), and can be made as well for other eqns. (31), (32), (33), and (34) without any
difficulty (see details in Bannister, 1974).

Other constants in equations mentioned above have also eco-physiological meanings;
β, b, and n are related to photoinhibition and the effects of changes in these constants
on curve approximation have been shown by Vollenweider (1965), Fee (1969), and Parker
(1974). Actual evaluation for these constants for photoinhibition has to wait until fundamental mechanisms for photoinhibition become clearer from physiological and biochemical laboratory experiments. Knowledge of how these constants, including ϕ, k,
β, b, and n, actually change with environmental conditions and algal populations will
have to be based on many P vs. I curves taken from different areas and at different
seasons; such data may eventually provide suitable P vs. I curves, without having to
make actual measurements of photosynthesis.

As has already been mentioned in this section, photosynthetic rate changes diurnally
(see also Section 3.1.2). For example, evening rates at optimum light intensity may
drop as low as one-tenth of that measured during the early morning hours (e.g. Doty,
1959). In some waters this decrease may be a function of diurnal nutrient depletion
(e.g. in the tropics) or of saturation of photosynthetic dark reactions.

Apart from the physiological depression in photosynthetic response, the overall diurnal response of photosynthesis is dependent on changes in the light intensity. In general,
the diurnal change of incident light intensity on a horizontal plane can be given by
empirical equations (Vollenweider, 1965; Ikushima, 1967). According to Ikushima, for
a fine day

$$I_t = I_{max} \cdot \sin^3 (\pi/D)t \tag{38}$$

where I_t (expressed as a unit of photosynthetically available light energy or illumination) is the light intensity at a given time, t, on a given day and I_{max} is the light intensity when the altitude of the sun is highest at local apparent noon; D (hr) is the day length. If the relationship of light penetration in a water column (Section 3.1.2) is put into the I_{max} of eqn. (38), light intensity at a given depth and time can be calculated for a fine day as

$$I_{t \cdot d} = I_{0 \cdot max} \cdot \sin^3 (\pi/D)t \cdot e^{-kd} \tag{39}$$

where $I_{0 \cdot max}$ is the light intensity at the surface for the highest altitude of the sun, k is the average extinction coefficient of photosynthetically available light (m^{-1}), and d is the depth in meters. For overcast days the equations given above should be modified by taking the $\sin^2$ instead of the $\sin^3$; if I_{max} is reduced to account for cloud cover this will effectively reduce the total radiation while keeping the same general shape of the diurnal curves.

For combining a P vs. I curve (physiological response) and light conditions (environmental response), eqn. (39) is inserted into the I of eqn. (30) or (31) and an equation for calculating gross photosynthesis at a given depth (d) and time (t), $P_{g \cdot t \cdot d}$ is obtained. For example, from eqns. (30) and (39), it is possible to write

$$P_{g \cdot t \cdot d} = P_{g \cdot t} \cdot \frac{a \cdot I_{0 \cdot max} \cdot \sin^3 (\pi/D)t \cdot e^{-kd}}{1 + a \cdot I_{0 \cdot max} \cdot \sin^3 (\pi/D)t \cdot e^{-kd}} . \tag{40}$$

Instead of using empirical equations [eqns. (38) and (39)], actual measurements of solar radiation can be put directly into an equation [e.g. eqn. (31)] for a P vs. I curve if they are available (Section 3.1.3), and then the following calculations can be continued.

By integrating eqn. (40) with time ($t = 0 \rightarrow D$), daily gross photosynthesis (usually in mg C/mg Chl a/day), at a given depth (d) is calculated as

$$P_{g \cdot d} = \int_0^D P_{g \cdot t \cdot d} \cdot dt . \tag{41}$$

By integrating eqn. (41) with depth ($d = 0 \rightarrow \infty$), daily gross photosynthesis (usually mg C/day/m^2) is calculated as

$$P_g = \int_0^\infty P_{g \cdot d} \cdot dd . \tag{42}$$

For determining the growth of phytoplankton under natural conditions, net photosynthesis should be considered instead of gross photosynthesis. Net photosynthesis can be calculated by subtracting respiration, R(mg C/mg Chl a/hr), from gross photosynthesis as follows:

Hourly net photosynthesis, $P_{n \cdot t \cdot d}$;

$$P_{n \cdot td} = P_{g \cdot t \cdot d} - R . \tag{43}$$

Daily net photosynthesis, $P_{n \cdot d}$;

$$P_{n \cdot d} = P_{g \cdot d} - 24R \tag{44}$$

where respiration rate of phytoplankton is assumed to be constant throughout the day. The depth at which $P_{n \cdot t \cdot d} = 0(P_{g \cdot t \cdot d} = R)$ or, $P_{n \cdot d} = 0(P_{g \cdot d} = 24 R)$ is called the hourly and daily compensation depth, respectively. The hourly compensation depth will change during the day and be maximum at noon and zero during darkness; the 24-hr compensation depth will change with season. Unfortunately, it is not often clear in the literature which compensation depth is being reported.

Daily photosynthesis obtained above is the daily rate when 1 mg Chl a/m^3 is distributed homogeneously in the euphotic zone; this is not the actual daily photosynthesis of the water column since the amount of chlorophyll will vary in time and space. In order to obtain the actual photosynthesis, $\mathbb{P}_{g \cdot d}$, the $P_{g \cdot d}$ must be multiplied by the amount of phytoplankton biomass, S (usually expressed as mg Chl a/m^3):

$$\mathbb{P}_{g \cdot d} = P_{g \cdot d} \times S \tag{45}$$

$$\mathbb{P}_{n \cdot d} = P_{n \cdot d} \times S. \tag{46}$$

$\mathbb{P}_{n \cdot d}$ is sometimes referred to as ΔN, which represents the daily net increase in standing stock, N. Hourly photosynthesis in a P vs. I curve is sometimes expressed on the basis of a unit volume of water, such as mg $C/m^3/hr$. In such a case, however, the curve can only be applied to the depth from which the sample for the measurement of P vs. I was taken, and at all depths, only in the season when the chlorophyll a is distributed homogeneously, such as during active vertical mixing. When chlorophyll a is not uniformly distributed, the expression for photosynthesis per unit of chlorophyll a must be obtained by multiplying the photosynthesis by the actual amount of chlorophyll a at different depths in the water column. Finally the depth integration of $\mathbb{P}_{g \cdot d}$ can usually be done with a planimeter, or by graphic integration.

In determining an average growth rate, μ, for the phytoplankton over short intervals, the biomass (usually expressed as mg C/m^3) is taken initially as N_0. The measured daily increase of phytoplankton, assumed to represent daily net photosynthesis (ΔN) on a unit carbon base instead of chlorophyll a, is then added to the biomass after a day's growth (t). For transforming daily net photosynthesis $P_{n \cdot d}$ (mg C/mg Chl a) into ΔN, $P_{n \cdot d}$ must first be converted to cellular carbon by multiplying by the amount of chlorophyll a in the water [eqn. (46)]. Then

$$\Delta N = N_t - N_0 \tag{47}$$

where N_t is the biomass after t days and the average growth rate is calculated as

$$\mu = 1/t \cdot \log_e [(N_0 + \Delta N)/N_0]. \tag{48}$$

The average growth rate is often expressed in terms of 'doublings of algal biomass per day'. For this purpose the base of the logarithm in eqn. (48) should be replaced by 2 instead of e (2·7183).

Average growth rates of phytoplankton in the euphotic zone for various regions are shown in Table 19. The growth rates, measured in an experimental bottle, represent the potential rate under given environmental conditions on a particular day. The actual growth rate *in situ* may be smaller or larger than the potential rate. Smaller *in situ* growth rates may result from losses of phytoplankton cells in the water column due to sinking, drifting, dying, zooplankton feeding, vertical mixing, and the movement of phytoplankton themselves. Larger *in situ* growth rates may result in nutrient-depleted environments since the exclusion of animals from the incubation bottles may decrease nutrient regeneration.

Of these effects, the most pronounced in temperate latitudes is the depth to which the water column is mixed. Thus, under conditions of storm activity, or tidal mixing

in coastal areas, it is apparent that phytoplankton cells may be mixed down below
the euphotic zone and that the time that the cells spend in the aphotic zone before
being mixed back to the euphotic zone, may be sufficient to result in no net production;
the period of darkness below the aphotic zone having resulted in a loss through respiration of the carbon gained through photosynthesis in the euphotic zone. In this sense
the euphotic zone is defined strictly as the depth to which the photosynthesis of a
plant cell is equal or greater than its respiration; the depth of the euphotic zone is
known as the compensation depth (D_c) and the light intensity at this depth is the compensation light intensity (I_c).

TABLE 19. AVERAGE GROWTH RATE OF PHYTOPLANKTON IN THE EUPHOTIC ZONE FOR VARIOUS REGIONS. TEMPERA-
TURES INDICATED ARE FOR THE SURFACE OR THE AVERAGE IN THE MIXED LAYER (AFTER EPPLEY, 1972)

Location	Temp. (°C)	Growth rate (doublings/day)		Reference
		measured	max. expected*	
Nutrient-poor waters				
Sargasso Sea	—	0·26	—	Riley *et al.* (1949)
Florida Strait	—	0·45	—	Riley *et al.* (1949)
Off the Carolinas	—	0·37	—	Riley *et al.* (1949)
Off Montauk Pt.	—	0·35	—	Riley *et al.* (1949)
Off Southern California				
July 1970	20	0·25–0·4	1·5	Eppley *et al.* (1972)
Apr.–Sept. 1967	12–21	0·7 av.	0·9–1·6	Eppley *et al.* (1972)
Nutrient-rich waters				
Peru Current				
Apr. 1966	17–20	0·67 av.	1·5	Strickland *et al.* (1969)
June 1969	18–19	0·73 av.	1·4	Beers *et al.* (1971)
Off S. W. Africa	—	1·0 av.	—	Calculated from Hobson (1971)
Western Arabian Sea	27–28	>1·0 av.	2·4	Calculated from Ryther and Menzel (1965)

*From eqn. (29) (Section 3.1.4) assuming the maximum growth rate, μ, will be one-half the value calculated
as expected if daylength is 12 hr and μ is directly proportional to the number of hours of light per day.

The depth to which plants can be mixed and at which the total photosynthesis for
the water column is equal to the total respiration (of primary producers) is known
as the 'critical depth'. The concept of a critical depth was first suggested by Gran
and Braarud (1935) and developed into a mathematical model by Sverdrup (1953). The
following description of the model is taken from Sverdrup's original paper, with a few
modifications. In particular, Sverdrup reduced the incident solar radiation (I_0) by a
factor of 0·2 to allow for absorption of the longer and shorter wavelengths of light
in the first meter of sea water. In shallow water columns this may be too large a
factor and it is suggested instead that the incident solar radiation should be reduced
by a factor of 0·5 to allow for the absorption of non-photosynthetic ultraviolet and
infrared radiation in the first few centimetres of water. An average extinction coefficient
for light (400 to 700 nm) penetrating the rest of the water column should then be used;
the latter value may be rather larger than the value used by Sverdrup, especially if
the model is used in coastal waters.

The model is illustrated in Fig. 34. If I_0 is the surface radiation and k is the average
extinction coefficient, then the compensation light intensity (I_c) is related to the compen-

sation depth (D_c) as

$$I_c = 0.5 \cdot I_0 \cdot e^{-kD_c}. \tag{49}$$

At the compensation depth the photosynthesis of a cell (P_c) is equal to its respiration R_c; above this depth there is a net gain from photosynthesis $(P_c > R_c)$ and below it there is a net loss $(P_c < R_c)$. However, as the phytoplankton cells are mixed above and below the compensation depth they will experience an average light intensity $(\bar{I})$.

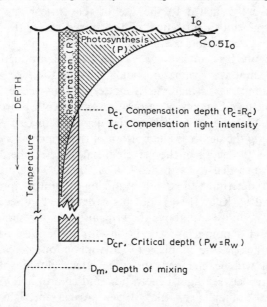

FIG. 34. Diagram showing the relationship between the compensation and critical depths, and the depth of mixing. (For explanation, see text)

The depth at which the average light intensity for the water column equals the compensation light intensity is known as the critical depth, D_{cr} [i.e. the depth at which photosynthesis for the water column (P_w) equals the respiration for the water column (R_w)]. The relationship of the critical depth to the compensation light intensity can be obtained by integrating eqn. (49) and dividing by the critical depth to obtain the average (compensation) light intensity for the water column. Thus

$$\bar{I}_c = 0.5 \cdot I_0 \int_0^{D_{cr}} \frac{e^{-kD} \cdot dD}{D_{cr}}$$

$$\bar{I}_c = \frac{0.5 \, I_0}{kD_{cr}} (1 - e^{-kD_{cr}}). \tag{50}$$

From eqn. (50) it is possible to determine the critical depth for any water column, knowing the extinction coefficient (k), the solar radiation (I_0) and assuming some value for the compensation light intensity (see Section 3.1.2). Where D_{cr} is large the equation reduces to

$$D_{cr} = \frac{0.5 \, I_0}{\bar{I}_c k}. \tag{51}$$

From the explanation given above, it is apparent that if the critical depth is less than the depth of mixing, no net production can take place since $P_w < R_w$. This is illustrated as the condition in Fig. 34 where the depth of mixing (D_m), measured to the depth to the bottom of the principal thermocline, has been drawn as being greater than the critical depth. However, if the critical depth is greater than the depth of mixing, a net positive production will occur in the water column ($P_w > R_w$) and conditions for the onset of phytoplankton growth will have been established.

Several conditions are attached to the use of the critical depth model; these are (i) that plants are uniformly distributed in the mixed layer, (ii) there is no lack of plant nutrients, (iii) the extinction coefficient of the water column is constant (this in fact has to be determined as an average value), (iv) the production of the plants is proportional to the amount of radiation, and (v) respiration is constant with depth. The last of these assumptions is made as a matter of convenience and there is in fact no ecological data on the constancy of respiration with depth. Furthermore, some difficulties may be encountered in establishing the depth of mixing; in Fig. 31 this has been simplified as being the depth to the bottom of the principal thermocline. However, in other environments the depth of the principal halocline or the depth at which there is a maximum change of density with depth ($\partial \varphi / \partial z \times 10^3$, Sverdrup et al., 1946) may be better employed to determine the depth of mixing. A similar approach to the critical depth model was used by Riley (1946) and this is discussed in Section 4.2.2.

Once the mixing zone moves up above the critical depth, the standing stock of phytoplankton increases and a number of conditions for the critical-depth model may be no longer valid. Further, as the biomass of phytoplankton increases, light conditions in the water column will be changed and the water column will become stratified. Empirical relationships between light penetration and chlorophyll a concentrations have been given by several authors (e.g. Riley, 1956a; Aruga and Ichimura, 1968). In the former reference an equation is given relating the average extinction coefficient (k_e, m^{-1}) to the chlorophyll a concentration (C, mg Chl a/m^3) for natural phytoplankton community as follows.

$$k = 0.04 + 0.0088\,C + 0.054 \cdot C^{2/3}. \tag{52}$$

The constant (0·04) in the equation is close to the average extinction coefficient for visible light in clean ocean water (0·0384) (see Table 20). The original equation was derived from the actual field observations carried out in the western North Atlantic. The equation does not specifically distinguish between phytoplankton and other materials and it may also include some extinction due to particulate and dissolved materials which are not directly related to chlorophyll (Riley, 1975). Nevertheless this is still a useful empirical formula which can be used to estimate the average extinction coefficient in natural water.

Figure 35 shows diagrammatic examples of phytoplankton growth in a stratified water column. At time, t_0, just after active vertical mixing of water, phytoplankton is distributed homogeneously in the water column but the in situ photosynthesis per unit phytoplankton biomass (mg C/mg Chl a/day) at each depth is different, being inhibited at the near surface by high light intensities and then decreasing from a subsurface maximum due to light attenuation. Assuming that the losses of phytoplankton at a given depth (from sinking, zooplankton grazing, etc.) are similar, the biomass of phytoplankton after a certain period can be estimated approximately by using eqns. (47) and (48). Because of the difference in the in situ photosynthesis per unit phytoplankton biomass at $t = 0$, the *in situ* growth rate of phytoplankton, μ will be different at each depth. Thus the profile of phytoplankton biomass and photosynthesis (called the 'productive structure'— Ichimura, 1956b) will change with time $t = 1$, $t = 2$, etc. As the phytoplankton biomass increases at the sub-surface photosynthetic maximum, the average extinction coefficient of light will also increase, and self-shading will occur (see Aruga, 1966). Accordingly, the daily compensation depth and the maximum in phytoplankton growth becomes shallower (Ichimura, 1956a). Providing there are sufficient nutrients, the pattern of photosynthesis and the standing stock of phytoplankton will finally maximize at the surface as a thin layer, and a bloom will occur ($t = 2$). As nutrients become exhausted in the surface layers, however, the depth of the maxima in the phytoplankton biomass

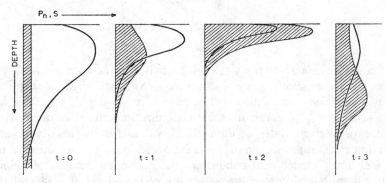

FIG. 35. Schematic changes in phytoplankton biomass (S) and daily net photosynthetic rate (P_n) after three time intervals (t) in stratified water (S or ▨, usually expressed in mg Chl a/m^3; P_n or ——, usually expressed as mgC/mgChl a/day.

and the primary productivity deepen ($t = 3$); the latter conditions occur in stratified water columns in temperate latitudes during the summer or may be found to prevail generally in tropical and sub-tropical waters throughout the year. However, other factors can modify the overall effect of light penetration on the production of the water column. Thus the feeding activity of zooplankton strongly affects the standing stock of phytoplankton and more detailed information of this subject is given in Section 4.2.2. In other cases, organic substances stimulate growth of phytoplankton (e.g. humic substances in sea water are reported to stimulate the growth of dinoflagellates—Prakash and Rashid, 1968). In some tropical areas motile phytoplankton, such as dinoflagellates (Eppley *et al.*, 1968), can use nutrients from the aphotic zone by diel migrations between near surface and deeper waters. A further method of overcoming nutrient limitation in tropical waters is through nitrogen fixation (e.g. Goering *et al.*, 1966), which sometimes gives rise to large surface blooms of *Trichodesmium*.

The quantity and quality of light in a water column is affected by the depth of the water column, the presence of dissolved coloured substances, and scattering due to suspended particles, including the phytoplankton. The light attenuation in a given water column is approximated from a logarithmic equation [eqn. (24) in Section 3.1.2] which is governed by the extinction, kD. The extinction is a function of the total phytoplankton biomass ($\bar{C}$), the depth of water (D) and the total amount of other suspended and dissolved matter ($\bar{M}$), as follows (Sakamoto, 1966);

$$kD = k_1D + k_2\bar{C} + k_3\bar{M} \qquad (53)$$

in which k_1, k_2 and k_3 are average extinction coefficients (400 to 700 nm) of a *unit* amount or thickness of phytoplankton, water and other matter, respectively. Assuming the daily compensation depth is the depth at which there is 1% of the surface illumination,* then the relation between chlorophyll a in the euphotic zone ($\overline{C}$, mg Chl a/m^2), the depth of euphotic zone (Dc, in meters) and suspended and dissolved matter ($\overline{M}$, expressed in units of thickness, e.g. meters) can be expressed by the following equation:

$$0{\cdot}01 = e^{-(k_1 D_c + k_2 C + k_3 M)} \tag{54}$$

$0{\cdot}01$ can be approximately replaced by $e^{-4{\cdot}6}$, which gives:

$$4{\cdot}6 = k_1 D_c + k_2 \overline{C} + k_3 \overline{M} \tag{55}$$

then

$$\overline{C} = \frac{4{\cdot}6 - (k_1 D_c + k_3 \overline{M})}{k_2} \tag{56}$$

where $\overline{C}$ in eqn. (56) represents the total chlorophyll a in the euphotic zone of a given water column. If the euphotic depth is close to zero (i.e. the light absorption due to water and suspended and dissolved matter is negligible, $k_1 Dc + k_3\overline{M} \simeq 0$), $\overline{C}_{max} \simeq 4{\cdot}6/k_2$; where $\overline{C}_{max}$ represents the maximum amount of chlorophyll a in the euphotic zone of a given water column. Some values for the maximum amount of chlorophyll a in cultures are shown in Table 20. *Skeletonema* culture shows a maximum of 650 mg Chl a/m^2. Under field conditions, the maximum amount of chlorophyll a is usually less than that of pure cultures because of the effects of $k_3\overline{M}$ and sometimes $k_1 Dc$ on the increase of kDc. For example, in the Kurishio current, the maximum amount of chlorophyll a in the euphotic zone has been computed to be 250 mg Chl a/m^2. Lorenzen (1972) found a highly significant correlation between the thickness of the euphotic zone (using the 1% light level for the euphotic depth) and the total chlorophyll content. From extensive field observations in the Pacific and Atlantic Oceans (including upwelling regions and offshore areas) he derived an equation as follows:

$$\ln \overline{C} = 8{\cdot}85 - 1{\cdot}57 \ln D_c \tag{57}$$

in which $\overset{\cdot}{\overline{C}}$ is the total chlorophyll a in the euphotic zone (mg Chl a/m^2) and D_c the 1% light depth (m). The data used in determining this equation covered a wide range of values, i.e. D_c, 10–91 m; $\overline{C}$, 7–277 mg Chl a/m^2.

An assessment of light extinction by the three factors water, phytoplankton, and particulate and dissolved matter can be made using eqns. (55) and (57). This is done by adding the attenuation due to water ($k_1 D_c$) and the phytoplankton [$k_2\overline{C}$, where $\overline{C}$ is estimated from eqn. (57)], and subtracting this from the total attenuation of the euphotic zone, which by definition is the natural logarithm of $0{\cdot}01$ ($\simeq 4{\cdot}605$). The quantity left is equal to the extinction due to dissolved and particulate substances ($k_3\overline{M}$). The relative effect of these three factors in euphotic zones of different thickness (Fig.

---

*Compensation light intensities for marine phytoplankton are generally found to be in the approximate range of $0{\cdot}002$ to $0{\cdot}009$ ly/min; since these values are very approximately two orders of magnitude lower than average surface radiation values, the compensation depth is sometimes simply estimated from the depth at which there is 1% of the surface radiation. A further approximate estimate of this depth is based on the relationship between the depth of the Secchi disc and the extinction coefficient. If this is assumed to be $k_e = 1{\cdot}7/T$ (p. 92.), then $0{\cdot}01 = e^{-kD_c}$ and $kD_c = 4{\cdot}6$ substituting for k with the Secchi depth (T), $Dc = (4{\cdot}6/1{\cdot}7)$. $T \simeq 2{\cdot}7T$. Thus twice the Secchi depth is sometimes employed as an estimate of the euphotic depth.

36) is reflected in the ratio $k_x X/kD_c$ where $k_x X$ is the attenuation of each factor with the thickness of the euphotic zone D_c as determined above, and kD_c is total attenuation throughout the euphotic zone. From any set of data, the thickness of the euphotic zone (D_c) and average extinction coefficient in the euphotic zone $(k, \text{ m}^{-1})$, one can estimate the possible light extinction due to water $(k_1 D_c)$, phytoplankton $(k_2 \overline{C})$, and other materials $(k_3 \overline{M})$ from Fig. 36. In springtime in the subarctic Pacific, for example, D_c and k' are 55 m and 0·080 (m^{-1}), respectively, then respective $k_1 D_c$, $k_2 \overline{C}$, and $k_3 \overline{M}$ can be estimated as 0·040, 0·037, and 0·003 (all these have a dimension of m^{-1}).

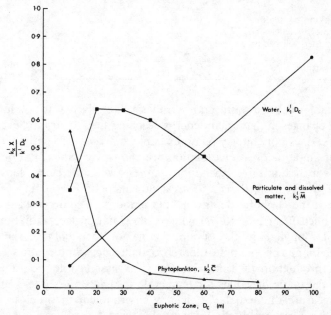

FIG. 36. Partitioning of total light attenuation $(k'D_c)$ into the fractions attributable to water $(k'_1 D_c)$, phytoplankton $(k'_2 C)$, and particulate and dissolved matter $(k'_3 M)$. See eqn (53) in text (from Lorenzen, 1972, 1976).

It is clear that $k_x \overline{C}/kD_c$ due to phytoplankton shows a logarithmic decrease with increasing thickness of the euphotic zone. The fraction due to particulate and dissolved matter $(k_3 \overline{M}/kD_c)$, which includes 'yellow substance', has a maximum around 20 ~ 30 m. This indicates that in high productivity areas generally characterized by thin euphotic zones, and if not influenced by land runoff or tidal mixing, most of the light is attenuated by phytoplankton ($t = 2$ in Fig. 35). After a period of time, the phytoplankton concentration is decreased by grazing or a decline of available nutrients, etc., and the euphotic zone deepens. It is at intermediate euphotic zone depths (20 − 30 m) that an increase in the attenuation of light by particulate and dissolved matter occurs, probably as a residue to a past plankton bloom ($t = 3$ in Fig. 35). Later, the euphotic zone attains maximum depths (>60–70 m). At this time, most of the light is attenuated by water, with only a very small fraction attributed to the phytoplankton. The decline in the fraction of attenuated light attributed to particulate and dissolved materials probably results from the euphotic layer. For example, particulates may be ingested by zooplankton, decomposed by bacteria, or sink into deeper layers. A similar treatment of light attenuation in the sea will be found in Tyler (1975); in this reference the author distinguishes between light absorbed by chlorophyll a and phaeophytin.

TABLE 20. AVERAGE LIGHT EXTINCTION COEFFICIENTS (400 TO 700 nm) OF CULTURED ALGAE AND SEA WATER (FROM TAKAHASHI AND PARSONS, 1972)

Materials	k_2 (mg Chl $a/m^2)^{-1}$	Chl $a \cdot$ max (mg Chl $a \cdot m^{-2}$)	References
	k_2		
Chlorella sp.	0·0115	400	Tominaga & Ichimura, 1966
Scenedesmus sp.	0·0051	900	Aruga, 1966
Cyclotella meneghiniana	0·0048	960	Steemann Nielsen, 1962
Skeletonema costatum	0·0071	650	Tominaga & Ichimura, 1966
	$k_2 + k_3$		
Natural phytoplankton populations (including suspended particles and dissolved matter, Kuroshio)	0·0184	250	Aruga & Ichimura, 1968
	k_1 (m^{-1})		
Sea water (clear ocean)	0·0384		Jerlov, 1968

It is predictable that the daily photosynthesis in a given water column will change with sustained changes in the radiant energy reaching the sea surface, even if the amount of chlorophyll a does not change. As an example, differences in daily net photosynthesis under different amounts of solar radiation are shown in Table 21. For the calculation of results shown in this table, the photosynthesis vs. light curve shown in Fig. 29B (in Section 3.1.3) was used, the P_{max} was assumed to be 1 mg C/mg Chl a/hr and the relation between the euphotic depth and the amount of chlorophyll a in the euphotic zone was obtained from Fig. 36. If data on the total surface radiation and the total amount of chlorophyll a in the euphotic zone (or the depth of the euphotic zone) can be obtained, one can estimate the approximate net photosynthesis from Table 21. In calculating production in Table 21, P_{max} was assumed to be 1 mg C/mg Chl a/hr, which falls in the range of values for temperate pelagic waters (see Table 17). It would be better to measure the actual P_{max} in any environment in order to obtain a better prediction of the in situ productivity; in such a case the values for net photosynthesis in Table 21 have to be multiplied by the value of the actual P_{max}. In highly productive areas, P_{max} may be greater than 5 mg C/mg Chl a/hr. Thus the calculated net photosynthesis from Table 21 could be at least $1800 \times 5 = 9000$ mg C/m^2/day. This value is in general agreement with maximum in situ primary productivity values of 5 to 10 g

TABLE 21. DAILY NET PHOTOSYNTHESIS FOR VARIOUS AMOUNTS OF CHLOROPHYLL a UNDER DIFFERENT LEVELS OF IRRADIANCE (FROM TAKAHASHI AND PARSONS, 1972)

Radiation (ly/day)		Net photosynthesis (mg C/m^2/day)								
	Euphotic depth (M)	175	150	100	50	30	20	10	5	<1
	Chl a amount (mg/m^2)	1	2	10	52	100	135	190	220	260
100		1·6	3	16	83	160	220	300	350	420
200		2·6	5	26	140	260	350	490	570	680
300		3·3	7	33	170	330	450	630	740	860
400		4·0	8	40	210	400	540	760	880	1000
500		4·7	9	47	240	470	630	890	1000	1200
600		5·3	11	53	280	530	720	1000	1200	1400
700		5·8	12	58	300	580	780	1100	1300	1500
800		6·3	13	63	330	630	850	1200	1400	1600
900		6·8	14	68	350	680	920	1300	1500	1800

$C/m^2/day$. Primary productivity values under culture conditions may be two or three times greater than *in situ* (see also Ryther, 1963).

Based on the photosynthetic production under a unit surface area, energy efficiency of photosynthetic production (photosynthetic efficiency) in natural water can be estimated taking a ratio of photosynthetic production $\{mgC/(m^2 \cdot day)\}$. The photosynthetic efficiency obtained like this has a dimension of $mgC/g \cdot cal$. However, a true efficiency should be dimensionless. This problem may be overcome by multiplying the photosynthetic production, expressed with carbon units, by an empirical conversion factor for carbon to calories (e.g. $11 \cdot 4 \ g \cdot cal/mgC$; Platt and Irwin, 1973), which gives a dimensionless efficiency:

Photosynthetic efficiency

$$= \frac{11 \cdot 4 \ (g \cdot cal/mgC) \times \text{Photosynthetic production} \ \{mgC/(m^2 \cdot day)\}}{\text{Incoming solar radiation} \ \{g \cdot cal/(m^2 \cdot day)\}}. \tag{58}$$

Photosynthetic active radiation (PAR, approximately 50% less than total solar radiation on a sunny day) is sometimes applied for estimating the efficiency. Therefore one must be careful in considering the quality of radiation, as well as the dimensions of the efficiency, when photosynthetic efficiencies from several different sources are compared. According to the extensive study in St. Margaret's Bay, Nova Scotia, throughout a year (Platt, 1971), the photosynthetic efficiencies fluctuated between 0·02 and 0·9% being an average of 0·26% (per total radiant energy) in which chlorophyll *a* in the water column varied from 10 to 340 mg Chl a/m^2, being an average of 30 mg Chl a/m^2.

Platt (1969) has proposed an index k_b^d in (m^{-1}) which represents that contribution to the absorption (extinction) coefficient of light energy at any depth (d) which is due to photosynthetic production. This should be equal to the absorption of light by photosynthetic pigments multiplied by a factor measuring the efficiency with which light is utilized in the photosynthetic process. It was shown that this index, k_b^d, was very nearly equal to the photosynthetic efficiency.

3.1.6 THE KINETICS OF NUTRIENT UPTAKE

The two earliest ecological considerations in the uptake of nutrients by phytoplankton were firstly, that at low nutrient concentrations, the rate of nutrient uptake was found to be concentration dependent, and secondly that the total yield of phytoplankton was directly proportional to the initial concentration of limiting nutrient and independent of the growth rate of phytoplankton. Ketchum (1939a) demonstrated the first of these results; using the diatom, *Nitzschia closterium*, he showed that the uptake of nitrate was concentration dependent over an approximate range from 1 to 7 μg at N/l. Spencer (1954), using the same organisms, demonstrated the second result when he showed a linear relationship between the initial concentration of nitrate and total yield of cells over nitrate concentrations ranging from approximately 15 to 150 μg at/l. Ketchum's results also showed that another nutrient, phosphate, followed similar kinetics to the uptake kinetics of nitrate and that phosphate could be taken up in the absence of measurable nitrate; however, phosphate uptake was increased when the concentration of nitrate was increased.

Taking the original work of Ketchum a step further, Caperon (1967) and Dugdale (1967) showed that nutrient uptake could be described using Michaelis–Menten enzyme kinetics in which

$$v = \frac{V_\mathrm{m} S}{K_\mathrm{s} + S}, \tag{59}$$

where v is the rate of nutrient uptake, V_m is the maximum rate of nutrient uptake, K_s is the substrate concentration at which $v = V_\mathrm{m}/2$ and S is the concentration of nutrient. By plotting S/v versus S, a straight line is obtained with an intercept on the abscissa $-K_\mathrm{s}$. The constant K_s is believed to be an important property of a phytoplankton cell since it reflects the ability of a species to take up low concentrations of a nutrient and as such it may determine the minimum nutrient concentration at which a species can grow. Thus MacIsaac and Dugdale (1969) have shown that coastal phytoplankton communities generally have K_s value of $>1\ \mu\mathrm{M}$ for nitrate uptake, while oceanic communities have lower K_s values of $ca\ 0{\cdot}2\ \mu\mathrm{M}$.* In line with other biochemical studies on Michaelis–Menten constants it is also important to recognize that K_s values are temperature dependent (Eppley $et\ al.$, 1969b).

In the case of Si uptake by diatoms on the other hand, it was found that the apparent uptake kinetics are slightly different from those for other nutrients; the plot of Si uptake against the Si concentration does not start from zero Si concentration (Fig. 38). Paasche (1973) has presented an equation with a slight modification of eqn. (59) in order to express this type of uptake. His equation is

$$v = \frac{V_\mathrm{m}(S - S_0)}{K_\mathrm{s} + (S - S_0)}, \tag{60}$$

where S_0 is the threshold concentration of Si at which $v = 0$. According to laboratory batch culture experiments using five different species of centric diatoms, threshold concentrations varied between $0{\cdot}3$ and $1{\cdot}3\ \mu\mathrm{g}$ at Si/l being rather greater in species with a large K_s, with the exception of $Ditylum\ brightwellii$ in which S_0 was small and the K_s was large (Paasche, 1973). K_s values also varied between $0{\cdot}8$ and $3{\cdot}4\ \mu\mathrm{g}$ at Si/l depending on different species. A similar level of K_s for Si† ($2{\cdot}93\ \mu\mathrm{g}$ at Si/l) has been reported from silicon limited natural phytoplankton populations measured by the recently developed non-radioactive masspectrometric technique (Goering $et\ al.$, 1973).

It should be noted both from eqns. (59) and (60) and Figs. 37 and 38, that the kinetic approach of nutrient uptake is only valid for a condition under which nutrients are limiting. In waters with naturally saturating nutrient concentrations, it is not possible to measure K_s.

Under conditions of no nutrient stress, nutrient uptake rate shows a hyperbola-shaped response to light intensity which is similar to the response commonly observed in photosynthesis/light relation. Such nutrient/light relation can also be described by using the Michaelis–Menten equation, but the following two cases must be considered. The first point is that the nutrient/light curve quite often shows a positive value of nutrient uptake even at zero light intensity, because some nutrients (e.g. nitrate and ammonia) are known to be taken up in the dark at low rates, with ammonium being taken up more readily (Dugdale and Goering, 1967). The second point is that nutrient uptake may be depressed under bright light intensities (e.g. above 10 or 25% of surface illumination) in the same way as are observed 'photo-inhibition' in photosynthesis/light responses. MacIsaac and Dugdale (1972) have proposed an equation slightly modified

---

*$0{\cdot}2\ \mu\mathrm{M}$ nitrate $\equiv 0{\cdot}2\ \mu\mathrm{gat}\ \mathrm{NO_3^-}$.
†Precisely silicic acid $\{\mathrm{Si(OH)_4}\}$, in this case, which is a hydroxylated product of silicate.

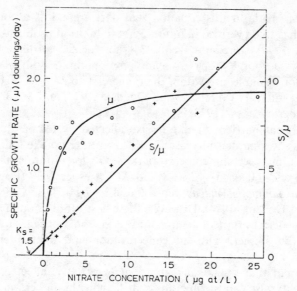

FIG. 37. Specific growth rate of *Asterionella japonica* as a function of nitrate concentration (redrawn from Eppley and Thomas, 1969).

from eqn. (59) in order to describe the nutrient uptake versus light intensity, in which they have taken into account the first point but not the second point mentioned above.

$$v = V_d + V'\left(\frac{I}{K'_I + I}\right) \tag{61}$$

where V_d is dark uptake, V'_m is the maximum velocity at nutrient and light saturation, K'_I is the constant of the equation and I is light intensity. This equation only describes uptake over the hyperbolic part of the curve but not for photoinhibition. There is

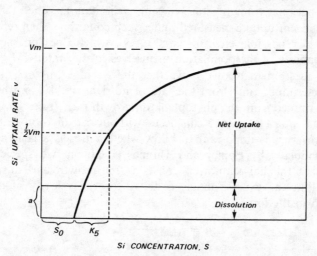

FIG. 38. Si uptake rate as a function of reactive silicate concentration in the medium (redrawn from Paasche, 1973) when the uptake experiments were carried out with silicate-depleted cultures; a correction for silica dissolution (*a*, generally 10 to 20% of the maximum uptake rate) is needed to drive the net uptake rates. S_0 is the threshold concentration.

also an assumption made in this equation that dark uptake is a constant at all light intensities, which has to be verified in future. Photoinhibition problems may be overcome by using Parker's (1974) expression* [eqn. (33)] for the second term on the right-hand side of eqn. (61). Values for K_I for nitrate and ammonium for some natural phytoplankton populations, for example, are within the range of 1 to 14% (*ca.* 0·004 ~ 0·06 ly/min) of the surface light intensity (MacIsaac and Dugdale, 1972) and ~2·3% (*ca.* 0·01 ly/min) for silicic acid (Goering *et al.*, 1973).

It is anticipated that nutrient uptake is also affected by temperature. Actually there are several reports showing that increases in K_s for nitrate uptake followed with increasing temperature within a certain temperature range; K_s for nitrate uptake of *Skeletonema costatum* were 0·0 and 1·0 μg at N/l at 8 and 28°C, respectively (Eppley *et al.*, 1969b), and those for *Gymnodinium splendens* were 1·0 and 6·6 μg at N/l at 18 and 28°C, respectively (Thomas and Dodson, 1974). However, there is still not enough information available to make a generalization of temperature effect on nutrient uptake in a manner similar to effects by ambient nutrient concentration and light intensity as described above.

It is now apparent that uptake of a nutrient (e.g. K_s) is affected by light intensity (and possibly quality also) and temperature, at least within a certain range of changes, as well as physiological properties of cell membranes. Furthermore, other nutrient(s) not considered here may also affect the uptake of the rate-limiting nutrients. Most past experiments using algal cultures done by many investigators were carried out under optimal or suboptimal environmental conditions. Under such circumstances (with no environmental limitation except for the nutrient being considered), K_s values obtained may reflect the physiological potential of the species or clones being used, and they can be compared with each other in that sense. However, in nature the situation may be rather different and it is hard to anticipate that there is only one parameter limiting nutrient uptake and that all others are optimal, even supposing that every different species of phytoplankton reacts the same way. For example, if a nutrient, which is the primary limiting factor, is introduced into a natural population in order to estimate K_s, some other parameter might start causing a limitation, other than the nutrient being studied. Nutrient uptake measured under such conditions then reflects the uptake characteristics under possible environmental stress, which does not necessarily reflect the potential capability of the population themselves in the water.

An important question to resolve is whether the K_s of a phytoplankton species in the sea can be determined only from the rate of nutrient uptake or whether the same value can be determined from the phytoplankton growth rate. It has been demonstrated experimentally that nutrient uptake can occur without cell division (Fitzgerald, 1968); ecologically, however, it is important to know whether this is a general phenomenon. From studies conducted by Eppley and Thomas (1969) on two species of diatom it was concluded that the half-saturation constants for growth and nitrate uptake are very similar so that the expression given above for the velocity of nutrient uptake can also be given in the form

$$\mu = \mu_{max} \frac{S}{K_s + S},$$ (62)

* $\quad v = V_d + V_m [I/K_I \exp(1 - I/K_I)]^{\alpha}$ (61a)

where α is a dimensionless constant. This equation also describes the saturated curve expressed by eqn. (61).

TABLE 22. HALF-SATURATION CONSTANTS FOR NITRATE AND AMMONIA UPTAKE

Phytoplankton species, clone or area	K_s (μgat/l) Nitrate	K_s (μgat/l) Ammonia	Reference
Oligotrophic, tropical Pacific	0·04 0·21 0·01 0·03 0·14	0·10 0·55 0·62	MacIsaac and Dugdale, 1969
Eutrophic, tropical Pacific	0·98		
Eutrophic, subarctic Pacific	4·21	1·30	
Oceanic species	0·1 to 0·7	0·1 to 0·4	Eppley et al., 1969b
Neritic diatoms	0·4 to 5·1	0·5 to 9·3	
Neritic or littoral flagellates	0·1 to 10·3	0·1 to 5·7	
Thalassiosira pseudonana Clone 3H Clone 7–15 Clone 13–1	1·87 1·19 0·38		Carpenter and Guillard, 1971
Fragilaria pinnata Clone 13–3 Clone 0–12	0·62 1·64		

where μ and μ_{max} are the growth rate and maximum growth rate, respectively, in units of time^{-1}. A plot of the above expression is given in Fig. 37 together with a plot of S/μ vs. S, which gives the intercept value on the abscissa of K_s. A summary of some K_s values are given in Table 22. From these data it was suggested by Eppley et al. (1969b) that there was a clear trend in K_s, being large in large cells and small in small cells; thus Coccolithus huxleyi, which is ca. 5 μm diameter, had a K_s of 0·1 μg at NO_3^-/l while Gonyaulax polyedra, which is ca. 45 μm diameter, had a K_s of >5 μg at NO_3^-/l. Similar trends were also found in silicic acid uptake by diatoms (Goering et al., 1973). It was also apparent from additional experiments conducted by Eppley et al. (1969b) that the ability to take up nitrate and ammonium ions differed between species and that some species could take up nitrate at lower concentrations than they could take up ammonia, and vice versa. Carpenter and Guillard (1971) showed that differences in K_s values were not confined to species but that clones of the same species had different K_s values depending on their environment. Clones isolated from low nutrient oceanic water had K_s values of <0;75 μgat/l while the same species taken from an estuarine region had a K_s of >1·5 μgat/l.

Differences in K_s values coupled with differences in the ability of species to reach their maximum growth rate at different light intensities, have been suggested by Dugdale (1967) and Eppley et al. (1969b) as being important factors in determining species succession in phytoplankton blooms. A schematic example is given in Fig. 39; in the first figure it may be seen that Coccolithus huxleyi reached its maximum growth at a low light intensity using ammonium ions and that its growth was approximately double that of the other species at a concentration of 0·5 μgat NH_4^+/l. However, in the second

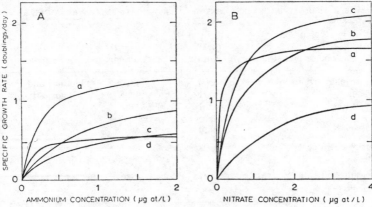

FIG. 39. Specific growth rate vs. ammonium and nitrate concentration at two light intensities, (B) approx. 4 times (A). (a) *Coccolithus huxleyi*, (b) *Ditylum brightwellii*, (c) *Skeletonema costatum*, (d) *Dunaliella tertiolecta* (redrawn from Eppley *et al.*, 1969b).

figure, at a higher light intensity, both *Skeletonema costatum* and *Ditylum brightwellii* grew faster than *C. huxleyi* at concentrations above 2μgat NO_3^-/l. Since specific growth rates are exponents it is quite apparent that even small differences, over a period of time, would lead to very large differences in the standing stock of a particular species. The validity of this approach to field studies has been demonstrated by Thomas (1970b) and Thomas and Owen (1971), who showed that it was possible to obtain good agreement between growth rates calculated from $^{14}CO_2$ uptake and from ammonium concentrations in tropical Pacific waters. However, this empirical justification for the determination of K_s and phytoplankton growth constants from field data is not wholly supported by experimental and theoretical considerations advanced by several authors. The relationship between growth rate and nutrient concentration described in the above equation depends on μ_{max} and K_s remaining constant; this condition is generally met during exponential growth when the nutrient content per cell remains relatively constant, such that

$$\frac{\text{uptake rate}}{\text{cell division rate}} = Q, \text{the nutrient content per cell.} \tag{63}$$

However, Kuenzler and Ketchum (1962) showed that the uptake of phosphorus by *Phaeodactylum* was not determined by the phosphate concentration in the medium but by the past history of the cells and their internal phosphorus content. The cell content of phosphorus varied between 2 and 66×10^{-15} moles/cell; Caperon (1968) showed a similar variability in the nitrogen content of *Isochrysis galbana*, from 2 to 40×10^{-15} moles/cell. Further it has been known from some of the earliest studies (e.g. Ketchum, 1939a) that nutrient uptake can take place in the dark when the cell division rate is zero. Eppley and Strickland (1968) have suggested that phytoplankton cells require a certain minimum content of nitrogen (or phosphorus) before cell division can proceed; for *Dunaliella tertiolecta* the minimal nitrogen per cell (Q_0) varied with light intensity between 0.7×10^{-15} and 1.4×10^{-15} moles/cell. Caperon (1968) showed that there was a hyperbolic relationship between the reserve nitrate content per cell (Q') and growth rate (μ), such that

$$\mu = \mu_{max} \cdot \frac{Q'}{A + Q'}, \tag{64}$$

where A is a growth constant. The value Q' used in the above equation represents the nitrate in the cell reservoir which is in addition to Q_0, the minimum amount of nitrate required for a normal cell at zero growth rate. Thus A represents the concentration of reservoir nitrate within the cell which is necessary to obtain a growth rate of $\mu_{max}/2$. In chemostat studies on vitamin B_{12} requirements, Droop (1970) showed that the growth rate could be expressed in terms of Q_0 and Q as

$$\mu = \mu_{max}\left(1 - \frac{Q_0}{Q}\right). \tag{65}$$

From these findings it is apparent that the ecological use of K_s values (Table 22) should be approached with caution. As pointed out by Eppley et al. (1969b), for example, the dinoflagellate, Gonyaulax polyedra, has a low growth rate and an apparently high K_s. However, this organism displays a diel migration in which it absorbs nutrients from 10 to 15 m during the night and swims to the surface during the day where light is available for photosynthesis and growth, but where the nutrients are much lower than at 10 to 15 m. Thus the uptake of nutrients can be independent of growth and the phenomenon of 'luxury consumption' (see Eppley and Strickland, 1968) may result in growth being proportional to the nutrient content per cell (e.g. Droop, 1968 and 1970) rather than the nutrient concentration in situ (e.g. Thomas, 1970b).

Nutrient-dependent differences in growth rates may account for the dominance of one species over another on the basis of their respective growth rates but as Hulburt (1970) has pointed out, there is little or no possibility of an abundant species monopolizing the nutrient supply and forcing a less abundant species to extinction. Each algal cell in a phytoplankton community may be visualized as the centre of a volume of water in which nutrient depletion decreases from the cell outwards; this volume may be referred to as the cell's nutrient zone. Hulbert (1970) based his conclusions on the fact that the nutrient zone surrounding a cell would have to overlap with that of another cell if one growth rate was to affect another. For cellular nutrient zones to overlap, Hulburt calculated that there would have to be at least 3×10^8 phytoplankton cells per litre; since most natural populations rarely exceed 10^6 cells/litre in open and coastal waters, there is generally no possibility of one nutrient zone overlapping with another. However, at cell concentrations of 10^9/litre, such as may occur in some estuarine and eutrophic environments, there is a possibility that dominant species will lead to the elimination of residual species and hence decrease the diversity of the system.

3.1.7 CHEMOSYNTHESIS

Some micro-organisms can satisfy their primary energy requirements by utilizing simple inorganic compounds, such as ammonia, methane, or nitrite, or elements, such as ferrous iron, hydrogen gas, or water-insoluble amorphous sulfur. All of the known organisms which comprise this (chemosynthetic) group are bacteria. Most of their inorganic substrates are derived from the decay of organic matter which is itself primarily formed through photosynthesis. Consequently, if one follows the origin of the energy used in chemosynthesis, chemosynthetic processes may not be considered as primary production. However, chemosynthesis usually involves carbon dioxide fixation and the primary formation of new particulate material. Thus chemosynthesis may be considered as a special case of primary production on the grounds of its trophic position in the marine food chain.

In the past it was believed that the inorganic substrates reacted directly with molecular oxygen to form oxidized end-products. However, Bunker (1936) showed that molecular oxygen was not directly involved in sulfur oxidation by thiobacilli. The fact that oxygen in the end-products originated from water resulted in a new concept of chemosynthesis, although in some cases it is still questionable whether the concept is adaptable to all cases of chemosynthetic activity. Thus reactions of the type suggested by Bunker involve dehydrogenations rather than oxidations. In general terms the chemosynthetic process can be expressed conveniently in three stages according to Bunker's concept (Gundersen, 1968).

(1) As the result of dehydrogenation, high reducing power is produced as follows.

$$n\text{AH}_2 + n\text{H}_2\text{O} \xrightarrow{\text{dehydrogenase}} n\text{AO} + 4n[\text{H}^+ + \text{e}^-] \tag{66}$$
$$\underset{\text{(Inorganic substrates)}}{} \qquad \underset{\substack{\text{(oxidized} \\ \text{end-product)}}}{} \quad \underset{\text{(reducing power)}}{}$$

where the symbol $[\text{H}^+ + \text{e}^-]$ is used as a synonym for the reducing power.

(2) A proportion of the reducing power is then utilized for energy production (adenosine triphosphate, ATP, synthesis) by being transferred through the cytochrome system to molecular oxygen. A second part of the reducing power is transferred to NAD (nicotinamide adenine dinucleotide) in order to produce reduced NAD (or NADH_2, reduced nicotinamide adenine dinucleotide). These relationships might be visualized as follows:

$$4[\text{H}^+ + \text{e}^-] + m\text{ADP} + m\text{P}_i + \text{O}_2 \xrightarrow[\text{system}]{\text{cytochrome}} 2\text{H}_2\text{O} + m\text{ATP} \tag{67}$$

$$2[\text{H}^+ + \text{e}^-] + \text{NAD} \longrightarrow \text{NADH}_2 \tag{68}$$

where ADP is adenosine diphosphate, and P_i is inorganic phosphate. Some anaerobes can use bound oxygen derived from inorganic compounds instead of free oxygen as shown in the above equation (e.g. sulphate-reducing bacteria).

(3) The ATP and NADH_2 are then used for the assimilation of carbon dioxide:

$$12\text{NADH}_2 + 18\text{ATP} + 6\text{CO}_2 \rightarrow \text{C}_6\text{H}_{12}\text{O}_6 + 6\text{H}_2\text{O} + 18\text{ADP} + 18\text{P}_i + 12\text{NAD}. \tag{69}$$

Thus the different inorganic substrates used by chemosynthetic bacteria are not merely the sole source of the organism's energy but also the sole source of their reducing power.

Depending on differences in the organic substrate (AH_2), chemosynthetic bacteria are classified into several groups, such as nitrifying, sulfur, hydrogen, methane, iron, and carbon monoxide bacteria. Table 23 shows some representative chemosynthetic bacteria inhabiting marine environments. The overall efficiencies of chemosynthesis during the growth of the bacteria (e.g. nitrifying or sulfur bacteria) can be expressed as the ratio between the total energy consumed in carbon dioxide assimilation and the energy liberated by the primary inorganic compounds during oxidation; these efficiencies are 6 to 8% (Gibbs and Schiff, 1960) although the efficiency may sometimes change drastically with different stages of cultures. For example, young cultures of *Nitrosomonas* gave values approaching 50%; in older cultures, the efficiency dropped to 7% (Hofmann and Lees, 1952). Most chemosynthetic bacteria require free oxygen as the electron acceptor in the second step described above. However, facultative or obligate anaerobic bacteria, such as *Thiobacillus denitrificans* and *Desulfovibrio desulfricans*, can use bound oxygen derived from nitrate or sulfate.

Among inorganic substrates available for the chemosynthetic bacteria, nitrogen and sulfur compounds are relatively abundant and widely distributed compared with the other reduced compounds in the pelagic environment of the oceans. Among reduced nitrogen compounds, ammonia may be present in concentrations up to *ca.* 5 μg at/l and nitrite at concentrations up to *ca.* 2 μg at/l (Riley and Chester, 1971). Actual occurrences of nitrifying bacteria, expressed in colonies per litre of sea water, were found to be <1 in the north Atlantic Ocean, <10 in the Pacific Ocean, *ca.* 10^4 in the Indian Ocean near an island, and *ca.* 10^6 in Barbados harbor (Watson, 1965; Hattori and Wada, 1971). Some of the strains of nitrifying bacteria isolated from marine waters are the same or similar to those from fresh water or soil, but others are peculiar to marine environments (e.g. see Watson, 1971). Nitrification by marine bacteria, using either ammonia or nitrite as a substrate, is reported to be more efficient at low (<0·1 ml/l) oxygen concentrations (Carlucci and McNally, 1969).

The quantities of reduced sulfur compounds in the marine environments are generally much greater than the quantities of reduced nitrogen compounds. Thiosulfate and polythionates may sometimes be present at levels from 0 to 100 μg at/l, the highest concentrations of these compounds being detected near to shore (Tilton, 1968). Sulphides are generally not detected within the analytical limit of <1 μg at/l in open ocean waters (Tilton, 1968). The number of colonies of sulfur bacteria (*Thiobacillus* spp.) have been found to range from 0 to 275 per 100 ml (Tilton *et al.*, 1967). These values for colony counts are several orders of magnitude smaller than those (10^3 to 10^4 per 100 ml) which were calculated by Tilton *et al.* (1967) on the basis of the amount of reduced sulfur compounds.

The reduced inorganic substrates are mainly produced through anaerobic metabolic processes. Consequently, anaerobic environments, which sometimes develop in fjords and estuaries, create favourable habitats for chemosynthetic bacteria. Under such conditions vigorous growth of certain species of chemosynthetic bacteria (usually sulfur, hydrogen, or methane bacteria) has been observed (Kuznetsov, 1959; Sokolova and Karavaiko, 1964).

It should be mentioned that an obligate anaerobic strain, *Desulfovibrio desulfricans*, has been isolated even from oxic ocean waters (Kimata *et al.*, 1955; Wood, 1958). In order to explain this phenomenon, Baas Becking and Wood (1955) proposed the idea of 'metabiosis' in which they suggested that several kinds of microorganisms co-exist on or in bacterial aggregates and conditions for the production of reduced substrates could occur within the aggregates. Thus the obligate anaerobic bacteria existing in oxic environments are probably present within the microcosm of a bacterial aggregate (Seki, 1972).

In the ocean, chemosynthetic activity can be estimated from carbon dioxide uptake in the dark using $NaH^{14}CO_3$. Compared with carbon fixation through photosynthesis, dark CO_2 uptake is usually small (i.e. less than 5% of the photosynthesis on a daily basis within the euphotic zone in pelagic areas). Dark CO_2 uptake is not entirely carried out by chemosynthesis, but heterotrophic processes by bacteria and algae (e.g. Wood and Werkman, 1935, 1936, and 1940) may result in the uptake of small amounts of CO_2 in the absence of light. Algal dependency on dark fixation of CO_2 is low (i.e. 3 to 5% of the total CO_2 fixed in the light) and this is about the same proportion of CO_2 fixed by aerobic heterotrophic bacteria when they are growing on an organic substrate (Sorokin, 1961 and 1966). However, among facultative autotrophic bacteria which belong to an intermediate metabolic type between obligate heterotrophs and

TABLE 23. SUMMARY OF CHEMOSYNTHETIC BACTERIA GROWING IN THE OCEAN AND THEIR INORGANIC SUBSTRATES

	Inorganic substrate	Oxidized end-product	Oxidizer	ΔF^* (k cal)	Ability to grow heterotrophically	Habitat
(1) Nitrifying bacteria						
Nitrobacter spp.	NO_2	NO_3	O_2	18		Soil, fresh, and sea waters
Nitrococcus mobilis WATSON & WATERBURG	NO_2	NO_3	O_2	18		Sea water
Nitrospina gracilis WATSON & WATERBURG	NO_2	NO_3	O_2	18		Sea water
Nitrosomonas spp.	NH_3	NO_2	O_2	85		Soil, fresh, and sea waters
Nitrosococcus oceanus (WATSON) comb. nov.	H_3 or NH_2OH	NO_2	O_2	85		Sea water
(2) Sulfur bacteria						
Thiovulum majas HINZE	H_2S	S		50	?	Sea water
Beggiatoa spp.	H_2S	S		50	+	Fresh and sea waters
Thiospira bipunctata MOLISCH	H_2S	S		50	?	Sea water
Thiothrix spp.	S	SO_4^-	O_2	119	?	Soil, fresh, and sea waters
Thiobacillus thioparus BEIJERINCK	$5/4\ S_2O_3^-$	$3/2\ SO_4^{--} + s$	O_2	112	+	Soil, fresh, and sea waters
Thiobacillus denitrificans BEIJERINCK	$5S$	$5SO_4^{--}$	O_2	112	+	Soil, fresh, and sea waters
(3) Hydrogen bacteria						
Hydrogenomonas spp.	H_2	H_2O	O_2	56	+	Soil, fresh, and sea waters
Desulfovibrio desulfricans (BEIJERINCK) KLUYVER & VAN NIEL	H_2	H_2O	SO_4	56	+	Soil, fresh, and sea waters
(4) Methane bacteria						
Methanomonas spp.	CH_4	CO_2	O_2		+	Soil, fresh, and sea waters
(5) Iron bacteria						
Grallionella spp.	$4FeCO_3$	$4Fe(OH)_3$	O_2	81	+	
(6) Carbon monoxide bacteria						
Sarcina bakerii SCHNELLEN	CO	CH_4	H_2		+	Soil, fresh, and sea waters

*Free energies per number of moles of electron donor indicated. Values reviewed by Gibbs and Schiff (1960) are mainly quoted here.

chemoautotrophs, there is a requirement of between 20 and 90% CO_2 during the oxidation of low molecular weight organic compounds, such as methane and formic acid; for obligate chemoautotrophs the requirement for CO_2 is close to 100%. Since in natural environments, microorganisms which depend on different kinds of substrates exist together, it is practically impossible to separate out and estimate the actual activity of chemosynthetic organisms except at the time of vigorous growth of any one species. However, a measure of dark CO_2 uptake is a useful measure of chemosynthetic activity in any environment. As an example, Seki (1968) studied seasonal changes in dark uptake of CO_2 in a small bay and showed that dark uptake of CO_2 near the bottom of the bay could be as high as 20% of the photosynthesis throughout the year and that during the spring dark uptake for the water column was sometimes 50% greater than photosynthesis.

Active chemosynthesis occurs in waters and sediments in which both aerobic and anaerobic environments exist in the same column. Sorokin (1964a and b) has studied the importance of chemosynthetic bacteria in the Black Sea, where a thick anaerobic zone exists below the depths of *ca.* 150 m in the central part of the sea. A summary of his finding is shown in Fig. 40. At depths shallower than 50 m, inorganic carbon is fixed through photosynthesis by algae, and the daily photosynthesis is reported to be *ca.* 350 mg C/m²/day during October. At the transition zone between 100 and 250 m, where environmental conditions are changing from aerobic to anaerobic, chemosynthetic fixation of inorganic carbon is predominant. An *in situ* maximum value for daily chemosynthesis was *ca.* 9 mg C/m³/day; chemosynthesis for the water column amounted to

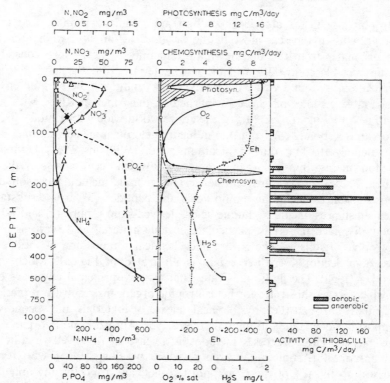

Fig. 40. Daily chemosynthetic production and some factors influencing it in the Black Sea (after Sorokin, 1964c).

approximately 200 mg $C/m^2/day$. This active chemosynthesis is carried out mainly by sulfur bacteria (aerobic and anaerobic *Thiobacillus*), and it is supported by a continual supply of reduced sulfur compounds, such as H_2S, S, and S_2O_3, from the anaerobic zone. The potential activity of thiobacilli in water samples taken from the depth of their maximum activity is very high compared with the *in situ* activity; by determining the oxidation rate of thiosulfate added to a water sample, a potential activity of 200 mg $C/m^3/day$ has been obtained.

Although the studies described above may appear to have limited geographical significance, it is probable in fact that similar microzonation in environmental factors can occur in many marine environments, especially in bays and estuaries. Also in all near-shore sediments where wave action does not disturb the benthos, a high chemosynthetic activity will occur in a depth zone of a few cm just below the sediment surface. If the transient zone of aerobic to anaerobic conditions comes up above the depth receiving a few per cent of the surface illumination, anaerobic photosynthetic sulfur bacteria, which require reductants (e.g. H_2S, S) for the hydrogen donor, grow vigorously. These organisms impart a purple or deep green colour to the sediment or water column in which they exist.

3.2 HETEROTROPHIC PROCESSES

3.2.1 THE ORIGIN OF ORGANIC SUBSTRATES

Heterotrophic organisms are dependent on an organic carbon source to provide energy for growth. In the sea, most of the organic carbon utilized by heterotrophic organisms originates from the marine biota, and only in near-shore coastal areas is there any appreciable contribution of organic materials from the land. Exceptions to this generalization are found where major rivers may influence the oceanic environment for a considerable distance off shore; in particular, the Amazon River, with an annual discharge of $6 \times 10^{12} m^3$ (or *ca.* 20% of the entire world wide river runoff), may contain sufficient organic carbon (2 to 10 mg/l) to influence heterotrophic activity over an oceanic area approximately $10^7 km^2$ during maximum runoff (Williams, 1968). Duursma (1965) recognized four main groups of organic compounds which occur as dissolved substances in sea water. These are (1) nitrogen-free organic matter, including carbohydrates, (2) nitrogenous substances, including amino acids and peptides, (3) fat-like substances, and (4) complex substances, including humic acids, derived from groups (1) and (2). In addition to these dissolved materials, particulate organic material also serve as a substrate for heterotrophic organisms. The nature and chemical composition of these substances are discussed in Section 1.4; their origin is illustrated in Fig. 41, which is adapted from Johannes (1968). The figure illustrates that the two principal pathways of organic materials which act as substrates for heterotrophic organisms result from the food chain of the sea, firstly through the physiological release of materials and secondly through the decomposition of plants and animals themselves. Autochthonous organic materials derived by physiological processes include the release of dissolved organic materials by phytoplankton; this subject has been reviewed by Fogg (1966), who records that in some cases, up to 50% of the CO_2 photosynthetically fixed by phytoplankton may be released as soluble organic carbon (Allen, 1956; Fogg, 1952). This process is sometimes called 'excretion' but should probably be called 'exudation'. The latter term is used by Sieburth (1969) and Sieburth and Jensen (1969) in referring to the loss of soluble organic material from seaweeds; the authors report that this loss may amount to 40% of the net carbon fixed daily. Further studies on the exudation of soluble organic carbon by phytoplankton (Eppley and Sloan, 1965; Hellebust, 1965) have shown that in general the release from healthy cells amount to 15% or less of the total carbon fixed; Watt (1966) has suggested a maximum loss of extracellular products of up to 30%. Two factors which appear to effect the release of soluble organics are the age of the cells and the light intensity. Hellebust (1965) and Anderson and Zeutzschel (1970)

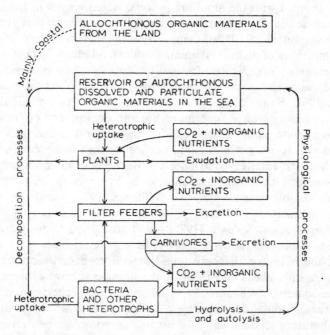

Fig. 41. Pathways of transfer and regeneration of organic substrates in an aquatic ecosystem (modified from Johannes, 1968).

have shown that exudation is greatest at high light intensities and among phytoplankton cells collected at the end of a plankton bloom.

The relationship between algal exudation and bacterial growth has been discussed by Bell and Mitchell (1972). From their own experimental results and from literature reports the authors conclude that bacterial growth may show a small increase during the logarithmic phase of algal growth but that a much greater maximum in the number of bacterial cells occurs when the algae reach a stationary growth phase and when there is some cellular lysis. The authors introduce the concept of a 'phycosphere' which is defined as a zone extending outward from an algal cell (or bloom) in which bacterial cells are stimulated by extracellular products of the algae. It appears that bacterial chemotaxis towards this zone can occur under experimental conditions but only when the organic material released by algae has reached relatively high concentrations from lysing algal cells.

It is quite apparent from these studies that since plants are the major producers of organic matter in the sea, any fraction lost through exudation will constitute an appreciable input to the organic reservoir in the environment. The type of compounds released by algae are known to include small amounts of amino acids and generally larger amounts of short chain acids (e.g. glycollic acid), glycerol, carbohydrates, and polysaccharides (Fogg et al., 1965; Watt, 1966).

The excretion of organic materials resulting from digestive processes of animals forms a second physiological pathway for the input of organic materials available for heterotrophic growth. In addition to the process of true excretion, animals may also release organic material directly from their bodies (e.g. Johannes and Webb, 1970), and this may in part be coupled with the moulting process, which results in the specific production of chitinous debris; for example, Lasker (1966) estimated that the loss of chitin by euphausiids amounted to 10% of their body weight every 5 days.

The natural mortality of plants and animals in the food chain forms the main pathway for the production of organic substrates by decomposition processes. Post-mortem changes in the permeability of cell membranes and the effect of autolytic enzymes (Krause et al., 1961; Krause, 1961 and 1962) may result in an initial loss of 15 to 50% of the total biomass of an organism. Soluble organic compounds released from dead organisms include amino acids, peptides, carbohydrates, and fatty acids.

Heterotrophs which utilize the organic particulate and soluble substrates produced by the food chain, themselves contribute to the organic reservoir of the sea, either by serving as food for filter feeding organisms or by hydrolysing substrates, and self-autolysis, to provide soluble organic materials. Williams (1970) found that soluble organic materials were assimilated by bacteria with an average efficiency of 67% for glucose and 78% for amino acids. Sorokin (1970) found that algal hydrolysates were utilized by natural populations of marine heterotrophs with an efficiency of ca. 45%. These efficiencies are much higher than are found in culture experiments (e.g. Parsons and Seki, 1970) where only about one-third of the organic substrate taken up is retained as new cellular material. Williams (1970) suggests that higher conversion efficiencies result with natural populations utilizing a variety of organic compounds, in contrast to the usual laboratory studies, where all organic carbon must be derived from a single organic compound. The largest fraction of organic carbon lost during heterotrophic activity is respired back to carbon dioxide, a process which also occurs throughout the food chain as organic carbon is transferred to higher trophic levels.

While information summarized in Fig. 41 shows the major pathways for the release of organic compounds in the heterotrophic cycle of the sea, there are in addition very specific organic compounds released in sea water which can modify the overall balance of organic materials in any one environment. Included among specific organic substances are growth factors and antimetabolites. As an example of the former, Burkholder and Burkholder (1956) and Starr (1956) showed that in coastal areas, vitamin B_{12} is produced by the bacteria; secretion of vitamins by some phytoplankton has also been demonstrated (Carlucci and Bowes, 1970). Since other species of phytoplankton require vitamins for growth, their presence can be assumed to affect the organic food chain. Antimetabolites have been detected in sea water, particularly in respect to the production of antibacterial substances by phytoplankton (e.g. Sieburth, 1964). More detailed studies on the release of antibacterial materials by phytoplankton (Duff et al., 1966) have revealed species specific differences and in at least one case (Antia and Bilinski, 1967) this had led to the identification of a lecithinase as being an anti-bacterial agent in the chrysomonad, Monochrysis lutheri.

3.2.2 HETEROTROPHIC UPTAKE

The two pathways of heterotrophic uptake shown in Fig. 41 are firstly through the plants (facultative heterotrophs) and secondly through the bacteria and other obligate heterotrophs, such as yeasts and moulds. The most important of these two pathways is undoubtedly the second, and the role of plants as autotrophic (or sometimes auxotrophic organisms) far exceeds their role as heterotrophic organisms.

The ability of marine bacteria to decompose a wide range of naturally occurring organic substrates has been widely demonstrated. These substances include the decomposition of chitin (Hock, 1940; Seki, 1965 a and b), cellulose (Waksman *et al.*, 1933), protein (Wood, 1953), and alginates (Meland, 1962). Jannasch (1958) showed that the presence of large numbers of bacteria in sea water was dependent on the concentration of dissolved organic matter; in the absence of soluble substrates, bacteria are found attached to particulate materials, such as inorganic particles and chitin. The heterotrophic activity of bacteria results in a number of complex transformations which are not implied in Fig. 41; these have been reviewed by ZoBell (1962) and include dissolution and precipitation of organic and inorganic compounds as well as the large-scale production of inorganic energy reserves (such as methane and hydrogen sulphide), when heterotrophic decomposition occurs under anaerobic conditions (see Fig. 64).

The heterotrophic activity of phytoplankton has been demonstrated in a number of specific cases. Out of forty-four pure cultures of littoral diatoms, Lewin and Lewin (1960) found that twenty-eight were capable of heterotrophic growth on glucose, acetate, or lactate media. Kuenzler (1965) found that only one out of thirteen species of phytoplankton could assimilate glucose for growth in the darkness. Antia *et al.* (1969) demonstrated that a photosynthesis cyptomonad, *Chroomonas salina*, could grow heterotrophically in the dark at high concentrations of glycerol but not at low concentrations. However, a variety of substrates including acetate, glutamate, and glucose stimulated growth of the organism in the light. Sloan and Strickland (1966) used radioactive carbon to determine the heterotrophic uptake of glucose, acetate, and glutamate by four species of marine phytoplankton. While some uptake was detectable, it was generally less than the dark uptake of radioactive carbon dioxide; the latter can be fixed non-photosynthetically in the dark through biochemical mechanisms, such as the Wood–Werkman reaction (Wood and Werkman, 1936 and 1940). At times the dark uptake of carbon dioxide may exceed the amount taken up by photosynthesis; this indicates that in such environments heterotrophic activity, resulting in the mineralization of organic matter, is in excess of autotrophic activity resulting in the production of organic matter (e.g. Seki, 1968, and Section 3.1.6).

The uptake of organic substrates at low concentrations has been demonstrated among a few species of coastal phytoplankton. North and Stephens (1971 and 1972) showed that *Platymonas* and *Nitzschia ovalis* could take up amino acids. In the case of *Platymonas*, for example, arginine, glycine, and glutamate were taken up at concentrations of *ca.* 5×10^{-6} moles/litre; the rate of uptake was increased 10-fold in nitrogen-starved cells. As a general conclusion, however, it appears that heterotrophic activity among the phytoplankton is limited to a number of species and that among these, some species may require very high concentrations of substrate in order to demonstrate heterotrophic activity.

Parsons and Strickland (1962b) introduced the use of radioactive organic substrates to a study of the kinetics of heterotrophic uptake in sea. The author's original experiments showed that the uptake of radioactive glucose and acetate by natural microbial populations in the sea could be fitted to the equation

$$v = \frac{V_m(S + A)}{K + (S + A)} \tag{70}$$

where v was the rate of uptake of the substrate, S was the natural *in situ* concentration of substrate, A was the concentration of the same substrate added to the water, V_m was the maximum rate of uptake and K was a constant, similar to the Michaelis–Menten constant for enzyme reactions (see Appendix to this section). Quantitatively v, in units of mg/m^3/hr, was obtained from the expression

$$v = \frac{cf(S + A)}{C\mu t}, \tag{71}$$

where c was the radioactivity of the filtered organisms (cpm), S and A were as defined above, C was the activity of $1\,\mu$Ci of ^{14}C in the counting assembly used, μ was the number of microcuries added to the sample bottle, f was a factor for any isotope discrimination between the ^{14}C isotope and normal carbon, and t was the incubation time in hours. By employing a reciprocal plot of eqn. (70), to obtain a straight line, the negative intercept on the x-axis becomes $-1/(K + S)$. If one could assume that $K \ll S$, the method would give a measure of the amount of natural substrate in sea water; alternatively, if $K \gg S$ the method would give a measure of the 'relative heterotrophic potential' based on K in mg C/m^3, which expresses the ability of the microbial population to take up the substrate. Wright and Hobbie (1965 and 1966) discussed the use of this method as applied to lake water and from their work it has been possible to draw a number of conclusions. By combining eqns. (70) and (71) and rearranging, the following form can be obtained:

$$\frac{C\mu t}{fc} = \frac{(K + S)}{V_m} + \frac{A}{V_m}. \tag{72}$$

When $C\mu t/fc$ is plotted against different concentrations of A (assuming A to be much larger than S), a straight line is obtained with a slope $1/V_m$, an intercept on the ordinate of $(K + S)/V_m$ and an intercept on the abscissa at $-(K + S)$. If an independent determination is made of S (e.g. by chemical analysis), the rate of uptake of the substrate under natural conditions can be determined, providing the organisms are complying with the relationship in eqn. (70). Three experimental qualifications governing the use of this equation are (i) that there must be no appreciable change in the microbial population during the experiment, (ii) that there should be no appreciable change in substrate concentration, and (iii) that careful temperature control is maintained. A plot of $C\mu t/fc$ versus A is shown in Fig. 42 from Wright and Hobbie (1965). From these results with natural lake water it is apparent that eqn. (70) applied over concentration range from 0 to 580 mg/m^3 of added acetate, but that above this concentration the relationship did not apply. In additional experiments, it was further shown by Wright and Hobbie (1965) that at high substrate concentrations a second uptake mechanism was involved which accounted for the departure from an asymptotic value of V_m predicated by eqn. (70). This was explained in terms of a passive transport of substrate due to diffusion at high substrate concentrations and the process could be expressed

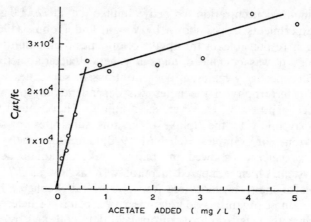

FIG. 42. Kinetics of acetate uptake by natural lake water samples at different substrate concentrations plotted according to eqn. (72) (redrawn from Wright and Hobbie, 1965).

in terms of a diffusion coefficient (K_d) as

$$K_d = \frac{V_1 - V_2}{A_1 - A_2},$$ (73)

where V_1, V_2 and A_1, A_2 are two rates of uptake and substrate concentrations, respectively. The difference in these two uptake mechanisms is illustrated in Fig. 43, curves A and B.

From these studies it is apparent that there are at least two transport systems for heterotrophic uptake: one showing dependence on Michaelis–Menten kinetics at low substrate concentrations and the other showing dependence on diffusion at high concentrations. Some indication of how low substrate concentrations can be in nature for the first mechanism, is given by the K values for two bacterial isolates obtained by Wright and Hobbie; these values were 5.0 mg/m^3 for acetate and 7 mg/m^3 for glucose. Since 90% of the maximum uptake velocity is obtained at *ca.* $10 \times K$, the active uptake mechanism may be assumed to operate maximally at substrate concentrations below 100 mg/m^3. The second mechanism for heterotrophic uptake appears to operate at about

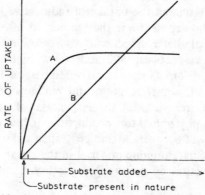

FIG. 43. Theoretical relationships between substrate addition and velocity of heterotrophic uptake showing (A) active transport, in which the transport system becomes saturated and (B) transport by diffusion in which the uptake velocity increases with substrate concentration. Note: added substrate is greatly in excess of natural levels.

10 times the maximum concentration for active uptake (i.e. at *ca.* $1.0 \, g/m^3$) and from some additional experiments it was shown by Wright and Hobbie (1965) that phytoplankton were largely responsible for the uptake of organics by the diffusion mechanism. From these findings it was concluded that, in general, bacterial activity in aquatic environments would effectively maintain concentrations of substrates which would be too low for algal heterotrophy, if the latter was dependent only on the diffusion of substrates at high concentrations. However, several authors have found that Michaelis–Menten kinetics may apply to the uptake of organic substrates by some species of coastal phytoplankton. For examples, Hellebust (1970) found that one species of marine diatom, *Melosira nummuloides*, showed an ability to take up amino acids through an active transport system which compared favourably (K as low as 7.7×10^{-6} m) with transport constants of many bacteria (see also North and Stephens, 1971 and 1972). At present, however, these organisms do not appear to account for heterotrophic activity in oceanic environments. This is indicated from field data collected by Williams (1970), who found that 80% of the heterotrophic activity in sea water samples from the Mediterranean and Atlantic Ocean would pass an 8-μm filter. This would eliminate a large part of the phytoplankton biomass. Particles of $<8 \, \mu$m would include all bacteria and most bacterial aggregates, other than those attached to large particles of detritus.

The hyperbolic curve (A) in Fig. 43 is shown passing through zero rate at zero substrate concentration. In practice it has been found (Jannasch, 1970) that zero growth may occur at some low but finite, or threshold, concentration. This phenomenon has been explained as a population effect in which a certain initial population of a particular species has to be present in order to modify the environment (e.g. redox potential) for the cells to multiply. The effect has been observed in chemostat studies but it may also be a common phenomenon under natural conditions in the sea. Kinetics accounting for a small departure from a zero intercept have been given by Jannasch (1970).

In some cases it has been reported to be difficult to establish any meaningful relationships between substrate concentration and uptake (e.g. Vaccaro and Jannasch, 1967; Hamilton and Preslan, 1970): these difficulties may be due to the existence of several competing heterotrophic populations or to the occurrence of more than one uptake mechanism operative over the same range of concentration. From Hellebust's (1970) data it is also apparent that a form of competitive inhibition of substrate uptake may be caused by the presence of similar substrates.

Sorokin (1970) has investigated the uptake of radioactive algal hydrolysates in water samples taken at different depths from the surface to 5000 m in the Pacific Ocean. The author found very little heterotrophic activity below 800 m and the maximum activity in tropical waters was between 400 and 600 m, which coincides with the depth of the oxygen minimum. From fairly extensive studies on the uptake of glucose and amino acids in the English Channel, Andrews and Williams (1971) concluded that heterotrophic processes in the area account for an uptake of organic material equivalent to 50% of the measured phytoplankton production. If this is converted into bacterial biomass with an efficiency of between 30 and 60%, then the contribution of heterotrophs to the formation of particulate organic material in the ocean would amount to between 15 and 30% of the production by autotrophic photosynthetic processes.

(Appendix to Section 3.2.2)

In a number of sections throughout this text reference is made to Michaelis–Menten kinetics. Caperon (1967) discussed the general applicability of this approach to the uptake of food materials by micro-organisms. From his review of the subject it is possible to write a general expression for the growth of a population of micro-organisms limited by food supply. The approach follows the original expression given by Michaelis and Menten for the action of enzymes on substrates.

The process of a population feeding may be described by the following equation in which C is the acquisition site for food, B is a food particle or unit quantity of food, $C \cdot B$ is an acquisition site filled by B, and P is a particle, or unit amount, of ingested food.

$$C + B \underset{k_2}{\overset{k_1}{\rightleftharpoons}} C \cdot B \overset{k_3}{\longrightarrow} C + P. \tag{74}$$

If $(C + B)$ to $(C \cdot B)$ is a reversible process it can be defined by constant K_m such that

$$K_m = \frac{[(C) - (CB)](B)}{(CB)} \tag{75}$$

or

$$(CB) = \frac{(C)(B)}{K_m + B}.$$

If the breakdown of (CB) to $(C + P)$ is largely an irreversible process, then the rate (v) for this process is given by

$$v = k_3(CB) \tag{76}$$

and

$$v = \frac{k_3(C)(B)}{K_m + (B)}. \tag{77}$$

The maximum rate (V_m) will be attained when the concentration of CB is maximized, which is when all the acquisition sites are filled [i.e. $(CB) = (C)$].

$$V_m = k_3(CB) = k_3(C). \tag{78}$$

By combining eqns. (77) and (78), the general Michaelis–Menten expression is obtained:

$$v = \frac{V_m(B)}{K_m + (B)}. \tag{79}$$

Under these conditions it is possible to relate growth to the rate of food intake. These conditions are (1) that the concentration of only one food is limiting population growth, (2) that a given amount of ingested food always results in the production of a fixed number of new individuals, (3) that there is no time lag in the response of growth rate to change in food concentration, and (4) that food once absorbed is ingested and not returned in any appreciable quantity as a food item. Under these conditions, growth, measured as production dn/dt per unit biomass (n) can be expressed as μ, the growth constant, where

$$\mu = \frac{1}{n}\frac{dn}{dt} \tag{80}$$

and μ and μ_{max} substituted for v and V_m in eqn. (79)

$$\mu = \frac{\mu_{max}(B)}{K_m + (B)}. \tag{81}$$

CHAPTER 4

PLANKTON FEEDING AND PRODUCTION

4.1 FEEDING PROCESSES

Among the zooplankton, methods of feeding may be broadly divided into filter feeding and raptorial feeding. In the former process a number of different mechanisms are employed to induce a flow of water against a screen which removes most of the particulate material in suspension. In the second process, animals actually seize individual prey items; the latter may either be large plants, such as individual diatom cells, or planktonic animals generally smaller in size than the predator. The two processes are not mutually exclusive and examples of both filter feeding and raptorial feeding can often be found in one species, especially among the planktonic crustaceans. On the basis of diet, zooplankton feeding may be herbivorous, omnivorous, or carnivorous; other divisions used by some authors are to describe the dietary requirements as being phytophagous, zoophagous, euryphagous, or detritivorous.

A general account of suspension feeding has been written by Jørgensen (1966), and an extensive study of the nutrition of abyssal plankton has been reported by Chindonova (1959); specific reviews on planktonic crustacean feeding have been written by a number of authors including Marshall and Orr (1955), Wickstead (1962), and Gauld (1966).

Among the planktonic protozoans, the feeding of the foraminifera and radiolaria is generally accomplished through the extrusion of pseudopodia from within the main body of the animal; a large number of pseudopodia may spread out into a reticulum and phytoplankton cells are captured on the surface and digested externally (MacKinnon and Hawes, 1961). In addition to capturing prey some of these animals may contain symbiotic algae known collectively as 'zooxanthellae' which may contribute directly to the animal's nutrition through releasing extracellular products· of photosynthesis (e.g. Khmeleva, 1967). A second large group of protozoa, the tintinnids, feed off a wide variety of particles including bacteria, detritus, flagellates, and diatoms; food is captured through a large aperture surrounded by cilia which induce a strong current of water towards the animal.

The position of zooplanktonic crustaceans, as by far the largest group of suspension feeders in the ocean, together with their ability to consume a great variety of prey, warrants special consideration of the feeding habits of these animals in the marine food chain. A schematic representation of the location of feeding appendages in a calanoid copepod is given in Fig. 44 adapted from Jørgensen (1966) and Gauld (1966).

124

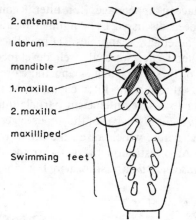

2.antenna

labrum

mandible

1.maxilla

2.maxilla

maxilliped

Swimming feet

FIG. 44. Diagram of ventral view of a copepod showing relative position of major appendages and flow of water (arrows).

While the details of these appendages vary considerably among the planktonic crustaceans, their general form and use includes a means of inducing water currents, filtering out organisms, seizing prey, and cutting or grinding food particles; the exact development of these processes is more or less specialized throughout the group. Thus in the illustration shown, a filter chamber is formed by the long setae of the (2) maxillae which project ventroanteriorly towards the labrum covering the mouth. Water currents are produced by a combination of vibrations of the (2) antennae, the mandibles, the (1) maxillae and the maxillipeds. The flow of water is as indicated in Fig. 44 and the removal of particulate material caught on the (2) maxillae is largely accomplished by a combining action of the (1) maxillae and the maxillipeds which pass the food forward to the mouth. The size range of particles retained by this type of filtering mechanism has been discussed by Gauld (1966) among others. The lower limit appears to be largely determined by the spacing of the setules which are small hair-like structures or bristles on the setae. In *Calanus finmarchicus* the setules of the (2) maxillae are spaced about $5\,\mu$m apart. Jørgensen (1966) has summarized reported measurements on a number of different copepod species including various life stages; the range of values for copepods was from $1.5\,\mu$m for *Pseudocalanus minutus* to $8.2\,\mu$m for *Metridia longa*. Similar structures on euphausiids showed a much greater range from about $7\,\mu$m for *Euphausia superba* up to $60\,\mu$m for *Bentheuphausia ambluops* (Nemoto, 1967). The largest particle which can be retained by the filter feeding process appears to be decided by the position of the maxillary setae in forming the filter chamber. Gauld (1966) has discussed the details of this structure in *Calanus finmarchicus* and has concluded that particles larger than $50\,\mu$m would be excluded from entering. It is quite apparent, however, that many filter-feeding zooplanktonic crustaceans are capable of feeding off diatom cells which are much larger than $50\,\mu$m diameter. This is accomplished through a second feeding mechanism in which large diatom cells are grasped by the animal and their contents sucked out (Beklemishev, 1954; Cushing, 1955). The process may be partly the same as that in which a zooplankter captures a smaller animal. However, the capture of large phytoplankters and small zooplankters may differ in respect to the predators' feeding response; in *Calanus hyperboreus* Conover (1966c) observed that *C. hyperboreus* could detect the movement of an *Artemia* nauplius which it actively pursued and caught.

The adaptation of crustacean appendages for filter feeding, raptorial feeding or a combination of both, has been discussed by a number of authors. Anraku and Omori (1963) showed that the (2) antennae, mandibles, (1) maxillae and maxillipeds of a predominantly herbivorous filter feeder were all well developed in terms of setae which increased the surface area of these appendages and therefore assured their efficient use in producing water currents through the filter chamber; setae of the (2) maxillae were also well developed in order to assure an efficient particle filter. In contrast, in a predatory species of copepod, *Tortanus discaudatus*, the appendages have few setae and instead there are modifications in structure which aid in the use of these appendages for seizing and holding a prey item. An illustration of two extremes in structural development in *C. finmarchicus* and *T. discaudatus* are shown in Fig. 45. Anraku and Omori (1963)

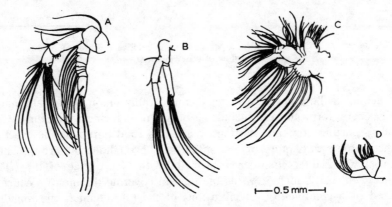

F IG . 45. Adaptation of crustacean appendages showing (2) antenna of (A) *Calanus finmarchicus* and (B) *Tortanus discaudatus* and (1) maxilla (C) and (D) of the same species, respectively (redrawn from Anraku and Omori, 1963).

also noted differences in the cutting edges of mandibles in different copepods and they describe a typical filter feeder as having grinding teeth while those of a typical predator have very sharp teeth. Between these two extremes, there are a variety of structures which enable some copepods to be omnivorous. Itoh (1970) has been able to summarize the difference between herbivores, omnivores, and carnivores on the basis of an 'Edge Index' (*EI*) derived from measurements of the cutting edges of the mandible. Calculation of the 'Edge Index' is illustrated in Fig. 46A and a plot of this value versus the number of cutting edges is illustrated in Fig. 46B. From the latter figure it is apparent that pelagic copepods may be divided into three groups, viz. Group I, with $EI \leq 500$ consisting of the Families, Calanidae, Eucalanidae, Paracalanidae, and Pseudocalanidae, which are largely herbivorous filter feeders; Group II, with $500 < EI \leq 900$, consisting of generally omnivorous feeders of the Families, Euchaetidae, Centropagidae, Temorida, Lucicutiidae, and Acartiidae, and Group III $EI > 900$ consisting of predominantly raptorial feeders of the Families, Heterorhabdidae, Augaptilidae, Candaciidae, Pontellidae, and Tortanidae. From earlier work on stomach contents of these animals it is sometimes possible to identify the raptorial feeders as being carnivores although the inclusion of stomach contents of a herbivorous prey makes it difficult to diagnose whether an animal is exclusively carnivorous.

In addition to extensive studies on the feeding of copepods, Nemoto (1967) has reported on differentiation in the body characters of euphausiids. Among euphausiids,

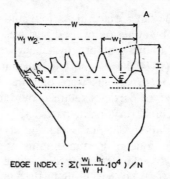

EDGE INDEX : $\Sigma(\frac{w_i}{W}\cdot\frac{h_i}{H}\cdot 10^4)/N$

FIG. 46A. Schematic representation of the 'Edge Index' on the cutting edges of the mandible, w_1, $w_2,\ldots,$ w_i represent the width of individual cutting edges and W the total width; similarly h_1, $h_2,\ldots,$ h_i and H represent the edge heights, respectively; N represents the number of edges (redrawn from Itoh, 1970).

raptorial feeders usually have very long 2nd or 3rd thoracic legs which terminate in grasping spines or small chelae; predominantly filter feeding euphausiids have thoracic legs of similar length and well-developed setae. In addition to modifications in the appendages of these animals, Nemoto examined differences in stomach structures and reported that herbivorous euphausiids had a cluster of spines along the posterior wall of the stomach while the same structures were largely absent from deep-water carnivorous species.

Among other major groups of zooplanktonic animals, the Coelenterata and Ctenophora are almost exclusively carnivorous and prey are collected largely through passive encounter with the animals' tentacles. However, Jørgensen (1966) has reported that at least one jelly fish is capable of suspension feeding and that the process of prey capture involved the adhesion of particles to the mucus surface of the animal, over which a surface current was induced by ciliary action. Another large carnivorous group of zooplankton are the Chaetognaths, or Arrow worms, which capture their food using specialized jaws. Among the planktonic molluscs, the Heteropods, the thecosomatus pteropod

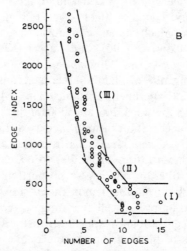

FIG. 46B. Relation between 'Edge Index' and number of edges of the mandible (redrawn from Itoh, 1970).

Limacina collects food on the flat surface of its foot and within its mantle cavity. The process of food collection is by adhesion of food particles on a mucus membrane which is carried forward into the animal's mouth by ciliary action (Jørgensen, 1966). Chindonova (1959) and Gilmer (1974) have reported that among food items recognized in pteropods, the most common organisms were diatoms, Radiolaria and Globigerinae. In the latter reference the size of particles retained by different species of thecosomatus pteropod was shown to include very small particles, ($<5\,\mu$m, including bacteria and detritus) and large particles, up to 800 μm (including crustacean larvae), depending on the size of the pteropod. The food of gymnosomatus pteropods (having no shells) has been reviewed by Lalli (1970) who indicates that these organisms rather specifically predate thecosomatus pteropods. It has been further indicated that the mouth parts of gymnosomatus pteropods are specifically adapted toward only one or two species of thecosomatus pteropod. Another specialized feeding arrangement between similar animals is found among the ctenophores; Greve (1971) has described the predation of one ctenophore (*Beroe gracilis*) on another (*Pleurobrachia pileus*). The retention of very fine particulate materials is apparent in some members of the Urochordata, which includes the salps and the appendicularians. Among the latter, *Oikopleura* is reported to retain particles down to 0·1 μm in size; this is achieved by ciliary induced currents passing water through a funnel-shaped mucus net, the end of which is twisted into a string which is continuously being consumed through the animal's oesophagus (Jørgensen, 1966).

The rate of filtering among microcrustaceans is broadly related to body size but can vary in any individual depending on such factors as temperature and food concentration. Gilmer (1974) showed that filtering rate of the cosomatus pteropods was an approximate log/log relationship with size from 4·3 to 55·7 ml/hr for animals between 0·03 and 15 mg dry weight. Such a relationship would appear reasonable if one considers that the weight of an animal will increase as the cube and the filtering surface will increase as the square of an animal's linear dimensions. From a summary of measurements made by Marshall and Orr (1955) and Jørgensen (1966) for copepods, it appears that the volume of water filtered by an animal may range from *ca.* < 1·0 to 200 ml/day. Cushing and Vucetic (1963) determined that maximum filtering rates for adult copepods could be as high as 1 litre/day. The latter figure appears high, however, when compared to the filtering rates of the much larger euphausiid crustaceans; these have been measured by Raymont and Conover (1961) and Lasker (1966) who report a range of values for *Meganyctiphanes norvegica*, *Thysanoëssa* sp., and *Euphausia pacifica* from *ca.* 1 to 25 ml/hr. The filtering rate of planktonic crustaceans may decline as food and concentration is increased. In experiments conducted by Mullin (1963) adult *Calanus hyperboreus* appeared to filter between 200 and 300 ml/day, decreasing linearly to less than *ca.* 50 ml/day as the concentration of food increased. Frost (1972) also has shown a decrease in filtering rate of *Calanus pacificus* from *ca.* 10 to 2 ml/copepod/hr over increasing phytoplankton concentrations from 200 to 600 μgC/l. In addition, however, Frost (1972) has shown that filtering rate is also proportional to the size of the phytoplankton cells; a semi-log relationship is demonstrated showing an increase in filtering rate from 4 to 9 ml/copepod/hr over a cell size range of *ca.* 10^3 to $10^5\,\mu$m^3. At very low phytoplankton cell concentrations it appears (Frost, 1975) that filtering rate may decline. Two possible effects of filtering rate on the intake of food are shown in Fig. 47. In the first illustration a constant maximal filtration rate at low prey concentrations gives an increased food intake with increased prey concentration; the filtering rate then

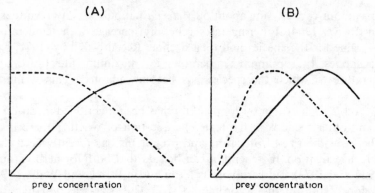

FIG. 47. Two possible effects of the filtering rate (---) on the food intake (——) at different prey concentrations. (A) Filtering rate constant or declining with increased prey concentration, (B) filtering rate increasing with prey concentration at low prey densities and decreasing rapidly at high prey densities.

declines at higher prey concentrations to give a maximal food intake. In the second illustration the filtering rate increases with prey concentration, reaches a maximum asymptotic value and declines at higher concentrations. Food intake in the second illustration is zero at some finite concentration of prey, reaches a maximum and then declines.

Among omnivorous microcrustaceans, the filtering rate can be greatly suppressed if the animal changes from filter to raptorial feeding; this has been observed by Lasker (1966) with the filtering rate of *Euphausia pacifica*, which was decreased by an order of magnitude, and in some cases eliminated, when nauplii were offered as prey in the presence of phytoplankton.

Over longer time periods the capacity to feed at any concentration must be governed by the animal's capacity to absorb its food. For example, the chaetognath, *Sagitta his-*

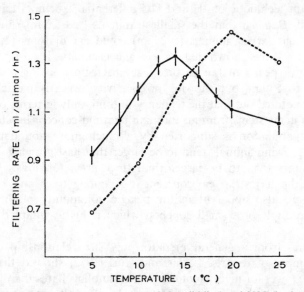

FIG. 48. Filtering rates of *Daphnia rosea* grown at 12°C (solid line) and 20°C (broken line) (redrawn from Kirby, 1971).

pida, had a maximum capacity for about 50 *Artemia* nauplii per day regardless of food concentration (Reeve, 1964). Filtering rates generally increase with temperature up to some optimal value for the species and then decline. Recently Kibby (1971) has shown that within a species, the optimum temperature for maximum filtration may depend on the temperature to which the species has become adapted. This is illustrated in Fig. 48.

The size of food particles eaten by zooplankton has been mentioned in various connections in this and other sections of the text. The question of whether there is a general relationship between the size of predators and prey items has recently been the subject of considerable investigation. In general, it can be said that both for herbivorous feeding (e.g. Frost, 1972 and 1974) and carnivorous feeding (e.g. Reeve and Walter, 1972) there is some selection of prey items according to size. This size-selection hypothesis has two properties: these are, firstly, that predators are generally larger than their prey and, secondly, that within the prey size range of a particular predator, the largest prey items will be selected when available (e.g. Mullin, 1963; Hargrave and Geen, 1970). Nival and Nival (1973 and 1976) formulated an equation for the efficiency of particle retention based on the mean distance between setules, the dispersion around the mean and the filtering surface. Using this index they showed that filtering efficiency increased from a maximum at 5 μm particles to a maximum at 12 μm particles during the maturation of *Acartia clausi* from Copepodite I to adults.

Boyd (1976) has pointed out that the interpretation of large particle 'selection' may be erroneous and that if one considers the mouth parts of a zooplankter as having sieve-like properties with a wide range of pore sizes, then large particles would be retained better than small particles for purely mechanistic reasons. However, one should also recall that the two modes of feeding among some zooplankters (i.e. raptorial and filter) may have different energy requirements at different prey concentrations. In this case a metabolic cost benefit 'electivity' may determine that seizing large prey items is generally more economic than filtering small prey items.

Using an elegant technique of ^{3}H and ^{14}C differential labelling, Lamport (1974) was able to show with *Daphnia* that the smallest animals (<1.5 mm) showed a preference for bacteria but that larger animals (>2 mm) could eat diatoms (28 μm) or bacteria (1.5 μm). Poulet (1974) has shown that *Pseudocalanus minutus* is an opportunistic feeder, taking advantage of peaks in the particle size spectrum (see Fig. 4) between particle sizes of 3.57 μm to 57 μm, but showing no electivity towards particles of <3.57 μm. Heron (1973) has pointed out that the tendency to think only in terms of large organisms eating small prey items stems from an Eltonian pyramid concept in which it is assumed that a predator population is supported by a much more abundant, faster-growing prey. Since slow-growing animals tend to be larger than fast-growing animals (e.g. Fenchel, 1974) predators tend to be larger than their prey. Obviously many exceptions to this occur in the terrestrial environment (e.g. among trees and insects). However, such exceptions are also apparent in the pelagic community of the sea and Heron (1973) gives an example of a small copepod which is highly adapted to predation on large salps.

Other departures from a general predator/prey size relationship can be found by example in a number of reports. Esaias and Curl (1972) showed that ingestion rates by copepods were lower for highly bioluminescent dinoflagellates than for species having a reduced capacity for bioluminescence. The uptake of small amounts of soluble organic material as food among filter feeding copepods is suggested in the work of Khaylor and Yerkohin (1971). Selectivity of prey towards male or female copepods has been demonstrated for two freshwater carnivores (Haly, 1970).

The problem of feeding in deep-water environments where food is scarce has recently been reviewed by Harding (1974) who suggests that three pathways are available for the transport of food to deep-water animals. These are zooplankton vertical migrations (Vinogradov, 1955), the 'rain' of detritus from the surface (Vinogradov, 1962), and the transfer of dissolved organic carbon into particulate matter (e.g. Riley, 1970; Parsons and Seki, 1970). While the first of these pathways is an active transport of animals to different levels in the upper water column, it is probable that it is not an effective mechanism below 1000 m where there is no evidence for diurnal migrations (Angel and Fasham, 1973). The question with both the other two mechanisms is that even if they are effective in producing particles at great depths, are these concentrations sufficient to support the feeding requirements of deep-sea zooplankton? From metabolic requirements for growth and maintenance the minimum particulate carbon concentration required for a copepod would be *ca.* 25–50 μg C/l (Parsons and Seki, 1970) which is below the general background level for deep-water particulate carbon (e.g. Menzel, 1967). However, it has been suggested (Parsons and Seki, 1970) that microlayers with high concentrations of particulate matter may exist in the deep ocean. Some evidence for this has been given by Wangersky (1974) who found anomalously high particulate organic carbon values in about 8% of his samples. These samples were also found to contain small zooplankton.

The suggestion contained in Harding (1974) that deep-water particulate material is formed *in situ* from soluble organic carbon has been further examined by Williams (1975). Using data from Williams *et al.* (1970) on the ^{14}C enrichment of deep water in the aftermath of atomic bomb testing, Williams (1975) has pointed out that both surface plankton samples and bathypelagic organisms are enriched with ^{14}C. In contrast, deep-water soluble organic carbon shows no such enrichment. This is taken as indicating a relatively rapid transportation of surface organic carbon to bathypelagic community; however, one species of bathypelagic fish analysed did not show any ^{14}C enrichment which indicates the possible presence of alternative links in the bathypelagic food chain.

The filtering rate (F) of a zooplankter over a given time period (t) can be measured as the decrease in phytoplankton cell concentration at the beginning (C_0) and end (C_t) of the experiment (Gauld, 1951). If the assumption is made that the concentration of cells decreases exponentially with time, then

$$F = \frac{v(\log C_0 - \log C_t)2\cdot303}{t},$$

(82)

where v is the volume of water per animal. A possible error in the use of this expression is that it assumes that F is constant over the period of the experiment. However, since the filtering rate is known to vary with the concentration of cells, the assumption that F is constant is only an approximation which will be true for small changes in $C_0 - C_t$. Control flasks, in which changes in C_0 can be measured, and gently stirred or rotated incubation chambers which assure an even distribution of cells, are necessary refinements in most experiments. A discussion of the methodology of measuring filtering rates has been given by Rigler (1971).

In conclusion it appears from what is known of feeding processes among the zooplankton that the smallest particles (*ca.* 0·1 μm) are retained by animals possessing a mucus net through which water is pumped by ciliary action. Larger particles (*ca.* >5 μm) are retained by filter feeding copepods and some other crustaceans while very large particles (*ca.* > 50 μm) are retained by raptorial feeding; the latter process includes the capture of many of the smaller zooplankton which serve as food for larger members of the plankton community.

4.2 SOME TROPHODYNAMIC RELATIONSHIPS AFFECTING THE PLANKTON COMMUNITY

4.2.1 FOOD REQUIREMENTS OF ZOOPLANKTON

The distributional needs for food among marine zooplankton are similar to any other representative of the animal kingdom. This is shown in Fig. 49 as a flow diagram in which the energy in the ration consumed is distributed as a flow chart into the

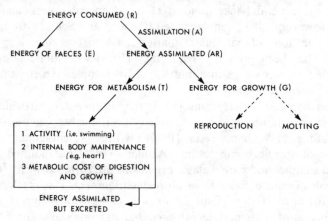

FIG. 49. Distribution of food energy in a marine zooplankter.

various life requirements of the animal. The major divisions of food energy (*R*) are first between how much is assimilated (*A*) and how it is excreted (*E*) and secondly, of the assimilated ration (*AR*), how much is used for various metabolic needs (*T*) and how much is used for growth (*G*). Growth is generally defined as an increase in body weight but among crustaceans a certain amount of growth tissue will be lost as moults, while the female of a species will also lose an appreciable fraction of her body weight in egg production. Among metabolic needs, movement, body maintenance and the metabolic cost of the extra effort required in digestion and growth are clearly identifiable. Further, the waste products of these metabolic activities must then be excreted; in high animals this excrement would be identifiable with urine but in small zooplankton it is extremely difficult to identify this loss.

An important point to recognize in Fig. 49 is that the fractions of energy found in any one section are not always the same for different species or for the same species, either under different feeding regimes or at different ages in the organism's life cycle. Thus, while we shall attempt to give values to some of these life requirements, there are actually 'feed-back' relationships which can alter the absolute fraction assigned to any one process [e.g. the fraction of food used for growth (*G*) is not a simple fraction of the amount of food consumed (R)]. These 'feed-back' relationships will also be discussed.

The food requirement of a zooplankter can be expressed as the sum of the major requirements for growth, metabolism and faeces (Richman, 1958).

$$R = G + T + E \qquad (83)$$

Since E can be expressed as $(R - AR)$ where A is the assimilation efficiency of the food, eqn. (83) can be written as

$$G = AR - T. \qquad (84)$$

The extent to which food is assimilated by a zooplankter is difficult to measure experimentally since there are technical difficulties in making quantitative collections of faeces from zooplankton, even under laboratory conditions. By assuming that the ash content of the food material was largely unabsorbed in the digestive process, Conover (1966a) derived a formula which allows for an estimation of the assimilation efficiency (A) in terms of (J), the ash-free dry weight to dry weight ratio in the ingested food and (L), the same ratio in a sample of faeces.

$$A = \frac{(J - L)}{(1 - L)(J)} \times 100(\%). \qquad (85)$$

Use of this formula avoids the difficulty of quantitatively collecting faeces and also permits measurement of assimilation efficiencies on samples of particulate material collected *in situ*. Conover (1966b) examined factors which affect the assimilation efficiency and found it to be largely independent of temperature, the amount of food offered, or the amount of food consumed. The assimilation efficiency was affected by the ash content of the phytoplankton food, however, and a simpler empirical relationship than the one described above was established, such that

$$A = 87 \cdot 8 - 0 \cdot 73X, \qquad (86)$$

where (X) is the percentage ash per unit dry weight of food.

The determination of assimilation efficiencies based on the ash content of food and faeces may be a useful method for determining organic carbon assimilation but it appears that the assimilation efficiency of elements, such as nitrogen and phosphorus, cannot be determined with any accuracy by this method (Kaczynski, personal communication). The quantities of these elements absorbed are much smaller than the quantities of organic carbon and small changes in the ash content of the food and faeces can greatly affect the calculated elemental assimilation efficiency. An alternative method which relies on the ratio of elements to each other (e.g. N to P) in the food and faeces is given by Butler *et al.* (1970).

Conover (1968) summarized his own data and a number of experimental reports in the literature and concluded that herbivorous zooplankton assimilated phytoplankton with 60 to 95% efficiency. This may reflect a maximum range of assimilation efficiencies since Gaudy (1974) has clearly shown that assimilation decreases with food intake. His results are given in Fig. 50 and show that herbivore assimilation efficiencies are much lower (10–20%) at high food intake than indicated by Conover (1968). For carnivorous zooplankton, assimilation is generally higher than for herbivores; this is explained by assuming that a carnivore can better digest an animal prey that is biochemically more similar to itself compared with a plant prey item. Thus, Conover and Lalli (1974) found

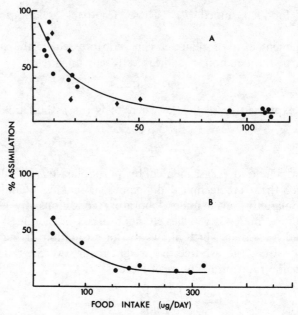

FIG. 50. Relationship between assimilation rate and daily food intake in (A) *Centropages typicus* (diamonds) and *Temora stylifera* (circles), (B) *Calanus helgolandicus* (redrawn from Gaudy, 1974).

that the carnivorous pteropod, *Clione limacina*, had an assimilation efficiency of $>90\%$; Cosper and Reeve (1975) showed by direct measurement that the chaetognath, *Sagitta hispida*, had an assimilation efficiency of 80%. Assimilation efficiencies may also vary with age; Mootz and Epifanio (1974) found that the pelagic larvae of the stone crab, *Menippe mercenaria*, showed an increase in assimilation efficiency from 40 to 85% during maturation from Stage I to megalopa, when fed on *Artemia* nauplii. Direct measurements of fecal material excreted by zooplankton (E) have been made in a few cases. For example, Hargrave (1970) measured the feces produced by a deposit feeding freshwater amphipod and found that the fecal material represented *ca.* 83% of the ration; this gives an assimilation efficiency of less than 20%.

Welch (1968) has derived an interesting relationship between assimilation efficiency and growth efficiency [K_2 eqn. (94)] for aquatic organisms, including copepods and fish. His relationship shows that assimilation decreases as growth efficiency increases. The interpretation of this result is that high assimilation must be associated with high energy expenditure and therefore low growth efficiency. For example, a carnivore spends more energy searching for and catching a prey than a herbivore spends in grazing. However, once having caught the prey, carnivore assimilation efficiencies are high— higher that is than herbivore efficiencies for reasons indicated in the paragraph above. It is important to recognize, however, that the relationship between assimilation and net growth efficiency (K_2) does not hold for gross growth efficiency (K_1); only ration (R), and not assimilated ration (AR), is considered in the latter.

Food used for metabolism within an animal (T) can be expressed as a function of the animal's body weight (W) such that

$$T = \alpha W^{\gamma} \tag{87}$$

or
$$\log T = \log \alpha + \gamma \log W.$$

TABLE 24. REGRESSION EQUATIONS OF
LOG SPECIFIC RESPIRATION RATE, R' (μl
O_2/mg BODY wt./hr) AND LOG BODY
WEIGHT EXPRESSED AS THE DRY WEIGHT
OF THE ANIMAL (mg/ANIMAL) DERIVED
FROM DATA IN FIG. 51 (IKEDA, 1970)

Species	Regression equation
Boreal	$R' = -0.169W + 0.023$
Temperate	$R' = -0.309W + 0.357$
Tropical	$R' = -0.464W + 0.874$

T can be measured in terms of an animal's respiration (e.g. μl O_2/animal/hr) and the value of α will be a function of the units which are used to express the animal's weight (e.g. wet weight, dry weight, etc.) as well as being influenced by environmental factors such as temperature. γ, on the other hand, has been found to be relatively constant (Zeuthen, 1970) and to reflect the internal metabolism of the organism; since there are certain overall similarities in cellular metabolism of a large variety of species, the relative constancy of γ appears logical. However, from a very extensive survey of different zooplankton organisms, including crustaceans, pteropods, chaetognaths, coelenterates, and polychaetes, Ikeda (1970) established that there was a significant difference between the value of γ for tropical, temperate, and boreal species taken from the tropical Pacific, the temperate Pacific southeast of Hokkaido, and the Bering Sea, respectively. The results of Ikeda's studies are shown in Table 24 and Fig. 51. These results, in which metabolism is expressed per unit weight of animal, clearly show that the smallest animals have the highest metabolic rate and that there are geographic differences, not only

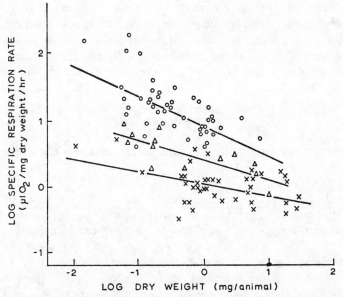

FIG. 51. Regression lines for planktonic animals from boreal ($\times$), temperate ($\triangle$), and tropical (O) waters drawn for log specific respiration rate (μl O_2/mg dry wt./hr) against log dry weight animal (mg)—(redrawn from Ikeda, 1970).

in the value of α, but also in the value of γ, which reflect an environmental difference in basic metabolic processes of animals from different areas.

For the same species of animal at different temperatures (i.e. within species variation), Comita (1968) determined a multiple regression equation for *Diaptomus* which gave the oxygen uptake (T in $\mu l/mg$ wt./hr) in terms of the temperature (t in °C) and the dry weight (W in mg) as follows:

$$T = 0.0364t - 0.3418 \log W + 0.6182. \tag{88}$$

The conversion of metabolism, which is generally measured as respiration in units of oxygen consumed, to biomass of food in terms of organic carbon is made by the following equations:

$$\text{mg } O_2 \text{ consumed per unit time} \times \frac{12}{32} \times RQ$$

$$= \text{mg C utilized per unit time;}$$

or

$$\text{ml } O_2 \text{ (NTP) consumed per unit time} \times \frac{12}{22.4} \times RQ$$

$$= \text{mg C utilized per unit time}$$

where RQ is the respiratory quotient, $+\Delta CO_2 / -\Delta O_2$, which may range in animals from 0.7 to 1.0 depending on whether fats or carbohydrates are being utilized for energy. Thus, in a copepod living off fat reserves, an RQ of 0.7 may be recommended, while a copepod feeding off phytoplankton may have an RQ closer to 1.0.

The food consumed by a zooplankter (R) can be expressed in terms of the amount of food available. Using a variety of fish, Ivlev (1945) determined that the quantity of food eaten increased with the concentration of food offered, up to some maximum ration which could not be further increased by increasing the concentration of food. From this observation it was stated that if the maximum ration is taken as R_{max}, then the relation between the size of the actual ration, R, and the concentration of prey, p, must be proportional to the difference between the actual and maximal ration such that

$$\frac{dR}{dp} = k(R_{max} - R) \tag{89}$$

where k represents a proportionality constant. Integrating, this expression gives

$$R = R_{max}(1 - e^{-kp}). \tag{90}$$

Sufficient evidence exists in the experimental data from a number of authors (e.g. Reeve, 1963; Mullin, 1963; Parsons *et al.*, 1967; Sushchenya, 1970, and references cited therein) to show that the general form of this relationship can be applied to zooplankton grazing. Two modifications to this expression have been proposed on the basis of additional experimental results. The first of these was shown by Ivlev (1961) to apply to the feeding of fish under conditions in which the effect of prey aggregation on feeding was to increase the ration obtained when the prey concentration was held constant. It is difficult to establish on the microscale of zooplankton whether this effect with fish is applicable to zooplankton feeding. It is apparent, however, that natural mechanisms which tend to concentrate prey cause patchy distributions of both plants and animals (see Section 1.3.5), and that from the observation given above, the occurrence of patchy distributions may increase the food available when the average prey concentration is quite low (Paffenhöfer, 1970). The second modification to eqn. (90) concerns the prey concentration at which feeding starts. From observations on zooplankton grazing conducted at sea (Adams and Steele, 1966; Parsons et al., 1967), as well as from laboratory experiments (e.g. Nassogne, 1970), it appears that grazing ceases at some minimum or threshold prey concentration, p_0. Equation (90) can be modified to include this value as

$$R = R_{max}(1 - e^{k(p_0 - p)}).$$ (91)

Threshold values for p_0 have been reported to range from 40 to 130 μg C/l depending on species of microcrustacea and food (Parsons and LeBrasseur, 1970). In other experiments, however, it appears that filtering continues down to zero prey concentration (e.g. Mullin, 1963; Paffenhöfer, 1970). The question of a prey threshold concentration has been further examined by Frost (1975) who found that as prey concentrations decreased, the copepod Calanus pacificus fed at a constant maximal rate until food intake fell to below about 15% of the maximal hourly ration. At this point, a significantly depressed feeding rate was observed. The actual concentration of phytoplankton at which this occurred (presumably a function of p_0) depended on the size of the phytoplankton species studied.

There is some discussion of the actual mathematical form of the feeding relationship given in eqn. (90) (Frost, 1974; Mullin et al., 1975). From the latter reference it is apparent that a rectilinear equation statistically best fitted experimental data. However, both the curvilinear eqn. (90) and a Michaelis–Menten equation [eqn. (79)] have been used to describe experimental results and neither could be statistically rejected in the study by Mullin et al. (1975).

The concentration of phytoplankton at which herbivorous copepods reach their maximum ration is almost certainly dependent both on the species of zooplankton and the species (or size of phytoplankton). Parsons and LeBrasseur (1970) generally found concentrations of food organisms greater than 300 μg C/l for copepods and up to 1000 μg C/l for euphausiids. Paffenhöfer (1970) showed that the copepod Calanus helgolandicus could attain its maximum ration, as determined from the largest sized animals, at food concentrations of between 100 and 200 μg C/l. Frost (1974) showed that the maximum ration for female Calanus pacificus was attained at between 100 and 300 μg C/l depending on the size of phytoplankton. It was found that small phytoplankton cells (11 μm) had to be approximately three times more concentrated in terms of biomass (μg C/l) than large cells (87 μm) for the animal to obtain the same ration (R_{max}).

A more specialized predator/prey relationship than that given by eqn. (90) has been described by Greeve (1972). In the feeding relationship between copepods and ctenophores, Greeve (1972) noted that adult copepods tended to destroy larval ctenophores but that adult ctenophores ate copepods. A further complication in this feeding relationship was that if there were too many copepods, then adult ctenophores themselves suffered a certain mortality because their tentacles were eventually broken by too frequent encounters with large numbers of copepods. Greeve's (1972) representation of these observations could be modelled as shown in Fig. 52. Here, the probability of ctenophore survival among young ctenophores is shown to decline with an increase in copepods. As ctenophores become older the probability of ctenophore survival in-

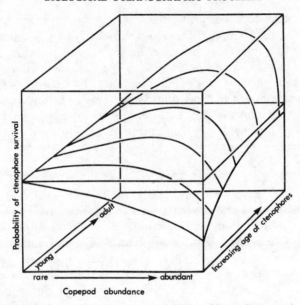

FIG. 52. Three-dimensional model showing the probability of ctenophore survival with increasing
abundance of copepods and age of ctenophores (redrawn from Greve, 1972).

creases with an increase in copepods abundance, but decreases again at high copepod
abundance.

The actual quantity of food eaten by zooplankter (R) depends on the type of prey
as well as its concentration; thus *E. pacifica* obtained 15% of its body weight per day
off a bloom of *Chaetoceros* but at equivalent concentrations of nanoplankton, *E. pacifica*
could only obtain approximately one-third of this amount of food per unit time (Parsons
et al., 1967). In this example it may be assumed that the small nanoplankters (8 μm
diameter) were much less efficiently filtered than the larger diatoms (32 μm diameter).
Prey selectivity may also determine the amount of food eaten. Conover (1966c) and
others have observed, for example, that some food particles will be rejected by zooplankton even after they have been seized during raptorial feeding. Also it has been shown
that within the same species of phytoplankton, larger food particles, consisting of paired
cells, may be selectively grazed over smaller single cells (Richman and Rogers, 1969).
In addition, although some phytoplankton cells may be eaten, their digestion by crustaceans may be ineffectual and they may pass through the animal as viable cells. This
was observed by Porter (1973) who showed that gelatinous green algae remained alive
after passage through the guts of *Daphnia*. The degree to which an animal selects a
prey over the natural abundance of different prey in the environment can be expressed
as an electivity index, such as that given by Ivlev (1961) where

$$E = \frac{r_i - p_i}{r_i + p_i} \tag{92}$$

and r_i is the relative proportion (e.g. percent) of a prey in the ration and p_i is the
relative proportion of the same prey in the water. Prey concentration may be expressed
as units of biomass or numbers of organisms and the choice of units will affect the
value of E. The value of E in the above expression ranges from -1 to $+1$, higher values
denoting a greater selection of a prey item over lower values.

In general it appears that food ingested (as indicated by R in experiments) has been found to range from a few percent up to *ca.* 100% of the body weight of zooplankton per day, with lower average values of 10 to 20% of the body weight per day for the larger crustacean zooplankton and 40 to 60% of the body weight per day for the smaller crustacean zooplankton (Bell and Ward, 1970; Sushchenya, 1970; Mullin, 1963; Parsons and LeBrasseur, 1970). From these references it is also apparent that prey densities at which zooplankton reach the asymptotic value of R_{max} appear to be in the range 200 to 1000 μg C/l, depending (as in the case of p_0) on the particular combination of predator and prey. Inhibition of zooplankton food intake at high prey concentrations appears in some experimental data. From Mullin's (1963) results it was shown that the number of diatom cells (*Thalassiosira fluviatilis*) ingested by *C. hyperboreus* increased over the range *ca.* 200 to 4000 cells/ml but then decreased to about 25% of their maximum intake over the range, 4000 to 8000 cells/ml. Diel periodicity in zooplankton grazing has been observed by Duval and Geen (1976). These authors found that several species of freshwater crustaceans exhibited bimodal maxima in feeding rates at dawn and dusk, with the lowest feeding at midday. The diel increase in feeding rates amounted to 6 times the midday minimum; a corresponding diel periodicity in respiration was on the average twice as great at dawn and dusk compared with midday.

Apart from the obvious need for major metabolites such as proteins, carbohydrates and fats, there are other properties of diets, as well as properties of the sea water in which animals reside, which may influence an animal's growth and survival. These properties are at present poorly defined but they are apparent from a number of examples. Thus Wilson (1951) showed that *Echinus* larvae survived better in some waters than others; Provasoli *et al.* (1959) showed that only one or two algae were satisfactory as food for *Artemia* and *Tigriopus* while a combination of several algae generally proved a much more adequate diet. Lewis (1967) showed that a mixture of trace metals improved the survival of *Euchaeta japonica* and that periodicity in the occurrence of natural chelators in seawater was a possible factor affecting *E. japonica* survival, *in situ* (Lewis *et al.*, 1971).

In eqn. (84) the food available for growth (G) is achieved when the gain in food (AR) exceeds the loss (T). This indicates that for a constant basal metabolism (T) a linear increase in growth can be achieved by increasing the ingested ration (AR). However, in a review of published data on the feeding of fishes, Paloheimo and Dickie (1965, 1966 a and b) found that increasing the ration (R) resulted in a decreased efficiency of food utilization for growth and that the relationship between gross growth efficiency (K_1) and ration was log-linear, such that

$$\ln K_1 = -a - bR \tag{93}$$

where a and b are constants, characteristic of the predator and its prey, and gross (K_1) or net (K_2) growth efficiency are defined as

$$K_1 = \frac{\Delta W}{R\,\Delta t} \quad \text{and} \quad K_2 = \frac{\Delta W}{AR\,\Delta t}. \tag{94}$$

Since $\Delta W/\Delta t$ represents growth per unit time it follows that

$$\frac{\Delta W}{R\,\Delta t} = e^{-a-bR} \tag{95}$$

or

$$\frac{\Delta W}{\Delta t} = Re^{-a-bR} \tag{96}$$

which is an expression of growth (G) in terms of ration (R). Substituting this into eqn. (84),

$$Re^{-a-bR} = AR - T$$

or

$$T = R(A - e^{-a-bR}). \tag{97}$$

The application of Paloheimo–Dickie equations to zooplankton feeding has been discussed by Conover (1968) and Sushchenya (1970) who have concluded from experimental data on zooplankton that the approach may have general application to the study of food requirements of zooplankton. Thus Sushchenya (1970) showed from Richman's data that there was a linear relationship between the logarithm of the gross growth efficiency and ration in the feeding of *Daphnia pulex* and that the maximum gross growth efficiency was 60%. Further it was possible to derive an equation [eqn. (97)] for *Daphnia* feeding on *Chlamydomonas* which closely fitted the experimental growth data. Sushchenya (1970) accumulated various values for the gross and net growth efficiency, and the food used for respiration (T) as a percentage of the food assimilated (AR); the values are reproduced in Table 25. From these results it is apparent that gross efficiencies vary over a wide range but the K_1 for most animals lies in the range of *ca.* 10 to 40%; K_2 values are less variable and show a range of *ca.* 25 to 55%, while the expenditure of food on metabolism generally lies in the range 40 to 85% of the food assimilated.

The significance of the values (a) and (b) in eqn. (93) has been discussed by Paloheimo and Dickie (1966b). Neither value was affected by temperature which is known to affect the level of metabolism. This was interpreted as meaning that temperature affected

TABLE 25. VALUES OF ENERGY COEFFICIENTS K_1 AND K_2, AND PERCENTAGE OF ENERGY EXPENDITURE FOR RESPIRATION (T), IN % OF AR (FROM SUSHCHENYA, 1970)

Species	t °C	K_1	K_2	T	Author
Artemia salina	25	18·5	23·6	76·4	Sushchenya, 1962
Artemia salina	25	13·0	26·5	73·5	Sushchenya, 1962
Artemia salina	25	9·0	27·4	72·6	Sushchenya, 1962
Daphnia pulex	20	13·2	55·4	44·6	Richman, 1958
Daphnia pulex	20	9·1	57·6	42·4	Richman, 1958
Daphnia pulex	20	4·8	56·9	43·1	Richman, 1958
Daphnia pulex	20	3·9	58·7	41·3	Richman, 1958
Calanus helgolandicus	10	48·9	52·5	47·5	Corner, 1961
Calanus helgolandicus	10	37·2	47·0	53·0	Corner, 1961
Calanus helgolandicus	10	41·9	46·2	53·8	Corner, 1961
Asellus aquaticus	14–19	28·3	40·4	59·6	Levanidov, 1949
Asellus aquaticus	14–19	23·3	33·3	63·7	Levanidov, 1949
Asellus aquaticus	14–19	20·6	30·0	70·0	Levanidov, 1949
Asellus aquaticus	14–19	18·7	26·7	73·3	Levanidov, 1949
Asellus aquaticus	14–19	16·3	23·3	76·7	Levanidov, 1949
Euphausia pacifica	10	7·1	7·4	92·6	Lasker, 1960
Euphausia pacifica	10	31·8	39·2	60·8	Lasker, 1960
Euphausia pacifica	10	14·2	14·9	85·1	Lasker, 1960
Euphausia pacifica	10	28·0	29·0	70·0	Lasker, 1960

the total rate of turnover of food but not the distribution of food among the various metabolic components. By contrast, changes in factors such as salinity and type of food (especially particle size) influenced both parameters.

From Fig. 45 it is apparent that the specific metabolism of a zooplankter decreases with body size. This gives rise to another general relationship if we consider also that the amount of energy used for growth will be large in young (and therefore small animals) compared with older (and larger) animals; as the latter become adults, the amount of energy used for growth reaches zero. Thus, at some point in an animal's life cycle there must be a period of maximum growth efficiency when the specific metabolism per unit body weight is lowest for a maximum in the growth rate. Experimentally, this was shown to be true by Makarova and Zaika (1971) who found that for *Acartia* maximum growth efficiency (K_2) of *ca.* 30% was achieved at approximately one-third

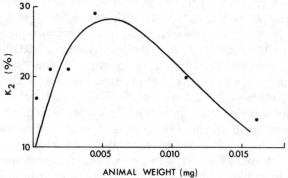

FIG. 53. Changes in the growth efficiency (K_2) during the growth (mg dry wt.) of *Acartia* (from Makarova and Zaika, 1971).

of the animal's adult body weight (Fig. 53). While these data apply to growth efficiency changes within the life cycle of a particular animal, it must also be considered that the same general relationship may apply to different species of different size. Thus, Parsons (1976) calculated that for animals growing at 7% of their body weight per day, growth efficiency (K_1) of a small (0·05 mg) copepod would be about half that of a large (5 mg) copepod, if all other factors were equal. This would appear to give a distinct advantage to the growth of large zooplankters. However, such an assumption is made without consideration to life cycles, food supply, reproduction, and a number of other factors. For example, as Taniguchi (1973) and Fenchel (1974) have pointed out, small zooplankton grow faster than large zooplankton and thus reach maturity sooner. This gives small zooplankton an advantage, especially, for example, in tropical waters where increases in food supply may be short-lived as a result of temporary upwelling in warm waters. Such a growth regime gives an advantage to small zooplankton maturation in short time periods because of their inherently more rapid growth rate. This growth rate is further accentuated in warm waters because the Q_{10} for growth exceeds that for metabolism by a factor of *ca.* 2 (e.g. Ivleva, 1970; Chang and Parsons, 1975). Another advantage in the life-cycle strategy of a small zooplankter is that more energy is devoted to reproductive tissue in small crustaceans compared with large crustaceans (Khmeleva, 1972). In contrast to these advantages for small animals in tropical environments, an advantage for large zooplankters in colder waters would be to grow slower but at the same time to be able to carry greater energy reserves through periods when food is virtually absent (i.e. during winter). As Taniguchi (1973) has pointed out, the size distribution of small zooplankton in the tropics and large animals in temperate waters can be explained in part by these metabolic and growth strategy considerations.

In further extensions of generalized relationships governing animal size and metabolic complexity (e.g. unicellular vs. multicellular organisms, and heterotherms vs. homio-therms), Fenchel (1974) has discussed the relative advantages and costs of different life strategies. From correlations drawn from a wide range of animal sizes at different levels of metabolic complexity, Fenchel (1974) reaches certain conclusions which may be of general interest to trophic relationships in the marine habitat. His first conclusion is similar to Khmeleva's (*loc. cit.*) conclusion that larger animals spend more of their assimilated energy on maintenance than productivity; this has the advantage of giving larger animals a more independent life style. Obviously metabolic complexity (i.e. from unicellular to multicellular organisms) is another means of achieving greater ecological independence. However, according to Fenchel this is only achieved with an approximate eightfold increase in the specific metabolic needs of multicellular organisms. Since the growth rate of a metazoan is approximately twice as great as a protozoan of the same size, the overall cost in terms of growth efficiency is approximately 4 times lower for metazoan growth compared with protozoan growth. Moving further afield, Fenchel (1974) shows that the metabolic cost of homiothermic existence is about 28 times greater than that of a heterotherm of the same size, while productivity of the latter is about 1·7 times lower. This indicates that growth efficiency of heterotherms should be about 16 times higher than homiotherms (e.g. compare growth efficiency for a fish of *ca.* 30% with that of a cow, *ca.* 2%). While these latter generalizations have drifted away from the subject of trophodynamics within a plankton community, they serve to tie together the ultimate extremes of life strategies available both in marine and terrestrial habitats (for the latter, see MacArthur and Wilson, 1967). Thus, in one extreme a highly successful strategy for survival is in being small, numerous, highly productive, but open to severe predation; in another situation being large, scarce, relatively unproductive but highly independent and consequently generally safe from predation is also a success-ful strategy. According to Sheldon *et al.* (1972) the actual biomass or standing stock of organisms occupying these two extreme positions, as well as positions in between, is about the same for the entire size spectrum of organisms from bacteria to whales in the world's oceans!

Several food budgets involving the terms G, AR, and T in eqn. (84) have been derived for microcrustaceans. Some results given in Table 26 from Petipa (1966) show that the growth and rate of metabolism of *Acartia* nauplii are 2 or 3 times greater than for stage V copepods, and that consequently the food intake per unit body weight of the young animals is considerably greater than for the older animals. These results are similar to results obtained for euphausiids (Table 27) which show that faster-growing animals devoted proportionally more energy to growth than metabolism, when com-

TABLE 26. DISTRIBUTION OF FOOD EATEN AS A% OF THE BODY WEIGHT IN *Acartia clausii* AT 17–25°C ASSUMING AN ASSIMILATION (A) OF 80% (FROM PETIPA, 1966)

	Dry weight (mg)	Daily growth (%)	Daily metabolism (%)	Daily ration (%)
Nauplii	0·000 09	20·3	98·3	148
Stage V	0·002 56	8·3	49·8	73

TABLE 27. DISTRIBUTION OF FOOD REQUIREMENTS IN A EUPHAUSIID FROM LASKER (1966)

Animal	Dry wt. (mg)	Growth Final wt. / Initial wt.	Time (days)	% food assimi- lated (A)	% food for Growth (G)	% food for moults (G')	% food for eggs (G")	% food for metabolism (T)	Location of study
E. pacifica	1·19	3·95	69	86	30	8·0	0	62	Laboratory
E. pacifica	1·65	1·16	63	78	6·2	7·0	0	86·8	Laboratory
E. pacifica	0·23	39·2	580	—	9·4	15·3	8·9	66·4	Natural population

pared to slower-growing animals. In the case of euphausiids, however, an appreciable fraction of up to 15% of the food is devoted to moults. This latter figure may be higher in euphausiids than in copepods. Corner et al. (1967) estimated <1% of the food assimilated by copepods was lost to moults; Mullin and Brooks (1967) showed that Rhincalanus lost ca. 6% of its body carbon as moults during its life cycle from egg to adult. The amount of tissue devoted to reproduction (Fig. 49) may be generally related to body size. Khmeleva (1972) suggests that 'generative growth' is proportionally higher in small compared with large crustaceans. From some recent data on the percentage of assimilated food available for reproductive tissue (i.e. eggs) it appears that in harpacticoid copepods this may be as high as 23% (Harris, 1973), in mysids 19% (Clutter and Theilacker, 1971), in copepods 12% (Corner et al., 1967) and in euphausiids this is seen (Table 27) to amount to less than 10%. However, in another species of euphausiid Nemoto et al. (1972) found that the carbon content of eggs in Nematoscelis difficilis was equivalent to 28% of the body carbon. Marshall and Orr (1955) showed that the total number of eggs laid by Calanus was related to the amount of food available. From these results it appears that while size may in part govern generative growth, other factors, such as temperature and food supply, must also influence the fecundity of zooplankton.

Vlymen (1970) has discussed the energy requirements for movement (swimming) among copepods. From a theoretical analysis of the problem he concludes that the energy cost of acceleration from rest to a given constant velocity is very small, requiring slightly more than 0·1% of basal metabolism. Further, that even during vertical migrations the rate of energy expenditure is slightly less than 0·3% of basal metabolic rate.

The principal product of excretion among crustacean zooplankton appears to be ammonia. However, the methodology of estimating how much ammonia is lost as assimilated excreta (Fig. 49) is extremely difficult. This is because it is difficult to separate soluble material that may come from fecal pellets as opposed to excreta from the products of metabolism. Butler et al. (1970) performed some experiments in which it was shown that the percent of body nitrogen excreted per day was found to vary seasonally from a high of ca. 10% in the spring to a low of ca. 2% in the winter. Phosphorus is also excreted as a product of assimilation and Butler et al. (1970) found that the ratio of N:P in soluble excreted material was ca. 5 in spring and ca. 6 in winter. The fraction of phosphorus excreted was a greater (ca. 5 to 25%) fraction of the body phosphorus compared with the body nitrogen. Some other reports in the literature indicate that much higher fractions of body nitrogen and phosphorus are excreted (e.g. Harris, 1973; Pomeroy et al., 1963). While these reports may be correct it also should be considered that excessively high excretion rates are difficult to accommodate in terms of known growth efficiencies unless there is a form of wasteful or 'superfluous' feeding (see Butler *et al.*, 1970, for discussion). *In situ* measurements of nitrogen and phosphorus excretion by *Euphausia pacifica* and *Metridia pacifica* have been made by Takahashi and Ikeda (1975). They showed that compared with the excretion rate at zero phytoplankton density, the estimated increase in excretion rate was as high as 5 times at 15 μg Chl a/l for *M. pacifica* and almost 9 times at 70 μg Chl a/l for *E. pacifica*.

Throughout the above discussion reference has been made to 'food' requirements and its metabolic distribution. Food may be better expressed in terms of energy units, and the following conversion factors can be used as general approximations:

1 ml of O_2 consumed $\equiv$ 5 g cal assuming,
RQ of approximately 1·0,
1 mg of food (dry wt.) $\equiv$ 5·5 g cal.

assuming an average dry weight composition of food as 50% protein, 20% fat, 20% carbohydrate, 10% ash, and caloric equivalents of 5·65 cal/mg for protein, 4·1 cal/mg for carbohydrate and 9·45 cal/mg for fat.

The effect of temperature on the metabolic rate (T) has been discussed earlier in this section. However, since temperature affects such overall processes as growth, reproduction, and the final size of organisms, it is appropriate to consider whether there is a general relationship between such complex biological processes and temperature.

McLaren (1963, 1965, and 1966) has discussed the general acceptability of empirical relationships between temperature and the rate of biological reactions with particular reference to the zooplankton community. From his discussions it appears that the two extremes in such approximations are firstly that biological reactions increase in rate two to threefold with a 10°C rise in temperature (the Q_{10}); this is described by McLaren as a clumsy approximation which may be assumed to have been adopted for its mathematical simplicity. At the other extreme, mathematically complex polynomial equations can be devised to fit almost any deviations in different groups of data. However, from his own data, as well as from a wide variety of published reports, McLaren concluded that the closest fit was generally obtained with Belehradek's empirical formula which is given as

$$V = a(t + \propto)^b \qquad (98)$$

where V is the rate of metabolic function, t is the temperature and a, b, and $\propto$ are constants. When other factors could be assumed not to be limiting (e.g. food supply), this function gave good descriptions of development rate, metabolic rate, and size (McLaren, 1963); a number of examples are shown in Fig. 54 A, B, and C (note that metabolic rate, V, has been expressed in these figures in terms of time taken to complete a process or length achieved as adults, i.e. as $1/V$ on the ordinate). Some physiological and environmental justifications for the use of this function have been discussed by McLaren in terms of the 'constants', a, b, and $\propto$. Thus the scale correction, $\propto$, is known as the 'biological zero' and expresses the temperature at which $V = 0$. The value $\propto$ appears to be positively related to environmental temperature and consequently has been shown to vary with latitude (and altitude in the case of terrestrial lake communities); the lowest values of $\propto$ being found in the most northerly populations. The constant, b, reflects the degree of curvature and represents the general dependence on temperature of all metabolic processes leading up to changes in the measured parameter, V. Thus when the same metabolic function (e.g. egg hatching) is measured among similar groups of organisms, the curvilinear response, b, should be the same for all groups.

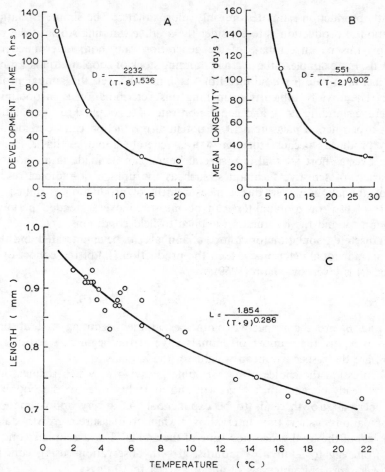

FIG. 54. Effect of temperature on biological development. (A) Time taken for hatching eggs of *Calanus finmarchicus*. (B) Mean longevity of *Daphnia magna*. (C) Mean cephalothorax length in female *Acartia clausi* (redrawn from McLaren, 1963).

The proportionality coefficient, *a*, is determined by the units in which *V* is measured; however, it is also apparent the value of *a* can be related to size in such a way as to indicate that it is an expression of a surface to volume restriction governing the exchange of gases across a membrane. This was illustrated by McLaren (1966) who showed a positive correlation between *a* and egg diameter for a variety of copepod eggs, corrected for differences in their yolk content.

4.2.2 THE MEASUREMENT OF PRODUCTION

Production is defined as the total elaboration of new body substance in a stock during a unit time, irrespective of whether or not it survives to the end of that time (Ricker, 1958). Production depends on the time interval over which it is measured, the presence or absence of predators, and the growth and natural death rate of the population. The combination of these factors is very difficult to measure *in situ* and consequently production estimates are only approximate.* The simplest relationship for the estimation of production is to consider the *average* standing stock of the population and multiply this by an estimate of the generation rate, both determined over short time intervals. For example, if the average standing stock of zooplankton is 1 g/m^2 and the average generation rate (or doubling time) is 1 month, then the annual production is 12 $g/m^2/yr$. The obvious difficulty in making this determination is in obtaining figures for the average standing stock and generation rate of a zooplankton population *in situ*.

A direct experimental measurement of potential production can be made in the case of primary producers because their growth is generally rapid so that a change in the population over a short interval (e.g. several hours) can be made in an isolated sample using a variety of sensitive techniques, such as the uptake of radioactive $^{14}CO_2$. In the case of animals, however, direct measurement of production is difficult although *growth rate* of fish, or zooplankton can be measured over extended periods of time in the laboratory and the information applied to field conditions.

Where a discrete population (or cohort) of animals can be enumerated and the average weight of an individual determined, then the production (P_t) of the cohort over a short time period (t) is given by Mann (1969):

$$P_t = (N - N_t) \times \frac{(\overline{W} + \overline{W}_t)}{2} + (B_t - B) \tag{99}$$

where B and B_t are the respective biomasses at the beginning and at time t, and where N and N_t are the number of animals alive at the beginning and at time t, and $\overline{W}$ and $\overline{W}_t$ are the respective mean weights of the animals.

In this expression the choice of a *short* time period over which to measure changes in W and N really depends on the growth and mortality rates of the population since over long periods growth tends to be exponential. As a very approximate guide for animal populations, a short time interval over which to measure a growth stanza should be 10% or less of the total generation time of the animal. For a zooplankton population maturing from egg to adult in 3 months, growth observations using eqn. (99) could be made over *short* time intervals of about a week to 10 days.

Equation (99) represents the production of the total population. That is the production removed by predators (and natural mortality) is the number of individuals lost per unit time, multiplied by the average weight of those lost, plus the increase in biomass of the surviving population. If the production removed by predators is determined over short time intervals and the loss totalled for the year, then the total production can be determined as the sum of the surviving population at the end of the year plus the sum of the total lost to predators, (i.e. $P_{\text{Total}} = P_{t_1} + P_{t_2} + \cdots + P_{t_i}$).

For a population of copepods having one generation per year the growth rate can be determined *in situ* from changes in the weight of two different stages and the time interval between the maximum numbers of each stage (Cushing, 1964; Parsons *et al.*, 1969). This is illustrated in Fig. 55 where the time interval between the maximum number of Stage I and Stage III copepodites is given as 44 days; the mean weight of the respective stages was 0·15 and 0·60 mg. Since growth over an extended period tends to be a logarithmic function of time, the growth rate per day can be approximated from the expression

$$\%\Delta W = 100[10^{1/t(\log W_2 - \log W_1)} - 1] \tag{100}$$

Note: Primary productivity estimates as measured by the $^{14}C-$ or oxygen techniques actually represents the *potential* production in the absence of sinking, grazing, etc.

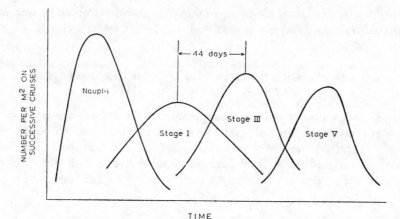

FIG. 55. Estimation of the time intervals in the maximum occurrence of different stages of a copepod having one generation (schematic representation of data from Parsons *et al.*, 1969).

where % ΔW is the percent increase in weight per day, W_1 and W_2 are the weights of Stage I and III copepodites respectively and t is the time interval. For the values given above, % $\Delta W = 3.5$ (Parsons *et al.*, 1969). From this growth rate the potential production of Stage I copepodites can be determined using $N_1 \times W_1$ (as the total biomass of the cohort) and determining $N_2 \times W_2$ from eqn. (100). The difference between the mean *in situ* biomass of copepods $(\overline{N_2 W_2})$ and the calculated value of $N_2 W_2$, will represent an estimate of the production lost to predators and natural mortality.

In a population of zooplankton which has more than one generation per year it is more difficult to determine production since several generations of the species may be present and the time of development for a single stage can be determined as in Fig. 47. However if the development time of a stage (or the individual daily growth) is found experimentally (e.g. by isolating stages and incubating them under *in situ* conditions of temperature and food supply), then the production of the population can be calculated as the sum of the individual stages. In this technique it may not be necessary to deal with the species in stages, and weight groups are in fact more satisfactory

TABLE 28. PRODUCTION OF *Acartia clausii* DURING THE SUMMER (FROM MANN, 1969)

Size groups (mg × 10⁻³)	Number per m³	Individual daily growth (mg × 10⁻³)	Daily production (mg × 10⁻³)
0–2·5	850	0·09	76·5
2·5–10	250	0·45	108·0
10–20	87	1·20	104·0
20–30	35	1·68	58·8
30–50	47	1·20	56·4
50–70	40	0·47	18·8
		Total	422·5

for the purposes of the calculation. An example of this determination was made by Greze and Baldina (1964) and is reported in Table 28 from Mann (1969).

4.2.3 THE PELAGIC FOOD PYRAMID AND FACTORS AFFECTING ITS PRODUCTION AND STABILITY

The flow of energy up the pelagic food pyramid is illustrated in Fig. 56. When the energy flow is considered from a point of view of groups of animals of a generally similar habitat, then these are referred to as *ecological levels* and the system is referred to as a *food chain* (right-hand side of Fig. 56). If successive steps in the food pyramid are considered, then these are referred to as *trophic levels* and the system is called

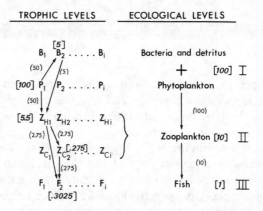

FIG. 56. Diagram of a pelagic food pyramid involving three ecological levels and five trophic levels (adapted from Ricker, 1968). B_1, B_2,..., P_1, P_2,..., etc., represent different organisms at each ecological level. Z_H represents herbivorous zooplankton and Z_c represents carnivorous zooplankton. The amount of energy available to be transferred is represented by the standing stock [5] and (2·75) represents the amount of energy transferred by a particular route assuming that all energy at one trophic level is available to the next but that it is transferred with only 10% efficiency. In this schematic food web it is also assumed that where a food resource is shared, 50% goes to each predator.

a *food web*. The amount of energy transferred up a food chain is not the same as that transferred by a food web. This is also illustrated in Fig. 56 where an arbitrary 100 units of primary production has been moved up the food pyramid through both a food chain and a food web. The details of this example calculation are given in the figure caption and the result obtained shows that approximately three times more energy reaches the fish community if one considers a model based on ecological levels compared with one based on trophic levels. Furthermore, the position of animals within a food web is not fixed throughout their lifetime; for example, copepodite stage *Oithona* are considered to be largely herbivores (Z_H) while Stage V *Oithona*, becoming adult, transfers to the level of primary carnivores (Z_C) (Petipa *et al.*, 1970).

From the above discussion it is apparent that the representation of the food pyramid in the sea as a food web in which there are many trophic levels is probably a more accurate representation of what is actually going on than the simpler representation of a food chain. However, for comparative purposes between different environments it is also apparent that the representation of a community as a food web could become highly complex. Thus some approximate comparisons between different communities may be made in terms of food chains. In this event it becomes important only to know how much energy is being produced by the primary producers, the number of ecological levels and the *ecological efficiency* (E); the last is defined as

$$E = \frac{\text{Amount of energy extracted from a trophic level}}{\text{Amount of energy supplied to a trophic level}}. \tag{101}$$

Slobodkin (1961) indicated that this value was about 10%. However, it is very important to know the correctness of this value since it can greatly influence final estimates of production in which man is interested. For example, Schaefer (1965) considered the effect of ecological efficiencies ranging from 10 to 20% on the production of fish at the fifth trophic level. The general form of this production (P) can be given as

$$P = BE^n \tag{102}$$

where B is the annual production at the primary trophic level, E is the ecological efficiency and n is the number of trophic levels. Starting with an annual production of phytoplankton of 1.9×10^{10} metric tons of carbon, the production of fish at the 5th trophic level (3rd carnivore; $n = 4$) was 1.9×10^6 and 30.4×10^6 metric tons of carbon at $E = 10\%$ and $E = 20\%$, respectively. Thus a doubling of the ecological efficiency can result in an order of magnitude increase in the production of a terminal resource.

However, the ecological efficiency as defined above does not take into account recycling processes which take place in nature, but which would not be apparent in a single predator/prey relationship. Thus a 10% transfer of material between trophic levels actually includes a 90% loss to the system. Part of this loss will be real, in that organic carbon will be respired back to CO_2 at each trophic level. However another fraction of the 90% loss will appear as organic debris and be recycled through the food chain (e.g. as indicated in Fig. 41 and discussed in Section 3.2.1). Further it is apparent that at least at the herbivore level, experimental data has recently been presented to show that the gross growth efficiency of copepods (K_1 see Section 4.2.1) from nauplius to adult may be as high as 30 to 45% (e.g. Mullin and Brooks, 1970). Ecological efficiencies will not approach these values for various reasons connected with life cycles, the variety of organisms included in natural trophic levels, predator/prey electivity and the need for wild organisms to spend more energy in hunting their prey. Nevertheless it is apparent that the value of 10% is probably too low for all trophic levels and that it is especially low as an ecological efficiency involving transfers from the first trophic level. For overall estimates it may be assumed that the ecological efficiency at the herbivore level is probably no less than 20% and that efficiencies at higher trophic levels are probably between 10 and 15%. From eqn. (102) it is also apparent that the number of trophic levels (n) occurs as an exponent and as such it can greatly affect production. In an examination of differences in fish production throughout the world, Ryther (1969) considered the number of trophic levels in three communities which may be described as 'oceanic', 'continental shelf', and 'upwelled'. Over large areas the authors suggested that oceanic communities have long food chains with low ecological efficiencies, the latter being determined by the three or four levels of carnivorous feeding. As an example of such a community the author gave the following food chain which is summarized here as a diagram:

Nanoplankton   →   Microzooplankton   →   Macrozooplankton   →
  (small flagellates)   (Herbivorous   (Carnivorous crustacean
                          protozoa)   zooplankton)

Megazooplankton   →   Planktivores   →   Piscivores
  (e.g. Chaetognaths   (e.g. lantern fish &   (e.g. squid, salmon &
  & euphausiids)   saury)   tuna)

The food chain is shown as a continuous flow of biomass from phytoplankton to fish but in fact salmon, for example, may also feed on euphausiids and squid; thus their position at the 6th trophic level would actually be better represented as being between the 5th and 6th trophic levels. For purposes of comparison with other environments, however, it is apparent that there are approximately five trophic levels leading to the production of a commercial species of fish. Further it was determined from the literature that the annual primary production of oceanic areas was generally low and $50 \text{ g C/m}^2/\text{yr}$ was considered to be an average value.

The second food chain given by Ryther was described as 'coastal' but should be more properly called 'continental shelf' since it may sometimes exist at considerable distance from the coast, such as in the Atlantic on the Grand Banks. This food chain may be represented diagrammatically as follows:

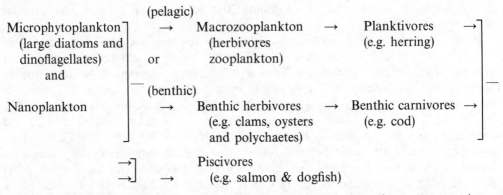

As in the oceanic food chains, the flow of food represented above is an average picture and may include such steps as carnivorous zooplankton in addition to those shown. Essentially, however, the food chain can be represented by three trophic levels, whether this is via the benthic or pelagic community. The total annual primary production of these areas was assessed as *ca.* $100 \text{ g C/m}^2/\text{yr}$.

Finally the food chain of areas in which there is persistent upwelling, such as off the coast of Peru or in Antarctic seas, could be represented by the following diagram:

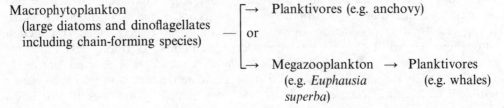

In this food chain it is known that certain fish, such as adult anchovy, may feed directly off phytoplankton, while another relatively short food chain may exist between eupha-usiids and whales. The total number of trophic levels in such environments was repre-sented by Ryther as one and a half. The total primary productivity of this community type was assessed at 300 g C/m^2/yr since the process of upwelling leads to a continual supply of nutrients.

TABLE 29. ESTIMATED FISH PRODUCTION IN THREE OCEAN COMMUNITIES
(ADAPTED FROM RYTHER, 1969)

Marine environment	Mean primary productivity (g C/m^2/yr)	Trophic levels	Efficiency (%)	Fish production mg C/m^2/yr
Oceanic	50	5	10	0·5
Continental shelf	100	3	15	340
Upwelled	300	1·5	20	36 000

In these three communities ecological efficiencies at each trophic level were assumed to be highest, when governed largely by phytoplankton/herbivore associations and low-est for communities in which there were secondary and tertiary carnivores. Consequently Ryther assigned a 10% overall efficiency to the oceanic food chain, a 15% efficiency to the continental shelf food chain, and a 20% efficiency to the food chain in upwelled areas. The resulting potential fish production is shown in Table 29. From these results it may be readily concluded that the upwelled areas should produce the largest fisheries and that the continental shelf areas and oceanic are of decreasing importance. This is borne out in part by the fact that the anchovy fishery of Peru is the largest fishery in the world, exceeding the total catch of an intensive fishing nation such as Japan; in contrast very few fisheries are conducted in the open ocean waters except where there is a continental shelf, such as in the northern hemisphere off the Grand Banks. The description of food chains given by Ryther (1969) may be open to certain criticisms as to details of his calculations (e.g. Alverson et al., 1970) but the general approach is illustrative of the importance of trophic relationships in assessing the productivity of the ocean.

In the previous paragraphs it will be noted that the size of phytoplankton cells appears to be a critical factor in determining the length of the pelagic food chain and the ultimate yield of fish to man. The next question to ask is, what factors govern the size of phytoplankton cells? Malone (1971 a, b) suggested that large phytoplankton cells only predominate during periods of positive vertical advection and when nitrate concentrations are above 1–3 μM. Further, he found some observational data supporting earlier experimental results that zooplankton selectively grazed large cells as opposed to the nanoplankton. Semina (1972) suggested on the basis of environmental data that the size of phytoplankton cells was governed by at least three factors, viz. (1) the direc-tion and velocity of vertical water movement, (2) the value of the density gradient of the main pycnocline, and (3) the phosphate concentration. Some of these and ad-ditional ideas were incorporated in an equation which attempted to describe cell size in terms of eight environmental and physiological variables (Parsons and Takahashi, 1973). While this equation appeared superficially correct, its use has been criticized (Hecky and Kilham, 1974; Malone, 1975) mainly on the basis of a lack of terms governing alternate possibilities, including grazing and horizontal advection. Laws (1975) has suggested that cell size is determined largely by three factors: phytoplankton growth rate, respiration, and sinking. By showing that large phytoplankton cells catabolize a smaller fraction of their biomass than small cells, Laws (1975) uses a model to demonstrate that the net growth rate of large cells may exceed that of small cells when the mixed layer is relatively deep. From these reports it appears that phytoplankton cell size must be linked ultimately to the physical, chemical, and biological processes but that it may still be too early to clearly define the causative agents.

Cushing (1971 and 1973) determined the ecological efficiency from environmental data by assuming that the amount of energy extracted (yield) from a trophic level was proportional to the production at that level. Consequently the ratio of production at one ecological level to the production at the next is an estimate of the ecological efficiency which Cushing (1971) called the *transfer efficiency*. By plotting the transfer efficiency between primary and secondary producers against the primary production, Cushing (1971) showed that ecological efficiencies probably declined with production (Fig. 57). Experimental support for this assumption is found in Paloheimo and Dickie (1966b) who showed that growth efficiency [K_1, eqn. (80)] declined with ration. The connection between growth efficiency and ecological efficiency is established through yet another type of efficiency used by biological oceanographers and that is the *ecotrophic efficiency* (E_c). This is defined as the fraction of the annual production of a trophic level which is consumed by predators (Ricker, 1968). In Fig. 48 it was assumed for simplicity that all of one trophic level was available for transfer to the next trophic level; in fact a certain fraction must remain in order to maintain the production. For example, if we assume that 10% of the primary production is needed to maintain production against other losses, then 90% of the primary production is available for consumption by secondary producers. Further, if this is consumed by the secondary producers with a growth efficiency (K_1) of 20%, then the ecological efficiency is 18%. The relationship between these three efficiencies can be summarized as

$$E = E_c \times K_1. \tag{103}$$

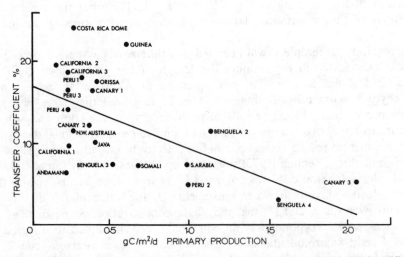

FIG. 57. Dependence of transfer coefficients on the intensity of primary production for different upwelling areas in different seasons (redrawn from Cushing, 1971).

Two further corollaries of eqn. (103) are that ecological efficiency (E) should decline with successively higher steps in the trophic pyramid and, secondly, that ecological efficiency should decline with the age of a population. The first of these corollaries comes from the fact that P/B ratios decrease with trophic position (Fig. 60) and therefore less production can be made available to the next trophic level (i.e. E decreases); the second corollary is indicated by the fact that K, as a measure of growth, must decrease with age.

Riley (1946) first considered quantitative relationships governing productivity at the primary and secondary levels in the marine food chain. Using data from Georges Bank, Riley (1947) considered two approaches; in the first several quantities were measured simultaneously and their interrelationships derived by some statistical method, such as by multiple correlation. In the second approach, a number of simplified assumptions were made, based on experimental data, and these were synthesized into a mathematical description of the events.

TABLE 30. PERCENTAGE CHANGE IN THE PHYTOPLANKTON CROP PRODUCED BY INCREASING THE VALUE OF EACH ENVIRONMENTAL FACTOR FROM ITS MEAN TO THE LIMIT OF ITS STANDARD DEVIATION (FROM RILEY, 1946)

	Sept.	Jan.	Mar.	Apr.	May	June
Depth	−1	60	−20	11	0	−1
Temperature	−9	−37	−31	−5	10	−26
P	7	74	−23	−24	6	−28
N	7	41	−57	1	−21	1
Zooplankton	−6	5	−1	−9	−31	−10

As an example of the first approach, Riley (1946) determined multiple regression equations between plant pigments, the depth of the water, temperature, the amount of phosphate, the amount of nitrate, and the abundance of zooplankton. Proportionality 'constants' in the equations varied with season and the effect of changing the mean value for each component by one standard deviation caused different seasonal effects which indicated the temporary importance of each factor. This is illustrated in Table 30 which shows that major changes in the phytoplankton crop were caused by increasing the nitrate concentration during March, phosphate concentration during April, and zooplankton during May. Thus it is apparent that no one factor exercises complete control over phytoplankton production and that the relationship between the few major components in Table 30 indicates a complex system in which one factor after another gains momentary dominance. Combining data from different seasons Riley (1946) developed a multiple regression equation in which

$$PP = -153t - 120P - 7{\cdot}3N - 9{\cdot}1Z + 6713 \qquad (104)$$

where PP was the phytoplankton crop in Harvey pigment units (not used today, but proportional to chlorophyll a), t was the temperature, P and N were phosphate and nitrate concentrations, respectively, and Z was the zooplankton standing stock. This equation reproduced the observed data with an average error of about 20% for all seasons.

More recent studies using various types of multiple-component analysis have been used to show the principal components governing plankton production in different environments. Williamson (1961) showed that out of eleven factors in a correlation matrix, four accounted for 83% of the variance in plankton abundance in the North Sea and of these, one factor related to the extent of vertical mixing in the water column predominated by accounting for 48% of the variance. Walsh (1971) showed that 66% of phytoplankton variation across the Antarctic Convergence could be predicted in terms of water mass structure, turbulence, light, silicate, and an index of heterotrophic conditions. However, as the latter author has concluded, definite statements about functional relationships between the biological and habitat variables cannot be made because of the inherent non-causal nature of correlation and regression analyses. A similar conclusion led Riley (1946) to consider a second approach to describing changes in a plankton community. In this approach he assumed that the rate of change in a phytoplankton community could be determined as the difference in reaction rates between processes of accumulation and loss of biomass (or energy) in the plankton population. Riley considered that the most important reactions could be summarized by an equation which can be given in a modified form as

$$\frac{dN}{dt} = N(P_h - R) - G \tag{105}$$

in which dN/dt was the rate of change in the phytoplankton population (N) having a photosynthetic rate (P_h), a respiration rate (R), and a rate of grazing by zooplankton (G). Riley (1947) considered that each term in the above equation would be subject to environmental influence but that a description of their effect could be given in some mathematical form. Thus the average light intensity ($\bar{I}$) to the depth of the mixed layer, D_m, can be given by the equation

$$I = \frac{I_0}{kD_m}(1 - e^{-kD_m}), \tag{106}$$

I_0 is the photosynthetic radiation at the surface and k is the extinction coefficient (for the derivation of this equation see Section 3.1.5). The photosynthetic response to this light intensity can then be obtained from a P vs. I curve (e.g. Fig. 28) which might be expressed in the form

$$P_h = P_{max}(1 - e^{\alpha(I_c - I)}) \tag{107}$$

where P_h is the photosynthetic rate at $\bar{I}$ and P_{max} is the maximum photosynthetic rate for the phytoplankton population, and α is a constant.

Since in eqn. (107) the compensation light intensity (I_c) is determined by the respiration (see Fig. 28), the term R in the eqn. (105) does not have to be treated separately. However, temperature (which was included in Riley's original equation as only affecting respiration) should be included as an effect on the photosynthetic rate. This has been discussed in Section 3.1.4 and using Eppley's empirical relationship between temperature and photosynthetic rate, a temperature adjustment can be included having the form

$$P_h = P_{max}\, e^{(at-b)} \tag{108}$$

where a and b are constants and t is the temperature.

On the basis of more recent findings (see Section 3.1.6) the nutrient limitation on phytoplankton growth can now be expressed in the form of Michaelis–Menten kinetics as

$$P_h = P_{max}\frac{[S]}{[S] + K_m} \tag{109}$$

where P_{max} is the maximum rate of photosynthesis at which an increase in the rate limiting nutrient concentration [S] no longer results in an increase in P_h, and K_m is the Michaelis constant for the uptake of the nutrient.

The grazing component, G, can also be expressed more appropriately than in Riley's original equation. Thus if $G = HR'$ where H is the standing stock of zooplankton and R' is the ration per unit time per animal, then from Section 4.2.1

$$R' = R'_{max}(1 - e^{-k'N}) \tag{110}$$

where N is the concentration of the phytoplankton population. Thus one attempt to construct a determinate model of phytoplankton productivity on the basis of Riley's classical paper might be written as

$$\frac{dN}{dt} = N\left[P_{max}(1 - e^{\alpha(I_c - \hat{I})})\left(\frac{[S]}{[S] + K_m} \right)(e)^{(at - b)} \right] - HR'_{max}(1 - e^{-k'N}) \tag{111}$$

where $\bar{I}$ is defined by eqn. (106) and all the variables affect dN/dt independently.

Further embellishments of this type of equation can be made by adding factors, such as might be required to account for the sinking rate of phytoplankton, or by using other expressions for the photosynthetic light response, such as are discussed in Section 3.1.5. In Riley's (1946) original model, which differs from eqn. (111), he was able to reproduce the essential features of the phytoplankton standing rock with approximately the same precision as was obtained by the statistical estimate given in his first approach eqn. (104). Riley (1947) extended his determinate equations to express changes in the herbivorous zooplankton population (dH/dt) as

$$\frac{dH}{dt} = H(gP - R'' - aC - D), \tag{112}$$

where H was the herbivorous zooplankton population, gP was the input from grazing on phytoplankton, R'' was loss from zooplankton respiration, aC was loss by carnivorous predation, and D was natural mortality.

Vinogradov and his colleagues (Vinogradov et al., 1972, 1973) have developed an ecosystem model in which they have tried to simulate the time changes in biomass of organisms at different trophic levels, inorganic biogenous elements (i.e. nutrients), and detritus. The model is intended to represent a natural water column and the possible co-reaction between each parameter (e.g. Fig. 58). In this model the animal community was classified into several different trophic categories as follows: protozoa (F_1), nauplii (F_2), small and large-herbivores (F_3, F_4), cyclopods (S_1), carnivorous calanoids (S_2), and chaetognath and polychaetes (S_3). This treatment makes the model more complicated than others (e.g. Riley's), but the final result is probably more realistic for the description of a natural ecosystem. In the Vinogradov model, the rate changes of nutrients (N), phytoplankton (P), and bacteria (B) were described in a manner similar to eqn. (105) but including also several other terms such as mortality, sinking and input and output due to physical water-mass transportation. The changing rate of zooplankton biomass was then described as follows:

$$\frac{\partial X_i}{\partial t} = u_i \Sigma C_p - r_i X_i - \mu_i X_i - \Sigma X_q \tag{113}$$

$$Cp = P, B, D, F_1 \ldots F_4, S_1 \ldots S_3$$

$$Xq = F_1 \ldots F_4, S_1 \ldots S_3$$

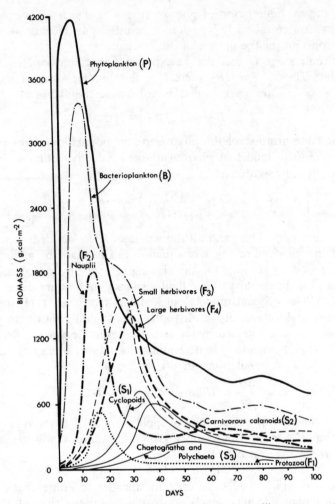

FIG. 58. Changes in time of the major components of a tropical upwelling ecosystem in the upper 200 m, predicted by a simulation model; for explanation see text (redrawn from Vinogradov *et al.*, 1973).

where X_i is the biomass of animal in the ith category of the trophic food web ($F_1 \ldots F_4$, $S_1 \ldots S_3$), in the water column, u_i food consumption efficiency of the animal, ΣC_p total food eaten, r_i respiration rate of the animal, μ_i mortality of the animal, and ΣX_q biomass loss of the animal due to feeding by other animals. The rate of detritus production was described as follows:

$$\frac{\partial D}{\partial t} = \Sigma(h_i + \mu_i)X_i - \Sigma d_i X_i + k\frac{\partial^2 D}{\partial Z^2} - w\frac{\partial D}{\partial Z} \tag{114}$$

where D is the concentration of detritus in the water column, h_i is undigested food of the ith zooplankters ($i = F_1 \ldots F_4$, $S_1 \ldots S_3$), d is the removal rate of detritus by ith zooplankters ($i = F_1 \ldots F_4$), k is the turbulent diffusion constant, w is the average sinking rate of detritus, and Z is the depth.

This approach has been applied to a tropical upwelling community (Fig. 58), in which the X-axis can also be read as the distance from the upwelling area as well as days.

The biomass of both phyto- and bacterio-plankton increases rapidly, followed by a development of herbivorous animals. Then phyto- and bacterio-plankton decrease in their biomass due to heavy grazing by herbivores, as well as possible nutrient deficiency. Predators develop their biomass even later. The peak height of organisms in each category generally diminishes at higher trophic levels. Fifty to sixty days after starting, the system comes to a semi-stationary state, characterized by low concentration of all live elements. The time and spatial changes predicted are supported by actual observations (e.g. Vinogradov *et al.*, 1970). Menshutkin *et al.* (1974) expanded Vinogradov's original approach, using data from the Sea of Japan, and taking into account the nekton (i.e. fishes and squid).

Many additional descriptive models using both approaches have been developed since Riley's (1946 and 1947) papers. Cassie's (1963) paper serves as a more elaborate example of recent developments in statistical analyses of plankton communities while Patten (1968) has reviewed models which synthesize empirical relationships between the major components governing the production of plankton.

The structure of a plankton community is often not so readily resolved into definite trophic levels as indicated by the use of Riley's models. More often there is a complex food web in which a species of plant or animal may belong to more than one trophic group at any time (e.g. as illustrated in Fig. 56). There are very few quantitative descriptions of such food webs but one has been attempted for the planktonic communities of the Black Sea (Petipa *et al.*, 1970); unfortunately the data presented in this study contain several inconsistencies but the approach serves as an example of a planktonic food web. There are also one or two similar studies in which authors have considered the distribution and flow of energy (or biomass) in other aquatic environments (e.g. the Thames river study by Mann, 1965; the Georgia salt marsh study by Teal, 1962).

In order to understand the energetics of the Black Sea planktonic communities it is necessary first to describe the trophic levels involved, and particular attention is given here to Petipa's description of the epiplanktonic community. The phytoplankton of this community consisted of autotrophic organisms (such as chlorophyll containing diatoms and dinoflagellates) and saprophagus organisms (principally *Noctiluca* which feeds by engulfing particulate food, 70 to 90% of which was reported to be detrital in origin). Thus the authors chose to consider these two groups of organisms as a single trophic level, based on the fact that both groups of organisms were involved in the primary formation of particulate material for filter feeding animals. The distribution of biomass (standing stock) within the phytoplankton community was over 90% in favour of the saprophagus *Noctiluca*. It appears from the data, however, that this organism had a very high death rate (42% per day) and that this was the chief source of detritus being utilized by filter feeders. In contrast the smaller autotrophic phytoplankton were producing one or two generations per day with a natural mortality of less than 5% per day. Herbivorous organisms formed the second trophic level and these included nauplii of copepods, some copepodite stages, *Oikopleura*, and the larvae of some benthic molluscs and polychaetes. Omnivorous organisms, which consumed both plant and animal plankton, formed the third level and consisted mostly of the later stages of the copepods, *Acartia*, *Oithona*, and *Centropages*. One primary carnivore, adult *Oithona*, was recognized as the fourth trophic level, while the fifth and sixth trophic levels were recognized as secondary carnivores (mainly *Sagitta*, which ate both herbivores and primary carnivores) and a tertiary carnivore (*Pleurobrachia*) which fed off all other zooplankton; for most calculations the two latter levels were considered

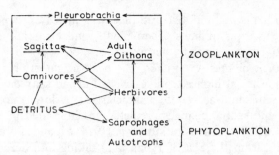

FIG. 59. Food web of the epiplankton community in the Black Sea during early summer (redrawn from Petipa *et al.*, 1970).

together as one. A diagram of this food web is shown in Fig. 59. While the figure represents fixed relationships described above it was also considered by Petipa *et al.* (1970) that certain organisms transferred from one level to another during their lifetime so that *Oithona*, for example, transferred to the primary carnivore level on becoming an adult; in addition, a number of early copepodite stages transferred from the herbivore to the omnivore level as they matured.

Petipa *et al.* (1970) considered that the diel rate of production of matter (P') at one trophic level was determined as the number of organisms eaten (G), the number of organisms dying per day (M), the number of organisms transferring to the next level per day (L) and ($B_1 - B_0$) the difference in standing stock at the end of the day. The best units in which to compare the activity of different levels in such a food web are units of energy (e.g. calories/m²/day) although actual measurements are usually made in terms of numbers or biomass of organisms which are assigned caloric equivalents. Experimental data on the six trophic levels described above were collected from an area of high stability in the Black Sea. Food composition and daily rations were obtained from gut contents; reproductive rate of the algae and the respiration of animals were obtained from samples incubated *in situ* and the total weight increment of animals was determined from the duration of stages at different temperatures and the known weight of each stage. The results of these measurements are shown in Table 31. From

TABLE 31. PRODUCTION OF THE TROPHIC LEVELS OF BLACK SEA EPIPLANKTON COMMUNITY (mg/m²/DAY) FROM PETIPA *et al.* (1970)

Trophic level	Amount of food eaten from a given level (G)	Death rate (M)	Difference between final and initial standing stock ($B_1 - B_0$)	Transport to the following level (L)	Production (P)	Standing stock of living organisms (B)	P'/B (%)
Primary producers and saprophages	478·5	12 593·8	5397·2	—	18 469·5	29 249·6	63
Herbivores	361·2	4·0	−6·3	156·0	515·0	792·0	65
Omnivores	67·0	1·2	20·0	82·0	170·4	193·5	88
Primary carnivores	53·5	1·3	27·4	—	82·2	172·2	47
Secondary–tertiary carnivores	33·0	2·8	−1·1	—	34·7	560·5	6·2

these data it is apparent that the phytoplankton community is predominated by the rather unusual death rate which has been attributed to the saprophage, *Noctiluca*. Further, it is apparent that both biomass and production show a general decrease going from the lowest trophic level towards the highest. Production per unit biomass (P'/B) would have been highest at the primary level if data on *Noctiluca* had been excluded. There is some evidence in the data that P'/B ratios decrease with increasing trophic levels, as might be expected for slower-growing organisms which have to spend an increasing amount of energy on capturing their prey.

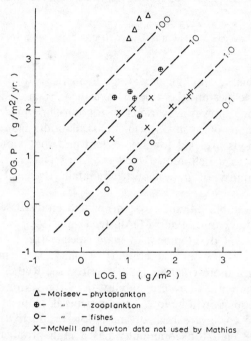

Δ – Moiseev – phytoplankton
⊕ –   "   – zooplankton
O –   "   – fishes
X – McNeill and Lawton data not used by Mathias

FIG. 60. Production (*P*) and biomass (*B*) relationships (data accumulated by Dickie, 1972 from various sources. Dashed lines indicate log intervals of constant *P/B*).

Dickie (1972) has attempted to summarize P'/B ratios for different trophic levels and their results are reproduced in Fig. 60. P'/B ratios in Table 31 are not directly comparable to data in Fig. 60 since the former are based on daily estimates and do not allow for an organism's life cycle which may impose a natural restriction on its growth for part of a year. Thus phytoplankton may continue to multiply throughout the year while more complex organisms have to grow to maturity, mate, and reproduce; these processes may require several years in higher organisms, including many fish. P'/B ratios on an annual basis indicate that at the three principal trophic levels, the respective ratios are *ca.* 300 for phytoplankton, 10 to 40 for most zooplankton, and 1 or less for most fishes depending on how long they take to complete their life cycle. The data also indicate that the annual average standing stock at different trophic levels may be very similar, but that low ecological efficiencies reduce the transfer of high productivity at low trophic levels, as material is moved up the food chain.

An important property of a food web is its stability. The simplest case is to consider a single predator and a single prey over the course of time (Fig. 61). If there are no

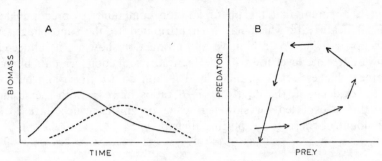

FIG. 61. (A) Predator (----)/prey (——) association with time in which (B) predator exhausts its
food supply and the population is extinguished (redrawn from Maly, 1969).

other restrictions on this system the predator will increase to a point where it runs
out of food and then declines and become extinguished. Such a system lacks stability.
Under natural conditions, however, various restrictions are placed on predator/prey
associations which tend to dampen out violent oscillations. Dunbar (1960) suggested
that in general oscillations are bad for any system and that violent oscillations may
be lethal. This has led to a definition (Hurd *et al.*, 1971) in which it can be stated
that stability is the ability of a system to maintain itself after a small external
perturbation.

Factors which tend to stabilize natural systems have been discussed by several authors
(e.g. Rosenzweig and MacArthur, 1963; McAllister *et al.*, 1972). MacArthur (1955) and
others have postulated that development of many species (i.e. a high diversity) is the
principal component in establishing community stability. Since more energy may be
required in establishing a more complex food web (e.g. see Ryther's three trophic relationships) it is also apparent that where stability has been acquired through diversity,
productivity per unit biomass will be low. Thus in general, tropical plankton communities and food webs tend to have a high diversity, low productivity, and high stability.
In contrast a temperate plankton community has a high productivity caused by one
or two species of zooplankton (e.g. *Calanus plumchrus* and *cristatus* in the subarctic
Pacific) and a low stability in which there are large seasonal fluctuations in production.
Some other important factors leading to the stabilization of a system are the imposition
of a limitation on the predator, other than its supply of prey; a hiding place for the
prey; periodic migration of the predator away from its food source; a threshold concentration below which the prey is not consumed by the predator; and the patchiness
of prey distributions. Examples of these stabilizing influences can be found in the aquatic
environment. Thus the diel migration of zooplankton into the euphotic zone may be
interpreted as adding to community stability (as well as being of some physiological
advantage to the zooplankton population, e.g. see McLaren, 1963). The possible occurrence of a threshold prey concentration in phytoplankton/zooplankton relationships
(e.g. Parsons *et al.*, 1967) and the ability of zooplankton, such as euphausiids, to change
from being carnivores to herbivores (Parsons and LeBrasseur, 1970) are further examples
of ways in which stability of a community may be maintained. On the other hand,
instability may be imparted on a plankton community by time delays (Dickie and Mann,
1972); one example of this is the maintenance of the barnacle community in the Firth
of Clyde which is highly dependent on the timing of *Skeletonema* bloom in association
with the release of barnacle larvae (Barnes, 1956).

An important observation regarding the stability of a system is based on the natural tendency for components of a system to be more variable than the system itself. Thus it is apparent from nearly all ecological oceanographic data (e.g. Riley's 1947 data on George's Bank) that individual components of a food web leading to production at a particular trophic level are more variable *in toto* than the variability in the production level itself. This has been expressed by Weiss (1969) and given by Dickie and Mann (1972) as a definition: viz. a system exhibits stability when the variance of the whole is less than the variance of the parts. This definition has some merit in that it can be expressed mathematically as

$$S = \frac{N_i V}{\Sigma(v_1 + v_2 + v_3 \ldots v_i)} \tag{115}$$

where S is an *index* of stability of a system having a variance V and a number of components (N_i) making up the system with variances of $v_1, v_2, v_3 \ldots v_i$. An advantage to the use of the above ratio as an index of stability is that it removes the problem of deciding on the relative size of perturbations as destabilizing effects on a system. Thus one system which shows regular oscillations over time of several hundred percent may be as stable as a system which shows practically no variation with time.

Patten (1961 and 1962b) considered the stability of a plankton community and its environmental parameters. The stability index evolved by Patten (1962b) was expressed as

$$S' = \frac{\sum_{j=1}^{m} \det P_j}{\sum_{j=1}^{m} (s/\bar{x})_j} \tag{116}$$

where

$$P_j = \begin{bmatrix} P_{id} & P_{ii} \\ P_{dd} & P_{di} \end{bmatrix} \tag{117}$$

which is the matrix of transition probabilites for the jth of m variables, P_{id} being the probability for a decrease in value of the variable following an increase, P_{ii} that of an increase following an increase and so on. $\det P_j$ is then determined as

$$\det P_j = P_{id}P_{di} - P_{ii}P_{dd} \tag{118}$$

and $(s/\bar{x})_j$ is the standard deviation divided by the mean of the jth variable. Equation (116) takes into account both the direction of changes and their amplitude. Thus if a parameter tends to show more of an increase than a decrease (or vice versa) the stability index will be negative; a large $s/\bar{x}$ ratio will further increase the value S' indicating a lack of stability. In comparing S' for components of the physical environment and plankton community, Patten found that the plankton (S'_p) was 6·7 times more stable than the environment (S'_e); this is not unreasonable since the plankton must to some extent absorb the 'shock' of environmental change and if $S'_p < S'_e$ the plankton community would collapse.

From these observations it is apparent that variability in the components of a system contribute to the systems stability which is the opposite process to that contributing to the instability of a machine (Weiss, 1969), or a highly controlled ecosystem, such as is attempted in aquacultural projects. In controlled operations, such as machines, the variance of each of the components add up to describe the variance in the performance of the total unit. Only by imposing maximum control over the components is it possible to predict the operation of the machine. In contrast, in systems the components vary widely but the system itself (if stable) may show only small (or if large, regular) oscillations. This contrast is brought out when attempts are made to simulate a system using a computer model which inadequately describes the buffering action of the natural environment. Thus small changes in a component of such models can be shown to produce practically any change in a population, however unrealistic such changes may be in a real ecosystem (e.g. McAllister, 1970). In contrast it appears that a natural aquatic ecosystem is made up of highly variable components which at some point in their oscillating states interlock to allow for a flow of energy up the food chain; the exact position of the interlocking point in any process being maintained by fluctuations about a mean, rather than by rigid control of an absolute value. One corollary of these observations is that a perturbation applied to the top of a food chain should have more effect on the nature of the food chain than if a perturbation of similar magnitude is applied to the bottom of the food chain. This may be seen in part by two large-scale perturbations applied to sockeye-producing lakes in British Columbia. In the first experiments (Foerster and Ricker, 1941; Foerster, 1968) 90% of the sockeye predators were removed from a lake with a resulting *ca.* 300% increase in the biomass of sockeye produced in the lake. In a second experiment (Parsons *et al.*, 1972; Barraclough and Robinson, 1972) nutrients added to a large sockeye-producing lake were increased by 100% which resulted in only about a 30% increase in the biomass of sockeye, due to depensatory components (particularly a temperature effect). While these two experiments are only very crudely comparable they serve to illustrate that the proportional increase in fish production was at least twice as effective when a perturbation was applied to the top, in contrast to the bottom, of an aquatic food chain. However, while production is the only consideration in this section, the removal of a predator as a means of raising production may not be economically sound if the same predator is the focal point of another interest, such as the sport fisherman.

Other work on lake communities has been helpful in obtaining an understanding of marine pelagic food chains. Since lakes are relatively discrete bodies of water, it is often easier to establish cause and effect relationships in the aquatic food web than in the sea where the exchange of water may remove organisms from the study area. Density-dependent relationships between the growth of planktivorous fish, and phytoplankton and zooplankton growth rates and abundance are particularly difficult to establish in the marine environment. In lakes, however, it can be demonstrated (e.g. Brocksen *et al.*, 1970) that the abundance of fish is generally related to the abundance of plankton. However, depensatory mechanisms are often apparent; thus in the case of sockeye-salmon-producing lakes, there is a general relationship between zooplankton production and sockeye-salmon production *between different* lakes. However, *within* a single lake, a large number of young fish may compete so heavily for a limited amount of food that the entire population is weakened and the number of fish surviving to become adults may be very much smaller than if fewer young fish were present initially. Another depensatory mechanism has been discussed by Kerr and Martin (1970). In a study of trout populations in Ontario lakes, these authors found that while there was a general relationship between phytoplankton productivity and trout production there was no relationship between trout production and the length of the food chain in different lakes, as implied by the general relationship in eqn. (102). Thus the biomass of large trout feeding as piscivores was similar to the biomass of small trout feeding as planktivores (other factors being equal, between lakes). This was explained by the fact that the large piscivorous trout expended less energy on feeding and were more efficient feeders than the smaller, planktivorous trout. While this depensatory mechanism may be valid in a relatively confined environment, it is probably less valid in marine ecosystems where herbivorous zooplankton are much larger and often more densely aggregated than in lakes.

Predation by planktivorous fish on the plankton community may also affect the structure of an aquatic food web. This has been illustrated by Brooks and Dodson (1965) in the case of a number of lake communities in the eastern United States. Under conditions in which there were few planktivorous fish, it was found that large zooplankton generally predominated. It was assumed that this was because large herbivorous zooplankton are generally more efficient phytoplankton feeders since they filter both large and small phytoplankton. However, in lakes where planktivorous fish were predominant, large zooplankton were selectively grazed by the fish; this markedly decreased the dominance of large zooplankton, allowing smaller zooplankton to flourish. Further, since the smaller zooplankton could not feed off large phytoplankton, the latter also tended to flourish. Thus the whole size structure of planktonic organisms in similar lakes was largely determined by the abundance of planktivorous fish.

CHAPTER 5

BIOLOGICAL CYCLES

IN THE preceding section the relationship between different organisms in the food web of the sea has been expressed without regard to life cycles. For unicellular organisms, such as phytoplankton and bacteria, this may be a useful simplification but it should not be interpreted as meaning that unicellular organisms do not have life cycles; many standard texts will describe life cycles of unicellular organisms including resting spore and auxospore formation, sexual reproduction, etc. (e.g. Fritsch, 1956). Rather, it means that the growth and decay of unicellular populations can often be accurately represented by a simple exponential function [e.g. eqn. (48)]. In the case of more complex organisms, however, the life cycle of the organism over a period of one or more years may not permit the use of simple kinetics. This is because life cycles of multicellular organisms often involve different strategies for which there is no simple mathemetical expression, other than for the period in which the population shows maximum growth. In order to gain some appreciation of these strategies, examples of life cycles of a number of zooplanktonic organisms are discussed below.

In addition to life cycles, it should also be apparent from the preceding sections that organic and inorganic materials are distributed throughout the food chain and that the mineralization of organic matter back to carbon dioxide and inorganic radicals serves, in itself, to assure a 'food' supply for autotrophic and chemosynthetic organisms. Thus the distribution of carbon and other elements tends to follow a cyclic pattern in which a single element may be present in several different phases (e.g. as a solid, gas, or in solution). In contrast to the recycling of materials, energy can only be used once. Mann (1969) has commented that while materials circulate through the biosphere, energy flows through the system in one direction, entering as light during photosynthesis and being lost as heat during respiration. In one sense this is true but the passage of energy through organic compounds into inorganic compounds (e.g. H_2S, NH_3^+, etc.) and back to organic compounds and the food chain represents a redistribution of energy which is cyclic.

Many attempts have been made to depict organic and inorganic cycles in aquatic systems (e.g. Strickland, 1965; Kuznetsov, 1968; Nakajima and Nishizawa, 1968; Riley and Chester, 1971). The closer these come to representing all the intricate possibilities involved in the flow and feedback of materials, the more complex they become. Unfortunately, this leads to difficulties in understanding their significance. In an attempt to clarify this situation, we have represented degrees of cyclic complexity by separating overall processes from some of the more detailed interdependent reactions.

5.1 ZOOPLANKTON LIFE CYCLES

Details of life cycles of some species of zooplankton will generally be found in text-books on invertebrate zoology. Such details are mostly concerned with morphological and physiological changes; the fact that life cycles also represent different strategies for survival in the pelagic environment of the sea is not generally emphasized and consequently a number of different examples are given here.

Reproduction among zooplanktonic crustaceans is generally unisexual, involving both male and female animals, although parthenogenesis is known among the cladocera and ostracods. The typical life cycle of a pelagic copepod has been described by Marshall and Orr (1955). From egg to adult the genus *Calanus* passes through six naupliar and six copepodite stages. Metamorphosis in the first few naupliar stages may take place in a matter of days and may not involve feeding. The six copepodite stages may be completed in less than 30 days (depending on food supply and temperature) and several generations of the same species may occur in the same year (i.e. an 'ephemeral' life cycle). The organism generally overwinters as a late copepodite stage in deep water and starts to breed again as an adult in the spring. The total number of eggs per female (or fecundity) depends on the food supply and may vary from <10 to several hundred. McLaren (1965) showed that for *Pseudocalanus* both egg size and body size of adults decrease with an increase in temperature so that late summer generations of the same species are smaller than early spring generations. Typical common species following this life cycle are *Calanus finmarchicus*, *C. pacificus*, and *Pseudocalanus minutus*. Because these species take advantage of new supplies of food to create new generations, they are often described as "opportunistic'.

A distinct variation in the above life cycle among common pelagic copepods is that exhibited by such copepods as *Calanus plumchrus*, *C. cristatus*, and *C. hyperboreus*. As an example, the life cycle of *Calanus plumchrus* is described by Fulton (1973). In this species the development of naupliar stages through to Stage V copepodite takes place in the surface waters from March to July (northern hemisphere). By June, Stage V copepodites start to migrate down below 300 m so that for the balance of the year (*ca.* 265 days) the organism remains in deep water. Maturation and reproduction occur at depth during the winter months and eggs are released to float towards the surface and hatch into nauplii during February to March. This vertical change during the animal's life cycle is known as 'ontogenetic migration'.

Another large group of zooplanktonic crustaceans of ecological importance in the total biomass of the oceans are the euphausiids. The number of developmental stages in the life cycle of euphausiids is greater than in copepods and consists of nauplius, metanauplius, calyptopsis, furcilia, cyrtopia, and adult forms. Breeding populations may occur within one years (e.g. *Euphausia pacifica*) or animals may continue to mature and breed in their second to third years (e.g. *Euphausia superba* or krill). With the latter species eggs appear from November to March (southern hemisphere) and sink to 500–2000 m. Larvae start to develop soon after the eggs are laid and ascent to the photic zone occurs during three stages from nauplius I, nauplius II to metanauplius. Independent feeding starts at the calyptopsis stage. Juvenile euphausiids overwinter at 150 to 200 m and mature further the following summer; adults occur after two full years and eggs are laid during the following Antarctic summer (November–March); thus, the complete life cycle of the krill involves a period of three summers and two winters (Marr, 1962).

Some species of zooplankton are 'oviparous' in that their eggs are released into the water column where they are free-floating and beyond the control of the parents (e.g. *E. superba*). In some other zooplankton (e.g. among mysids, isopods, and amphipods) eggs are laid into a modified brood chamber where they hatch to be liberated from the adult as young stages (i.e. the species are 'ovoviviparous'). Grice and Gibson (1975) showed that among some oviparous crustacean zooplankton, eggs which were laid in the fall remained viable on the bottom during the winter; the eggs hatched as nauplii to re-enter the pelagic community in the spring.

Mauchline (1973) showed that among mysids, egg production amounted to about 10% of the adult body weight regardless of the size of adult. However, meso- and bathypelagic species of mysids had fewer but larger eggs. Mauchline (1972) was unable to diagnose specific seasonal breeding cycles among a number of bathypelagic organisms but the adaptation to longer-lived species would indicate a slow, irregular breeding cycle which would maximize conservation of resources in a nutritionally sparse environment. In addition it appears that with deep-water zooplankton, brooding of young tends to be more common than among shallow-water animals. This extra care for the offspring may be another adaptation towards a general scarcity of food at great depths.

Among chaetognaths, sexual hermaphroditic reproduction occurs giving rise to free-floating eggs which may hatch in several days. Young chaetognaths are referred to as larvae but are essentially similar in form to the adults. The number of generations per year may range from several in tropical waters to a biennial life cycle in arctic waters. As with many other species, size is inversely correlated with temperature (Reeve and Cosper, in press; McLaren, 1966; Dunbar, 1962).

Among the coelenterates reproduction is commonly carried out by alternate generations (metagenesis) involving a sexual budding from a polyp stage and sexual reproduction in a free-floating jellyfish stage. In the true jellyfish (Scyphozoa) the polyp stage is often suppressed and although metagenesis occurs, members spend most of their life as free-floating jellyfish. In the Hydrozoa, only the medusan stage is planktonic and thus the hydrozoans may contribute only seasonally to the meroplankton. One or several generations of coelenterates may occur per year but where there is a hydroid (or polyp) stage, this is used for hibernation.

Pteropods reproduce as protandric hermaphrodites by sexual reproduction. Eggs are usually released as an egg mass or ribbon in which the eggs are embedded in mucus. As molluscs, the first larval stage is known as a veliger and these are generally reported from laboratory experiment to be herbivores. Veliger metamorphosis may occur in 10 to 20 days giving rise to larval forms similar in structure to the adults. Among gymnosomatous pteropods, carnivorous feeding starts after metamorphosis of the veliger stage (Lalli and Conover, 1973).

In addition to the examples given above, in continental shelf areas the life cycle of many benthic organisms will contribute larval forms to the plankton. Such larval forms will commonly include echinoderm and polychaete larvae, Veligers and cypris larvae of barnacles.

Figure 62 contains a summary of some of the different pathways in the life cycle of marine planktonic organisms. This is not intended as an exhaustive survey of various options but as an introduction to the subject. As Cole (1954) has commented, "the number of conceivable life history patterns is essentially infinite, if we judge by the possible combinations of features which have been observed. Every existing pattern may be presumed to have survival value under certain environmental conditions." From

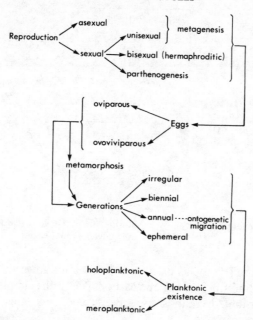

FIG. 62. A summary of some life cycles strategies among the zooplankton illustrating various options employed by different organisms in the pelagic environment of the sea.

this, one must conclude that there is a certain danger in employing over-simplified descriptions of growth processes, especially within higher trophic levels. Further, the deployment of different life cycle strategies is not unrelated to what has been called 'stability' in the previous chapter. Obviously there remains a wide gap between the real world biology of the sea and our ability to view it as a series of coupled processes that respond to physical and chemical forces.

5.2 ORGANIC CARBON AND ENERGY CYCLES

In a preceding section it has been noted (Section 2.4) that the largest fraction of organic carbon in the oceans exists as debris, either as organic compounds dissolved in sea water or as particulate organic detritus. It is also apparent from the utilization of organic substrates by heterotrophic organisms, as well as from the utilization of particulate detritus by zooplankton (e.g. Chindonova, 1959), that the reservoir of organic debris in the sea feeds back into the food chain. From these observations Riley (1963) has concluded that there is a flexible system of reversible reactions allowing organisms to draw upon the reservoir of organic carbon and replenish it in a variety of ways. This in turn has tended to stabilize the aquatic environment by providing a food source for living organisms over a longer period of time than that in which a single phytoplankton bloom can be sustained by the environment.

This system can be illustrated by Fig. 63, which has been numbered to show (1) the input into the system from the food chain, (2) the reservoir of organic debris which is assumed to have reached a steady state over a long period of time, and (3) the output from the reservoir which includes both the mineralization of organic matter to carbon dioxide and inorganic radicals, as well as the utilization of organic matter

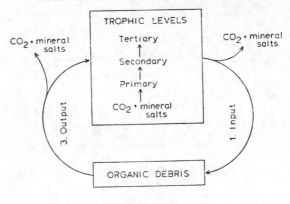

FIG. 63. A generalized organic carbon cycle showing the exchange of carbon between the food
chain and the organic debris of the sea.

by the food chain. Olson (1963) has discussed the kinetics governing the input of organic
material (L) to a standing stock of organic carbon (X). In the simplest case of a continual
input and with a constant decomposition rate, k(time^{-1}), changes in X with time can
be given by the equation

$$X = \frac{L}{k}(1 - e^{-kt}).\qquad(119)$$

In a situation where sufficient time has been allowed for the system to come to equilibrium
(t large), a steady-state value of X will be reached such that

$$X_{ss} = \frac{L}{k}.\qquad(120)$$

Minderman (1968) has criticized the use of this approach in terrestrial environments
on the grounds that the initial decay of fresh organic material is much faster than
can be accounted for by use of the kinetics given above (see also Section 2.4). In aquatic
environments, however, decomposition starts as the debris descends towards its final
place of accumulation; in shallow seas this will generally be the sediments while in
the deep oceans the largest accumulation of organic material is in the deep-water
column. Thus the initial rapid loss of organic material will probably have already
occurred when the input term (L) is assessed. Skopintsev (1966) using the above kinetics
[eqn. (119)] estimated that the one or two parts per million of soluble organic carbon
in deep ocean waters had accumulated over a period of several thousand years; a value
which is similar to the residence time of 3400 years found by radio-carbon dating (Williams *et al.*, 1969). Using the same approach, the turnover time of organic carbon in
a near-shore sediment was found to be about 30 years (Seki *et al.*, 1968).

The need to consider biological systems in terms of energy units instead of weight
has already been expressed at several points in this text. In practice, however, the expression of results in terms of weight per unit area is more easily conceived and compared
with everyday problems of agriculture, fisheries, and nutrition. However, the absolute
need to use energy units when discussing biological cycles emerges whenever a consideration is given to autotrophic and chemosynthetic processes in an environment. This
is illustrated in Fig. 64, which is modified from Sorokin (1969).

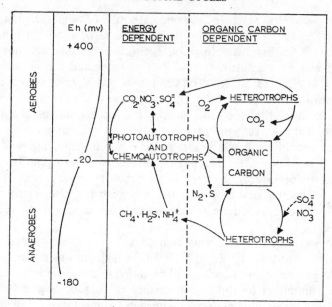

FIG. 64. Organic carbon and energy-dependent cycles in marine aerobic and anaerobic environments (modified from Sorokin, 1969).

The left-hand side of the figure is divided into aerobic and anaerobic environments; since oxygen disappears in anaerobic environments, the best division of the two environments is in terms of the 'oxidizing potential', or E_h, measured in millivolts. Generally most of the environments discussed in this text have been aerobic environments in which the oxygen concentration was adequate to supply all trophic levels. In the Black Sea, and in some fjords and lakes, however, anaerobic environments exist below the surface and it is at the interface of these two zones that a number of biological reactions may occur. The areas mentioned above are not large compared with the hydrosphere but it should be considered that certain near-shore sediments, which are not mixed by wave action, may contain very similar aerobic/anaerobic zones. The important aspect of an anaerobic zone wherever it occurs is that it represents a storehouse of chemical energy which may have some ability to feed back into the food chain; certainly this has already been demonstrated in meromictic lakes and the Black Sea (Sorokin, 1969). In these environments different groups of bacteria exist which can decompose organic material by using sulphate and nitrate as a source of oxygen, and in the process form reduced substances, such as CH_4, H_2S, and NH_4^+. The latter compounds can be utilized by other bacteria, some of which are strictly chemoautotrophs since they can use CO_2 as a source of carbon, and inorganic compounds as a source of energy. The extent to which CO_2 is taken up independently of photosynthesis has been studied by a number of authors (e.g. Romanenko, 1964 a, b; Sorokin, 1969) and a discussion of these processes is given in Section 3.1.7.

The top of Fig. 64 is divided into energy-dependent reactions and organic carbon-dependent reactions. In moving from either photoautotrophic or chemoautotrophic reactions, the cycle can either continue to be described in energy units, or quantities of organic carbon can be substituted (the latter are generally easier to measure in small amounts). Thus a photosynthetic organism receives its energy to grow from light, but once new organic material is formed, it can be expressed as so many calories of organic matter or in terms of dry weight of organic carbon, wet weight, and so on. All other steps in the food chain, represented by the 'organic carbon' box in Fig. 64, may also be expressed in terms of biomass. However, due to the very different inorganic chemical composition of marine organisms (including the water and ash content of plankton, e.g. Table 12), it is often advisable to retain units which express either organic carbon or a growth-limiting element, such as nitrogen, and to avoid units such as total wet or dry weight. Some authors recommended the exclusive use of energy units but it is apparent from Table 11 that the primary producers may at times be rich in energy compounds (i.e. carbohydrates and lipids) and low in nitrogen compounds (i.e. proteins) and vice versa depending on the availability of inorganic nitrogen. Thus two different metabolic cycles or food webs may exist in nitrogen limited (e.g. tropical) and nitrogen available (e.g. temperate) environments; the former being characterized by excessive amounts of energy rich compounds and the latter being an energy-starved but nitrogen-rich system.

Figure 41, in Section 3.2.1, is another representation of a carbon cycle; in this figure emphasis is given to the loss of organic material through various processes in the marine food chain. A large component which could be added to this cycle is the loss of materials to the benthic community by sinking. In certain areas, such as the Grand Banks, an important pathway for organic material is through sedimentation of phytodetritus, benthic filter feeders and from thence to demersal fish which inhabit such shallow areas in great abundance.

5.3 INORGANIC CYCLES

Many of the earliest studies on the inorganic micronutrients of the sea were concerned with the study of the phosphorus cycle (e.g. Harvey, 1957, and references cited therein). The occurrence of phosphorus in three different forms (i.e. particulate phosphorus, soluble organic phosphorus, and inorganic phosphate) over a period of a year in a coastal environment is shown in Fig. 21, Section 2.1. From this figure it is apparent that the biological phosphorus cycle of the sea involves the uptake of inorganic phosphate by phytoplankton during the summer in temperate latitudes. Phosphorus is then redistributed as particulate and soluble organic phosphorus, the latter resulting from the breakdown of cellular material, as well as from the release of organic phosphorus from living plants and animals. In tropical and subtropical latitudes seasonal cycles are far less well defined and the distribution of phosphorus may be assumed to be in a state of flux in which phosphorus from decomposition processes is utilized as it becomes available.

Pomeroy et al. (1963) and Johannes (1965) showed that zooplankton, and particularly marine Protozoa, could excrete phosphorus as organic compounds and as inorganic phosphate, in daily amounts in excess of their body phosphorus content. The excretion of phosphorus by zooplankton was shown by Martin (1968) to be inversely proportional to food abundance; the explanation for this was that phosphorus was used for storage products or egg production when food was abundant, but excreted due to a relative increase in body metabolism when food was scarce. A distribution of dietary phosphorus for *Calanus* during active feeding (April) has been given by Butler *et al.* (1970) as 17·2% retained for growth, 23·0% excreted as fecal pellets, and 59·8% excreted as soluble phosphorus. Thus the grazing activity of *Calanus* effectively returns more than 80% of the phytoplankton phosphorus to the environment. This process of phosphorus remineralization has been observed in nature. For example, Cushing (1964) showed that during 10 weeks of the spring phytoplankton/zooplankton bloom in the North Sea, inorganic phosphate did not decrease below 0·6 μg at/l. However, Antia *et al.* (1963) observed a decrease in phosphate to 0·1 μg at/l after 2 weeks of phytoplankton growth in the absence of zooplankton.

Both the excretion or organic phosphorus by phytoplankton and the uptake of phosphorus from organic phosphorus compounds for phytoplankton growth have been demonstrated by a number of authors (e.g. Kuenzler, 1965 and 1970). Studies on the remineralization of phosphorus from decaying phytoplankton in the absence of light (Antia *et al.*, 1963) showed that half the organic phosphorus could be released as reactive inorganic phosphate in a period of 2 weeks. Phosphorus lost to sediments through

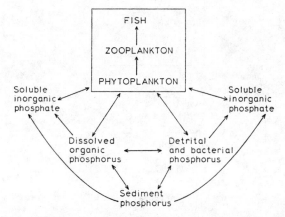

FIG. 65. Major pathways in the phosphorus cycle of the sea.

sinking as particulate phosphorus may be partly converted into insoluble mineral phosphates or recycled by the benthic animals and microflora.

A general summary of the major pathways in the phosphorus cycle of the sea is shown in Fig. 65. From this cycle and the observations given above it is apparent that the various forms of phosphorus are often readily exchangeable as metabolites. The rate of turnover, especially in the presence of zooplankton, indicates that phosphorus is generally available in the marine environment although the absolute concentration of phosphate may sometimes be sufficiently low as to determine the actual growth rate of the phytoplankton species.

Phosphorus is readily hydrolysed from organic compounds, either by hydrolysis at the alkaline pH of sea water or by phosphatases, which are hydrolytic enzymes present in many bacteria and on the surface of some phytoplankton, particularly those from environments low in inorganic phosphate. In contrast, organic nitrogen is cycled with much greater difficulty since its fixation, reduction, and oxidation back to nitrate requires the exchange of energy. Thus in the same experiment in which half the phytoplankton phosphorus was mineralized in 14 days, no mineralization of organic nitrogen was observed in a period of up to 75 days (Antia *et al.*, 1963). However, as in the case of phosphorus, zooplankton feeding greatly enhances the recycling of nitrogen, either as ammonia (e.g. Harris, 1959; Corner and Newell, 1967) or through the release of soluble organic nitrogen compounds (Johannes and Webb, 1965; Webb and Johannes,

1967 and 1969). A dietary nitrogen budget for actively feeding *Calanus* has been given by Butler *et al.* (1970) as 26·8% retained for growth, 37·5% excreted as fecal pellets, and 35·7% excreted as soluble nitrogen compounds.

In temperate waters the uptake of nitrate nitrogen follows a seasonal cycle similar to that for phosphate with the exception that the supply of nitrate in the surface layers often becomes completely exhausted during the summer months, while phosphate may continue to be present in low concentrations. In tropical and subtropical waters nitrogen availability is often regarded as the rate-limiting nutrient throughout the year. Dugdale and Goering (1967) have shown that ammonia is taken up by phytoplankton more

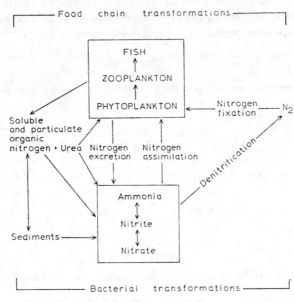

FIG. 66. Major pathways in the nitrogen cycle of the sea.

rapidly than nitrate; this was particularly noticeable in phytoplankton from subtropical compared with temperate regions. Urea may also serve as a nitrogen source for some phytoplankton and significant concentrations (*ca.* to 5 μg at/l) of this compound have been found in coastal and oceanic waters (Newell, 1967; McCarthy, 1970; Remsen, 1971). In tropical environments the fixation of molecular nitrogen by certain bacteria and more especially by blue–green algae, may be very important. Goering *et al.* (1966) measured the uptake of molecular nitrogen in waters containing *Trichodesmium* sp. and found a maximum rate of 0·32 μg N/l/hr. The loss of nitrogen through denitrification of nitrate appears to be an anaerobic reaction. However, in some aerobic environments. denitrification may occur in the presence of detrital particles; this is probably due to the formation of anaerobic microzones of bacterial activity within the particles (Jannasch, 1960).

The conversion of ammonia to nitrite and nitrate may proceed in any environment by a series of reversible reactions. These are largely carried out by bacteria but some phytoplankton may produce extracellular nitrite during the course of nitrate uptake. A general schematic summary of nitrogen conversions in the sea has been given by Vaccaro and Ryther (1959) and a modified version of their diagram is presented in Fig. 66. In this diagram the food chain, as the primary consumer of inorganic nitrogen, has been represented at the top, and the bacteria, as the primary agents in recycling nitrogen, are represented at the bottom. However, as in the case of phosphorus, nitrogen can be effectively recycled between phytoplankton and zooplankton under conditions of a plankton bloom which favours both zooplankton grazing and the growth of phytoplankton. (e.g. Takahashi and Ikeda, 1975). The uptake of different nitrogen substrates by phytoplankton also indicates that various forms of nitrogen may serve as nitrogen sources during photosynthesis. Antia *et al.* (1975) found that among twenty-six species of phytoplankton, 88% showed good growth on urea, 69% could grow on hypoxanthine, and 50% could grow on glycine.

Ammonia, particulate organic nitrogen, and soluble organic nitrogen from the food chain may be present in different proportions in different environments; the bacterial oxidation products of these reduced forms of nitrogen may also occur in different proportions, especially in depth profiles. In particular, the occurrence of a nitrite maximum at the bottom of the euphotic zone is often a well-defined phenomenon in thermally stratified surface waters. Nitrite can be produced either through the oxidation of ammonia (a dark reaction) or by reduction of nitrate (stimulated by light). It is possible that the ammonia pathway is carried out largely by bacteria while the nitrate pathway is mediated through phytoplankton. The presence of the nitrite maximum has recently been reviewed by Wada and Hattori (1971), who considered that it was due to nitrite production from ammonia and nitrate at a rate in excess of biological and physical processes which remove nitrite. The latter would include the biological uptake of nitrite precursors (i.e. nitrate or ammonia) in the euphotic zone above the nitrite maximum, and decreased nitrite concentration due to advective processes, particularly below the thermocline.

The cycling of other micronutrients in sea water is generally less well defined than in the case of nitrogen and phosphorus. Silicon, which is required in diatom metabolism for the formation of cell walls, follows a seasonal cycle of abundance in temperature waters. The lack of soluble silicate may to some extent determine species succession from a diatom to a flagellate community. Many biologically essential inorganic materials in sea water are present in very large amounts so that any cyclic process involved in their distribution is not important to the rate control of metabolic processes in the food chain. These nutrients include sodium, magnesium, potassium, calcium, sulphate, water, and carbon dioxide. In some cases elements may become limiting to phytoplankton growth due to their availability and not to their absolute concentration. Thus iron, particularly in sub-tropical and tropical oceanic waters, may become growth limiting in the absence of organic chelators, which makes it available for uptake by plants.

5.4 TRANSFER OF ORGANIC COMPOUNDS WITHIN THE FOOD CHAIN

Certain organic compounds are transferred within the food chain of the sea; this may also involve a cyclic process in which the compounds are recycled between different trophic levels but at present this has not been a subject of investigation. In the following section some examples are given of compounds which are transferred intact between different trophic levels.

The importance of preformed compounds in the nutrition of lower organisms has been well recognized in the vitamin requirements of certain algae. In a review of this subject Provasoli (1958) showed that thiamine, biotin, and vitamin B_{12} are often required by marine algae. Since these compounds are also required in a preformed state by higher organisms it is reasonable to consider their transfer up the food chain as a necessary part of food chain stability. The exact state of the preformed compound has been investigated in the case of vitamin B_{12} by Droop et al. (1959), who showed that various analogues and precursors of vitamin B_{12} could also serve to meet the requirement for this vitamin in some algae.

TABLE 32. FATTY ACID COMPOSITION OF *Chaetoceros, Artemia,* AND GUPPY OILS (WEIGHT PERCENT OF TOTAL ESTER) [FROM KAYAMA (1964)]

Fatty acid	Chaetoceros	Artemia	Guppy 17 ± 1°C	Guppy 24 ± 1·5°C
Shorter chain	trace	0·4	trace	
12:0	0·4	trace	0·2	trace
13:0	0·7	trace	trace	trace
14:0	13·0	4·8	1·5	0·9
15:0	1·8	1·5	trace	0·2
14:2	0·6	trace	0·6	0·5
16:0	18·1	11·6	22·9	36·0
16:1	47·9	44·9	15·9	8·9
16:2	2·7	trace	0·2	0·2
16:3?	4·0	1·7		0·6
16:4?	trace			0·5
18:0	0·5	1·9	8·2	9·8
18:1	8·7	18·4	18·3	15·0
18:2	1·7	0·7	trace	trace
18:3	trace	0·5	1·4	0·8
20:1		0·9		
18:4 & 20:2		0·8	0·3	trace
20:3			0·2	trace
20:4		trace	2·0	2·0
20:5		12·0	4·8	4·6
22:4			1·3	1·0
22:5			6·1	7·3
22:6			16·5	11·5

The transfer of an antibacterial substance in the food chain of penguins has been investigated by Sieburth (1961). The compound which was found to have strong antibacterial properties was acrylic acid; this is present in relatively large amounts in a common antarctic phytoplankton, *Phaeocystis* (a genus which is also found in the northern hemisphere). In the Antarctic this organism forms substantial blooms which are consumed by *Euphausia superba* or krill; the euphausiids in turn are one of the principal food organisms of penguins. Acrylic acid is transferred to the penguin where it causes bacteriological sterility in the anterior segments of the gastrointestinal tract.

The transfer of fatty acids up the food chain from phytoplankton to fish has been indicated by Williams (1965) and studied experimentally by Kayama (1964) and Lee et al. (1971). Kayama (1964) fed a diatom, *Chaetoceros* sp., to brine shrimp which were then fed to guppies. Fatty acid composition of the lipids from the three trophic levels were then examined and the result for C_{12} to C_{22} fatty acids are shown in Table 32. These results show a transfer of 14:0, 16:0, 16:1, and 18:1 fatty acids, but an apparent synthesis of polyunsaturated fatty acids by *Artemia* (20:5) and guppies (22:5 and 22:6). However, from Table 32 it is also apparent that C_{20} and C_{22} polyunsaturated fatty acids occur among some phytoplankton, in which case they are probably transferred intact up the food chain (e.g. Lee *et al.*, 1971). Hinchcliffe and Riley (1972) fed different phytoplankton diets to brine shrimp (*Artemia*) and found that the levels of saturated acids were comparatively constant regardless of the food organism used. They showed also that polyunsaturated fatty acids occurred in their different phytoplankton diets and that they were transferred to the brine shrimp. However, some fatty acids occurred in higher proportions in the brine shrimp than in the algal food (e.g. oleic acid, as shown also by Kayama, 1964). Ackman *et al.* (1970) studied the fatty acids and lipids in north Atlantic euphausiids and showed differences between samples which could be explained in terms of diet. As a general conclusion, therefore, it appears that under natural conditions, where animals feed on a broad range of food, there will be qualitative similarities but some quantitative differences between the fatty acid composition of aquatic animals and their diet.

From studies on hydrocarbons in algae it appears that an important biological substance found in many algae is a 21:6 hydrocarbon, all-*cis*-3, 6, 9, 12, 15, 18-heneicosahexaene or 'HEH'. This substance is present to the extent of a few thousandths of a percent of the cell weight of algae and is also found in small quantities in zooplankton and higher organisms including oysters, herring, and the basking shark, among animals investigated (Blumer *et al.*, 1970). It has been suggested (Youngblood *et al.*, 1971) that this compound is involved in some way in a biochemical function of the reproductive cycle in both marine plants and animals. The evidence for this is at present circumstantial but it appears that polyunsaturated hydrocarbons are highest in algae during exponential growth, they are concentrated in the reproductive structure of at least one benthic algae, and in the specific case of HEH there is some correlation between the predominance of HEH to other hydrocarbons in algal foods and the relative proportions of male animals hatched from *Calanus helgolandicus* when fed different algae. The general chemistry of hydrocarbons in the marine food chain also appears interesting from the fact that different species of algae appear to have quite specific hydrocarbon ratios and that these differ from those synthesized by copepods and present in mineral oils. Zooplankton contain complex mixtures of C_{19} and C_{20} isoprenoid alkanes and alkenes while mineral oils contain many isomers including large amounts of cyclic compounds, but no olefins. Since hydrocarbons are relatively indestructible in the food chain of the sea, it is apparent that transfer routes within the food web as well as the source of oil pollutants might be discovered through hydrocarbon chemistry (Blumer *et al.*, 1971).

CHAPTER 6

BENTHIC COMMUNITIES

6.1 DISTRIBUTIONS OF BENTHIC ORGANISMS

6.1.1 TAXONOMIC, HABITAT AND SIZE GROUPS OF BENTHOS

When organisms live in, on, or are occasionally associated with aquatic sediments, their mode of life is referred to as benthic and collectively they form the 'benthos'. Bacteria, plants, and animals from all phyla are represented, but, unlike the plankton, sizes of bottom-living organisms span several orders of magnitude. Truly benthic organisms, as opposed to those which temporarily come to the bottom, are considered to be sessile and relatively inactive. Organisms such as seaweeds, encrusting sponges, corals, barnacles, and some echinoderms are firmly anchored for the duration of their life. Burrowing worms and molluscs demonstrate sedentary behaviour but movement within and over the bottom also occurs. In contrast, some benthic animals such as certain molluscs, crustaceans, and demersal fish species often undergo extensive vertical and horizontal movements away from the bottom. The distinction between a pelagic and benthic mode of life cannot be made for such organisms.

Benthic organisms differ from planktonic ones in their adaptation for an association with a substrate. While a planktonic existence necessitates small size and a specific gravity close to that of sea water, a sedentary life permits a variety of size, shape and density. Seventy-five per cent of the total number of marine species live on firm substrates (rocks, coral reefs), 20% occur on sandy and muddy bottoms, and only 5% of the total are planktonic (Thorson, 1957). Calcareous shells, elongated stalked and branched body forms and the development of appendages (cilia, bristles) and body musculature which enable movement over, into, and through sediment are characteristic of benthic organisms.

A characteristic feature of all sediments is their vertical zonation into a surface oxidized layer and a subsurface reduced zone where dissolved oxygen is depleted (see Section 6.2.1). Vertical distributions of sediment fauna and flora clearly reflect these sharp vertical gradients in chemical conditions (Fenchel, 1969). The organisms themselves produce and utilize metabolites which maintain chemical profiles. Bacteria are most important in this regard. They are the most numerous and ubiquitous organisms present in sediments and their metabolic activities largely control the chemical nature of sedimentary environments. Groups of bacteria isolated from marine sediments are generally similar to those in water in antibiotic sensitivity, degradative and fermentative capability and cation requirements (ZoBell, 1946; Wood, 1965; and Stevenson *et al.*, 1974). Besides the common genera of bacteria which occur in water (Section 1.1), *Flavobacterium* and *Achromobacter* are often present in sediments. Both gram negative and gram positive cell types can be present and cell size has been observed to be larger than in bacteria isolated from water (Stevenson *et al.*, 1974). Most motile sediment bacteria possess polar flagella and rod and coccoid cell forms predominate.

Bacteria using oxygen as a hydrogen acceptor to oxidize organic compounds (aerobic heterotrophs) occur in surface sediment layers to depths determined by the penetration of dissolved oxygen. Subsurface sediments are usually anoxic and anaerobic bacteria use various organic and oxidised inorganic compounds as hydrogen acceptors. Bacteria capable of anaerobic respiration (fermentation) and chemosynthesis characterize these sediment layers (ZoBell, 1946; Bass Becking and Wood, 1955; and Fenchel, 1969). When reducing conditions extend to the sediment surface where light is present, sulfurbacteria can occur in patches—termed 'sulfureta' since energy transfers involving sulfur predominate (Lackey, 1961; Fenchel, 1969).

Fungi have been thought not to be numerically important in marine sediments but recent observations demonstrate that microbial flora may be dominated by fungi at certain times of the year (Litchfield and Floodgate, 1975). Marine waters subject to freshwater input may receive fungal spores through drainage. Humic materials present in mud sediments could provide an available carbon source to these micro-organisms.

When water is sufficiently shallow, light reaches the sediment surface and micro-benthic algae and macrophytes grow, often in great abundance. Pomeroy (1959), Grøntved (1960), Meadows and Anderson (1968), Steele and Baird (1968), and Gargas (1970) document the presence of diatoms in surface sediments which receive light. Benthic microflora can be very diverse with all major classes present in the phytoplankton represented. Besides photosynthetic bacteria, cyanophyceans, cryptomonads, euglenoids (green and colourless), dinoflagellates, and diatoms are often found in well-structured vertical distributions corresponding to the presence of light and oxidation-reduction conditions (Wood, 1965; Fenchel and Straarup, 1971). Attached macrophytes (Thallophytes—seaweeds) and emergent vascular plants (sea grasses) are often abundant in shallow-water littoral and intertidal areas provided exposure to water movement is not great (Mann, 1972a).

Benthic fauna, unlike the zooplankton, is not dominated by any particular class of animals. Representatives of all phyla occur but a particular habitat may be characterized by a few dominant species. Thus, species associations described below have been used to describe bottom community types (Thorson, 1957). An obvious example of dominance exists in coral reefs, oyster banks, and mussel beds. These are extreme cases and usually a diverse and heterogeneous fauna, both in species abundance and size, exists. The question of why some species are common and others are rare remains as an unanswered ecological problem.

It has been apparent since the voyage of the *Challenger* that trawl and dredge samples from the deep sea contain numbers of benthic animals in inverse proportion to depth (Sanders and Hessler, 1969; Rowe, 1971b). Faunal changes associated with water depth have been classified by a system of vertical faunal regions based on the distribution of particular genera and species (Menzies *et al.*, 1973). These authors calculated an 'index of distinctiveness' (the percentage of genera or species not held in common by any two sampling points along a depth gradient) on the basis of isopod crustacea. 'Intertidal,' 'shelf', 'archibenthal' (zone of transition), and 'abyssal' faunal groups are

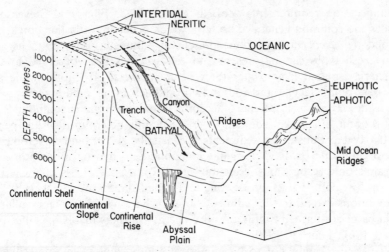

FIG. 67. Horizontal and vertical zonation in a hypothetical cross-section of a transect across the western Atlantic. Geological features of the ocean floor are illustrated in conjunction with terms commonly used to differentiate water column depth layers. Depth and distance relations are not drawn to scale (redrawn from Heezen *et al*, 1959).

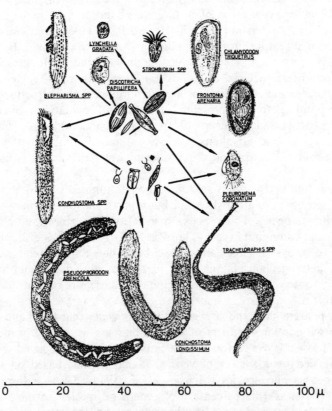

FIG. 68. Examples of microbenthos: bacteria, flagellates, diatoms and herbivorous ciliates in surface sublittoral sands (from Fenchel, 1969).

terms usually used to indicate vertical zonations in the sea (Fig. 67). The zone of transition (continental slope and rise) is also referred as the 'bathyal' zone. Trenches and
canyons in the abyssal region which may drop to below 6000 m form the 'hadal' zone.
These regions had distinctive faunal assemblages perhaps reflecting reproductive isolation which exists between populations at great depths.

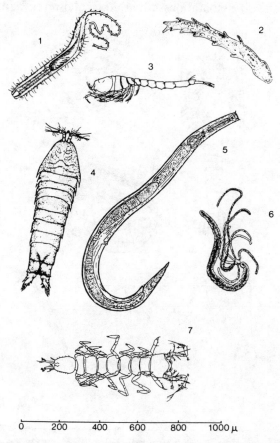

FIG. 69. Examples of meiobenthos: gastrotrich [*Urodasys* (1)], mollusc [*Pseudovermis* (2)], cumacean
[*Eudorella* (3)], harpacticoid copepod [*Paramphiacella* (4)], nematode [*Ethmolaimus* (5)], coelenterate
[*Halammohydra* (6)], isopod [*Microjaera* (7)]. From Eltringham (1971), Hyman (1951), Smith (1964)
and Marcotte (1974).

Benthic animals are also generally described on the basis of their position in the
sediment relative to the surface and their size. 'Infauna' are animals of any size which
live within the sediment utilizing either interstitial space between sediment particles,
burrows, or tubes. 'Epifauna' live at or on the sediment surface. Sessile and slowly
moving epifauna have been differentiated (Petersen, 1913). Many of these latter species
frequently enter the water column. Shrimp and scallops, for example, make extensive
movements by swimming off the sediment surface. Gametes are often released freely
into the overlying water during reproduction, or migration of the reproducing animals
themselves occurs. Tube-dwelling polychaetes may leave the sediment during such
spawning migrations.

Size separation of benthic fauna, like that for planktonic organisms, is operationally useful. Mare (1942) first suggested that benthic fauna be divided into 'microfauna' (1–100 μm) (bacteria, protophytes and protozoans excluding the foraminifera), the 'meio-fauna' (100–1000 μm) (foraminifera and small metozoans-nematodes, turbellarians, and juvenile macro-invertebrates) and the 'macro or mega fauna' (> 1000 μm) (Figs. 68, 69, and 70). Usually metazoans passing through a 500-μm sieve are considered as meiofauna (McIntyre, 1969). The size separations are arbitrary, however, and the groupings have

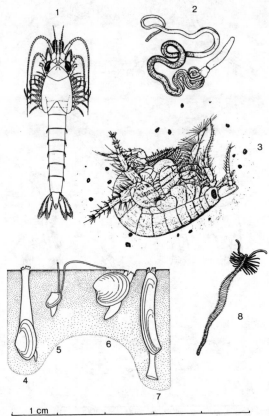

Fig. 70. Examples of macrobenthos: mysid [*Heteromysis* (1)], hemichordate [*Saccoglossus* (2)], amphipod [*Bathyporeia* (3)], molluscs [*Mya* (4), *Tellina* (5), *Mercenaria* (6), *Ensis* (7)], polychaete [*Dodecaceria* (8)]. (From Smith, 1964 and Nicholaisen and Kanneworff, 1969).

no biological meaning. All feeding types, from selective feeders on specific bacterial cell types to omnivores and carnivores, are represented in all size categories of benthic organisms. 'Demersal' (bottom-living) fish which prey on all size groups usually enter the benthos as larval stages. Most demersal fish species remain at or just above bottom (as epifauna) but some browse and bury themselves in the sediment surface.

6.1.2 DIVERSITY AND SPECIES ASSOCIATIONS IN BENTHIC COMMUNITIES

6.1.2.1 Indicator species and community associations. Descriptions of species composition in benthic communities are of two major types. Broad sampling over large geographical areas has been used to define species composition and distribution in terms of dominant (usually macrofauna) species. Studies by Petersen (1913), Thorson (1957), and Nesis (1965) are examples of this type. The regular recurrence of a few common species over wide areas is used to name a community in terms of one or two of the most abundant species. Alternatively, studies such as those by Jones (1956), McIntyre (1961), Sanders (1956), Lie (1968), Mills (1969), and Parker (1975) describe and enumerate benthic fauna on a similar basis but within one localized area. These surveys have indicated limitations in describing benthic communities on the basis of a few dominant 'indicator' species.

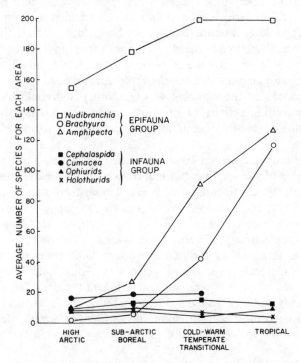

FIG. 71. Average number of species of different groups of bottom invertebrates from equally large coastal areas in different latitudes. Redrawn from Thorson (1957).

Thorson (1957, 1958) presented the concept of 'parallel communities' which implied that similar animals will be found in association wherever similar environmental conditions exist. He believed that where similar selective forces and responses (predation and settlement, for example) occurred that similar groups of more or less closely related species would always occur. The accumulation of evidence for latitudinal gradients in diversity created difficulties in applying the concept, however. An increasing gradient in diversity (species number) occurs in benthic communities, as in others, as one moves from the arctic to the tropics (Fig. 71). Tropical and deep sea benthos are highly diverse, and usually have no clear dominant species (Sanders and Hessler, 1969).

While some benthic communities may be characterized in terms of a few dominant species, the current view regards the community as an abstraction from a series of species distributions along gradients (Mills, 1969). Mills demonstrated that the abundance of smaller less numerous species along a transect of an intertidal sandflat was not correlated with peak numbers of abundant species. There was thus little justification for assuming that dominant species were functionally related to other less abundant species. This interpretation corresponds to Jones' (1950) classification of benthic communities as ecological units separable by physical (temperature, oxygen, sediment type) and biological interactions. The schemes assume that groupings of organisms are caused by both physical and biological factors.

Descriptions of some benthic species associations in terms of dominants remains convenient, however, particularly in certain habitats. Activities of single predatory species may greatly modify the physical and biological nature of some communities. Paine (1969) has termed these 'keystone species' since their activities and numbers determine community structure, integrity, and stability. Starfish which feed on a variety of prey, especially those capable of monopolizing rock shores, demonstrate that through predation one species can determine overall community composition and structure. Fager (1964) and Mills (1969) describe the stabilizing effect which tube-building polychaetes (*Owenia fusiformis*) and amphipods (*Ampelisca abdita*) can have on fine-grained sediment. The structure imposed on the sediment was temporary and could have been physically disturbed by the feeding activity of rays and snails, respectively. Both numbers and species composition in benthic communities thus may be regulated by the activities of a single species.

Species other than macrofauna have been used to indicate associations of benthic organisms. Analyses of species abundance and diversity of meiobenthic copepods above and below the Atlantic continental shelf break demonstrate unique assemblages in the deep sea (Coull, 1972). Shelf samples were dominated by few species while those from below 1000 m were different and more diverse. Wieser (1960) and Warwick and Buchanan (1970) were able to differentiate communities on the basis of nematode meiofauna in Buzzards Bay and off the Scottish coast. Differences in species composition corresponded to differences in sediment grain size. Ankar and Elmgren (1975) report a similar dependence of species dominance on sediment texture. Macrofauna associations predominated sandy sediments in the northern Baltic while meiofauna (ostracods and nematodes) dominated mud bottom communities. Benthic foraminifera have also been used to characterize bottom community associations (Schafer, 1971).

6.1.2.2 Indices of species associations.

Species-number relations, thought to indicate aspects of community structure, have provided a rationale and focus for collections of benthos as for other organisms. Abundance and importance (ecological function) need not be related, however, and calculations of indices of diversity provide only one measure of potential interactions between organisms.

Two different approaches to measures of diversity in community ecology have already been described for planktonic organisms (Section 1.2). Empirical observations, like Sanders' (1968) rarefaction method, are considered to be independent of *a priori* assumptions about the structure of the community studied, although random spatial distributions are assumed. The method, dependent on the shape of the species abundance curve rather than the absolute number of specimens per sample, does not measure any biological property of the community. Nor does it account for relative spatial distributions of species components.

Hurlbert (1971) proposed a measure of the probability of interspecific encounters which has parameters with more direct biological interpretation. Assuming random movement, all individuals in a community can encounter every other individual. Of the $(N) (N - 1)/2$ potential encounters of N individuals, $\Sigma_i (N_i) (N - N_i)/2$ encounters involve different species and

$$\Delta_1 = \sum_{i=1}^{s} \left(\frac{N_i}{N}\right)\left(\frac{N - N_i}{N - 1}\right)$$

$$= \left(\frac{N}{N - 1}\right)\left[1 - \sum_{i=1}^{s} \left(\frac{N_i}{N}\right)^2\right] \tag{121}$$

represents the proportion of potential encounters that are interspecific where N_i = number of individuals in the ith species in the community (or collection), $N = \Sigma_i$ N_i = total number of individuals in the community, and S = the number of species in the community. Δ_1 ranges from 0 to 1 and expresses the probability that any two individuals encountered at random in a community will belong to different species. If an encounter is non-lethal (that is, it does not involve predation or cannibalism), subsequent encounters are possible and the probability corresponds to

$$\Delta_2 = 1 - \sum_{i=1}^{s} \left(\frac{N_i}{N}\right)^2. \tag{122}$$

Hurlburt's attempts to develop diversity or 'species composition' parameters with biological interpretations also permit estimates of species richness = the number of species present in a collection of specified number of individuals,

$$E(S_n) = \sum_{i=1}^{s} \left[1 - \frac{\binom{N - N_i}{n}}{\binom{N}{n}}\right]. \tag{123}$$

This describes the expected number of species in a sample of n individuals selected at random from a collection containing N individuals and s species. Besides cumulative plots of n versus $E(S_n)$ for comparison, as in Sanders' rarefaction method, the parameter quantifies the expected encounters, for example, of a predator which enters a community and in a specified time encounters n individuals at random. The value could serve as an index of prey availability and should be proportional to a predator's searching rate.

Although Hurlbert's parameters have some biological meaning, data required to measure the probability of encounters will be difficult to obtain. Information on habitat size and movement of organisms thought to share that space will be required. Few past studies of benthic communities can provide such information. Measures of small-scale spatial distribution of benthic fauna (discussed in Section 6.1.3.2), however, suggest that intra-specific aggregations might be used to quantify the partitioning of habitat space in benthic species. Encounter probabilities might then be calculated assuming random movement of individuals within that space.

Another weakness of indices of species diversity based on relative numbers of individuals per species or sample is that abundance *per se* may be unrelated to relative energy consumption. Numbers provide but one measure of potential interspecific interactions and they do not specify in any way the number of energy pathways. 'Trophic diversity' was calculated by MacArthur (1955) using eqn. (5) to describe community stability where P_i referred to the proportion of a community's food energy which passed through the ith pathway in a food web. Information content (H') in this usage refers to the uncertainty of predicting the pathway of a given unit of food energy in a food web. Alternatively, other measures, such as reproduction or respiration per species, may be informative. Banse *et al.* (1971) ranked macrofauna species on the basis of respiratory demand and demonstrated differences in relative importance of the same species ranked according to numbers and biomass.

Hurlbert (1971) suggested that the importance of a species can be defined by the sum, over all species, of changes ($\pm$) in production which would occur on removal of a particular species.

$$\text{Importance of } j\text{th species} = \sum_{i-1}^{s} (P_{i,t=1} - P_{i,t=0}) \tag{124}$$

with P_i = production of the ith species before ($t = 0$) and after ($t = 1$) removal of the jth species. This index requires that production of *all* species affected by the removal of a particular species be measured—a substantial requirement.

The alternative approach to direct calculation of species abundance curves (Sanders, 1968), information statistics (Pielou, 1966), or other indices (above) calculated from observed relative abundance of species in a collection involves representing aspects of species composition in terms of parameters fitted to theoretical models. The use of logarithmic functions of species abundance or rank are examples. Numerous studies have demonstrated that indices determined by these methods are sample-size dependent and the fit of actual data to theoretical distributions is seldom perfect (Sanders, 1968; Hurlbert, 1971). In addition, there is no reason why actual distributions should conform to a theoretical distribution.

One advantage of using parameters of theoretical distributions as descriptive statistics of benthic communities lies in the fact that quantitative measures of variance of predicted distributions may be derived. Some objective judgement of goodness of fit is thus provided. Also, use of certain distribution statistics permits separation of diversity into two components—'species richness' and 'unevenness'—each of which can be quantitatively estimated (Gage and Tett, 1973).

Lie (1969) demonstrated that abundance of infauna benthic species in Puget Sound was distributed according to a truncated log-normal distribution. The method involves converting numerical abundances of species into logarithms, ranking these and estimating the medium abundance ($\bar{a}_R$), the total number of species sampled (s^*) and the standard deviation of the distribution of species abundances (σ). Gage and Tett (1973) used the same method and found, except for rarer species, a reasonable fit to the distribution of collections of macrobenthos in two Scottish sea lochs. They further interpreted the log-normal parameters in ecological terms—$\bar{a}_R$ = the abundance of the typical species in a community (those neither common nor rare), s^* = the total species number and σ = 'unevenness' (the inverse of 'evenness' and 'equitability').

In an earlier study, Gage (1972) analysed the same data set using Sanders' rarefaction method. There was a significant positive correlation between stations ranked by rarefaction diversity and by values of s^* (equivalent to 'species richness'). Additionally, a positive correlation existed between s^* and median sediment particle diameter and values of σ increased with distance up the lochs. Tett (1973) also observed increased 'unevenness' in phytoplankton species abundance at the head of two of the same lochs using similar methods. Increased salinity fluctuations and reduced exchange at the head of the narrow fjord-like lochs were thought to contribute to unevenness among species abundance. Decreased values of s^* with distance up the loch for phytoplankton did not occur in the macrobenthos, however. Thus, on the basis of parameters of the log-normal distribution, factors affecting benthic community structure do not necessarily affect phytoplankton communities in a similar way.

Various types of numerical analyses have been applied to benthic survey data as an alternative approach to identifying species associations. All such methods use statistical criteria to evaluate the frequency of occurrence and/or abundance of species and to quantify associations as groups through various types of 'similarity analyses'. Co-occurrence is assumed to indicate interaction between species within groups and thus these methods provide an objective form of community classification. As Mills (1969) has pointed out, however, co-occurring species may be independently distributed and thus structural groupings are not necessarily functional ones.

Numerical methods have two aims: to group stations as to faunal similarity and to compile species lists characteristic of each group. When surveys are conducted over large geographical areas, a heterogeneous collection of organisms results. Both species composition and abundance changes and single samples contain only a small fraction of the total number of species. Analysis of such data on the basis of species presence or absence alone is adequate.

Two similarity coefficients have been applied to characterize species groups when samples include heterogeneous types of benthic communities. Jaccard's coefficient is the number of species common to both samples expressed as a percentage of the total number of species in both samples:

$$JC = \frac{c}{a + b + c} 100 \qquad (125)$$

where a is the number of species in sample A only, b is the number of species in sample B only, and c is the number of species common to both samples. Czekanowski's coefficient which includes measures of species abundance in each sample is:

$$CC = \frac{2c}{a + b} 100 \qquad (126)$$

where the same notation is used to express the number of species common to both samples as a percentage of the mean number of species in both samples. As in planktonic populations, logarithmic transformation of the abundance of common species gives a better measure of abundance than do actual numbers (Cassie, 1962a) and coefficients may be calculated on this basis (Field and McFarlane, 1968). Dendrograms derived from matrices of coefficients can be used to indicate levels of similarity between groups of stations (Day et al., 1971).

Numerous studies have compared the effectiveness of both coefficients in separating benthic faunal groups. Jaccard's coefficient does not separate samples differing in relative abundance (Field and McFarlane, 1968) but Czekanowski's coefficient based on log-transformed abundance of common species separated four faunistic zones at increasing depth across the continental shelf off North Carolina (Day et al., 1971). Field (1971) was able to separate species-poor groups of benthos in a turbulent zone from deeper water groups in areas not subject to wave disturbances by this coefficient. Changes in fauna corresponded to changes in sediment parameters.

Fager (1957) suggested an index of affinity, termed 'recurrent group analysis', between species in a series of collections based on the geometric mean of the proportion of joint occurrences corrected for sample size. Only presence or absence records are required. The grouping of zooplankton on this basis has been described (Section 1.1). The method has the advantage that levels of affinity can be specified. Connections between groups are determined in terms of the number of species in each group which show affinity with species in other groups. Data on occurrence of demersal fish along trawl stations off the coast of Africa demonstrated the existence of up to twenty-one recurrent groups which could be marshalled into major assemblages (communities) corresponding to water depth (Fager and Longhurst, 1968). Analysis based on joint occurrences of groups along transects formed two assemblages which corresponded to differences in sediment or bottom type.

Multivariate statistical methods have been used in numerical studies of benthic organisms. These methods assume a multivariate normal distribution and raw data is usually logarithmically transformed (Cassie and Michael, 1968). Two types of analyses have been used. Principal component analysis assigns stations or species coordinates on axes representing the major components of variance. Canonical correlation is analogous to multiple regression analysis where the maximum correlations between variates are computed for each vector (variable). Chardy *et al.* (1976) have summarized relations between these and other methods which fit a set of points with given weights and distances into a subspace of defined dimension. Cassie and Michael (1968) and Hughes *et al.* (1972) have used these methods to demonstrate the proportion of variation in occurrence of certain benthic macrofauna attributable to various components. In both studies, approximately 50% of the variance was associated with differences in sediment type.

Numerous clustering or sorting techniques have been used to consider hierarchical associations between groups of benthic species. Comparison of sum-of-squares agglomeration (Field, 1971; Hughes *et al.*, 1972) separates classes into a dendrogram on a probabilistic basis. Macrofaunal groups in both studies were identified which were associated with different depth zones and bottom type. Vilks *et al.* (1970) used an index of association based on chi square (χ^2) values to construct a hierarchy of benthic Foraminifera groups in an Arctic marine bay. Nearshore-shallow (50 m) and offshore-deep (250 m) water assemblages could be differentiated.

All of these analyses provide an objective method for separating species groups which often confirm subjective classification (Fager and Longhurst, 1968). Their value lies in the quantitative association which they demonstrate between co-occurring species. Groups which occur in particular habitats may be assumed to have similar environmental requirements. It is not surprising that sediment-related properties are common factors upon which groups of benthic species can be differentiated.

6.1.2.3 Colonization and successional changes.
Experimental manipulation by the removal or addition of species, or the colonization of artificial substrates, allows consideration of diversity-causing mechanisms. Studies by Paine (1966, 1969, 1971) demonstrated that diversity (measured as species abundance) decreased when predatory starfish were removed from rocky intertidal areas. These simple exclusion experiments showed that microscopic species could be greatly modified by the presence of a single species which was itself unimportant as an energy transformer. Predation by these species can enhance coexistence between the species by preventing monopolization of some limited resource. For attached benthos, two-dimensional space is a major limiting factor as demonstated by measurable competitive intraspecific interactions which occur when pre-

dators are absent (Connell, 1961; Paine, 1966). Similar studies to test these ideas in the deep sea have not been carried out.

Experiments to evaluate ecological dominance of macroscopic attached algae are described by Dayton (1975). Competitive advantages were discernible for dominant species under various conditions of light and exposure to wave action. Sea urchins were the only herbivore to over-exploit most algal species. Two important carnivores were capable of clearing large areas of urchins, resulting in patches in which algal succession followed. Breen and Mann (1976) describe similar interactions between sea urchins, *Laminaria* (a subtidal seaweed), and lobsters. These keystone species are not only important in determining community structure, however. They are also major energy pathways within the benthic community (Miller and Mann, 1973).

The rate and patterns of species colonization and subsequent development of community structure in habitats initially devoid of species form a body of ecological research which has only recently been extended to include marine benthic communities. Studies of fouling communities, resulting from wartime interest in the condition of ship bottoms, attempted to interpret patterns of colonization and species composition of organisms which were altered by chemical and physical characteristics of surfaces on which attachment occurred. Sieburth *et al.* (1974) used scanning electron microscopy to demonstrate the effect of substrate type on the sequence of microbial colonization. Little information exists about the predictability of these colonization events from a species point of view, however.

When solid substrates, in the form of compressed refuse or wood, are placed in the sea, they become sites of aggregation of various organisms both (mobile and attached) which otherwise remain dispersed. Turner (1973) observed that boring pelecypods attacked wooden blocks within 3 months. Micro-organisms also rapidly colonize nutrient substrates (Jannasch and Wirsen, 1973). The opportunistic formation of aggregations in the deep sea was nicely demonstrated by the use of baited cameras (Dayton and Hessler, 1972). Dispersed scavengers were rapidly concentrated and remained associated as long as food was available. These simple observations show that deep-sea benthic communities, despite their apparent stability, may be structured by patterns of disturbance, but they provide no information of successional changes over time which follow.

Changes in benthic communities with time will be discussed in the following section. However, disturbance through physical erosion during storms may induce recolonization of redeposited or freshly exposed sediment surfaces. Reineck (1968) and Ewing (1973) have demonstrated erosion of sediments at depths greater than 30 m during heavy storms but subsequent short-term changes in benthic species composition following such catastrophes have not been documented. Johnson (1972) postulated that local disturbances of the sediment-water interface could operate to maintain diversity in benthic communities.

Oxygen depletion can occur when circulation patterns prevent advection of water into stagnant basins. The disappearance and recolonization of benthic fauna in the Bornholm Basin of the Baltic was correlated with oxygen depletion and subsequent replacement (Leppäkoski, 1973). High densities of many previously unrecorded species occurred 2 years after oxygen-rich water re-entered the area. Such phenomena could be used to follow changes in benthic communities during recolonization and thus establish quantitative structural relationships between species. Rosenberg (1973) has documented such a succession in benthic macrofauna in a Swedish fjord following closure of a sulfite pulp mill.

6.1.3 ABUNDANCE IN SPACE AND TIME

6.1.3.1 Sampling statistics. Marine benthos, like all other organisms, are seldom distributed uniformly. Even relatively homogeneous areas of sediment can contain dispersed patches of fauna and flora. Problems of sampling such heterogeneous associations are not unique to benthic ecology. Thus it is not surprising that techniques to quantify spatial distribution patterns developed by botanists, geographers, and geologists have been applied to measure certain geometrical properties of distribution patterns (Pielou, 1975). The ecological relevance of geometric distribution patterns may be obscure, however, especially when organisms are compared which vary in size and mobility. Pattern recognition is also strongly dependent on the size of the sampling unit (quadrate size) (Elliott, 1971; Fager, 1972). It is thus impossible to quantitatively assess non-random distribution patterns for all organisms in a community by use of a single sample size.

Inefficiency in collection gear operation and variability in organism numbers due to non-random distribution are separate problems which combine to hinder the accurate assessment of the abundance of benthic organisms. Holme and McIntyre (1971) reviewed collection methods commonly used to sample the benthos. Photography and television permit enumeration of large surface-dwelling invertebrates and fish (Machan and Fedra, 1975). Although large areas may be surveyed, specimens cannot usually be collected. Quantitative collections are usually made by grabs, coring tubes, dredges, or trawls. Whether diver or line operated, the efficiency of all samplers is problematic. Particular substrates or groups of organisms usually require the use of a specific sampler. Once collected, sediments must be sieved, and the organisms removed for counting or otherwise enumerated. No quantitative estimates of abundance are possible without assessment of extraction or counting efficiency. Even when efficient collection methods are used, it is common to observe considerable variation in both species composition and abundance in replicate samples from the same sampling areas. Explanations for aggregated distributions of benthic organisms will be considered below, but patterns of variation can often be due to spatial differences in sediment properties or hydrographic conditions. Often, however, heterogeneity occurs over a small area where these factors are thought not to vary significantly. Tests which provide information on the spatial distribution of organisms can then be used to assess quantitative differences between samples.

Elliott (1971) has summarized statistical measures commonly used to assess spatial distribution patterns of benthos. Tests for various types of frequency distribution: positive binomial ($s^2 < \bar{x}$), Poisson series ($s^2 = \bar{x}$), negative binomial ($s^2 > \bar{x}$) and normal distribution (with transformations) can be used as models for analysing samples from a population. Frequency distributions fitted to these models permit definition of population dispersion in mathematical terms.

A random distribution is usually an initial hypothesis and early studies of macrofauna communities (Clark and Milne, 1955) suggested that this was the general pattern of distribution for most species. However, these authors and others noted that detection of non-randomness depends greatly on the size of sampling units. If these are large or small relative to the size of clumps and clumps are regularly or randomly dispersed, then dispersion appears random. Thus, even small sampling units (500 cm^2) will only detect non-randomness for certain species if there are more than a few individuals in each clump.

The simplest and most widely used test for randomness, the variance-to-mean ratio, or 'index of dispersion' (I) has been described (Section 1.3.1) where

$$I = \frac{\text{sample variance}}{\text{theoretical variance}} = \frac{s^2}{\bar{X}} = \frac{\Sigma(x - \bar{x})^2}{\bar{x}(n - 1)} \tag{127}$$

where s^2 is the variance, $\bar{x}$ the arithmetic mean, and n the number of sampling units. The ratio will tend towards unity for samples from populations with a random distribution. The expression $[I(n - 1)]$ gives a good approximation to χ^2 with $n - 1$ degrees of freedom and thus the significance of departures of the ratio from unity can be tested (Elliott, 1971).

The choice of sample size in analyses of distributions of benthic organisms is critical since it affects the numbers sampled. Counts with low means cannot be reliably fitted to a Poisson distribution. Thus, the smaller the mean for an aggregated species, the more nearly will numbers in samples conform to the Poisson distribution (Gage and Geekie, 1973). Clark and Milne (1955) also observed that evidence of aggregation may disappear in smaller sampling units because the mean per unit is reduced.

A further difficulty with the use of the dispersion index is the dependence of the variance on the mean. With larger and larger sampling area, values for both variance and mean increase but not necessarily in a linear manner. Angel and Angel (1967) and Rosenberg (1974) attempted to standardize for this effect of sample size (and number) by calculating coefficients of dispersion for increasing numbers of sampling units. Both studies demonstrated that I values increased for all species when sample unit size increased. However, the former workers compared these curves with those generated due to changing values of n alone by:

$$1 \pm 2\left(\frac{2n}{(n - 1)^2}\right). \tag{128}$$

These curves were used as baselines or limits against which the significance of aggregation could be tested. Three of the eleven common species analysed were aggregated at all sample unit sizes, five were randomly distributed, and three had fluctuating distributions. Unless a range of sample sizes was used, multiple random sampling was concluded not to be a suitable method for quantifying distribution patterns of benthic populations.

Gage and Geekie (1973) used logarithmically transformed values of variance and mean to calculate coefficients of dispersion for each species in sixteen samples taken at various stations in different Scottish sea lochs. Slope values of regression curves which compared coefficients for each species provide a comparative measure of overall community aggregation. Fauna in samples from muddy sand generally tended to be more aggregated than those in samples from soft mud bottoms.

Few studies have attempted to quantify the distribution of benthos on a within-sampling unit basis (i.e. by subcoring). Rosenberg (1974) concluded that benthic macrofauna populations were aggregated at least on a scale larger than $600 \, cm^2$ (his minimal sample size) since values of I increased when multiple samples were combined. As discussed above, however, this increase would be expected on the basis of increased numbers alone. Cores covering $46 \, cm^2$ were collected by Angel and Angel (1967) and the three species which demonstrated aggregation did so down to the smallest sampling block size (2 cores = $92 \, cm^2$). For smaller organisms, such as meiofauna, McIntyre (1969) suggested that numbers be assessed per $10 \, cm^2$.

Jumars (1975 a, b) tested the assumption that macrofauna species in 0.25-m² box cores from 1200 m off California were randomly dispersed by subsampling twenty-five contiguous samples of 100 cm² each. Dispersion coefficients were calculated for each species to test for strong aggregation of individuals within species. These indices were then summed to calculate a 'total dispersion chi-square' which was partitioned into two components—a 'heterogeneity' value which expressed the proportion each species comprised of the total number of individuals per sample and a 'pooled' value reflecting the total number of individuals per sample. Since contiguous subcores were sampled, a 'joint method' was also used which considered the number of adjacent presences or absences in particular species.

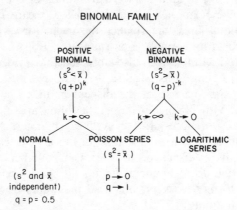

FIG. 72. Relationships between frequency distributions used as models for the three possible relationships between the variance and arithmetic mean of a population. Redrawn from Elliott (1971).

Jumars observed that few species showed aggregation either between or within cores of 100 cm² area. If all species were considered together as replicates, intraspecific aggregation could be detected between cores, but uniform dispersion predominated within cores. Environmental structure, either physical or biological, was assumed to have a 'grain' smaller than 100 cm² and it was concluded that in the stable deep-sea, organisms themselves provide a major source of environmental variability. Relations between the scale of variation and organism size will be considered below.

An alternative to testing for goodness of fit to a random distribution, when the variance exceeds the mean, is to use a negative binomial distribution as a model. This probability series is given by:

$$(q - p)^{-k} \qquad (129)$$

where $p = \bar{x}/k$ and $q = 1 + p$. $\bar{x}$ is the arithmetic mean and k is not the maximum possible number of individuals possible per sampling unit, but is related to the spatial distribution of organisms in bottom samples. $1/k$ is thus a measure of clumping. As the exponent approaches infinity a Poisson series results and as it approaches zero the distribution converges to the logarithmic series. The relationship between normal, Poisson, and logarithmic distribution series is illustrated in Fig. 72.

Negative binomial distributions have been found to adequately describe distribution patterns of some groups of microfauna in subtidal sediments (Gray and Rieger, 1971), while other groups were normally distributed. Gärdefors and Orrhage (1968) used paired samples taken at a constant and known distance from each other to assess macrofauna aggregation. A method of calculating numbers and ratio of patches by 'tieline' sampling was used but results depend on the form and size of aggregates relative to the distance between double samples. Populations of ophiuroids and polychaetes were not randomly distributed and negative binomial curves described observed distributions.

A final group of tests used to measure departures from random patterns of distribution in benthic organisms involve distance measurements between individuals. Indices based on 'nearest-neighbour' measurements require that organisms be stationary and visible (Elliott, 1971).

Levinton (1972a) examined spatial patterns of distribution of a small infaunal deposit-feeding mollusc (*Nucula proxima*) in the laboratory by using X-radiography. Animals were added to homogenized sediments in shallow trays (56–225 cm^2) and their position recorded on X-ray film after 10–89 days. Distances from each animal to its nearest neighbour and from a randomly located point to the nearest animal were recorded. If organisms are randomly arranged, the mean distances of the latter measurement would equal the distance from an individual to its nearest neighbour. A multiple-tube sampler (36 contiguous cores of 17 cm^2 each) was also used to provide a field comparison. *Nucula* showed no tendency to aggregate in experimental trays and random spatial patterns occurred over different elapsed times. There was a slight tendency for field samples to be aggregated, but the total area sampled was up to twice that in trays. Environmental heterogeneity could have induced non-random patterns at this scale.

Harvey *et al.* (1976) calculated cumulative frequency distributions of r^2 values, where r was measured as the distance from each individual to its nearest neighbour, to compare settlement of a bryozoan and an annelid polychaete on plastic discs. Distributions were compared with those expected assuming random spacing. At low densities, the polychaete (*Spirorbis*) showed aggregated settlement, but at high densities spacing apart occurred. The bryozoan (*Alcyonidium*) showed aggregation at all densities with no tendency for spacing.

The spatial dispersion of a population is a fundamental and characteristic property of a species which reflects environmental pressures and behaviour patterns. In addition, population density cannot be adequately assessed without some effort to estimate errors in population parameters (mean and variance). The choice of sample (quadrate) size is critical in this respect. A small sampling unit is usually more efficient than a larger one when contagious distributions occur (Elliott, 1971). However, organism size and scales of obvious environmental heterogeneity affect the choice. In general, clumped distributions occur but the size of sampling units may affect our ability to quantify their size and extent.

Comparisons of organism density from area to area are desirable and numbers per square metre of sediment are usually calculated on this basis for macroscopic species. McIntyre (1969) has suggested that meiofaunal density be assessed for 10 cm^2 sediment surface area as most samples are taken by cores of this area or less. Angel and Angel (1967) concluded that a range of sizes of samplers would be required to adequately assess abundance and patterns of distribution for different species. However, both Hargrave and Peer (1973) and Rosenberg (1974) felt that a quantitative evaluation of abundant macrofauna was obtained by multiple (5–10) random sampling within one area to sample a total area of 0·5 to 1 m^2. Saila *et al.* (1976) calculated that as few as three 0·1-m^2 grab samples would permit an estimate of the five to seven most abundant macrofauna species to a precision of $\pm 50\%$ of the mean with a probability of 90%.

6.1.3.2 Small-scale variation in horizontal and vertical distribution. Principal factors which cause local non-random aggregations of benthic organisms can be grouped as follows:

1. Physical/chemical effects (sediment grain size, oxidation-reduction state, dissolved oxygen, organic content, and light).
2. Biological effects
   (a) food availability and feeding activity,
   (b) predation effects and removal of certain species,
   (c) reproductive effects on dispersal and settlement,
   (d) behavioral effects which induce movement and aggregation.

Variations in numbers or biomass in benthic species which occur over small distances usually reflect effects of at least one of these factors.

Apparently uniform sediment can be composed of a variety of microhabitats of varying size. Lackey (1961) demonstrated this for protozoa and microflora populations in restratified sediments. Aggregations occurred apparently in response to or through facilitating heterogeneity on a microscale of sediment physical and chemical characteristics. Arlt (1973) and Olsson and Eriksson (1974) observed horizontal distribution patterns of natural meiofauna populations which indicated significant differences in abundance between 1–10-cm^2 cores taken adjacently. Schafer (1971) quantified the patchy lateral distribution of living benthic Foraminifera in shallow coastal waters. Species proportions and total numbers in replicate samples were most variable in samples from shallow water where turbulence and sediment movement could occur more frequently. Significantly higher numbers of living specimens were associated with sand-ripple crests and troughs at different times suggesting hydrodynamic transport across the sediment surface. Environmental gradients for these organisms thus appear to occur over distances on the order of a few centimetres.

Studies of benthic organism distribution which have concentrated on small localized areas and attempted to describe and enumerate as many species as possible indicate that different scales of variation exist for different organisms. Level bottom communities identified by Thorson (1957) where a few species are characteristically dominant over a large area seldom exist in shallow water (Parker, 1975). Heterogeneity in habitat features, such as differences in sediment texture, prevent the development of well-defined community associations over broad areas. It has been possible, however, to demonstrate that recognizable benthic species assemblages occur on specific sediment types. A classification scheme (Fig. 73) is generally used to describe differences in sediment texture by reference to the proportion of silt, clay, and sand. Commonly used size ranges of particles in each group are given in Table 33. The total amount (as weight, volume, or number) of particles in various size classes of a deposit may also be measured and size-frequency distributions compared. The scale of size class interval ('grade scale') chosen for grouping is important for comparison of results from different studies. Sheldon (1969) noted that geometric grade scales suggested by Wentworth and Atterberg which grade by factors of 2 and 10 respectively are not readily comparable. Logarithmic transformation of the Wentworth scale to $\phi = -\log_2 d$ where d = diameter (mm) does not physically alter the grade scale. Sheldon suggested that successive grades based on particle volume changes by a factor of 2 could be generally useful since this combines features of both binary and metric diameter scales.

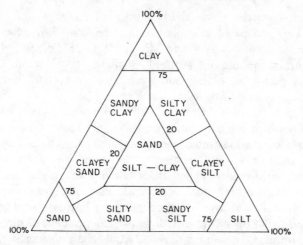

FIG. 73. Classification scheme for sediment texture according to percentage composition of silt, clay, and sand (redrawn from Eltringham, 1971).

Jones (1950) made an early attempt to classify marine benthic communities on the basis of sediment texture. He demonstrated (Jones, 1956) that muddy sand contained a fauna recognizably different and of higher biomass than other grades of deposit in a nearby area. Sanders (1956, 1958, and 1960) identified a soft-bottom polychaete (*Nephtys*)-mollusc (*Yoldia*) community in Long Island Sound on the basis of sediment texture. Deposit feeders were most numerous in areas of fine sediments with a high organic and clay content. Filter feeders, on the other hand, reached an optimum abundance in sediments with a median grain size of 0·18 mm. Substrate movement and optimal amounts of suspended matter were thought to interact to determine optimum sediment texture for the establishment of populations of filter feeders. Bloom *et al.* (1972) made similar comparisons and identified species assemblages of intertidal infauna in which numbers of deposit and filter feeding species were inversely related. The predicted optimal grain size for filter feeders was supported. Hughes *et al.* (1972) demonstrated that 46% of the variance in occurrence of polychaetes and echinoderms in a coastal bay

TABLE 33. COMMONLY USED SEDIMENT PARTICLE SIZE
NOMENCLATURE

		Size (mm)
Gravel	Boulders	> 500
	Cobbles	25–500
	Pebbles	10–25
	Fine gravel	2–10
Sand	Very coarse sand	1–2
	Coarse sand	0·5–1·0
	Medium sand	0·250–0·500
	Fine sand	0·100–0·250
	Very fine sand	0·050–0·100
Silt	Coarse silt	0·020–0·050
	Medium silt	0·005–0·020
	Fine silt	0·002–0·005
Clay	Clay	< 0·002

was associated with an area of soft mud. Numerical methods described above have also identified species groups of macrofauna and demersal fish which correlate with differences in sediment texture (Cassie and Michael, 1968; Field, 1971; Fager and Long-hurst, 1968). In contrast to these observations, Buchanan (1963) found that macrofaunal communities off the east coast of England were poorly correlated with sediment type.

Meiofauna species, because of their small size, might be expected to be more sensitive to changes in sediment texture. McIntyre (1969) reviewed aspects of the ecology of marine meiobenthos which indicated that characteristic fauna occur in sand and mud deposits. Wieser (1960) demonstrated that particular nematode species occurred in muddy sediments. Warwick and Buchanan (1970) confirmed these observations and demonstrated decreased diversity in nematode fauna with an increase in the silt–clay fraction. They hypothesized that more ecological niches are present in sandy habitats. Gage and Geekie (1973) also observed that fauna on current-swept muddy sand was more aggregated than that on soft mud sediments. The presence of more spatial organi-zation in faunal groups in non-mud deposits could be considered indirect evidence for greater niche diversification.

The proportion of silt and clay (in addition to the absolute amounts of either size fraction) is of importance for the distribution of many organisms since porosity and interstitial space are directly controlled by the relative abundance of different sized sediment particles. Driscoll and Brandon (1973) observed that the distribution of a selective deposit-feeding mollusc (*Macoma tenta*) was directly related to the silt:clay ratio. Parker (1975) noted that silty-coarse sand generally supported an order of magni-tude more animals per unit area than silty-clay sediment. These effects may reflect optimum conditions that exist in deposits which are stabilized, yet permit exchange of dissolved nutrients and gases.

Sediments which differ in grain size also differ in numerous properties of significance for organisms. Food availability, discussed below, is related to sediment particle size for many deposit-feeders. Also, sediment porosity, or 'interstitial space' (voids between sediment particles), is critical for small organisms living within the sediment. Fenchel (1969), Webb (1969), and Gray (1974) discussed numerous biologically significant physi-cal properties of marine sediments. Many of these factors are directly controlled by water movement through the sediment which depends on the size, shape, and degree of packing of particles.

Measurements of sediment water content (by drying) or pore volume (the amount of water necessary for saturation) have been used as indices of available space within sediments (Jansson, 1967). Porosity (total pore space) is not necessarily related to average grain size, however, and permeability (the rate of flow from a constant head pressure) may be a more representative measure of spatial relations of particles within sediments (Morgan, 1970). Frazer (1935) suggested that in systematically packed spheres:

$$P = kD^2 \tag{130}$$

where permeability (P) varies directly with the square of the diameter of a sphere (D). Morgan tested this relationship and found that it adequately described permeability through various grades of natural sand except with coarse material (over 2000 μm diameter).

Webb (1969) and Morgan (1970) have also discussed spaces which theoretically exist when spheres of equal diameter are packed. Frazer (1935) suggested that the diameter ratio of smaller spheres capable of passing through 'throats' between larger spheres ('critical ratio of entry') varied between $0.15D$ for tight packing and $0.41D$ for loose packing (where D is the diameter of the large sphere). Similarly, a 'ratio of occupancy' existed as the ratio of the diameter of a sphere too large to pass between larger spheres but capable of existing within a void without disturbing packing and the diameter of the larger spheres. This ratio varied between $0.44D$ and $0.72D$ for tight and loose packing respectively. Morgan (1970) demonstrated that maximum diameters of inter-stitial amphipods (*Pectenogammarus*) capable of burrowing and inhabiting various grades of substrate corresponded to calculated diameters of 'throats' between sediment particles. Amphipods had a preference for substrates which permitted free movement through voids.

Direct pore-size measurements on sections of resin-impregnated sediments have indi-cated that interstitial space may be partly filled by small particles. Williams (1972) observed that this occurred and that the abundance of major faunal groups was related to pore-size distribution in four different sediment types. Positive correlations existed between animal size and pore diameter for all groups of interstitial fauna. Although total biomass of interstitial fauna was the same in all grades of sediment, the estimated volume of animals/voids was higher in samples which contained silt.

Several authors have attempted to compare the quantitative importance of benthic micro- and macrofauna in various types of sediments. Thorson (1966) and Fenchel (1969) reviewed many of these studies. Wigley and McIntyre (1964) observed that the relative quantity of macrobenthos compared to meiobenthos was similar in marine sediments from different areas. The overall ratio of macro- to meiobenthos was 1:70 by numbers and 24:1 by weight. Numbers were generally higher on fine sandy sediments but macro-fauna biomass was highest on fine deposits. Although these authors found no clear relation between these ratios or absolute abundance and sediment grain size, and com-parison with other studies (Table 34) confirms this, numbers of meiofauna and macro-fauna are generally much more variable than biomass estimates. Sanders (1960) con-sidered numbers a more valid estimate of community size for benthic fauna than biomass since less than 1% of fauna numbers in Buzzards Bay sediment constituted over 50% of the total weight. However, ratios of biomass of macrofauna:meiofauna are much less variable than numbers (Table 34) suggesting that different communities with different taxonomic composition may conform to some uniformity in faunal size composition.

Fenchel (1969) estimated metabolic rates of different size groups of benthos to provide an alternative comparison of relative importance. Microfauna, particularly ciliates, pre-dominated metabolism in sediments containing fine grain sizes. Nematodes were respon-sible for the bulk of animal respiration in sand finer than 100 μm, while other meiofauna (harpacticoids, ostracods, rotifers) were dominant in coarse sediment (<250 μm). In com-paring the relative importance of different size groups of benthic fauna, however, Fenchel noted the apparent absence of animals in the 10^{-5}–10^{-4} g weight category which marked the separation of macro- and meiobenthos. The lack of the size group was not thought to be due to inefficient extraction methods, but rather to reflect the division between macrofauna large enough to burrow and microfauna small enough to live interstitially. Thus, while descriptions of pelagic communities on the basis of particle size (Section 1.1) may demonstrate elements of community structure due to predator–prey relation-ships (Sheldon *et al.*, 1972) similar classification of benthic organisms (by size or weight) may relate more to the physical structure of the habitat.

No quantitative studies of marine sediments have been made which enumerate all organisms. Sanders (1960), Fenchel (1969), and McIntyre (1969) have discussed problems

TABLE 34. PROPORTIONS BETWEEN VARIOUS SIZE GROUPINGS OF BENTHIC FAUNA ON A NUMERICAL, FRESH WEIGHT AND ESTIMATED METABOLIC RATE PER SQUARE METRE BASIS IN SEDIMENTS FROM DIFFERENT AREAS

Animals weighing more than 10^{-4} g but less than 5 g each are considered as macrobenthos infauna (Wigley and McIntyre, 1964; Fenchel, 1969). Ratios refer to macrobenthos:meiobenthos (and ciliates when considered separately)

Location sediment type	Proportions			Reference
	Numbers	Biomass	Metabolism	
English Channel				
(coarse silt 20–50 μm)	1:60	90:1		Mare, 1942
Buzzards Bay				
(clayey silt 2–50 μm)	1:30–1:140	33:1		Wieser, 1960
Fladen Grounds, North Sea				
(coarse silt 20–50 μm)	1:40	40:1		McIntyre, 1961
Loch Nevis				
(clayey silt 2–20 μm)	1:100	15:1		McIntrye, 1961
Martha's Vineyard				
(coarse sand 1470 μm)	1:35	14:1		
(fine medium sand 250 μm)	1:180	18:1		Wigley and McIntyre, 1964
(fine sand 130 μm)	1:170	22:1		
(coarse silt 40 μm)	1:770	37:1		
Ålsgårde Beach				
(fine sand 175 μm)	1:40:1500	190:1·5:1	2·7:1:1·4	
Helsingør Beach				
(fine sand 200 μm)	1:28:3000	3·9:1·6:1	1:3:8	Fenchel, 1969
Nivå Bay				
(medium sand 350 μm)	1:10:50	170:10:1	4:2:1	

resulting from the lack of standard methods. This particularly applies to estimates of sediment microflora (bacteria and algae) abundance. Bacteria predominate all benthic communities in both numbers and metabolic importance. Fluorescence staining and direct microscopic counts (Wood, 1955), colony counts on nutrient plates, or chemical methods specific for cell wall muramic acid (Moriarty, 1975) have demonstrated that densities *per unit volume* usually exceed those in sea water by several orders of magnitude (ZoBell, 1946). Only recently, however, have densities been calculated per unit area to allow comparison between different substrates. Marine silt, sand, and pebbles have comparable densities (100–400 μm^2 per cell) (Dale, 1974). This approaches Mare's (1942) estimate of 3×10^{11} cells/m^2 in marine mud. Hargrave (1972d) replotted ZoBell's (1946) data and demonstrated an inverse logarithmic relation (slope of -1.0) between average sediment particle diameter and bacteria numbers determined by plate counts. Dale (1974) observed a similar relationship for direct counts of bacteria on intertidal sediment grains. Over 80% of the variance in abundance was accounted for by difference in particle grain size alone.

While accessible space in sediments is an important feature for all sizes of organisms in sediments, high permeability also permits a greater influx of water containing dissolved nutrients, oxygen, and particulate matter. Steele *et al.* (1970) suggested that drainage in littoral areas and pumping due to wave action were dominant methods of introducing water into interstitial systems. Riedl and Machan (1972) used thermistor probes to provide the first direct measurements of hydrodynamic flow patterns within marine sediments. Their observations on intertidal high-energy beaches demonstrated flow patterns which varied in a predictable although complex way with tidal cycle. Interstitial currents were variable and reached high velocities (up to 2 mm/s) relative to the modal body length of the fauna (1.3 mm). Maximum water movement by wave action and tidal pumping through sediments would be expected on exposed beaches, however, but similar although less intense flux may occur in deeper water (Riedl *et al.*, 1972).

Water movement through sediment is of primary importance in determining oxygen supply to benthic systems. Oxygen uptake by aerobic heterotrophic organisms, particularly micro-organisms, which are numerous and have a high weight-specific metabolism, rapidly depletes dissolved oxygen within surface sediment layers. A high rate of supply is necessary to prevent anoxic conditions. Despite the supply of oxygen, in all but

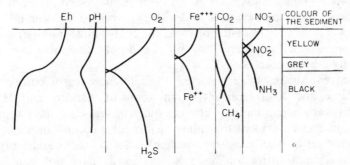

FIG. 74. Diagrammatic vertical profiles of chemical conditions within marine sediments. Redrawn from Fenchel (1969).

exposed beaches, a reduced sulfide-containing zone underlies the oxidized layer in marine sediments (Fenchel, 1969; Fenchel and Riedl, 1970). Partial oxidation of organic matter results in the accumulation of reduced compounds which diffuse upwards towards the surface. The boundary between oxidized and reduced conditions is operationally measurable as a discontinuity in vertical profiles of oxidation-reduction or sulfide ion potentials using platinum or silver electrodes (Whitfield, 1969). Since the sulfur cycle of bacterial conversion of sulfate to hydrogen sulfide often predominates anaerobic decomposition in marine sediment (Section 6.4.1.1), redox potentials and sulfide activity may be correlated and used jointly to indicate the presence or absence of reducing conditions.

Numerous studies have shown that the absence of oxygen, or presence of sulfide, is a significant condition affecting the distribution of bottom organisms (Theede *et al.*, 1969; Fenchel and Riedl, 1970). Wieser and Kanwisher (1961) related the vertical distribution of nematodes in a salt marsh to oxygen content and oxidation-reduction (Eh) potentials within the sediments. Fenchel (1969) reviewed previous studies and demonstrated that both microfauna and microflora in sediment could be characterized by their vertical zonation relative to oxidation-reduction potentials and associated chemical conditions (Fig. 74). Striking discontinuities occur at the interface between oxidized and reduced sediment layers similar to those observed across the interface of aerated–deoxygenated water in stagnant fjords, the Black Sea, and unmixed lakes. The thickness of the oxygenated layer, or conversely the upward extent of sulfide in sediment, can be measured as the depth of the 'redox discontinuity'. Its depth is determined by a balance between processes which move dissolved oxygen into surface sediments and those which consume it. Since the oxidative state of sediments reflects and results from the interaction of numerous physical, chemical, and biological (particularly by micro-organisms) effects, Eh measurements provide a useful integrative parameter of ecological conditions within sediments.

Sediment organic matter, either as a nutrient source or an attractive factor which induces settlement, has been related to the distribution of benthic organisms. Organic matter may enhance bacterial activity in localized areas of sediment which can result in an accumulation of partially oxidized metabolites and a lowering of Eh potentials (ZoBell, 1946; Wood, 1965; Hallberg, 1973). Distinction must be made, however, between non-living organic matter (debris) in sediments and that associated with organisms, particularly bacteria. Since Newell's (1965) studies with deposit-feeding molluscs (Section 6.4.2), organic matter in micro-organisms is generally viewed as being the food supply for invertebrates which ingest sediment. Much of the total organic content in deposits occurs as refractory compounds such as lignin or cellulose which are not considered to be digestible by most invertebrates. Bader (1954) reported a linear relation between sediment pelecypod (mollusc) populations and the acid-soluble (non-refractory) fraction of sediment organic matter. There was no significant relation between population densities and total organic content.

Many subsequent studies have observed a similar absence of correlation between faunal abundance and total sediment organic content. Sanders and Hessler (1969) pointed out that variation in biomass between different areas is usually so much greater than variation in organic content that other factors must control distribution patterns. Little information exists concerning small-scale variations of sedimentary organic matter. Studies of fauna microdistribution discussed previously have not been accompanied by measurements of organic content. Bottom sediments in the deep sea contain only a fraction of 1% of organic carbon (Riley, 1970). If deposition occurs as fine particulate matter, a rather homogeneous distribution pattern on a small scale should exist which would not conform to differences in faunal abundance in closely spaced samples (Jumars, 1975 a, b).

Despite the desirability of differentiating available and non-assimilable fractions of sediment organic matter for more meaningful comparisons with fauna distributions, several authors have observed linear correlations between biomass of certain inverte-brates and total sediment organic carbon and nitrogen (Newell, 1970; Longbottom, 1970). The comparisons are usually confounded, however, by significant inverse correla-tions between sediment particle size and organic content. Thus, regression lines for biomass on particle diameter are almost identical to those for organic matter (Longbot-tom, 1970). No quantitative evaluation of the effects of organic content *per se* separate from associated changes in particle size on abundance of deposit-feeding benthos appears to have been made.

Organic compounds in sediments and associated with solid surfaces attract benthic organisms. Experiments to test for substrate selection by various benthic invertebrate species (Gray, 1966, 1967) indicated that organic films produced by bacteria attached to particle surfaces can render sediments attractive. Organic films of different origin can also be differentiated. Numerous studies of settling behaviour by planktonic larvae of benthic invertebrates (oysters, mussels, and barnacles) have demonstrated correlations between settlement activity and the presence of acidic proteins or protein–carbohydrate complexes which may be species specific (Meadows and Campbell, 1972; Larmen and Gabbott, 1975).

Interrelations between sediment grain size, organic content, and oxidation-reduction conditions as they affect total numbers (by direct counts) of bacteria in sediments were quantified by Dale (1974) (Fig. 75). Water motion (waves and tides) was considered to be the dominant influence affecting accumulation of organic matter, grain size, and levels of oxygenation (Eh potentials) in sediment. Partial correlation coefficients indicated that strong correlations between bacterial numbers, carbon, and nitrogen were reduced but still significant when grain size was controlled. Thus, not all of the covariance between bacterial numbers and organic content was attributable to a bacteria:grain size relationship alone. The lack of a significant correlation between Eh and bacterial numbers probably reflects the diverse relationships which micro-organisms have with this variable (Bass Becking *et al.*, 1960).

Many studies have considered the effect of other abiotic factors which can affect the distribution of benthic organisms. Tolerance and preference experiments have demonstrated that with respect to salinity two broad classes (euryhaline and stenohaline) of response exist (Jansson, 1968). Variation in this factor is greatest in estuarine waters

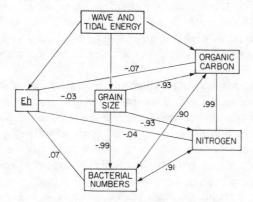

FIG. 75. Inter-relationships indicated by correlation coefficients for linear regressions between total bacterial numbers, sediment organic carbon and nitrogen, oxidation-reduction potentials and assumed effects of water movement within marine intertidal sediments. Redrawn from Dale (1974).

where brackish surface water may periodically reach organisms normally exposed to higher salinities. Gage and Tett (1973) considered that stress induced by reduced and unpredictable (fluctuating) salinity with distance along fjord-like inlets might affect pelagic larval stages of benthos. Since altered salinity conditions exist mainly near the surface, however, effects after settlement would be minimal. In any case, changes in factors like salinity and temperature would not be expected to have localized effects which would induce changes in small-scale patterns of distribution.

Light is a factor of considerable importance to the vertical distribution of some benthic organisms. Sediments which lie within the photic zone (intertidal to 150 m in tropical oceans) usually contain both algal and photosynthetic bacterial populations which are vertically highly stratified. Intact pigments and various breakdown products are also preserved within sediment layers (Orr *et al.*, 1958; Vallentyne, 1969). Patterns of vertical distribution result from a combination of effects. Although many types of algae and bacteria in sediment are capable of restricted movement (up to several millimetres) in response to light and tidal changes, living cells may occur to several centimetres depth (Taylor, 1964; Fenchel, 1969). Pigmented cells in deep aphotic sediments could be displaced from the surface by sediment mixing through physical or biological reworking.

Fenchel and Straarup (1971) measured light penetration and the vertical distribution of various photosynthetic organisms in shallow-water marine sediments. While several other studies have documented the presence of photosynthetic pigments in sediments

(Taylor, 1964; Steele and Baird, 1968: Pamatmat, 1968), no previous investigation had compared relations between adsorption of different wavelengths of light and pigment distribution. A microphotometer (1 × 1 mm) with filtered white light was used to demonstrate that long-waved (near infrared) light penetrated farthest in pigment-free quartz sand. Light intensity was reduced to 50% of surface values at 0·76 mm. Photosynthetic bacteria show maximum absorption at these wavelengths and their vertical distribution is restricted to sediment layers where reduced compounds (H_2S, H_2, or various reduced organics) exist close enough to the surface to permit exposure to light. In deep water infrared light is absent due to absorption within the water column and only shorter wavelengths reach the sediment surface. Fenchel and Straarup observed that blue light was attenuated the most in bleached sand probably reflecting increased refraction and scattering expected with shorter wavelength. The depth of the photic zone (light intensity $< 1\%$ of surface values) in silt and very fine sand was about half that observed in coarse sand. The vertical distribution of living algal cells and chlorophyll a corresponded and indicated maximum light penetration in the upper 6 mm of shallow-water sand sediments.

While abiotic conditions may affect distribution patterns on a small scale, the benthos itself alters physical and chemical features in sediments. These effects may induce heterogeneity on a scale comparable to habitat dimensions of the organisms themselves. Reworking, through digging and burrowing, and feeding activity (tube building, particle selection, fecal production) creates structures which locally modify sediments. Fager (1964), Mills (1967), and Young and Rhoads (1971) described examples of the consolidating effect of tube-building macrofauna on sediment. Increased surface area, both external and internal, provide attachment surfaces for other organisms which can serve as a food source. This will be discussed (Section 6.4.2) but irrigation and particle selection by deposit-feeding fauna enhances the growth of digestible micro-organisms (Hylleberg, 1975). Fecal pellets are usually enriched with organic matter in such organisms (Hylleberg and Gallucci, 1975). Certain invertebrates may also stimulate bacterial growth by mucus production during feeding (Newell, 1970; Fenchel et al., 1975). The spatial extent of such enrichment would correspond to the amount of feeding activity which for many deposit-feeding invertebrates is proportional to body size (Hargrave, 1972c).

Burrowing and feeding activity of benthic fauna may also lead to less consolidated sediments. Mills (1967) described the sediment-sorting effect of feeding by a tube-dwelling, selective deposit-feeding amphipod Ampelisca abdita. While tube formation stabilized sediment and fecal pellets consolidated ingested fine material, feeding increased the proportion of fine particles at the sediment surface, leading to increased sediment instability within aggregations. Rhoads and Young (1970) suggested that 'trophic amensalism' be used to describe organism interactions whereby the feeding of one species makes the habitat unsuitable for another. The process was demonstrated by the unstable sedimentary conditions created by deposit-feeders which restricts suspension feeding and attachment by sessile epifauna. Rhoads (1973; 1974) and others have noted that biologically reworked sediments are usually more easily suspended than consolidated deposits not exposed to bioturbation. Rowe et al. (1974a) attributed high rates of erosion at the upper end of Hudson canyon to excavation activity of large decapod crustaceans. Holothurian populations of only a few individuals per square metre increased water content, vertical sorting to 20 cm, and produced topographical relief (Rhoads and Young, 1971). In such locations, the activities of a single large animal may dominate the physical structure of the sediment for smaller more abundant organisms. Even when the abundance of large fauna is reduced, as in the physically constant deep sea, organisms themselves provide a major source of environmental variability (Jumars, 1975 a, b, 1976).

The role of predation as a structural force in benthic communities is demonstrated by results of exclusion experiments and observations concerning the existence of keystone species in certain habitats (Section 6.1.2.1). Dayton *et al.* (1974) observed that space monopolization in Antarctic epibenthic sponge populations was controlled by asteroid and nudibranch predation. Mann (1973) described interactions between lobsters, sea urchins, and kelp in a marine coastal area which demonstrated destruction of seaweed beds by sea urchin grazing. A balanced predator–prey relationship between lobsters and sea urchins which can be upset by exploitation of lobster stocks was hypothesized (Mann and Breen, 1972; Breen and Mann, 1976). Similar balanced relationships may exist in the deep sea. Dayton and Hessler (1972) advanced the hypothesis that high faunal diversity on the sea floor is maintained by predictable disturbance through cropping. This permits the coexistence of species which share the same limited resources (thought to be food supply). Jumars (1975b), however, disagreed with the suggestion that predators in the deep sea operate in a homogeneous environment. He viewed predation as one of many organism-induced effects which create a highly variable environment. Levinton (1972b) also considered competitive interactions to be important between coexisting deposit-feeders. Fager (1964) observed the destruction of a dense bed of tube-building polychaetes (*Owenia*) possibly due to feeding activities of bat rays. Levings (1973) demonstrated directly that dominance structure in a benthic macrofauna community in a shallow marine bay was altered by fish predation. Absolute numbers and relative abundance of macrofauna species changed following the onset of spring feeding by the flatfish *Pseudopleuronectes*. Changes in community structure induced by predation were superimposed over changes due to natural seasonal life-cycle patterns (Levings, 1975).

Griffiths (1975) reviewed studies which demonstrated that prey size and abundance are dominant factors controlling prey 'availability' to predators. He concluded that many invertebrates and larval vertebrates eat prey as encountered while adult vertebrates feed to maximize energy gain. Even 'passive' deposit-feeding invertebrates may demonstrate size-dependent food uptake. *Phascolosoma* apparently ingests sediment non-selectively but size of ingested meiofauna (nematodes, copepods) was positively correlated with sipunculid size (Walter, 1973). Prey size frequency distribution and spatial distribution are thus major factors determining susceptibility of many organisms to predation.

Fauna which crop micro-organisms also enhance bacterial turnover rates and nutrient cycling in sediment. Hargrave (1976) demonstrated that fecal pellets produced by a variety of benthic invertebrates are foci of increased oxygen uptake due to enhanced bacterial growth. Barsdate *et al.* (1974) observed that protozoa grazing on detrital bacteria increased mineralization rates and phosphorus cycling. Cropping could have a localized effect in selection for rapidly growing micro-organisms at sites of more intensive feeding.

Thorson (1957) suggested that aggregations of many benthic species which produce pelagic larvae reflected settlement in restricted areas of suitable substrate as a result of favourable hydrographic conditions. Recent studies have directly linked the abundance of some species of macrobenthos to settlement success. Peer (1970) observed a large aggregation of the tube-dwelling polychaete (*Pectinaria*) which did not reappear in subsequent years (Peer, pers. comm.). Buchanan *et al.* (1974) documented recruitment failure of polychaete (*Ammotrypane*) and mollusc (*Abra*) populations which were accommodated by increased abundance of other species. Eagle (1975) also observed dramatic variation in community structure which appeared to reflect settlement success. Non-selective deposit-feeders produced an unstable substrate which was unsuitable for spat or juveniles while filter feeders and selective surface feeders did not prevent juvenile settlement. Interactions of dominant species through adverse effects of feeding on larval and juvenile survival restricted dominant animals to single-year classes. Such environmental modification, which may account for non-overlapping distributions and the formation of infaunal suspension-feeding and deposit-feeding assemblages, is consistent with Sanders' (1958, 1960) and Rhoads and Young's (1970) observations of specific animal-sediment relations. Woodin (1976) reviewed evidence of adult–juvenile macrobenthos interactions through competition for space or food. Interference through sediment destabilization was considered a major factor in maintaining separation of assemblages of burrowing deposit feeders, suspension feeders and tube-building infauna.

Buchanan *et al.* (1974) classified species of dominant benthic macrofauna on the basis of fluctuations in abundance from year to year. They differentiated

1. volatile species (alternate between high and low densities),
2. opportunistic species (high reproductive potential capable of rapid increase in numbers in response to elimination of previously abundant species), and
3. conservative species (stable population densities unaffected by fluctuations of volatile species).

Opportunistic species tend to be of small body size ($< 500 \mu$m) which permits a rapid increase in numbers (Mills, 1967). Fenchel (1974) demonstrated that the intrinsic rate of natural increase per day was inversely related to body weight for a variety of organisms. Thus, micro-organisms (bacteria, protozoa) are capable of rapid increases in abundance when local conditions favour growth. Larger metazoans have much lower reproductive potentials.

Spacing between individual benthic organisms, particularly sedentary species, has been related to intraspecific competition for space. Knight-Jones and Moyse (1961) reviewed early studies which indicate how spacing behaviour during larval settlement serves as an adaptation to prevent overcrowding. Hargrave (1972d) and Hargrave and Phillips (1977) discussed spatial densities of micro-organisms attached to surfaces. Bacteria cover only a few percent of natural particulate surfaces in various types of deposits and although density can be increased through nutrient enrichment, cell distribution tends to be over-dispersed. Holme (1950) found over-dispersion of the surface deposit-feeding mollusc *Tellina* which suggested separation due to foraging activity in a territory around each individual. Crowded populations were more randomly distributed. Mobile deposit-feeders like *Nucula*, which may not benefit from territoriality, also tend to have random distribution patterns (Levinton, 1972a).

Fenchel (1975) suggested that distribution patterns of local subpopulations of four species of mud snails (Hydrobiidae) in a complex estuarine environment resulted from interactions of habitat selection, dispersal and colonization rates, interspecific competition and extinction. The relation between body size and particle size was the same for three of the four species which were of similar size when they occurred separately. However, when stable populations coexisted, the size-frequency distribution of the species differed sufficiently to permit feeding on different sediment grain sizes. Character displacement was thought to reduce competition by permitting partitioning of a limited food resource. Such a view of dynamic interactions between marine benthos and their biotic and abiotic environment contrasts Thorson's (1957) concept of fixed and stable organism assemblages. It also confirms Mills' (1969) suggestion that a range of types of association occur from closely associated species groups to loosely integrated aggregations due to co-occurrence. Small-scale patterns of distribution in coexisting species may reflect the nature and degree of association in such groupings.

6.1.3.3 Geographic variation.

A century ago it was commonly believed that no life existed in any ocean below 550 m. The idea, termed an 'azoic hypothesis', originated from dredgings in the Mediterranean in 1841 by Edward Forbes (Merriman, 1965). It was described by Spratt and Forbes (1847) as follows: "As we descend deeper and deeper in this region, its inhabitants become more and more modified, and fewer, indicating our approach to any abyss where life is either extinguished, or exhibits but a few sparks to mark its lingering presence". Dredgings by Sars, Thompson and Carpenter off Norway and Scotland, the recovery of animals attached to cables raised from the Mediterranean and bottom samples collected during the voyage of the *Challenger* (1872–6) demonstrated that organisms lived at least to 5500 m depth (Mills, 1973). The fauna collected, albeit inefficiently, was diverse, widely distributed and generally of small size with many carnivorous species. The conclusion that these features were adaptations to a limited food supply in the deep sea (Honeyman, 1874; Murray, 1895) has only recently been substantiated.

Wherever light sufficient for photosynthesis reaches sediments or solid substrates, attached and epibenthic algae and macrophytes occur. Kelps are found below low-tide level on almost all temperate rocky shorelines while sea grasses occur in more sheltered areas of fine sediment (Mann, 1972a). Mangrove swamps are also extensive in tropical latitudes throughout the world. Organic matter produced by these plants in both particulate and dissolved form enriches all coastal waters and provides a detritus supply that may extend to the deep sea (Menzies and Rowe, 1968). The importance of light, substrate, and nutrient supply for production of benthic plants will be considered below (Section 6.4.1.1).

The effect of nutritional factors on geographic distributions of numbers and biomass of marine benthic fauna has been demonstrated by qualitative (trawls, dredges) and quantitative (cores and grabs) sampling. Sample collections at different depths and distances along transects from shore (Sanders *et al.*, 1965; Rowe, 1971b) and at widely spaced stations in the open ocean (Zenkevitch, 1963; Sokolova, 1972) have demonstrated decreased macrofaunal numbers and biomass with increased water depth. Macrobenthos biomass off the Pacific coast of the U.S.A., for example, varies between 20 and 200 g/m^2 (fresh weight) at depths between 400 and 1200 m. Average values of 1·4 g/m^2 occur below 2000 m and 0·2 g/m^2 below 4000 m (Vinogradova, 1962). Sokolova (1972) differentiated eutrophic and oligotrophic oceanic regions on the basis of differences in macrobenthos feeding type (Fig. 76). Eutrophic conditions in peripheral and equatorial regions result in a dominance of deposit-feeding macrobenthos in sediments which contain excess assimilable organic matter. Oligotrophic areas occupy central areas of the three oceans where low concentrations of more refractory sedimentary organic matter results in a predominance of suspension feeders. The extent of these regions correspond, in a general way, to the distribution of primary production on a global scale (Section 1.3.2).

Depth and the amount of food material added to sediments from overlying water are inversely related since decomposition removes utilizable energy during sinking (dis-

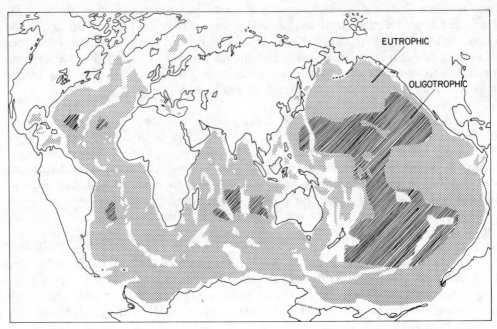

FIG. 76. Trophic regions in the deep sea (depths greater than 3000 m) inferred from biomass of deposit-feeding or suspension feeding macrobenthos in trawl hauls. Redrawn from Sokolova (1972).

cussed in Section 6.4). Depth *per se*, or some factor related to it, however, seems to be the most critical overall factor limiting benthic biomass. Sanders and Hessler (1969) observed that changes in faunal composition over a vertical distance of 100 m were equivalent to those which occur horizontally over thousands of kilometres. Also, abrupt faunal discontinuities tend to occur at shallow depths (100–300 m). These changes are usually not paralleled by large changes in total sediment organic content perhaps indicative that this is a poor measure of sediment nutritive value. Changes in fauna biomass below a few hundred metres depth, however, do generally correspond to the vertical distribution of other parameters which decrease rapidly with increasing depth.

Several studies have demonstrated that organic carbon and total particulate volume decrease rapidly in the upper 200 m (Hobbie *et al.*, 1972), but vertical distributions (plotted arithmetically) appeared constant with depth below 200 to 800 m (Riley, 1970; Gordon, 1971; Menzel, 1974). Sheldon *et al.* (1972), however, observed that total particle volume decreased with depth when data were plotted on logarithmic axes. Wangersky (1976) and Gordon (1977) also compared numerous depth profiles of particulate carbon and nitrogen from different areas on this basis but the rate of decrease with depth was not quantified. Vinogradov (1968), Rowe (1971b), and Rowe *et al.* (1974b), on the other hand, used exponential and semi-logarithmic relations to linearize depth-dependent decreases in macrozooplankton and macrobenthos biomass (numbers and weight). Johnston (1962) also used an exponential curve to describe vertical distributions of fish and plankton biomass.

Vinogradov (1968) emphasized that generalized curves for vertical distributions of zooplankton biomass from the surface to maximum depths may transect several depth zones. Depending on the water mass, surface, intermediate midwater, and abyssal sectors may be differentiated and a continuous decrease with depth does not occur. Despite these discontinuities, linearized profiles permit comparison of depth-related decreases. Hyperbolic, logarithmic, and exponential curves may describe depth profiles equally well. A power function, however, provides the best fit to depth profiles of various measures (Table 35). Even though only a few data points were used to calculate some curves, slope coefficients show that the rate of decrease of planktonic and benthic biomass is much greater than that of parameters related to suspended particulate matter. Depth-related decreases in macrobenthos thus appear to be more closely linked to suspended living biomass than to total particulate matter. The similarities may also indicate that food sources for zooplankton and benthos are the same (Rowe *et al.*, 1974b).

Sanders and Hessler (1969) noted that macrobenthos density decreased rapidly on the North Atlantic continental slope (the region of greatest rate of depth change, Fig. 67) and thus depth, rather than distance from land, appears to determine food availability to benthic organisms. Differences in primary production in surface waters are thought to be slight in this region and thus a lower food supply must account for reduced biomass with depth. Regional differences in primary production, however, can

TABLE 35. INTERCEPT (a) AND SLOPE (b) COEFFICIENTS OF LINEAR REGRESSION CURVES COMPARING $\log_{10}$ VARIOUS PARAMETERS (x) WITH $\log_{10}$ DEPTH (y)

Variable	Depth range (m)	No. of data points	a	b	Location and reference
Particle volume (ppm)	1–4500	47	0·63	−0·35	Various stations in North Atlantic and South Pacific (Sheldon *et al.*, 1972)
Total organic carbon (μg/l)	35–4000	6	9·33	−0·37	Central Pacific (Station Gollum) (Gordon, 1971)
Particulate organic carbon (μg/l)	25–3000	5[b]	105·4	−0·52	North Atlantic (Gordon, 1977); Central Pacific (Gordon, 1971)
Particulate nitrogen (μg/l)	25–3000	5[b]	14·7	−0·56	North Atlantic (Gordon, 1977);
	25–4000	6[b]	46·9	−0·70	Central Pacific (Gordon, 1971);
	0–4000	5[b]	642·0	−0·69	South Atlantic, Pacific
	0–4500	5[b]	180·0	−0·48	(Wangersky, 1976)
DNA (μg/l)	25–3000	9	8·8	−0·55	North Atlantic (Gulf Stream) (Holm Hansen *et al.*, 1968)
ATP (mμg/l)	10–1025	10	3810	−1·09	California coast.
	0–2400	19	10177	−1·30	Peruvian coast (Holm Hansen, 1970b)
'Living carbon' (ATP equivalent as percent of total)	10–1025	10	599	−0·844	California coast (Holm Hansen *et al.*, 1966)
Net mesoplankton biomass (mg/m³)	25–6000	8	5383	−1·21	Tropical Pacific (oligotrophic);
	25–7000	9	40044	−1·39	Tropical Pacific (eutrophic);
	50–7300	9	62332	−1·18	Southwest Pacific;
	25–6000	9	84313	−1·16	Subarctic;
	25–3500	8	4150	−1·12	North Atlantic (Sargasso Sea) (Vinogradov, 1968).
	126–6225	9	13875	−1·12	Peru–Chile Trench (Frankenberg and Menzies, 1968);
Macrobenthos	200–3900	8	36·7	−1·70	Gulf of Mexico, North Peruvian coast (Rowe, 1971b)
biomass	200–6300	8	2547	−1·27	
	30–4950	22	62098	−0·58	Northwest Atlantic shelf to
(g/m²)	550–2080	5	$1·5 \times 10^7$	−1·24[a]	abyssal plain (Rowe *et al.*, 1974b)

[a] Continental slope stations only.
[b] Means derived for depth intervals.

account for some variation in benthic biomass. Rowe (1971b) observed that average biomass was higher in regions of higher phytoplankton production. Hargrave and Peer (1973) also noted a direct linear correlation, which was most significant during the spring (March–May), between benthic macrofauna biomass and average chlorophyll *a* concentration per unit volume in the photic zone (Fig. 77). Lower correlation coefficients were obtained when the comparison was made with average chlorophyll standing crop

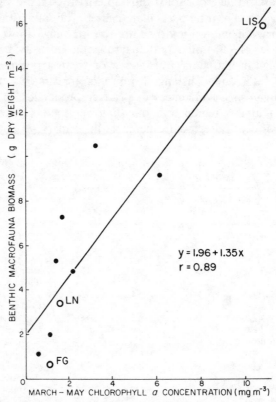

FIG. 77. Comparison of benthic macrofauna biomass and average chlorophyll *a* concentration per unit volume in the photic zone during March–May in various areas. Points refer to sampling locations in the Gulf of St. Lawrence and in coastal inlets in Nova Scotia. Open circles from previously published data collected in the North Sea (Fladen Grounds), Loch Nevis and Long Island Sound. Redrawn from Hargrave and Peer, 1973.

during summer and fall months. The relationship implies that in inshore coastal areas bottom fauna biomass may be determined primarily by sedimentation of organic matter produced during a spring bloom.

Nesis (1965) observed that the biomass of macrobenthos on continental shelves in the Pacific and North Atlantic was maximum in areas like the Grand Banks where shallow depths ensured direct input from phytoplankton production. However, Lie (1969) found no correlation between depth and standing crop off the Washington coast, although mean biomass offshore was less than half that of Puget Sound where primary production was considerably higher. The lack of a significant effect of depth on density when sampling is restricted to continental shelves (> 200 m) suggests that at these shallow depths factors not necessarily related to depth affect biomass. Often, maximum densities occur at the outer edge of a continental shelf which may be regions of increased sedimentation or enhanced primary production (Rowe and Menzies, 1969; Carey, 1972).

The impingement of an oxygen minimum layer onto a continental slope, or reduced oxygen levels in deep basins with restricted circulation, may limit total biomass and species composition in benthic communities. The zonation of benthic meiofauna in the northern Baltic is clearly related to the availability of dissolved oxygen in overlying water (Elmgren, 1975). Abundance and diversity decreased with depth with only nematodes tolerant of permanently reduced oxygen concentrations below 100 m depth. Frankenberg and Menzies (1968) also observed that benthos off Peru subject to oxygen stress was dominated by a few species of a small size. High densities of organisms may occur in such areas, however, which could reflect an ability to use excess assimilable organic matter accumulated in sediments not well supplied with oxygen (Richards and Redfield, 1954).

Studies demonstrating decreased macro- and meiobenthos density, measured by numbers and biomass, with increasing depth and distance from shore have been reviewed by Thiel (1975). He observed that meiofauna numbers per unit area decrease at a slower rate than those for macrofauna with increasing depth. Thus, the average size of benthic metozoans decreases with depth. A general reduction in body size in organisms in the deep sea or areas subject to oxygen depletion may reflect a morphological adaptation to limitations in growth imposed by restrictions in food and oxygen supply. Energy expenditures by small-bodied organisms are primarily for growth and reproduction with little used for maintenance metabolism (Fenchel, 1974). There is thus a distinct energetic advantage for small-bodied organisms when food availability is low as in the deep sea.

Besides the observations of macrofauna distribution mentioned above (Sokolova, 1972), zoogeographic boundaries have also been identified for meiobenthic fauna. Tietjen (1971) differentiated a shallow-water (50–500 m) fauna consisting of numerous taxa from deep-water (800–2500 m) assemblages with only nematodes and foraminifera present in large numbers. Changes in distribution of both total meiofauna and nematodes were correlated with changes in sediment grain size and bottom water temperature reflecting the effect of the Gulf Stream. Coull (1972) also separated meiobenthic harpacticoid copepods into shelf, lower slope, and abyssal groups on the basis of faunal affinities (rarefaction method) within and between areas. Differences in faunal assemblages observed along the continental shelf appeared to reflect different hydrographic conditions between areas which extended to abyssal regions. Deep-sea groups were not homogeneously distributed over wide geographic regions.

Photographic and submersible surveys have indicated that demersal fish and megafauna (holothurians, echinoderms, amphipods) are often important members of benthic communities. Trawl and grab samples do not adequately sample these large mobile animals and little truly quantitative data on abundance is available (Thiel, 1975). Animals are attracted to baited traps (Dayton and Hessler, 1972) indicating that dispersed individuals can become concentrated. High-density aggregations of some animals, like echinoderms, apparently occur naturally. Ophiuroid densities of 3 to 15/m^2 were observed in the San Diego Trough (Barham et al., 1967). Pronounced aggregations of sea urchins which move as their food supply is depleted occur in shallow (Mann, 1973) and deep water (Grassle et al., 1975). Generally, however, megafauna densities of a few individuals per square metre are common, which may be about 100 times more abundant than demersal fish (Rowe et al., 1975b). As with macrofauna, species composition of megafauna varies markedly with depth (Grassle et al., 1975).

6.1.3.4 Temporal variation. Massé (1972) identified three time-related changes which affect the abundance and biomass of benthic organisms:

1. short-term changes correlated with altered hydrodynamic conditions or feeding activity of large organisms (physical disruption);
2. seasonal changes related to reproduction and recruitment;
3. long-term changes resulting from the successful recruitment of a previously non-abundant species.

Storm-induced water motion and excessive salinity or temperature changes can drastically alter community structure in near-shore areas (Ziegelmeier, 1970). Such disruptions maintain communities in an early successional state characterized by small organisms with a rapid growth rate. Rachor and Gerlach (1975) attributed reduced macrofauna density and species abundance during winter to the passage of heavy storms across the German Bight during autumn. Regular seasonal changes in benthic species assemblages have been observed by Muus (1967) and Levings (1975) where the appearance of predators which utilized abundant prey species in the spring paralleled an increase in temperature. Bodiou and Chardy (1973) only observed seasonal changes in harpacticoid species groups at 10 to 15 m. No seasonal associations occurred at 20 m where abiotic factors had less variations.

Benthic communities may be physically or biologically controlled depending upon the degree to which abiotic factors vary (Sanders, 1968). In boreal waters temperature is considered to control the onset of breeding in marine invertebrates (Thorson, 1954), but thereafter predation or interspecific competition for space may regulate species abundance and biomass. The importance of reproduction as a dispersal mechanism for benthic fauna is also demonstrated by the large number of species which have planktonic larval stages (over 80% in temperate and tropical areas with a slightly lower proportion in arctic and deep sea) (Thorson, 1950). Species with short pelagic lifetimes usually produce a few larvae from eggs rich with yolk and recruitment does not vary greatly from year to year. Larvae with a long planktonic phase, however, are usually produced in large numbers and survival depends on the presence of an adequate food supply. Physical transport away from areas where parental populations exist by advective currents can be responsible for gene flow between populations large distances apart (Scheltema, 1971). Benthic fauna in the deep sea reproduce throughout the year and only a few species, typical of shallow-water areas, display synchronous annual reproductive cycles (Rokop, 1974). In contrast, Bourget and LaCroix (1973) demonstrated that the settlement of benthic epifauna on hard substrates in a subarctic estuary occurred only during a short summer period after which peak densities declined as a result of competition for space, sediment accumulation, and hydrodynamic factors. In contrast, constant physical conditions in the deep sea, deep-water polar regions, and tropics result in communities which are biologically accommodated. Seasonal variations in species composition may be small compared to those occurring in temperate waters and changes which do occur appear due to predation effects which prevent monopolization of space by single species (Dayton *et al.*, 1974).

Few studies of benthic communities have been of sufficient duration to quantify long-term changes occurring between years. Rachor and Gerlach (1975) concluded that despite frequent disruptions due to storms, the macrofauna species abundance and density has remained substantially the same at 25 m depth in the North Sea during the last 25 years. In contrast, Vilks (1967) calculated that 50% of the variation in living benthic foraminifera could be attributed to temporal changes while only 25% was due to spatial differences within a single sampling area. Lie (1969) observed reduced macrofauna bio-mass at four stations in Puget Sound sampled three years apart and, although the species composition of the fauna remained unchanged, the relative dominance of species varied (Lie and Evans, 1973). Masse (1972) also observed large changes in density of macrofauna on fine sand sediments in the Mediterranean during 4 years. Differences were greatest for small-bodied short-lived species which appeared subject to more year-to-year variation in recruitment success. Communities dominated by large long-lived species demonstrated more constant biomass levels since reproductive success or failure in any one year has little effect on total biomass. Eagle (1975) also noted that a single year-class structure led to dramatic variations in population densities of deposit-feeding species. This would account for observations discussed in Section 6.1.3.2 which document the rapid decline in species of polychaetes and molluscs following a period of successful settlement. Buchanan et al.'s (1974) observations that the total number of species and their production was similar over a 4-year period while the total number of individuals more than doubled indicates that while biomass and production levels are maintained between years, the proportions due to various species may change.

Stephen (1938) concluded that peaks in recruitment of the mollusc Tellina every three to four years over twelve years were usually correlated with above average tempera-tures. However, Trevallion et al. (1970) observed that plaice predation on Tellina siphons could cause reduced growth and interfere with reproduction. The effect was greatest when mollusc density was high. A threshold in fish predation existed such that during years of low population density, plaice consumed alternate food sources thus permitting increased Tellina population growth. The functional relation between predator and prey thus induced a density dependent recruitment.

Breen and Mann (1976) have documented the long-term decline in subtidal kelp (Laminaria) due to extensive grazing by sea urchins (Strongylocentrotus). Experimental removal of urchins permitted recolonization by a succession of algal species which did not occur in control areas (Paine and Vadas, 1969). Mann (1973) suggested that sea urchin populations may be limited by predators such as lobsters which thus prevents severe grazing. Once above a certain density, however, urchins can completely destroy a kelp bed and even a small residual population may prevent algal recolonization by all but encrusting species. The long-term effect is to restrict kelp to exposed sites where water movement prevents sea urchin attachment while the urchins themselves dominate large barren areas of solid substrate. These events may represent a successional phase in a predator–prey oscillation or they may lead to a stable situation which only permits kelp to grow in refuge areas. In either case it demonstrates, as Paine (1969) suggested, that benthic species associations may be drastically altered on a long-term basis by the feeding activity of a single species.

Changes in species composition and abundance in sediments subjected to natural or man-made stress serve as useful natural experiments for considering factors which determine the structure of benthic communities. Bagge (1969) observed that macrofauna diversity in the Gullmar Fjord which has received sulfite pulp waste for 80 years has decreased near the head to the fjord during the last 30 years. The zone of maximum fauna abundance has also moved 1·5 km farther seaward during this time. Selective effects of low oxygen and the softening of the bottom in the innermost part of the fjord were related to changes in macrofauna species dominance at different stations. Epibenthic and suspension-feeding species were intolerant of increased organic input while some burrowing invertebrates increased in abundance under these conditions. The boundaries between species groups in the fjord were not stable but changed seasonally as some species expanded their range.

Pearson (1975) has also documented changes in benthic macrofauna species induced over 11 years by the addition of pulp-mill waste into a Scottish sea loch. Fluctuations in the amount of effluent discharge were related to successional changes in macrofauna. A progressive elimination of species occurred as enrichment proceeded with a compensatory increase in numbers in surviving species. Excessive organic deposition resulted in anoxic conditions which could only be tolerated by annelid species. Rosenberg (1973) observed that fluctuations in the composition and distribution of benthic communities in deep basins in the Baltic were due primarily to variations in the oxygen content of deep water. Water exchange and oxygenation of anoxic sediments permitted immigration of large numbers of species not previously present. These populations quickly reached high biomass levels but the long-term stability of the species associations was unassessed.

The inundation of near-shore embayments by drifting macroscopic algae can induce anoxic conditions which dramatically alter benthic community size and composition. Watling (1975) identified three structural phases (re-establishment, larval recruitment, and summer decline) in an estuarine subtidal deposit-feeding community subject to such stress. Five dominant species were compared on the basis of their total abundance, percentage occurrence, diversity, equitability, and dominance. Fager's (1972) scaled standard deviation of diversity and dominance were directly related and each followed a repeatable seasonal cycle which was thought to reflect the degree to which resources were being shared by the deposit-feeding species in the community. Changes in the Shannon-Weaver function (H) and average within-sample diversity during larval recruitment implied a decrease in average niche width and niche overlap during periods of larval recruitment.

It is useful to distinguish adjustment stability (some measure of how rapidly a community returns to its original state) and persistence stability (the constancy of community structure over time) (Margalef, 1969). For example, Peterson (1975) compared benthic macrofauna communities (primarily bivalve molluscs) in two lagoons over 37 months. The lagoon with the lowest species diversity was composed of species whose populations exhibited a high temporal variability (thus a low stability of component species) but the proportionate community composition of each lagoon varied to an equal degree through time (equivalence in stability of community composition). Stability at the species level and stability at the community level are thus not equivalent and the degree of difference is determined by the interdependence among species. The lagoon with the lowest species diversity was populated by species which had a high level of synchrony in their natural temporal fluctuations and thus community composition was less variable than would be expected if all species fluctuated independently. Peterson suggested that community persistence stability was a more conservative community property than stability at the species population level.

6.2 CHEMICAL COMPOSITION

6.2.1 CHEMICAL COMPOSITION OF SEDIMENTS

Since benthic organisms affect and are themselves affected by the chemical composition of bottom deposits, it is necessary to consider briefly the chemical characteristics of marine sediments as they relate to biological processes within benthic communities and overlying water. Hayes (1964) discussed earlier studies with sediments from this point of view and Berner (1971), Broecker (1974), and reviews edited by Riley and Skirrow (1975) provide current comprehensive summaries of aspects of geochemical processes in sediments. In all but well-flushed coarse sediments, concentrations of biologically important nutrients (silicate, nitrate, and phosphate) are usually present in concentrations in interstitial water which are high relative to those in overlying water and increase with depth (Fanning and Pilson, 1974). Under conditions of advection and vertical mixing dissolved nutrient salts may diffuse and be distributed into overlying water. Rittenberg *et al.* (1955) and Goering and Pamatmat (1971) have considered regeneration of nutrients from sediments as a mechanism for supply to phytoplankton in overlying water on this basis. The major transport phenomenon for exchange at the sediment surface is thought to be diffusion and models for diffusive flux have been proposed (Bouldin, 1968; Morse, 1974). Diffusion across a stagnant benthic boundary layer (up to 1 cm thick) may control transport across the sediment–water interface into a layer where turbulent mixing becomes important (Morse, 1974).

Aspects of chemical exchange between sediments and water have long been considered important in lakes (Mortimer, 1971), but only recently have marine sediments been regarded as more than a sink for nutrients. Rittenberg *et al.* (1955) concluded that less than 1% of the annual nutrient requirement of phytoplankton could be supplied by regeneration from sediments. Their observations, however, were made in deep anoxic basins off southern California. Rowe *et al.* (1975a) measured fluxes of ammonia, silicate, and phosphate out of bottom sediments in shallow inshore waters, on the Atlantic continental shelf, and off the Spanish Sahara. In these well-mixed areas, over 35% of the nitrogen estimated to be required for photosynthesis could have been supplied by flux out of the sediments. Sediment pore water nutrient concentrations were several orders of magnitude higher than those in overlying water and the gradient between sediment and water decreased offshore. Thus, the importance of sediment nutrient regeneration may reflect the degree of coupling between pelagic and benthic systems which in turn should reflect water column depth and circulation patterns. For example, Smetacek *et al.* (1976) observed that interstitial water is flushed out of coarse-grained sediment by gravity displacement due to changes in density of bottom water. Both diffusion and hydraulic pumping may be important mechanisms which transport soluble nutrients out of sediments especially on continental shelves where internal waves and advective currents impinge on the bottom.

The existence of high concentrations of dissolved inorganic ions in sediment pore water results from the metabolic activities of bacteria and other micro-organisms. Baas Becking *et al.* (1960) identified pH and oxidation-reduction potential boundaries which are characteristic for various types of micro-organisms which occur in natural marine sediments. pH generally decreases with depth. High values, in excess of 9, may occur in shallow-water sediments (Pomeroy, 1959; Wieser *et al.*, 1974) due to photosynthesis of epibenthic algae while low values with depth reflect CO_2 production or sulfate reduction by bacteria. These processes may be highly stratified in sediments and distinct pH minima may correspond to the redox discontinuity layer (Fenchel, 1969, Fig. 74). Depth-related changes in pH vary with sediment type (ZoBell, 1946) but the range of values is restricted. Precipitation of sulfides prevents the occurrence of values below 6·9 (Ben-Yaakov, 1973) while the upper limit is controlled by buffering of the carbonate system. Pomeroy (1959) has discussed temporal variations in pH in shallow marine sediments which reflect metabolic activities of photosynthetic organisms in the top layer of illuminated sediments.

ZoBell (1946), Fenchel (1969), and Fenchel and Riedl (1970) discussed the importance of H_2S formation by sediment sulfate-reducing bacteria (such as *Desulphovibrio*). The oxidation-reduction potentials in sulfide-rich marine sediments may be controlled by the sulfide–sulfur half cell. Both Berner (1963) and Fenchel (1969) observed a linear relation between Eh and pS^{--} for a variety of natural H_2S-containing sediments and artificial sulfide systems. Jørgensen and Fenchel (1974) found that 50% of the total mineralization of organic matter in sand microcosms enriched with *Zostera* detritus was due to sulfate reduction. Initially most sulfide was precipitated as FeS but with time an increasing fraction remained in solution and diffused to the sediment surface where elemental and organic sulfur accumulated in bacterial plates. Both Hallberg (1968) and Jørgensen and Fenchel (1974) observed that insoluble sulfides could be formed at localized sites to produce reduced microniches within oxidized surface sediment. Thus, while oxidized and reduced sediment layers are generally separated vertically (Fig. 74), localized concentrations of organic matter in aerated surface layers can create hetero-geneous redox conditions which permit a close association of aerobic and anaerobic micro-organisms in oxidized sediment. Iron sulfides may be transformed by biological and inorganic processes to pyrite (Berner, 1963; Hallberg, 1968) which is the dominant form of sulfur in oxidized as well as reduced sediments (Kaplan *et al.*, 1963). Heavy metals in hydrogen sulfide-rich layers are not necessarily all present as sulfides. They may be adsorbed by clay minerals and organic matter, as Hallberg (1974) observed a positive correlation between these variables and concentration of several metals.

The presence of biological structures on the inner and outer surfaces of manganese nodules suggests that accretion of manganese around basaltic fragments may be caused in part by micro-organisms. Some Foraminifera species build tubes which incorporate manganese micronodules and Greenslate (1974) suggested that at least one species may actively concentrate the element during test construction. Thus, whereas most planktonic Foraminifera use calcium carbonate and radiolarian species use silica for test construction, some benthic species may use manganese.

Few accurate measures of dissolved gases in sediment pore water have been made due to contamination and degassing during sampling. Methods recently developed (Martens, 1974) have confirmed Reeburgh's (1969) observations that argon and nitrogen gases are present in surface sediments and vary seasonally with concentrations near those of overlying water. Argon and nitrogen concentrations decreased with sediment depth below 25 cm while methane concentrations increased below this depth. The changes were thought to reflect mixing to 25 cm and the selective removal of nitrogen by stripping caused by release of bubbles of methane. Methane is usually only detectable in anoxic sediments when anaerobic fermentation of carbohydrates occurs (Hutton and Zobell, 1949). Martens and Berner (1974) observed that methane does not occur in anoxic sediments until about 90% of the sulfate is removed by sulfate-reducing bacteria. Upward diffusion of methane and its production in sulfate-free microenvironments could explain the coexistence of methane and sulfate in the upper layers of anoxic sediment.

Sediments within the photic zone may be colonized by populations of epibenthic algae which photosynthesize and enrich surface sediment layers with dissolved oxygen. Heterotrophic respiration by organisms and chemical oxidation of reduced inorganic and organic compounds predominates all sediments, however, and unless water move-ment maintains a high rate of supply, dissolved oxygen is only present within a thin layer of oxidized sediment (Fenchel, 1969; Fenchel and Riedl, 1970; Hargrave, 1972a). Oxygen consumption and CO_2 production in sediments can affect concentrations of these gases in overlying water. Low dissolved oxygen in stagnant marine basins reflects consumption at both the sediment surface and within the water column. The degree of oxygen depletion depends on basin morphometry (the total amount of oxygen present and the ratio of sediment surface-to-water volume), the rate of deposition of organic matter, and the amount of downward and horizontal water exchange which can supply oxygenated water. Despite problems of interpretation, measures of oxygen uptake by undisturbed sediment cores are used as an index of metabolism by benthic communities (Pamatmat, 1975). Relations between this measure and the true metabolic activity (both aerobic and anaerobic) in sediments will be considered in Section 6.4.3.

Sediment organic matter exists in both dissolved and particulate form in concentrations which per unit volume exceed those in overlying water. As in sea water, organic compounds in sediments are complex mixtures which may be derived from terrestrial and aquatic sources and it may exist in either original or altered form. Studies of the origin, composition, and fate of sedimentary organic matter are of interest to chemists, geologists, and biologists and they form subject matter for the field of organic geochemistry.

Bordovsky (1965) reviewed biological and physical transformations which modify organic substances accumulated in marine deposits and Eglinton and Murphy (1969) edited a collection of papers which provides a comprehensive review of developments in the field. For example, Moore (1969) discussed changes which may occur through microbial attack of organic matter in sediments. Differences in the relative abundance of amino acids, carbohydrates, and lipids reflect diagenetic changes due to microbial utilization, transformation, and solubilization. The identification of gross molecular composition, however, gives no information concerning rates of these changes. Some organic compounds are useful biogenic markers since their origin is known and they are preserved in deposits. Plant pigments (chlorophyll derivatives, carotenoids, and porphyrins) can accumulate (Orr et al., 1958) and these compounds must originate from deposited phytoplankton or macrophyte debris. Cyclohexanes and numerous aromatic hydrocarbons found in petroleum do not usually exist in organisms and their presence may indicate petroleum residues in surface sediments (Farrington and Meyer, 1975).

Waksman (1933) first identified the presence of humic substances and the refractory nature for biochemical degradation of many marine sediments. These compounds are high molecular weight products condensed from polyphenolic and proteinaceous material which may be derived from lignin of higher plants or synthesized by microbial activity. They are generally characterized by being acid- and alkali-insoluble and termed 'crude fibre' (Strickland and Parsons, 1968). These stable compounds can account for 30 to 50% of the total organic matter in sediments and the proportion may increase with depth indicative of sediment accumulation (Kato, 1956) or a terrestrial origin (Pocklington, 1976). A large proportion of humic-like compounds could account for the often observed lack of correlation between bulk organic content and either fauna biomass (Bader, 1954; Kato, 1956; Sanders and Hessler, 1969) or oxygen consumption by stirred sediment (Kato, 1956). Little of the accumulated organic matter in marine deposits is available for metabolic utilization.

Total organic matter in sediments is measured as the percentage weight loss on combustion (550°C) or by elemental analysis of carbon after correction for inorganic carbonates. In shallow (near-shore) coastal areas up to 30% of sediment weight may be organic and carbon can account for up to 50% of this amount. Open ocean sediments, however, usually contain less than 1% organic carbon but the amount follows a variety of patterns with depth and distance from land (Sanders and Hessler, 1969). Organic matter in sediments under upwelling regions (off Peru) and in deep anoxic basins may exceed 4% in organic carbon (Rowe, 1971b) indicative of preservation which occurs under reduced conditions. Nitrogen concentrations are an order of magnitude less than those of organic carbon since most marine sediments have C:N ratios in excess of 10. An inverse relation also usually exists between sediment particle size and organic content which reflects the high surface area for organic adsorption in fine-grained deposits (Bader et al., 1960; Hargrave, 1972d). Fine-grained sediments also usually occur in depositional areas where high sediment organic content may result simply by accumulation.

Seasonal cycles of sediment organic matter may occur, particularly in shallow water where production results in periods of increased organic input. Volkmann and Oppenheimer (1962) found organic matter in sediments of a shallow marine lagoon to be minimal (0·1%) and more refractory during winter. Maximum organic concentrations (3·7%) occurred in coarse sediments near shore during summer possibly reflecting production by photosynthetic algae. Organic matter is usually less concentrated and more refractory with increased sediment depth but both temporal and vertical differences may be less in offshore sediments due to the considerable decomposition which has occurred during deposition (Kato, 1956; Sanders and Hessler, 1969).

The relative proportion of dissolved and particulate organic matter in interstitial water has not been extensively studied. The few observations which exist (Starikova, 1970; Karl et al., 1974) indicate that pore water may contain concentrations of dissolved organic carbon an order of magnitude above those in sea water (up to 25 mg/l). High concentrations immediately above the bottom and decreases with sediment depth imply active transport and/or consumption across the sediment surface. Stephens (1967) considered the uptake of dissolved organic matter of low molecular weight as a nutrient source for various marine invertebrates. Glucose and twelve neutral and acidic amino acids occurred in interstitial water of a mud flat from trace amounts to concentrations of $2·5 \times 10^{-5}$ M. Experiments with labelled compounds indicated accumulation of these compounds from solution at these low concentrations by several soft-bodied invertebrates. The metabolic importance of dissolved organic compounds for organisms in sea water and sediments remains unknown, however. Bacterial uptake and physical adsorption by surface-active organic and inorganic material should maintain low concentrations. Accumulation in sediments could indicate the refractory nature of the material. Also, filters (usually 0·45 mμ or 0·8 mμ) are used to separate dissolved and particulate fractions. While operationally necessary, separation by dialysis membranes or molecular sieve filtration would be more informative in characterizing the degree to which compounds are actually in true solution.

6.2.2 BENTHOS

Studies of the chemical composition of macroscopic attached benthic algae, macrophytes, and detritus derived from these sources indicate that, in contrast to phytoplankton (Section 3.2), carbohydrates predominate other constituents. Over 50% of dry weight of Spartina exists as carbohydrates and while protein content of detritus derived from dead Spartina may reach 20%, living plants contain less than 10% protein on a dry weight basis (Odum and de la Cruz, 1967; Hall *et al.*, 1970). Carbohydrates may also accumulate seasonally. Harrison and Mann (1975) observed that *Zostera* leaves lost organic carbon but not nitrogen during the fall. Himmelman and Carefoot (1975) also observed distinct late summer–winter maxima (up to 3·5 kcal/g dry weight) in calorific values of three seaweed species; however, the composition of major metabolites was not determined.

The relative proportions of fats, carbohydrates, and proteins and caloric content in detritus and benthic algae are of obvious importance to benthic consumer organisms. While Paine and Vadas (1969) concluded that food preference reflected more the availability than the caloric value of benthic algae for a variety of invertebrate herbivores, nutritional quality does affect the rate at which calories can be obtained (Himmelman and Carefoot, 1975) and feeding avoidance may occur for protein-deficient food sources (de la Cruz and Poe, 1975). Seasonal and geographic differences in chemical composition are also of importance for commercially harvested species of benthic algae. Polysaccharides like carrageenan may form up to 60% of the dry weight of species such as *Chondrus* with concentrations varying seasonally and in response to nutrient conditions. Buggeln and Craigie (1973) have reviewed aspects of the biochemical composition of this commercially important species.

Fatty acids and hydrocarbons, particularly n-alkanes with odd-carbon-number chains, are synthesized by both plants and animals. The specificity of these compounds for a particular class of phytoplankton has been mentioned (Section 3.2). Jeffries (1972) also observed that salt marsh grasses have a terrestrial pattern of fatty acids rich in 16–18 carbon chains while fish (*Fundulus*) and a detritus-feeding shrimp (*Palaemonetes*) have a marine pattern dominated by long-chain poly-unsaturates. The relative distribution of various fatty acids in detritus and digestive tracts of these organisms was used to infer the proportion of detritus and animal tissue in their diet. Ackman (1965) noted the occurrence of an unusually high proportion of odd-numbered fatty acids in the detritus-feeding mullet and suggested a dietary source from algae. However, differences in lipid and fatty acid composition characteristic for two species of oysters could not be substantially altered by feeding with phytoplankton containing different fatty acids (Watanabe and Ackman, 1974). Thus, although some hydrocarbons may not be modified during passage through a food web (Farrington and Meyers, 1975), some species rapidly convert unusual fatty acids to a species-specific composition.

The proximate composition of benthic crustaceans generally corresponds to that of zooplankton (Section 3.3). Water usually accounts for 70 to 80% of wet tissue weight which is slightly lower than values for planktonic animals (Table 12). Lipid content reflects age, reproductive and feeding conditions but values seldom exceed 15% of dry weight (1 to 3% wet weight). Ansell (1975) found that carbohydrate (mostly glycogen) (Ansell and Trevallion, 1967) and lipid content in the bivalve mollusc *Astarte* decreased during winter following spawning. Values for tissue carbohydrate (15%) and lipid (5.9%) were maximum prior to spawning during late summer when caloric content was also maximum (4·85 cal/g). Moore (1976), on the other hand, observed high lipid content (20 to 25% of dry weight) in five species of benthic crustaceans and neither this nor ash content (20%) varied seasonally. Optimal feeding conditions may have caused the accumulation of lipids in these species. Ackman and Cormier (1967) observed that lipid reserves decreased from 4·8 to 1·2% (wet weight) during 2 weeks of starvation of the periwinkle *Littorina*. Ansell and Sivadas (1973) also noted that starvation resulted in rapid losses of carbohydrate, lipid, and protein in the bivalve *Donax*. These changes in tissue biochemical composition suggest that the relative proportion of the three organic fractions could serve to indicate the nutritional state of organisms under natural conditions.

Concentrations of phytoplankton pigments and derivatives have been measured in benthic fauna to index the utilization of algae as a food source by bottom-feeding invertebrates (Fox *et al.*, 1948; Ansell, 1975). Absorption and fluorescent spectroscopy were used to compare spectral characteristics (peaks) in tissue extracts and phytoplankton thought to serve as a food source. Caution is necessary in interpreting spectra, however, since material synthesized by invertebrates themselves may interfere by absorption or fluorescence at similar wavelengths. An alternative approach to identifying energy sources for benthic invertebrates has been the measurement of the potential for enzymatic digestion. Hylleberg (1972) compared the carbohydrase activity in twenty-two species of marine benthic invertebrates. Hydrolysis of laminarin, glycogen, and amylose occurred in all species but was greatest in crustaceans. Little hydrolysis of oligosaccharides or structural polysaccharides occurred. Differences in the relative activity of carbohydrases between three coexisting species of deposit-feeding molluscs, however, could indicate a quantitative enzymatic response to divergent food sources (Hylleberg, 1976). Polysaccharides such as cellulose may be hydrolysed by cellulase enzymes present in some benthic species. Wildish and Poole (1970) identified the carboxymethyl-cellulase component of cellulase in the hepatopancrease of the amphipod *Orchestia*. At least part of the cellulase activity was attributed to the presence of symbiotic bacteria.

Separation of proteins from tissue homogenates using gel electrophoresis serves as a tool for quantifying the amount of genetic variation in natural populations. Selective enzyme assays may be used to identify activity associated with particular protein bands on electrophoretic gel. Banding patterns are characteristic for each species and they can be used to test for genetic differences between populations. Tracey *et al.* (1975) used these methods to confirm that lobsters (*Homarus*) from various areas off the eastern coast of North America were genetically similar but subdivided into isolated inshore and offshore populations. Differences in particular protein (or enzyme) banding patterns may also occur within a species, however, and the variation can be used as a measure of genetic diversity (polymorphism). Doyle (1972) observed the frequency of a polymorphic esterase isoenzyme in the brittle-star *Ophiomusium* in the deep sea using these techniques. Contrary to the idea that selectively significant genetic variation should decrease with depth, since physical conditions appear to become progressively more stable (Sanders and Hessler, 1969; Grassle and Grassle, 1974), variability of the esterase isoenzyme was high. Banding patterns (gene frequencies) were similar over horizontal distances up to 200 km along isobaths but varied considerably over 8 km in a direction normal to this. Differences in the frequency of occurrence of the isoenzymes across isobaths implies a depth-related polymorphism of this enzyme system. Similar investigations of genetic polymorphism at two enzyme loci in eight species of *Macoma* indicated that environmental heterogeneity rather than temporal environmental variability may select for polymorphism (Levinton, 1975). Genetic variability thus exists in the deep sea as well as at shallower depths although the selective value of variation in esterase isoenzymes is unknown.

Measurements of nucleic acids (RNA, DNA) have been used as an index of growth in cell suspensions made from a variety of organisms. Sutcliffe (1969) observed a direct relation between growth rate and RNA concentrations in a variety of invertebrates. The relationship suggested that just as chlorophyll concentration may serve as an index of phytoplankton photosynthetic potential, the ratio and absolute concentration of RNA and DNA might measure the capacity of cells for protein synthesis (growth). Pease (1976) reviewed evidence that nucleic acid concentrations and growth rates of organisms are related and attempted to compare RNA/DNA and protein synthesis in oysters during one year. The RNA/DNA ratio was directly related to growth rate for a specific oyster year class and to the rate of protein synthesis during the preceding month. Uniformity in the ratio for adductor muscle from all sizes of oyster was thought to reflect the constant composition and function of this tissue while differences in whole-body analyses could have been due to the presence of gonads at certain times in specific size classes. The study indicated the need for improved methods of tissue homogenization prior to nucleic acid extraction and a more adequate measure of protein synthesis in animals which may show little growth over short periods.

All organisms dehydrogenate substrates and subsequently form high-energy ATP during transfer of hydrogen and electrons to acceptor compounds. Reduction of artificial electron acceptors, such as 2,3,5-triphenyltetrazolium chloride (TTC), has been used as a measure of dehydrogenase activity by sediment micro-organisms (Pamatmat and Skjoldal, 1974) since only viable cells are thought to be capable of enzymatic reduction of TTC to formazan. Pamatmat and Bhagwat (1973) discussed problems in applying the method to estimate *in situ* microbial metabolism in sediments. Non-enzymatic reduction of TTC was not observed but further studies demonstrated that reducing sugars and other reduced compounds caused chemical reduction (Pamatmat, 1975). Wieser and Zech (1976) applied measures of electron transport activity ($NADH_2$-dependent dehydrogenase transformation) to carbonate sediments. Over 80% of the activity was associated with sand grains indicating that attached microflora and fauna rather than interstitial organisms dominated energy flow through the sediment.

ATP has also been used as an indicator of living biomass in benthic organisms. While the ATP content of bacteria and algae depend on nutritional state and metabolic activity (Holm-Hansen, 1970a), average content in a variety of aquatic micro-organisms is approximately 0·4% of cell carbon (Hamilton and Holm-Hansen, 1967). Comparison of ATP and total sediment carbon demonstrated that microbial biomass would account for less than 1% of organic carbon in surface salt marsh sediment and values decreased with sediment depth (Christian *et al.*, 1975). Concentrations of 140 to 1600 ng ATP/ml wet sediment were observed in surface sediment from 50 to 1000 m while samples from 6000 m contained approximately 2 ng ATP/ml (Hodson *et al.*, 1976). Pamatmat and Skjoldal (1974) also observed a positive correlation between TTC reduction and ATP concentration in marine sediment cores which would be expected if dehydrogenase activity and ATP concentration are present in micro-organisms in a constant ratio in proportion to metabolic activity.

6.3 BENTHIC AUTOTROPHIC PROCESSES

6.3.1 PHOTOSYNTHESIS AND PRODUCTION BY BENTHIC MACRO- AND MICROFLORA

Macroscopic and microscopic algae and photosynthetic bacteria develop extensive populations in and on sediments and solid substrates which receive light. Highest biomass and production are observed in salt marshes, intertidal sediments and in littoral areas up to 20 to 30 m deep where clear water permits adequate light penetration.

Although production by benthic macro- and microflora may occur in benthic communities to the compensation depth for phytoplankton (approximately 1% of surface radiation), the significance and magnitude of this source of production relative to that of phytoplankton may increase as depth decreases.

The magnitude of production by benthic microflora was first suggested by high cell counts and pigment concentrations observed in intertidal sediments (Pomeroy, 1959). Plant pigments occur in most aquatic sediments but usually these are accumulated degradation products (pheophytins, pheophorbides) of pigments originally present in phytoplankton or macrophytes (Orr *et al.*, 1958; Vallentyne, 1969). Shallow-water marine and lake sediments, however, may also contain a variety of undegraded pigments present in algae (chlorophyll *a*, *c*, diatoxanthin, diadinoxanthin, fucoxanthin, carotene, phycocyanin) and bacteria (bacterio-chlorophylls from green and purple bacteria) (Taylor and Gebelein, 1966; Fenchel and Straarup, 1971). Absorption spectra of intertidal sediments extracted with organic solvents and water (Fig. 78) indicate differences in peak absorp-

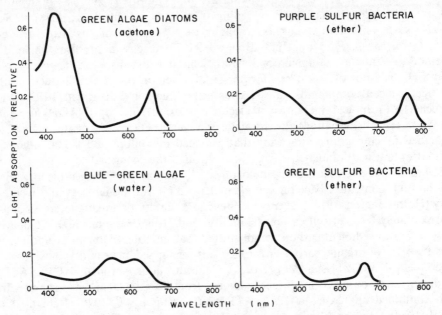

Fig. 78. Absorption spectra of extracts of different groups of sediment microflora (redrawn from Fenchel and Straarup, 1971).

tion for cells of different microflora groups and they demonstrate the difficulty in distinguishing between algal and bacteriochlorophylls. Absorption spectra cannot be used to distinguish pheophytin pigments because of spectral similarities to undegraded chlorophyll. Sedimentary chlorophyll concentrations are thus usually calculated by measuring light extinction at 665 nm and assuming a specific absorption coefficient (91 l/g cm— Fenchel and Straarup, 1971) or by relating the extinction to arbitrary units (Vallentyne, 1955; Pomeroy, 1959). Chromatographic separation permits the identification of specific pigments (Taylor and Gebelein, 1966) and spectrofluorimetric techniques have been used to quantify chlorophyll *a* and pheophytin in sediments (Pamatmat, 1968) and sedimented material (Ansell, 1974; Webster *et al.*, 1975).

The vertical distribution of benthic algal and bacterial photosynthetic activity corresponds to the marked attenuation of light which occurs in the upper few mm's of sediment (Section 6.1.3.2). Perkins (1963) observed that moist mud 1·0 and 2·0 mm thick reduced light intensity to 2% and 0·4% of incident illumination. In rinsed quartz sand, 1% of red and infrared light penetrates to 4 and 4·8 mm, respectively, while penetration only occurs to 3 mm in undisturbed estuarine sediment (Fenchel and Straarup, 1971). These authors also found bacteriochlorophyll a (present in purple sulfurbacteria) down to 6 mm depth with maximum concentrations between 2 and 3 mm. Decreases in sulfide concentration due to these photosynthetic bacteria were restricted to the upper 3 mm of sediment cores (Blackburn et al., 1975). Attenuation coefficients for light in sediments depend on grain size and wavelength with long-waved light penetrating deeper than short-waved radiation and fine-grained deposits attenuating light more than coarse sediments (Taylor, 1964; Fenchel and Straarup, 1971). Values between 0·52 mm^{-1} and 1·39 mm^{-1} were observed. Unlike the water column, where light extinction is primarily a function of phytoplankton and other suspended matter, reflection, refraction, scattering and light absorption by sediment particles (non-biological absorption) account for the rapid attenuation of light with sediment depth. Fenchel and Straarup (1971) estimated that a maximum of 20% of infrared, red, and blue light absorption could be attributed to plant pigments in intertidal mudflat sediments. The actual efficiency of light absorption depends on the density and vertical stratification of photosynthetic organisms within surface sediment layers.

Values of approximately 100 mg chlorophyll a/m^2 are common in intertidal sediments and the amount decreases with water depth and increasing amounts of wave action (exposure) (Steele and Baird, 1968; Fenchel and Straarup, 1971; Estrada et al., 1974). Moss (1968) reviewed measures of chlorophyll a content in algae attached to rock ('epilithic'), sand grain ('epipsammic'), and larger plant ('epiphytic') surfaces. Although concentrations up to 2000 mg/m^2 occur under conditions of high nutrient supply and on stable substrates, values from 10 to 100 mg/m^2 are typical for a variety of surfaces. Fenchel and Straarup (1971) noted the similarity of this range to that observed for standing crops of phytoplankton in lakes and the ocean when integrated per m^2 for the photic zone. These values are below theoretical maximum chlorophyll a concentrations of 400 to 800 mg/m^2 calculated to attenuate light to 1% of surface values (Steemann Nielsen, 1962). This is not surprising in sediments since particles themselves absorb and scatter most of the incipient radiation. When physically stable conditions exist, however, photosynthetic organisms may exist in thin layers, concentrated so as to maximize the efficiency of light absorption. In these cases, the microorganisms do not compete with their substrate for light and much higher chlorophyll concentrations may be achieved.

Photochemical processes during photosynthesis depend on light absorption by photosynthetic pigments while enzymatic conversion of organic compounds produced depends on enzyme concentrations and temperature. Photosynthetic organisms adapt to varying conditions of light and temperature by altering the relative proportions of photosynthetic enzymes and pigments. These effects of light and temperature on photosynthesis have led to the development of two different methodologies to estimate production by benthic microflora. In situ light and dark exposure is used with undisturbed sediment surfaces to estimate gross and net production (Pomeroy, 1959; Pamatmat, 1968) while incubation of stirred samples removed from specific depth layers and incubated under artificial light provides a measure of potential production when corrected for chemosynthetic uptake of label (Grøntved, 1960; Gargas, 1970). In situ light measurements are used to extrapolate these measurements to rates expected under field conditions. ^{14}C-techniques have also been used to estimate production of macrophytes (UNESCO, 1973). These methods offer sensitivity when production is low and when stirred samples are used, production by attached and free-living algae can be separated and both the light response and vertical distribution of microflora can be directly measured (Gargas, 1970; Hunding, 1971). Measures of gas exchange across light and dark undisturbed sediment surfaces provide a simple and rapid estimate of the relative importance of phototrophic and heterotrophic activity under *in situ* conditions. Hunding and Hargrave (1973) compared estimates of benthic primary production on a sandy beach measured by *in situ* and laboratory ^{14}C methods and demonstrated that both methods gave similar measures of the magnitude of production.

TABLE 36. PRIMARY PRODUCTION BY ATTACHED MARINE EPIBENTHIC ALGAE AND MACROPHYTES[a]

Location/source	Technique	g C m^{-2} yr^{-1}	Author
Epibenthic algae			
Georgia salt marsh	O_2, CO_2	200	Pomeroy, 1959
Delaware salt marsh	O_2	38–99	Gallagher and Daiber, 1974
Massachusetts salt marsh	^{14}C	(shaded) 106 (unshaded) 165	Van Raalte *et al.*, 1976
Intertidal sand flat	O_2	143–226	Pamatmat, 1968
Intertidal sand flat	O_2	0–325	Riznyk and Phinney, 1972
Intertidal mudflat	^{14}C	31	Leach, 1970
Estuarine subtidal	^{14}C	116	Grøntved, 1960
	^{14}C	90	Marshall, 1970
Sea and marsh grasses			
Thalassia beds	O_2	520–640	Westlake, 1963
Spartina (Georgia)	O_2	257–897	Teal, 1962
(North Carolina to Novia Scotia)	cropping	130–256	Mann, 1972b
(Massachusetts)	cropping	1100–2300[b]	Valiela *et al.*, 1976
Mangrove swamps			
Florida (net prod.)	O_2 (+ litter)	400	Mann, 1972b
Kelps			
Laminaria (N.S.)	cropping	1900	Westlake, 1963
(England)	cropping	1225	Bellamy *et al.*, 1968
(N.S.)	blade renewal	1750	Mann, 1972b
Macrocystis	cropping	400–820	Clendenning, 1971
Littoral seaweeds			
Fucus	O_2	<3000[c]	Kanwisher, 1966

[a] Only studies which provide annual estimates are included. Mann (1972b) and Van Raalte *et al.* (1977) provide extensive comparisons of macrophyte and microalgal production rates respectively in different areas.
[b] Includes above and below ground biomass and assumes 35% by weight as C.
[c] Calculated from hourly rates as net oxygen production by a dense stand of fronds covered by filamentous brown and red seaweed epiphytes.

Production by marine macrophytes (seaweeds, sea grasses) which grow between high water and a depth of approximately 30 m has been assessed by harvesting (cropping) to estimate seasonal increments in biomass (Westlake, 1963; Bellamy, 1968). A direct measure of tissue growth in the kelp *Laminaria* was provided by Mann (1972b) by punching holes in blades and measuring their relative position with time. Rapid tissue growth at the base of blades compensated for erosion from the tip and thus measures of standing crop alone would have underestimated production which had occurred. Despite the variety of techniques used, comparison of estimates of annual primary production by macrophytes and attached microbenthic algae (Table 36) indicates that carbon fixed per unit area in these communities may equal or be up to an order of magnitude greater than phytoplankton production. While phytoplankton production predominates as a source of synthesis of new organic matter in the open ocean, macrophytes and benthic algae make significant contributions to coastal regions. Only a fraction of this production may be grazed directly—almost all of it is thought to enter the water column as dissolved and particulate organic matter (Mann, 1972a). Shallow water and high advection, characteristic of coastal areas, ensure that a large fraction of this organic debris enters detritus-based food webs within sediments.

High levels of productivity by emergent and submerged macrophytes and microalgae in intertidal and shallow neritic areas, which per unit area are among the most productive communities in the world (Odum, 1971), are thought to be sustained by a high rate of nutrient supply (Mann, 1973). Wind, estuarine and tidal mixing ensure a supply of nutrient-rich water by upwelling and by transport of nutrients regenerated within sediments. The importance of nutrients for production by benthic plants was demonstrated when inorganic salts, added to stimulate phytoplankton production, were found to enrich sediments and cause increased growth of epibenthic algae (Marshall and Orr, 1948). The addition of sewage sludge, urea, and other nitrogen fertilizers enhanced production of salt marsh grasses two- to three-fold (Valiela and Teal, 1974). These and other studies (Valiela et al., 1976) have established that organic nitrogen, rather than nitrate or phosphorus, is responsible for increased production. Nitrate or ammonia added to sediments could be lost through nitrification in aerobic layers and denitrification in reduced anoxic subsurface sediment. Also, phosphate and nitrate may be present in deposits in concentrations sufficiently high so as not to be limiting. Nitrogen-fixing blue-green algae and bacteria also exist in salt marsh mud (Wood, 1965) and provide a source of organic nitrogen which is directly utilized by higher plants and algae (Jones, 1974). Nutrient enrichment of salt marsh sediments, however, did not increase chlorophyll concentrations per unit area in epibenthic algal populations (Estrada et al., 1974). Production by benthic algae was only increased by fertilization during early spring before a canopy of grasses (Spartina, Distichlis) had formed (Van Raalte et al., 1976). Thus, light availability rather than nutrient supply appears to limit standing crops and production of benthic algae.

Pamatmat (1968) concluded that incident radiation was of primary importance in determining seasonal patterns of production by epibenthic algae on an intertidal sandflat. Pomeroy (1959) and Hargrave (1969b) also noted linear correlations between gross oxygen production and light, but other factors (tidal exposure and temperature) also significantly affected production rates. Few measurements have actually quantified the effect of light intensity on benthic primary production but studies with epibenthic algae (Taylor, 1964; Gargas, 1971; Hunding, 1971) and macrophytes (Kanwisher, 1966; McRoy, 1974; Brinkhuis et al., 1976) have established two characteristic responses: benthic phototrophic organisms demonstrate maximum photosynthesis over a wide range of light intensities, thus light saturation occurs at very low light intensities (curve type 2, Fig. 30) and inhibition of photosynthesis seldom occurs even at high light levels. Burkholder et al. (1965) observed that planktonic algae generally show lower I_k values than benthic algae, but seasonal changes due to adaptation and differences in species composition make comparisons difficult. Gargas (1971) observed that I_k values for epibenthic algal communities decreased from 21 klx during summer to 4 klx in winter which followed changes in incident light. Also, I_k values for algae at various depths in sediment cores were equivalent to those for surface populations and they similarily declined with seasonal reductions in illumination. This suggests that benthic algal cells may not remain stratified at one depth or that when light is low, maximum photosynthesis may be maintained by prevailing light intensities (Hunding, 1971).

Taylor (1964) observed maximum carbon fixation by epibenthic diatoms receiving 12 g cal/cm²/hr (14% of incident radiation). Littoral seaweeds reach photosynthetic saturation at light intensities between 10% and 50% of incident levels (Kanwisher, 1966; Brinkhuis et al., 1976). The rate of bacterial sulfide oxidation also reaches maximum values at 2 klx (approximately one to two orders of magnitude less than full sunlight under a shallow layer of water) (Blackburn et al., 1975). Thus, all exposed sediments and those covered by shallow water in temperate and tropical areas receive light in excess of that required to saturate photosynthesis (Pomeroy, 1959). The absence of substantial photoinhibition even under full sunlight (Taylor, 1964; Pamatmat, 1968) and the concentration of cells at the sediment surface suggests that, as in seaweeds, growth may proceed until reduced by self-shading (Kanwisher, 1966). Adaptation for maximum utilization of light is also demonstrated by estimates of photosynthetic efficiency. Values may reach 3% of incident light photosynthetically fixed when light intensity is less than 0·15 g cal/cm²/min but lower efficiencies (0·1% to 1%) characteristic of phytoplankton occur at high illumination (Pomeroy, 1959; Hargrave, 1969b). Pamatmat (1968) noted diurnal changes in gross photosynthetic efficiency with higher values tending to occur at low light levels.

Marked vertical stratification, or zonation, which occurs in benthic algal and bacterial communities can in part be related to changes in the vertical distribution of light quality as well as quantity. *Chondrus* exhibited maximum photosynthesis under red light (Mathieson and Prince, 1973). Photosynthetic sulfide oxidation was also stimulated by long wavelength light and red light had a greater effect than infrared light (Blackburn et al., 1975). This would be expected if green sulfurbacteria (peak absorption near 660 nm) were more important in oxidizing sulfide than purple sulfurbacteria (peak absorption near 770 nm) (Fig. 78). Maximum concentrations of bacteriochlorophyll *a* also occurred between 2 and 3 mm in estuarine sediment which corresponded to the depth at which red and infrared light decreased to 1% of surface values (Fenchel and Straarup, 1971). The observations suggest that some photosynthetic organisms may concentrate at depths which maximize the efficiency of light absorption or at least correspond to the depth of penetration of light characteristic for specific photosynthetic pigments. Pomeroy (1959) and Brown et al. (1972), however, discussed vertical migrations of motile benthic algae (usually diatoms) which concentrate in layers only a few cells thick at the sediment surface. Such movements expose cells to maximum light intensities. Pamatmat (1968) failed to observe any time-related changes in profiles of chlorophyll concentration in intertidal sandflat sediment during a tidal cycle and so vertical migration of cells may not always occur.

Calculation of net photosynthetic production requires that respiration be known. While this is usually assumed to be light-independent (Section 4.1.2), there is increasing evidence for photorespiration during photosynthesis (Tolbert, 1974). In contrast to studies which demonstrate increased respiration in the light (Gregory, 1971), respiration appears to be replaced by a different process of CO_2 production during photosynthesis in some species of marine benthic algae (Brown and Tregunna, 1967). Net photosynthesis cannot be calculated by assuming a light-independent respiration rate if respiration is inhibited during photosynthesis. If respiration is assumed to be between 10% and 30% of gross photosynthesis (as dark O_2 uptake measurements indicate) (Pomeroy, 1959; Odum, 1971), however, the compensation point at which photosynthesis and respiration are equal for epibenthic diatoms is about 0·3% of midday surface radiation (Taylor, 1964). Thus, cells at 4 mm depth in fine sand sediment are near their compensation point while cells at 2 mm depth photosynthesize at more than 90% of their maximum capacity. The compensation point for the seaweed *Chondrus* occurs at less than 0·5% of full sunshine (Kanwisher, 1966) and this permits photosynthesis to occur to 20 m depth on a sunny day (Mathieson and Prince, 1973).

Calculations of annual primary production by ^{14}C or O_2 methods are usually based on measurements made throughout the year with periodic experiments of a few hours duration on different days. Diurnal changes are thus estimated by extrapolation from hourly rates. Burkholder *et al.* (1965) used a relationship between light and photosynthesis to estimate daily production by epibenthic algae. Pamatmat (1968) and Hargrave (1969b) calculated multiple linear regressions and substituted average daily values for parameters which accounted for a significant part of the variance in measures of oxygen production to make annual estimates. Endogenous daily rhythms in both respiration and gross oxygen production have been observed, however, which imply that in some types of benthic shallow-water communities, respiration may be controlled by epibenthic primary production.

Pamatmat (1968) observed an endogenous tidal rhythm in oxygen uptake and photosynthesis in settled intertidal sand samples held in the laboratory in continuous darkness or in alternating periods of light and darkness. Rates of photosynthesis and respiration were in phase and depressed during times which corresponded to low and high water in the field. Increased rates occurred during flood and ebb tidal periods. The maximum variation in photosynthesis (10 to 55 ml O_2/m^2; hr) and respiration (3 to 21 ml $O_2/m^2/hr$) in these samples corresponded to approximately 50% of the total range of values observed seasonally when bell jars were used to cover undisturbed sediments during tidal exposure. Other studies document different time-related changes. Beyers (1963) repeatedly found an evening burst in oxygen uptake in laboratory sediment–water microcosms even though they were exposed to continuous light. Gallagher and Daiber (1973) did not observe any diel variation in respiration of sediment cores from a *Spartina* salt marsh but an endogenous photosynthetic rhythm with a tenfold amplitude and maximum oxygen production at midday existed under constant light.

Diel patterns of oxygen production by epibenthic algae incubated under *in situ* conditions would be expected due to the effect of light and temperature on photosynthesis. However, increased oxygen uptake by undisturbed benthic communities during the late evening and night, observed by Hargrave (1969b) and Hunding (1973), light-induced increased respiration in salt marsh sediments (Gallagher and Daiber, 1973) and different linear relations between photosynthesis and illumination during different parts of the day (Hunding, 1973), indicate that complex relationships exist between heterotrophic and photosynthetic activity. Hunding (1973) suggested that where inorganic nutrient supply is low, epibenthic algal growth during the day may deplete nutrients. Organic substances could be released in increasing amounts during the afternoon to progressively stimulate microbial respiration. Alternatively, algal and bacterial cells could adapt to changing light conditions during the day or alter their position in the sediment thereby affecting the ratio of production:respiration. Measurements which demonstrate the appearance of extracellular photosynthetic products in sediments during the day and independent estimates of algal and bacterial respiration are needed to clarify these relationships.

Jassby and Platt (1976) compared a variety of empirical mathematical formulations for photosynthesis/light curves for phytoplankton and determined that a hyperbolic tangent function

$$P^B = P_m^B \tanh (xI/P_m^B) \tag{131}$$

where P^B is the instantaneous production rate normalized to chlorophyll biomass B, P_m^B is the specific photosynthetic rate at optimal illumination, x is the slope of the light-saturation curve at low light levels and I is the intensity of available light, best described data from various natural marine phytoplankton populations. The expression has the advantage that, unlike I_k, it describes the photosynthetic response of organisms to light by two parameters—x which is a function of photochemical reactions and P_m^B which reflects temperature-dependent enzyme reactions. Thus, it should be possible to describe the influence of temperature, nutrients, and adaptation on photosynthesis by these two parameters. Unfortunately, no study has normalized measures of benthic primary production to chlorophyll biomass and light-saturation curves have not been compared on this basis. When this has been accomplished, it will be possible to quantitatively compare the response of photosynthetic planktonic and benthic organisms to light.

6.4 BENTHIC PRODUCTION

6.4.1 MECHANISMS OF ORGANIC SUPPLY

6.4.1.1 Photosynthetic, heterotrophic and chemosynthetic production. Light and imported organic matter provide all energy requirements for production by benthic organisms. The processes of synthesis through oxidative and reductive pathways which these energy sources drive are interrelated and their relative importance depends on the availability of light, oxygen, and various organic and inorganic hydrogen acceptors (Fig. 79). The stratification of these processes, which may exist in a water column between aerobic and anaerobic conditions (Section 3.1.6), forms a major structural feature in almost all sediments since light and oxygen are only supplied across the sediment surface (Fenchel, 1969; Fenchel and Riedl, 1970).

The elemental cycles of carbon, nitrogen, and sulfur within sediments and across the sediment–water interface are due largely to the metabolic activities of autotrophic and heterotrophic bacteria (Brock, 1966). In aerated sediments as in the water column, where turbulence and diffusion maintain an adequate oxygen supply, heterotrophic bacteria decompose organic matter into oxidized inorganic compounds with oxygen as the terminal hydrogen acceptor. In subsurface anaerobic sediments, however, microorganisms use both organic and inorganic ($SO_4^=$, NO_3^-, CO_2) compounds as hydrogen acceptors. If organic substrates are used (fermentation), fatty acids, alcohols, and other simple organic compounds are produced. Utilization of inorganic hydrogen acceptors produces reduced inorganic compounds (CH_4, NH_3, H_2S) which diffuse upwards and serve as substrates for chemoautotrophic bacteria at the oxidized interface. The energy released by oxidizing these compounds is utilized to reduce CO_2 to carbohydrate in interstitial and overlying water. As Kuznetsov (1958) and Sorokin (1964a) pointed out,

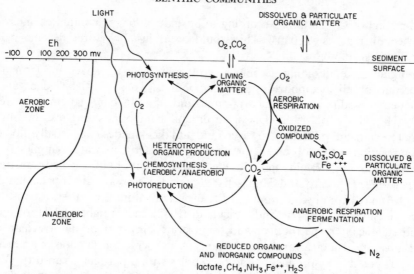

FIG. 79. Interrelations between photosynthetic, heterotrophic, and chemosynthetic processes which occur in sediment. Photosynthesis and photoreduction only occur in the presence of light. *Aerobic metabolic processes*: heterotrophic production (oxidation of reduced simple organic compounds with possible reduction of external CO_2); photosynthesis (reduction of CO_2 to carbohydrates using H_2O and light); aerobic respiration (reduction of oxygen to water with organic compounds as electron donors); aerobic chemosynthesis (oxidation of CH_4, H_2S, NH_3, Fe^{+++}, H_2 to form organic carbon compounds by fixation of external CO_2). *Anaerobic metabolic processes*: anaerobic respiration (oxidized inorganic end products of aerobic decomposition used as hydrogen acceptors for the oxidation of organic matter); fermentation (organic compounds used as hydrogen acceptors to produce CO_2, H_2O and reduced organic compounds lactate, glycollic acid, H_2S, NH_3); photoreduction (reduced compounds in the presence of light used to reduce CO_2 to carbohydrates with H_2S, SO_3, S, H_2 or reduced organics serving as a hydrogen donor); anaerobic chemosynthesis (oxidize inorganic compounds H_2, H_2S, Fe^{++}, $NO_2^=$ and use energy to reduce CO_2 to carbohydrates) (redrawn from Fenchel, 1969, with modifications).

these transformations resemble primary production since dissolved substances are converted into particulate organic matter. When reduced sediments are exposed to light (as in shallow water where organic enrichment leads to highly reduced conditions) photosynthetic bacteria utilize reduced inorganic and organic compounds to photosynthetically reduce CO_2. Fenchel (1969, 1971) has reviewed the trophic significance and vertical zonation of these processes in marine sediments.

When algae are not present, bacterial chemosynthesis which involves the flux of CO_2 has been measured as dark $^{14}CO_2$ uptake (Sorokin, 1965). Aerobic heterotrophic microflora generally fix only a few percent of CO_2–C in the dark, but chemosynthetic bacteria which oxidize simple reduced organic compounds may obtain 30% to 90% of their carbon supply from this source. Chemoautotrophic bacteria, like *Desulfovibrio* and *Thiobacillus*, utilize only CO_2 as a carbon supply. Thus, although assimilation of CO_2 does not measure total chemosynthetic production, if corrected for heterotrophic uptake it does provide a comparative index of metabolic activity for certain groups of microorganisms. Just as ^{14}C dark fixation expressed as a percentage of total carbon fixation (in light) increases with water depth as photosynthesis decreases, dark CO_2 assimilation relative to photosynthesis by benthic microflora (attached and freeliving within sediments) may change seasonally and increases with sediment depth (Gargas, 1970; Hunding, 1971). Absolute measures of chemosynthesis, however, may not increase with depth since lower numbers of algae with increasing sediment depth would also increase the ratio of dark:light CO_2 fixation. According to Sorokin (1965), maximum chemosynthesis occurs when Eh reaches -20 mV. This would occur just below the aerobic/anaerobic interface (Fenchel, 1969).

Micro-organisms undergoing aerobic heterotrophic growth, anaerobic fermentation, or photoreduction all metabolize organic compounds. Thus, ^{14}C-labelled organic substrates added to sediments are incorporated and, if fully oxidized, may be respired as $^{14}CO_2$ (Harrison et al., 1971). Dissolved organic compounds may also be physically adsorbed to sediment particles, however, and uptake kinetics are usually not simple first-order reactions often observed for algae and bacteria in water samples (Parsons and Strickland, 1962; Wright and Hobbie, 1966). Addition of labelled substrates to open ocean water (Vaccaro and Jannasch, 1966) and deep-sea sediments (Jannasch and Wirsen, 1973) have not always produced the expected saturation curve response typical of earlier measures by Parsons and Strickland (1962) and Wright and Hobbie (1966). Either no significant respiration of added substrate occurred or an irregular uptake of added substrate at increasing concentrations was observed. Induction of a kinetic-like response was possible by pre-incubation of samples with a particular substrate (Vaccaro and Jannasch, 1966; 1967) and Williams and Gray (1970) observed that the amount of induced activity was proportional to an increase in substrate concentration. Either inducible enzyme systems could be present or selective growth through enrichment could occur particularly in prolonged incubations. In either case, the observations indicate an ability of natural microbial populations to respond to increased substrate concentrations by increased metabolic activity. The measures do not necessarily quantify in situ metabolism.

Hall et al. (1972) demonstrated that temperature, nutrient availability, and the nature of the organic substrate affected the diffusion rate and uptake of dissolved organic compounds in lake sediments. Up to 80% of ^{14}C-glycine was respired within 2 hr when added to stirred surface sediment but acetate and glucose were mineralized to a lesser extent (10–25%). Also, when undisturbed sediment cores were used and a range of concentrations of substrate added, uptake mechanisms were not saturated, even at concentrations as high as 1 mg/l (interstitial values equalled 50 μg/l). Wood and Chua (1973) similarly observed that respiration amounted to 5 to 30% in 2 hr for various carbohydrates with no relation between uptake rate and glucose concentration in stirred sediments. Jones and Simon (1975), however, achieved saturation at concentrations >100 μg/l in undisturbed cores and stirred sediment with uptake equivalent to 400 to 600 μg/m^2/hr. Uptake of glucose by stirred autoclaved mud amounted to 10% of rates in fresh samples. Turnover times, while dependent on substrate type, are usually higher (minutes) in sediments than in water (hours) (Hall et al., 1972; Wood and Chua, 1973) as would be expected on the basis of microbial density per unit volume. Rates are also greatly affected by stirring. Uptake and mineralization can be increased by one to three orders of magnitude by mixing and dilution (Hall et al., 1972; Jones and Simon, 1975) and thus measurements with stirred sediment cannot be representative of those which occur across an undisturbed interface.

The assessment of microbial growth or metabolic transformation of energy sources in natural populations is a central problem in aquatic microbiology (Jannasch, 1969). Metabolic regulation, the interdependence of different groups of micro-organisms for essential nutrients and growth conditions and selective conditions imposed by incubation prevent the extrapolation of data obtained from cultures to natural mixed populations. In situ measures of growth under natural conditions are required. Brock and Brock

(1968) described microscopic autoradiographic techniques which permit a quantitative assay of uptake of labelled substrate by micro-organisms. Exposure of a culture of filamentous marine bacteria (*Leucothrix mucor*) to tritiated thymidine indicated that dividing cells synthesizing DNA accumulated the label in a linear fashion and that 1% of the cells were labelled in 0·002 generations. Generation times of 660 to 685 min were observed in natural populations while values of 94 min occurred in pure cultures growing epiphytically. Watt (1971) and Stull *et al.* (1973) have used similar methods to quantify species specific incorporation of $^{14}CO_2$ during phytoplankton photosynthesis. Generation times varied from over 1000 hr for some diatom species to less than 2 hr for several species of green algae. These techniques have not been widely applied to studies of microbial growth on solid surfaces. The conversion of grain counts to radioactivity requires many assumptions and quantitative results may be difficult to obtain (Knoechel and Kalff, 1976). Evidence that metabolic rate (oxygen uptake) and the degree of attachment of bacterial cells to particles is related to dissolved nutrient concentrations and the nature of the substrate could be assessed by these methods (Jannasch and Pritchard, 1972).

Pütter (1909) first suggested that dissolved organic matter in water might serve as a nutrient source when directly absorbed by metazoan organisms. Little experimental evidence for heterotrophic utilization of organic substances thought to be truly dissolved existed until studies summarized by Stephens (1967) demonstrated the uptake of various labelled dissolved organic compounds by numerous soft-bodied marine and estuarine invertebrates. Pogonophora species, which possess no internal digestive system, accumulated small organic molecules (glucose, amino acids, fatty acids) which could only occur through external body surfaces (Southward and Southward, 1974). These authors also calculated that at concentrations which exist in sea-water and sediments ($10^{-9} - 10^{-5}$ moles/liter), some benthic species could accumulate sufficient quantities to meet respiratory requirements. Weight specific uptake of amino acids varied inversely with body weight in the polychaete *Nereis* and logarithmic relations between substrate concentration and uptake rate existed for nereids and pogonophores (Southward and Southward, 1972).

Methodological problems persist in interpreting measures of uptake of dissolved organic compounds, however. Physical adsorption or uptake by micro-organisms on external surfaces must be controlled through the use of sterile and non-living organisms. Also, the addition of labelled compounds to filtered water provides no assurance that particulate matter is not present. Debris may be formed during filtration and even the smallest pore size of membrane filter passes particles (Sheldon *et al.*, 1967; Riley, 1970). When these variables are controlled, the incorporation of a labelled organic compound in body tissue from dilute solutions and the evolution of $^{14}CO_2$ indicate active accumulation and metabolism of dissolved substances. The relative importance of this heterotrophic uptake for nutrition remains in doubt, however. Excretion of various dissolved organic compounds by aquatic invertebrates can be comparable to respiratory energy loss (Hargrave, 1971; Field, 1972) and Johannes and Webb (1970) demonstrated a net release of amino acids by the polychaete *Clymenella*. Southward and Southward (1974) reviewed previous studies and concluded that the nutritive value of epidermal uptake by animals capable of particulate feeding was unknown. Pogonophores, however, do apparently exist only on absorbed soluble organic matter.

Dissolved organic carbon in sediment pore water decreases in concentration with depth in the upper 10 to 15 cm of abyssal sediments (Starikova, 1970; Karl *et al.*,

1976). Maximum bacterial numbers and metabolic activity also occurs in surface sediment layers (Wieser and Zech, 1976) and thus the turnover of these soluble organic compounds is probably maximum near the sediment surface. Lewin and Lewin (1960) and Andrews and Williams (1971) have also measured heterotrophic uptake of low molecular weight organic compounds by numerous species of planktonic and attached algae. The ability of many benthic algal species to remain viable despite prolonged burial could be due to heterotrophic utilization of dissolved organic compounds in interstitial water.

High carbon:nitrogen ratios in sediments and suspended organic detritus imply that, relative to carbon, nitrogen is often in short supply. Various micro-organisms which fix and transform nitrogen provide an additional source of supply of this essential element for aquatic production. Painter (1970) reviewed biological transformations of nitrogen which occur in water and soil. Nitrogen fixation (the synthesis of cellular nitrogenous compounds from elemental nitrogen) is carried out by both aerobic and anaerobic micro-organisms which are either free-living or in symbiotic associations.

Type	Micro-organisms capable of nitrogen fixation in sediments	
	Aerobic	Anaerobic
Heterotrophic	Azotobacter Pseudomonas	Clostridium, Aerobacter Pseudomonas, Bacilluspolymyxa
Autotrophic	—	Methanobacterium
Photosynthetic	Nostoc, Calothrix Anabaena	Chromatium, Achromobacter Chlorobium
Symbiotic	Puccinella rhizosphere	

Increases in Kjeldahl-N or NH_4-N have been used as a measure of N_2 fixation but ^{15}N incorporation provides a more sensitive and reliable method (Painter, 1970). More recent techniques depend on the ability of N-fixing micro-organisms to reduce acetylene to ethylene, which is measured by gas chromatography. Jones (1974) used this method to measure annual rates of nitrogen fixation in various zones of a salt marsh. Values ranged from 0·4 g N/m^2 on bare mud to 46 g N/m^2 in pools which contained blue-green algae (Nostoc). ^{15}N-labelled products of nitrogen fixation by micro-organisms were transferred to the roots and leaves of higher plants in the salt marsh. Nitrogen fixation by micro-organisms (primarily Desulfovibrio) in estuarine sediments reached maximum rates (1·07 ng N/g dry sediment/hr) under aerobic conditions during summer (Herbert, 1975). Incubations with anaerobic cores resulted in higher maximum rates of fixation (1·84 ng $N/g/hr$) and the addition of glucose to cores increased both aerobic ($\times 3$) and anaerobic ($\times 2·5$) rates.

Many micro-organisms capable of nitrogen fixation possess respiratory systems which use elements other than oxygen as a terminal electron acceptor. Thus, in anoxic sediments, sulfate may be reduced to sulfide by anaerobic chemoautotrophic bacteria such as Desulfovibrio which oxidizes H_2 by reducing $SO_4^=$. Similarly, Thiobacillus denitrificans oxidizes H_2S by reducing NO_3^-. Energy obtained by these processes is used to reduce CO_2 to carbohydrates and no other organic compounds are required for growth. The oxidation of organic matter by these anaerobic pathways, particularly in organically rich sediments, is often dominated by the sulfur cycle (Fig. 80). Over 50% of the total mineralization of Zostera detritus added to sand-filled aquaria was due to sulfate reduc-

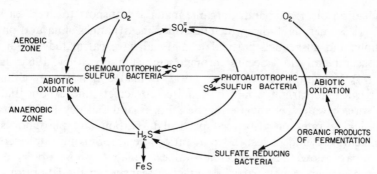

FIG. 80. Sulfur cycle in sediments (redrawn from Blackburn *et al*, 1975).

tion which decreased from 80 to 25 nM S/cm^3/day as the system became more oxidized during 7 months (Jørgensen and Fenchel, 1974). H$_2$S and FeS increased during mineralization and some of these compounds were oxidized abiotically and by chemo- and photoautotrophic bacteria which developed as a sulfuretum at the sediment surface. Oxidation rates of 0·1 mM S$^=$/l/min were only 10% of maximum rates under optimum light and sulfide concentrations (Blackburn *et al.*, 1975).

A succession of sulfur bacteria in artificial sulfureta appears to reflect competition for sulfide (Jørgensen and Fenchel, 1974). Since reducing conditions and sulfide initially only existed in the artificial systems below the sediment photic zone, chemoautotrophic white sulfur bacteria (such as *Beggiatoa* and *Achromatium*) catalysed the biological oxidation of sulfide and formed a dense plaque on the sediment surface. As sulfide concentrations increased within the photic layer, however, photoautotrophic purple (*Chromatium, Thiopedia*) and green (*Chlorobium*) sulfur bacteria developed in sequential layers overlying deposits of FeS. Layers of purple and green sulfur bacteria can become sufficiently dense to restrict the upward diffusion of sulfide. Reduced sulfide supply causes oxidation of stored elemental sulfur to sulfate. Destruction of bacterial plates follows in reverse order to their formation.

Carbon, nitrogen, and sulfur cycles in sediments are intimately related (Hallberg, 1973). Photosynthetic algae may participate in sulfide oxidation and sulfate reducing and photosynthetic sulfur bacteria may fix nitrogen. Autotrophic methane oxidizing bacteria are also able to oxidize ammonia to nitrite in amounts equivalent to methane oxidized (Hutton and ZoBell, 1953). Martens and Berner (1974) have demonstrated, however, that methane production and sulfate reduction are mutually exclusive processes in anoxic sediments. Methane production does not occur until dissolved sulfate is totally consumed. The interdependence of these processes ensures that if oxidation of organic matter is incomplete, a series of end products are produced which in turn serve as substrates for other metabolic pathways. The diversity of metabolic processes ensures continued bacterial transformation of sedimentary organic material regardless of oxidation-reduction conditions.

6.4.1.2 Sedimentation, resuspension and advection.

The import of dissolved and particulate organic matter through the water column is the only energy source for benthic communities not exposed to light. Despite measurements of the uptake of dissolved compounds by undisturbed sediments (Hall *et al.*, 1972; Jones and Simon, 1975), the relative importance of heterotrophic utilization of dissolved organic matter is unknown. Particulate sedimentation as a 'rain of detritus' has been thought to be the primary mechanism of organic and inorganic material transport to deep-sea sediments (Fournier, 1971). Numerous studies which described the dynamics of natural phytoplankton populations and suspended particulate material have calculated that a relatively small and constant daily loss occurs through sedimentation (Riley, 1965; Jassby and Goldman, 1974). Nakajima and Nishizawa (1972) estimated that a daily elimination of 2 to 4% of particulate carbon would be required to achieve the exponential decrease observed in surface waters of the Bering Sea. Burns and Pashley (1974) measured *in situ* vertical sinking rates of suspended particulate carbon and phosphorus in Lake Ontario using closed cylinders suspended at various depths. Sinking rates were between 0·1 and 2·5 m/day below the thermocline—values similar to those observed by Riley (1970) for aggregates and small particles settled in seawater in the laboratory.

Sinking rates determined in still water cannot be related directly to natural conditions and Burns and Pashley's data indicate that settling velocities of suspended particles are variable and dependent on physical and biological conditions. Negative settling velocities above the thermocline are consistent with the idea that turbulence in surface water reduces loss and may even permit dense particles such as diatoms to remain in suspension (Lund, 1959; Smayda, 1970a). Riley (1970) also agrees that suspended particulate matter may not truly sink but only have an apparent downward movement due to advection. Sedimentation may only be affected by actual particle sinking velocity at depths where advection becomes unimportant and in water where horizontal and vertical turbulence exists, sedimentation probably reflects conditions of turbulence more than settling of particles (Murray, 1970).

Kranck (1973, 1975) observed that particles in sea water flocculate into aggregates of characteristic stable size distributions dependent on the grain size of inorganic particles. Apparently, grains flocculate until all particles have approximately equal dynamic transport speed and are no longer in contact with each other. These processes significantly affect sedimentation since particle mass and settling velocity (flux) are increased. The effect may be most significant for small particles. Size fractionation has demonstrated that only 2% of total organic carbon in open ocean water is in particles larger than 0·8 μm 25% is colloidal and the remainder is 'dissolved' (Sharp, 1973). Of particles larger than 1 μm in surface water, 58% are between 1 and 10 μm and 30% are between 10 and 95 μm (Mullin, 1965). Light scattering and Coulter Counter determinations have also demonstrated the overwhelming abundance of particles <2 μm (Jerlov, 1968; Beardsley *et al.*, 1970). The cumulative number of particles (N) larger than a given diameter (d) can be described by a negative exponential equation (hyperbolic function)—

$$N = k\,e^{-md} \tag{132}$$

or a power function—

$$N = k\,d^{-m} \tag{133}$$

where k and m are constants (Bader, 1970; McCave, 1975). On linear scales the relations both ressemble a hyperbola but the 'fit' of the equations to actual data often only applies over a small diameter range (1–100 μm, for example) as an approximation.

McCave (1975) calculated negative exponential (hyperbolic) distributions of suspended particle number and volume from data provided by Sheldon and previous studies and found slope values between −2·4 and −3·6 for suspended material below 200 m. Density was assumed (with empirical evidence) to decrease with increasing particle size and volume histograms were converted to mass distributions. Stokes velocities, calculated for each particle size class, indicated that most of the particle flux (settling velocity

× mass) occurs as large particles. Suspended material <32 μm appears to constitute a 'background' of particulate matter which does not contribute substantially to downward flux due to its low settling velocity (0·01–1 m/day). Distribution may be controlled by horizontal advection. Particles between 90 and 362 μm diameter, however, may settle at rates between 10 and 100 m/day. The assumption of a steady state in suspended material throughout the water column necessitates a continuous aggregation of small particles in surface layers to maintain a continuous supply of these large rapidly sinking particles. The rarity of large particles in a small water volume prevents adequate assessment of their abundance by current sampling methods which use water bottles.

McCave (1975) reviewed numerous observations which support the hypothesis of accelerated sinking of particles in the sea. Localization of biogenic compounds under areas of high surface production, the presence of short half-lived radionuclides in deep-sea invertebrates and similarities between atmospheric dust and seabed sediments all imply transport of particles from the surface on a time scale of weeks to months. High settling velocities (10^2–10^3 m/day) of euphausiid and zooplankton fecal pellets (Table 2) and their resistance to physical breakage suggest that grazing activity could provide a supply of rapidly sinking material. Fecal pellets are abundant in material deposited in marine bays (Steele and Baird, 1972; Hargrave et al., 1976) and Seki et al. (1974) attributed the formation of anoxic bottom water in Tokyo Bay to the deposition of zooplankton feces. Heyraud et al. (1976) have also estimated that fecal pellets are the main route by which ^{210}Po is eliminated by the euphausiid Meganyctiphanes norvegica and that these play an important role in removal of ^{210}Po from surface layers.

Particulate material suspended above the sediment and throughout the water column may not only be derived from sinking remains of phytoplankton and zooplankton. Resuspension of bottom sediments and lateral transport as turbidity layers or concentration in turbidity maxima may occur in estuaries, coastal areas and the deep ocean. Postma (1967) discussed sediment transport processes which prevent loss of suspended matter from near-shore areas at rates expected from removal of coastal water. Settling and scour lags can combine with tidal movement to cause residual transport towards a coast and density differences due to river input may concentrate materials in turbidity maxima. Retention and sedimentation of fine-grained particles also increase due to flocculation. Suspended matter may remain concentrated as a nepheloid layer above the sediment or become detached where currents flow across steep bottom gradients. Turbidity layers separated from the bottom and extending at the same depth into the slope water column were found to be associated with thermal gradients as small as 0·1°C (Drake, 1971). Turbidity currents, a gravity current with an excess of specific weight due to suspended sediment, can also arise from submarine slumps and the influx of water across a sill (Heezen et al., 1955; Sholkovitz and Soutar, 1975). The thickness of the turbid layer and its horizontal extent will depend upon the mean current flow, turbulence, and the settling rate of suspended material. Relations between erosion, transportation, deposition, and inorganic sediment grain size indicate that deposition occurs at all particle sizes unless velocity exceeds 15 cm/s (Postma, 1967). Clay and silt size particles at the sediment surface, however, are usually incorporated into an organic matrix which may result from sediment reworking by deposit-feeding invertebrates (Johnson, 1974). These aggregates probably have different sedimentary properties from mineral grains of comparable size.

Wangersky (1974, 1975) and Gordon (1977) observed variability in suspended particulate carbon in the deep sea which indicated local small-scale concentrations from four to five times average background values. Particle 'clouds' were not associated with bottom nepheloid layers and they could not be attributed to macro- or microzooplankton. Aggregations existed on a scale smaller than 1 m (the size of the sample bottles) and they occurred in both horizontal and vertical directions with concentrations of organic carbon equivalent to those observed in surface waters during phytoplankton blooms (50–100 μg C/l). The persistence of these enriched concentrations is unknown but they must provide a significant energy source for deep-sea organisms. Fournier (1971) also observed pigmented cells (1–15 μm dia.) throughout the aphotic zone of various oceans. Maximum abundance occurred between 300 and 500 m, concentrations decreased with increasing depth and similar cells were observed in the guts of benthic and pelagic invertebrates (Fournier, 1973). Profiles of other parameters associated with suspended particulate matter also show depth-dependent decreases (Section 6.1.3.3) which suggests some consumption during deposition assuming no horizontal advection.

Conditions of water movement (both advective and turbulent) can critically affect measures of sedimentation. Collections of sedimenting material are usually made by exposing closed-bottom containers clamped to a rigid wire at various distances above the sediment (Edmondson and Winberg, 1971). Different designs of trap have been used but most collectors have a length greater than the mouth opening to reduce internal turbulence. Cylinders and funnels of various size collected similar amounts of deposited material in a stratified lake basin (Kirchner, 1975), however, in areas where water movement occurs, unrealistically high sedimentation rates may result because collectors impose an artificial interface. Large differences in deposition rates occurred in various types of trap exposed simultaneously under non-stagnant conditions (Johnson and Brinkhurst, 1971b). Traps designed as shallow trays (Håkanson, 1976), which are placed directly on the bottom and subject to scouring and removal of settled material, may provide a realistic measure of actual deposition.

The importance of trap configuration and placement for assessing particulate deposition onto sediment surfaces is apparent in studies which have measured vertical profiles of sedimentation (Steele and Baird, 1972; Ansell, 1974; Webster et al., 1975; Davies, 1975) (Table 37). The quantity of material trapped generally increased with depth. Although resuspension is usually implied from the presence of chlorophyll degradation products, inorganic material and high carbon: nitrogen ratios, in situ heterotrophic and fecal production or the existence of different water masses at different depths could also account for the observations. High sedimentation rates measured close to the sediment surface in shallow water can be attributed to resuspension through wind and tidal action (Oviatt and Nixon, 1975), but the process may contribute to deposition whenever traps are suspended close to a sediment surface. Attempts to correct for resuspension require that the origin of sedimenting material at a specific depth be known (Gasith, 1975) or assume a linear profile of sedimentation with depth (Steele and Baird, 1972; Davies, 1975). Neither method is applicable, however, when non-vertical supply occurs. Anomalous sedimentation patterns (high or low values between successive sampling depths) observed by Ansell (1974) and Hargrave et al. (1976) in marine bays indicated that material trapped at different depths was often of different origin.

Decomposition of sedimented material in traps has been neglected as an aspect of sedimentation measurements which could cause apparent differential sedimentation at different depths. Bacterial growth could alter organic carbon and nitrogen content in sedimented material and plant pigments would degrade over extended collection periods. Seki et al. (1968), by use of a bacterial growth bioassay, found that only a few percent of the carbon content in sedimented material was utilized in 3 weeks and exposure time is generally assumed not to influence calculated sedimentation rates (Edmondson and Winberg, 1971). Johnson and Brinkhurst (1971), on the other hand, demonstrated that freshly settled material lost 15% to 25% of its organic content in 6 days. The difference in results may reflect rapid losses which occur immediately after sedimentation when labile organic matter is degraded. Preservatives added to evaluate the magnitude

TABLE 37. ANNUAL CARBON DEPOSITION IN DIFFERENT MARINE ENVIRONMENTS

Location	Bottom depth (m)	Trap depth (m)	Sedimentation (g C m^{-2} yr^{-1})	Reference
Baltic	25	24	40	Zeitschel, 1965
Departure Bay	32	30	200	Stephens et al., 1967
Southampton Water				
Calshot	2·5	2	376	
Marchwood	5·7	5	1683	Trevallion, 1967
Netley	5·7	5	339	
Loch Ewe	25	18	30	Steele and Baird, 1972
Loch Etive				
E 6	40	36	247	
E 24	20	18	82	Ansell, 1974
Loch Creran C 3	40	24	262	
Loch Thurnaig	30	17	28	Davies, 1975
St. Margaret's Bay	15	13	118	Webster et al., 1975
	70	65	134	
	62	20	57	
	62	30	69	
Bedford Basin	62	40	77	Hargrave et al., 1976
	62	50	85	
	62	60	91	
La Jolla Bight	18·3	16·5	1214	Hartwig, 1976

of possible loss indicated that nitrogen was mineralized in material collected over 3 months in the surface waters of a meromictic lake (Matsuyama, 1973). The addition of formalin and chloroform did not alter the amount of organic matter trapped over various time intervals in lakes (Edmondson and Winberg, 1971). Hartwig (1976) added mercuric chloride to sediment traps. While this prevented fouling and colonization of traps by fauna, bacterial respiration in sedimented material was not eliminated and prolonged exposure (more than 21 days) did not significantly alter the percentage organic carbon present. If decomposition of material sedimented is not measured, trapped material collected after a few days or weeks exposure may best be considered as the supply of relatively stable organic matter residual after bacterial attack.

Previous studies of sedimentation in coastal marine areas and lakes (Table 37) have shown a variety of seasonal patterns. A bimodal distribution in sedimentation occurred in Departure Bay (Stephens et al., 1967). Phytogenous material was deposited during May–July, 2 months after the spring bloom, and terrigenous material sedimented during October–December, the time of peak runoff. Maximum carbon deposition occurred during winter in St. Margaret's Bay, possibly reflecting detritus supply from seaweeds (Webster *et al.*, 1975). Other studies showed seasonal patterns which indicated that resuspension occurred at certain depths during the year. In Bedford Basin, for example, high daily deposition rates during summer and fall corresponded to phytoplankton production (Hargrave *et al.*, 1976). Ansell (1974) observed a similar seasonal pattern in two Scottish sea lochs. Sedimentation of plant pigments in Bedford Basin, however, was maximum during winter when phytoplankton production was minimal and deposition of chlorophyll *a* increased with depth. Even though degradation could have occurred during collection, increased sedimentation in deep water must have resulted from horizontal transport mechanisms. The correlation between macro-benthos biomass and chlorophyll standing crop during spring (Hargrave and Peer, 1973; Section 6.1.3.3) implied that sedimentation during and following a spring bloom was an important source of organic matter. Hargrave (1975), however, compared results of eight different studies and observed no constant relationship between timing of peak periods of supply and sedimentation. Resuspension and factors related to water column stability could have been important factors which influenced sedimentation in many of the shallow water areas considered.

In the absence of direct measures of sedimentation in the deep sea, indirect observations have been used to infer the supply of organic matter. Many studies have suggested that 70% to 90% of the organic matter produced in the ocean is mineralized within the upper layers of the water column (Riley, 1951; Hargrave, 1973). Riley (1970) calculated that 75% to 80% of the carbon fixed by phytoplankton in the Sargasso Sea was consumed by zooplankton and bacteria between the surface and 900 m. Of the remaining particulate carbon, 80% was consumed between this depth and the bottom (4000 m). Thus only a few percent (1–2 g $C/m^2/yr$) of the carbon fixed by phytoplankton could reach the bottom. Other calculations for various regions in the Pacific Ocean and Black Sea estimated that between 5% and 10% of organic matter produced at the surface reaches depths below 2000 m to 3000 m (Bogdanov, 1965; Deuser, 1971). As a contrast, Riley (1956b) calculated that between 30% and 40% of primary production (60–80 g $C/m^2/yr$) would reach the sediments in Long Island Sound—an amount consistent with observations in other coastal waters (Table 37).

High organic input to coastal sediments relative to that in the deep sea would be expected on the basis of higher levels of primary production and shallow water depth. Also, if production and decomposition processes in surface layers are closely linked with only a small fraction of production actually deposited, sedimentation should be proportional to suspended organic matter concentration and inversely related to water depth (Ohle, 1956). Steele and Baird (1972) and Ansell (1974) found little relation between sedimentation rates and concentration of suspended material. However, Hargrave (1975a) compared annual particulate organic carbon supply and sedimentation in various lakes and marine bays ($n = 8$) with different total and mixed-layer depths. Organic carbon sedimentation (S) (g $C/m^2/yr$) was linearly related to particulate carbon supply (C_s) (g $C/m^2/yr$) and inversely related to mixed layer (Z_m) (m) (thermocline) depth. There was no significant relation between sedimentation and total depth but only a narrow range of values (12–65 m) was represented in the data. Since mixed-layer depth and carbon supply were independently related to sedimentation, a single equation was derived which related sedimentation to the ratio of carbon supply:mixed-layer depth:

$$S = 4 \cdot 9 + 3 \cdot 9 \left(\frac{C_s}{Z_m} \right).$$

(134)

Sedimentation, expressed as a percentage of carbon supply, was thus directly related to mixed-layer depth by the equation:

$$\left(\frac{S}{C_s} \cdot 100\right) = 73 \cdot 4 - 2 \cdot 8 Z_m. \tag{135}$$

Despite the lack of data from open ocean areas and the assumption that sedimented material is all derived from estimated carbon supply, the relations suggest that for a given carbon input at the surface mineralization occurring in the upper mixed-layer determines the amount of organic deposition. In areas where a thermocline exists between 10 m and 15 m, 30% to 40% of the carbon supply is sedimented. Where a mixed-layer depth exceeds 25 m, carbon input to the bottom amounts to a few percent of the supply.

6.4.1.3 Vertical migration, carcass deposition and macro-detritus. Riley (1951) and Vinogradov (1955) proposed predator–prey links between migrating zooplankton which would result in a rapid step-wise downward transport of material by interlocking cycles of vertical migration. Food ingested at shallow depths can be egested in deeper water and both horizontal and vertical transport can occur at rates dependent on descent rate and deep current flow. The rate of ascent and descent of the deep scattering layer on the order of several m/min (Section 1.3.4) indicates that these organisms may also facilitate rapid downward movement. In addition, evidence from baited cameras (Dayton and Hessler, 1972) and the gut contents of deep-sea demersal fish (Clarke and Merrett, 1972) indicates that certain predators feed on carcasses. Many species of fish have mouths adapted to ingest only large pieces of food and the numbers of scavengers attracted to bait suggests that efficient location mechanisms exist. Fragments of macrophyte debris, wood, and *Sargassum* found in deep-sea sediments indicate additional sources of organic matter which may be supplied as large particles (Menzies and Rowe, 1968; Schoener and Rowe, 1970). The relative importance of these sources of organic debris relative to that supplied as zooplankton feces or finely dispersed material has not been evaluated in the deep sea. The daily input of macro-detritus (algal fragments, worm tubes) per m^2 to near-shore sediments in La Jolla Bight was two orders of magnitude less than fallout of finely dispersed debris (Hartwig, 1976). Estimated resuspension accounted for 90% of the organic carbon deposited, however, and thus large particles of organic debris were a significant source of carbon input to these sediments.

6.4.2 FEEDING PROCESSES

6.4.2.1 Mechanisms. Yonge (1928) and Nichol (1960) classified invertebrate and vertebrate feeding habits on the basis of whether nutrients were obtained as a liquid or as small or large particles. This operationally useful separation of food sources avoids the necessity of assigning animals to specific trophic levels—a particularly difficult choice for benthic animals which often have catholic diets. Extensive reviews that differentiate liquid and particulate feeding (Jørgensen, 1966; Newell, 1970; Pandian, 1975; Conover, 1977) summarize feeding mechanisms and nutritional physiology in various marine organisms. Examples of both modes of nutrition exist in the benthos.

Direct utilization of dissolved organic matter occurs in benthic plants, microorganisms, and internal parasites of vertebrates and invertebrates. Absorption occurs through external surfaces and compounds are rapidly incorporated into cellular material. The apparent uptake of dissolved organic compounds by soft-bodied marine invertebrates, however, is still a topic of controversy since excretion of large amounts of similar substances also occurs (Section 6.4.1.1). Net uptake of dissolved material across external body surfaces has not been demonstrated under field conditions and since excretion is a little studied parameter in energy budget calculations, the physiological significance of the phenomenon for metazoans remains unknown. The reciprocal transfer of soluble organic compounds between symbiotic algae and host organisms (various corals, anenomes, turbellarian flat worms, molluscs), however, does illustrate the direct utilization of dissolved organic material by coexisting organisms. Conover (1977) has reviewed studies which demonstrated the rapid movement of dissolved organic compounds into and between symbiotic organisms. Again, it is seldom clear how essential soluble materials are for nutrition. Corals, for example, obtained sufficient energy by particle feeding when prey density exceeded minimal levels (Coles, 1969), yet intact coral heads appeared to meet only a small part of their respiratory requirements by capturing zooplankton (Johannes *et al.*, 1970). Perhaps pathways of nutrition in these organisms are regulated depending on the relative availability of dissolved and particulate organic matter.

Benthic animals which ingest small particles do so by one of three mechanisms which may be selective or non-selective:

1. filter feeding;
2. browsing (rasping);
3. deposit feeding.

To some extent, macrofauna which feed by different means are stratified within benthic communities. Filter feeders usually remove suspended material from water over the sediment, browsing organisms scrape material from surfaces or the sediment interface and deposit feeders ingest sediment particles directly. The feeding methods are not mutually exclusive. *Scrobicularia*, a deposit-feeding mollusc, also obtains particles through filtration by holding its inhalant siphon above the sediment surface when not directly ingesting sediment particles (Hughes, 1969). Also, it was earlier thought that these methods of feeding were generally non-selective. Filter feeders, like bivalve molluscs, barnacles, and sponges which use cilia and setae to create water currents and cause fine particles to collect on feeding surfaces where mucus bands carry food to the digestive tract, were thought to take what came to them (Hoar, 1966). Annelids, polychaetes, and echinoderms which lack piercing and sucking mouthparts and ingest sediment were also considered to feed indiscriminately. Conover (1977) discussed recent studies of invertebrate feeding which demonstrate that food-particle gathering is seldom automatic or indiscriminate regardless of the method.

Suspension feeding molluscs use gill surfaces for both gas exchange and food particle collection. Particles between 2 μm and 8 μm are usually retained with 100% efficiency but patterns of beating of latero-frontal cirri between adjacent gill filaments can be altered (Dral, 1967; Vahl, 1972). Feeding and respiration are also linked, as demonstrated by the effect of current speed and body size on pumping rate. Walne (1972) observed a power relation between water flow and filtration in five species of bivalves which demonstrates the importance of water movement past sedentary filter feeders. Pumping, as a metabolic process, is also related to body weight by a power function (Winter, 1969; Ali, 1970; Walne, 1972) but exponents of regressions of body weight on filtration rate are highly variable (-0.3 to -0.8). If pumping was directly related to respiration, a standard metabolic weight exponent of -0.2 to -0.35 would be expected. Although Conover (1977) cautions against comparisons when different methods were used to obtain data, regressions do permit calculation of weight-specific pumping rates. Comparison of various bivalve species on this basis suggested that mobile, rapidly growing species (scallops) had weight-specific filtration rates an order of magnitude greater than more sedentary species (Winter, 1969). Ali (1970) similarly observed lower pumping per unit weight by oysters and a rock-boring mollusc (*Hiatella*) in comparison to other species. In addition, Hughes (1969) found that the square root of gill area was linearly related to shell length in four species of molluscs, thus, pumping rates per unit gill area could be used as a standardized measure of filtration rate. When animals of the same shell length were compared, *Scrobicularia*, primarily a deposit feeder, had the lowest pumping rate per mm^2 of gill surface. The three species with the highest gill area-specific pumping rate (*Venus, Mytilus, Cardium*) filter clear water from above the sediment surface. *Scrobicularia* and *Mya*, which inhale water and suspended matter at the sediment surface, had lower rates on this basis.

Invertebrates which feed by scraping substrates (plant surfaces, rocks, and sand grains) may ingest particulate matter directly or absorb fluids and soft tissue by piercing and sucking. Various meiofauna such as tardigrades have tube-like mouths with stylets to pierce plant cell walls and a muscular pharynx to suck out cell contents (Swedmark, 1964). Snails and limpets scrape surfaces through rotation of a radula and their feeding action in intertidal areas often keeps rock surfaces free of epilithic algae (Castenholz, 1961). Moore (1938) first observed that the size of the feeding area (y as cm^2) and volume of the limpet *Patella* (x as cm^3) were linearly related by the equation:

$$y = 0.0125x \tag{136}$$

which can be calculated from his data. Castenholz (1961) did not measure feeding area, but he did demonstrate that volumes of littorine snails greater than 0.2 cm^3/dm^2 and limpet volumes over 0.8 cm^3/dm^2 kept rocks diatom free during summer. Since natural densities in tide pools exceeded these levels, feeding areas must have overlapped with rock surfaces being continually rescrapped depending upon duration of immersion. Browsing by sea urchins also prevents colonization of rocks by macroalgae (Paine and Vadas, 1969a; Breen and Mann, 1976).

Deposit-feeding organisms may ingest particles from on or under the sediment surface by a variety of methods (Newell, 1970). Some holothurian species, for example, appear to consume plant and animal debris from rock, sand, and mud surfaces as epibenthic feeders (Yingst, 1976) while other species feed on subsurface sediment and apparently accumulate interstitial meiofauna selectively (Walter, 1973). Selective feeding has been clearly demonstrated for various molluscs which feed on and within sediments. *Macoma* and *Scrobicularia* (bivalves) rotate their inhalant siphon to suck in the top millimetre of sediment from a circular area around the siphon-tube opening (Hughes, 1969; Hylleberg and Gallucci, 1975). Initial intake appears indiscriminate, but separation occurs on the gills and palps where coarse organic and inorganic particles and small particles low in organic matter are ejected as pseudo-feces. Hughes (1973, 1975) examined the mechanism of sorting in the bivalve *Abra* and found that the upper size limit of particles entering the mantle cavity was determined by the diameter of the inhalant siphon. Material ingested was usually smaller than that taken into the mantle cavity indicating sorting by pallial organs.

Behavioural, morphological, and physiological differences permit the coexistence of many deposit-feeding species. When related *Hydrobia* species (surface deposit feeding gastropods) occurred together, Fenchel (1975) observed that a narrower range of particle sizes was ingested than when each species occurred separately. Character displacement apparently caused a size separation which led to differences in the range of particle sizes ingested. *Hydrobia* also coexists with a burrowing amphipod (*Corophium*) and whereas *Hydrobia* generally consumes particles between 60 μm and 300 μm diameter, *Corophium* selects smaller particles and ingests a greater proportion of bacteria (Fenchel et al., 1975). Similarly, burrowing lugworms selectively ingest organically rich fine-grained sediment (Hylleberg, 1975) and polychaetes (*Pectinaria*) select particles in direct proportion to their body size (Whitlatch, 1974). Methods of selective feeding by burrowing species differ, however. *Abarenicola* uses its proboscis to excavate a water-filled pocket with coarse sand accumulated at the bottom and fine particles kept in suspension. *Pectinaria* uses palps to dig at the base of a vertically facing tube and ciliated grooved tentacles independently select particles near the mouth. Electivity indices for such organisms are often difficult to calculate since there may be a vertical gradation in particle size and individuals of different size may feed at different depths (Whitlatch, 1974). A similar problem exists in assessing feeding activity in numerous fish and crustaceans which feed on and within sediments. Often, while they are predators, inorganic and organic debris accounts for the bulk of their gut contents and material ingested in one location may be carried to another when the animals move (Sheldon and Warren, 1966). Odum (1970) also demonstrated that detritus-feeding fish (*Mugil*) either browse epiphytic algae or consume macroscopic debris and select for fine particles, depending on their relative availability.

All of these studies demonstrate discrimination in feeding which permits some degree of food-resource partitioning based on particle-size selection. In contrast, selective feeding benthic micro- and meiofauna are much more highly specialized with different kinds and sizes of micro-organisms (bacteria and algae) serving as specific food sources for particular species (Fenchel, 1969). The high diversity of microfauna in sediments probably reflects this specificity of food type (Fenchel et al., 1975). A lack of food specialization to this degree in many detritus and deposit-feeding invertebrates, however, is balanced by the development of alternative feeding methods. For example, *Ampelisca*, a tube-building amphipod, is a predator, but when food is unavailable, antennae are used to scrape up surface sediment (Kanneworf, 1965). Newell (1970) described feeding by many benthic invertebrates like *Nereis*, a polychaete, which generally filters water through a secreted mucus bag, but inorganic sediment particles and other organisms may also be ingested. Similarly, *Hydrobia*, which usually ingests individual mineral grains, may browse flat surfaces and perhaps trap particles in mucus in the surface film of water (Fenchel et al., 1975). Since switching of feeding methods appears to depend on food availability, experimental measures of food intake may be of limited value for extrapolation to nature.

Benthic animals which ingest large particles are generally carnivores. While some sedentary predators exist in the benthos (attached coelenterates, some polychaetes), most are mobile organisms which use visual and chemical senses to perceive prey. Epifauna are generally more vulnerable than infauna, but any prey may be swallowed whole, crushed, or drilled and partially consumed depending on the feeding method of the predator. Limits to ingested prey size are often dependent on predator size. For example, the carnivorous mollusc *Navanax* can only consume prey up to a certain maximum size and during growth the ability to manipulate and swallow small prey is lost (Paine,

1965). Many invertebrate predators (starfish, shell-drilling molluscs) only ingest a portion of their prey and demersal fish often browse on extended siphons of buried molluscs (Trevallion *et al.*, 1970; Tyler, 1971). While many of these predators are morphologically and behaviourally adapted to feed on selected prey species, often food selection is inferred from identification of prey in gut contents. Sympatric species-pairs of skates, for example, utilized many of the same species but in different proportions and with differences in dominance of epifauna and infauna (McEachran *et al.*, 1976). The differences largely reflected the availability of different prey in the benthic communities in which the species-pairs occurred. Levings (1973) also concluded that flatfish predation in a shallow-water benthic community was opportunistic with large vulnerable macrofauna being the primary food supply.

6.4.2.2 Food availability. Energy uptake by organisms, either as light, soluble nutrients or particles, has often been related to the abundance of the energy source. Light-photosynthesis curves (Section 3.1.2), substrate and nutrient uptake by microorganisms (Section 3.1.3), particle retention by suspension feeders (Section 4.1), and predation by both invertebrates and vertebrates, for example, has been represented by rectilinear or curvilinear equations which have threshold values where energy uptake becomes saturated. Many of these response curves have a shape similar to a rectangular hyperbola and such a formulation was used to describe light-photosynthesis relations (Baly, 1935) and population growth of micro-organisms (Monod, 1949). Caperon (1967) observed that the relation adequately described both nitrate uptake and photosynthesis-light responses in laboratory cultures of phytoplankton. Jassby and Platt (1976), however, found the equation inadequate to describe light response in natural phytoplankton populations. Uptake of substrates by mixed populations of micro-organisms and metazoans in sediments is also often not described by Michaelis–Menten kinetics (Section 6.4.1.1). Presumably interactions and conditions exist in these natural planktonic and benthic communities which prevent the occurrence of idealized response curves.

There have also been frequent attempts to use the hyperbolic equation or related hyperbolic functions to describe food supply and feeding in predator–prey systems. While these may suitably represent predation based on a two-phased feeding habit of searching and handling individual prey items before ingestion, they are generally not applicable to the process of filter feeding where food is gathered and processed simultaneously (Lehman, 1976). A more realistic model of selective feeding dependent on the relative abundance, size, and digestibility of particles was suggested. Boyd (1976) similarly pointed out the necessity to describe suspension feeding in mechanistic terms if observed particle size selection by suspension feeders is to be given a quantitative basis.

Jørgensen (1966) observed that the response of suspension feeding invertebrates to changes in particle number and type was variable, but that in general filtration declined at high particle concentration and ceased entirely when clogging impeded ciliary or appendage movement. Winter (1969) proposed a seven-step behavioural response in molluscs to changes in particle concentration. Food intake by mussels and oysters, for example, reached a plateau above a certain concentration while other species either demonstrated no threshold or decreased ingestion at high particle density (Conover, 1977). Lam and Frost (1976) similarly identified three phases in the feeding response of adult *Calanus* to differences in food concentration. Functional relations between filtering rate and food concentration for each phase of feeding behaviour permitted calculation of net energy gain expected dependent on food density.

Many studies of suspension feeding have indicated that food-particle quality as well as its size and abundance influences ingestion. Conover (1977) cited numerous examples of adaptation of feeding rates which appear to maximize useful ration. Thus, when suspension feeders consume insert or less nutritious particles. greater ingestion rates may ensue. Emlen (1966, 1973) proposed that such adaptation demonstrated a foraging strategy for selection of food on the basis of its energy content. Paine and Vadas (1969b) and Carefoot (1973), however, found no relation between consumption rate and calorific content of macrophytic seaweeds ingested by sea urchins and they concluded that texture and availability were more important criteria for choice. Many other predators also seem less dependent on total energy content of prey than on their accessibility (Conover, 1977). Carnivorous starfish (*Leptasterias*) provide an exception to these observations since when moderately energy-rich prey disappeared during winter, rarer species with a high energy content were selectively consumed (Menge, 1972).

Detritus and deposit-feeding species may also alter feeding rates in response to changes in food quality. Since Newell's (1965) observations on the feeding of the molluscs *Hydrobia* and *Macoma* on sediment particles. it has been assumed that micro-organism attached to particle surfaces, rather than non-living organic debris, provides the main food source for deposit-feeding invertebrates. Hargrave (1970), Fenchel (1972), Kofoed (1975), and Yingst (1976) presented evidence of high ingestion rates by deposit-feeding species with little or no assimilation of non-living organic detritus, while micro-organisms were assimilated often with a high degree of efficiency. Observations by Adams and Angelovic (1970) which suggested the assimilation of eel grass detritus, independent of associated micro-organisms, by three estuarine invertebrates were derived from experimental conditions which did not permit an unambiguous interpretation of the source of respired $^{14}CO_2$. A more direct assessment of the role of non-living organic debris can be provided by exposing organisms to labelled substrates such as cellulose and lignin under sterile conditions (Hargrave, 1970b; Kofoed, 1975). It must be realized, however, that the origin and chemical treatment of these materials provides substrates unlike those present in natural sediment and detritus samples. The extent of digestion of these materials in laboratory preparations may not reflect the availability of natural substrates.

These studies have suggested that egested material from deposit feeders, which contains refractory indigestible organic matter, may be colonized by micro-organisms and again become suitable food. Increases in nitrogen content have been used to document the colonization process (Newell, 1970). Often, however, selective ingestion concentrates organically rich material and elemental composition in feces, even after assimilation of the digestible fraction, exceeds that present in unsorted sediment thought susceptible to ingestion. Hylleberg (1975) described the effect of selective ingestion and the production of enriched fecal material by the lugworm *Abarenicola* and suggested that food supply was enhanced by feeding. The concept of 'gardening' in deposit-feeding invertebrates was also implied by the effects which benthic amphipods had in reducing detrital particle size, thereby increasing surface area for microbial growth (Fenchel, 1970), and in altering metabolism of benthic microflora (Hargrave, 1970a). Measures of oxygen consumption by fecal material may provide some measure of changes in nutritive value due to microbial colonization of egested material (Hargrave, 1976).

Few studies of sediment interface or deposit-feeding organisms have provided quantitative comparisons of feeding rates and food supply. Starvation and reduction in food quality can cause increased feeding and assimilation in some species (Calow, 1975) and thus adaptation to altered food supply can occur. Measures of egestion, however, might be used to infer feeding strategies. A review of feeding habits and assimilation in various marine organisms led Conover (1977) to conclude that food specialization resulted in selective ingestion and a high percentage of assimilation. Detritus and deposit feeders, on the other hand, process large amounts of material and assimilate only a small fraction. Since competitive interactions for a limited food supply are thought to characterize most benthic deposit feeding populations (Levinton, 1972b), selection should favour the development of feeding methods which maximize energy accumulation (Emlen, 1973). Models of optimal diet for predators stress the importance of size and abundance as factors which control a prey's availability and energetic value to a predator (Schoener, 1971; Pulliam, 1974). Deposit-feeding species, however, do not pursue prey and maximum energy gain is realized by sorting and selectively feeding on energy-rich particles or by maximizing the amount of food consumed. Besides the widespread occurrence of selective feeding, a general tendency towards the latter strategy is implied by the dependence of sediment turnover on body size in various deposit-feeding invertebrates (Hargrave, 1972c). While a single curve common to all species could not be derived, the data conformed to a power function ($b = 0.68$) relating body size and ingestion in terrestrial arthropods. The relation suggests that the rate of food passage is controlled towards an upper limit determined by body size (or gut volume). This is to be expected if energy intake is maximized.

6.4.3 ENERGY TRANSFORMATION AND ELEMENTAL BUDGETS

Measures of energy flow and elemental cycling through individual organisms, populations, and communities have been made with the hope that patterns exist which would provide a common basis for comparison of different ecological systems (Mann, 1969). Lindeman's trophodynamic model has provided a paradigm for many of these studies but observations of mixed diets and the apparently ubiquitous role of detritus (non-living debris and attached fauna and flora) in both pelagic and benthic food webs has made evaluation of pathways and transfer efficiencies under natural conditions difficult to assess (Strickland, 1970; Pomeroy, 1974). Laboratory measurements necessitate assumptions for extrapolation to nature and often balanced budgets are obtained by calculating unmeasured terms by difference. Despite these problems, budgets provide an organized way of describing how an organism partitions ingested energy or materials. Insofar as such studies permit growth and reproduction to be predicted, the comparisons may indicate mechanisms of control of population size and community structure.

IBP handbooks edited by Ricker (1971) and Holme and McIntyre (1971) summarize the separate processes which constitute the passage of energy and materials into and out of all heterotrophic organisms by an expanded version of eqn. (83),

$$R = P + G + T + E + U. \tag{137}$$

This equation partitions R, total food material or energy intake, between P, that incorporated into biomass (growth) and G, gonads and T, the amount used in metabolism and E, that released as undigested feces and U, excreta (secretion as mucus, urine, moults). Differentiation of gonad production from growth may be important seasonally particularly in organisms which store energy for gonad maturation at a later time (MacKinnon, 1972). Also, although few studies have separated soluble excretory products from non-digestible feces, this permits evaluation of A_b, the energy or elements absorbed or assimilated, where

$$A_b = R - E = P + G + T + U. \tag{138}$$

This is not equivalent to A_r, the retention of energy or elements by an organism, since soluble and particulate excretory losses occur. Thus,

$$A_r = R - (E + U) = P + G + T. \tag{139}$$

The difference between measures of absorbed and retained energy or elements has been stressed by Johannes and Satomi (1967) and several studies with benthic inverte-brates have demonstrated that the release of soluble excretory products occurs at rates which equal or exceed loss through respiration (Hargrave, 1971; Field, 1972; Kofoed, 1975). Kleiber (1961) further differentiated 'net energy gain' as retained 'metabolizable energy' corrected for the energy lost in food conversion (specific dynamic action) and 'surplus energy' $(P + G)$ which remained after metabolic costs of maintenance and ac-tivity. These distinctions are necessary if patterns of energy allocation in animals are to be quantified.

Attempts to construct short-term energy or elemental budgets for individual organisms have generally demonstrated that metabolic losses (as respiration and excretion) account for a large portion of assimilation (Conover, 1977). Little measurable residual energy or material may be stored for growth over a short time but exceptions can occur in young animals or reproductively active adults when growth and biomass accumulation are rapid. Few generalizations are possible since conditions under which feeding and respiration are measured, and the usual assumption that unmeasured excretory losses are small or a constant proportion of assimilation, do not permit realistic assessment of rates of energy or material transfer. In the first instance, assimilation may be difficult to accurately determine. When the amount of food ingested is known and feces are quantitatively collected, changes in weight or chemical content permit direct calculation of A_b from eqn. (138). This method is used widely to quantify feeding by fish (Ivlev, 1961) but it requires holding animals under laboratory conditions and the provision of an unnatural food supply. The measurements are often difficult for deposit-feeding invertebrates; however, where egested material cannot be quantitatively recovered, it may be mixed with uningested food particles and little weight change occurs if only micro-organisms are removed from ingested debris. When comparisons have been made, however, individual dietary components or specific elements are assimilated with a greater efficiency than total organic matter (Conover, 1977). Proteins and nitrogen are generally assimilated more efficiently than carbon, carbohydrates, or lipids, but digestibi-lity depends greatly on the nature of material offered as food. The principal component assimilated by the plant-feeding nudibranch *Aplysia* varied with food supply (Carefoot, 1967). Similarly, Hargrave (1970b) observed poor assimilation of non-living vascular plant tissue by the amphipod *Hyalella* and a high and variable assimilation of epiphytic algae. Kofoed (1975) observed that the fraction of ingested carbon respired and excreted by the snail *Hydrobia* varied when sterile hay or hay and bacteria were provided as food.

Radioisotopic methods can provide sensitive and specific measures of transport of elements from food into organisms (Sorokin, 1968). However, unless the specific activity in food and feces are directly compared over a time period sufficiently short to permit uptake and release processes to be differentiated, instantaneous rates cannot be properly calculated (Conover and Francis, 1973). Determination of removal and accumulation rates of specific elements is possible when homogeneous labelled food is used and recycling quantified. The use of ^{14}C-labelled food also permits quantitative measures of $^{14}CO_2$ respired and the importance of soluble excretion can also be evaluated (Johannes. 1964b; Hargrave, 1971; Kofoed, 1975). These studies demonstrated that soluble excretion of phosphorus or carbon by benthic amphipods and snails amounted to more than 30% of amounts assimilated. Peters (1975) similarly observed that food uptake by *Daphnia* measured by ^{32}P accumulation of 4 hr would be under-estimated by 53% if soluble phosphorus loss was ignored. Conover and Francis (1973) emphasized these and other errors which can occur when isotopic methods are improperly used to estimate feeding and assimilation rates.

Radioisotopic methods have also been used to estimate food uptake and elimination under natural conditions. Kevern (1966) calculated the amount of ^{137}Cs which would have to be ingested by young carp living in a pond contaminated with the isotope if an equilibrium body burden was to be maintained. The isotope was assumed to be in equilibrium in the fish and the pond environment and uptake by the carp was thought to be primarily by ingested food. Thus, with a steady state,

$$RA = Q_e k \qquad (140)$$

where R, the rate of ingestion of ^{137}Cs/day, and A, the fraction consumed assimilated, was balanced by Q_e, the body burden corrected for k, the fraction eliminated/day. When the concentration of ^{137}Cs in the food, D, was known, the rate of ingestion of isotope could be converted to the rate of food ingestion, R', since

$$R' = R/D. \qquad (141)$$

Thus the rate of ingestion of a specific dietary component required to maintain an equilibrium body burden was

$$R' = \frac{Q_e k}{AD}. \qquad (142)$$

Consumption of more than one dietary component was considered by measuring concentrations of ^{137}Cs in different food items (algae and detritus), D_i, by determining the assimilation of each food source, A_i, and by observing the fraction of each item in the diet. F_i, whereby

$$R' = \frac{Q_e k}{\sum\limits_{i=1}^{n} A_i D_i F_i}. \qquad (143)$$

Thus the rate of food ingestion required for maintenance metabolism was estimated when the fish had more than one source of diet. The quantity of the ith item was the product $R'F_i$. In addition, new tissue resulting from growth was assumed to be in equilibrium with concentrations of elements in old tissue and since the rate of growth of fish was known, the quantity of ^{137}Cs deposited and the increase in body burden, Q, were calculated.

Elimination (k) of isotope results from physical decay (negligible for ^{137}Cs) and biological removal from different metabolic pools. An exponential loss was assumed to occur from all pools and rates were determined by measuring decreased ^{137}Cs activity in fish fed labelled diets. The amount of isotope remaining at any time, A_t, was expressed as

$$A_t = A_0 \, e^{-kt} \text{ or } A_t/A_0 = e^{-kt} \tag{144}$$

where A_0 was the initial amount assimilated from a single feeding and t time. The length of time for 50% elimination (biological half-life), T_b, was thus

$$\frac{A_t}{2A_t} = e^{-kT_b} \text{ or } l_n(0.5) = -kT_b \tag{145}$$

and

$$k = \frac{-0.693}{T_b}. \tag{146}$$

Since elimination occurs from metabolic pools with different turnover times, k values were estimated for different time periods by semi-logarithmic analyses of A_t/A_0 on t.

These methods may be applied only where chronic supply of an isotope permits the assumption of a steady state and provides measurable concentrations. Even then, variable concentrations in different food items permits a wide range of estimates of feeding rate (Kevern, 1966). Where it can be used, however, the method provides estimates of the quantity of each diet item that is consumed for both maintenance metabolism and growth as well as gross intake under natural conditions. Thus, Kevern calculated that young carp consumed 5·3 g wet weight of food/day (3·9% of their mean body weight) consisting of 2·4 g detritus and 1·6 g algae for maintenance and 0·8 g detritus and 0·5 g algae for growth. Other studies have followed the elimination of ^{65}Zn from snails and plaice under field conditions (Mishima and Odum, 1963; Edwards, 1967). Once non-equilibrated label was eliminated, the long-term metabolic loss for free-living animals was inversely related to body size and temperature and paralleled oxygen uptake in snails and both studies demonstrated metabolic losses at least twice as high as those measured in laboratory studies. In subsequent experiments, Edwards et al. (1969) found that only plaice feeding at high ration levels achieved comparable metabolic rates. The observations support Winberg's view that active metabolism may be double that measured as 'standard' rates in the laboratory (Conover, 1977).

A third group of methods used to measure assimilatory uptake of material by aquatic organisms involves comparing the ratio of inert compounds to nutrients in food and feces. Conover's (1966) ash-ratio method (Section 4.2.1) was developed on this basis. While the technique does not require the quantitative recovery of fecal material for estimating assimilation, selective feeding or changes in ash content during passage through the gut prevents its use, although corrections for the latter problem are possible (Conover, 1977). Hargrave (1970a) and Lasenby and Langford (1973) achieved comparable assimilation rates by gravimetric and ash-ratio methods for an amphipod and mysiid feeding on epiphytic material and Daphnia respectively. However, mysiids selectively accumulated organic debris from detritus samples and Conover's method could not be used. Other studies have applied the method with varying degrees of success for different benthic organisms feeding on both artificial and natural diets (Conover, 1977).

In the more general inert indicator method, a non-digestible additive, such as Cr_2O_3, is mixed with a food supply. Percent assimilation is calculated from

$$A_b = \left[1 - \frac{N:R}{N:E} \right] . 100 \tag{147}$$

where $N:R$ and $N:E$ are the ratios of marker, N, in food, R, and feces, E. Calow and Fletcher (1972) used a double labelling method with the inert tracer ^{51}Cr (as $^{51}CrCl_3$) which is not absorbed and ^{14}C to label bacterial and algal cells offered as food to surface browsing snails. Attempts were made to ensure that both labels were homogeneously distributed in diatoms offered as food and both tracers were assumed to move uniformly through the gut and maintain a stable ratio in feces. Corrections for slight absorption of ^{51}Cr were applied and no leaching from feces was observed. Thus, the assimilation efficiency was calculated as

$$A_b = \left[1 - \frac{^{51}Cr : ^{14}C \text{ food}}{^{51}Cr : ^{14}C \text{ feces}} \right] . 100. \tag{148}$$

Assimilation efficiencies, in excess of 80%, were comparable to those estimated from the difference between ^{14}C lost from food discs and ^{14}C appearing in feces. The technique has the advantage that total fecal production need not be observed and fecal material derived from food can be distinguished from that derived as intestinal secretions.

Conover (1977) reviewed many studies which have attempted to quantify some or all budget terms of eqn. (137) for various aquatic organisms (Table 38). Calories, carbon, nitrogen, phosphorus, and sulfur have been used as a basis for calculation but few studies have estimated all terms as a check on the accuracy of measurements. Also, no study has calculated an energy and elemental budget simultaneously for individual organisms. Thus, although carbon is most often measured, growth, as protein synthesis, may be better assessed by measurement of nitrogen. Also, while carbon and calories are lost directly as CO_2 and heat during respiration, other elements may be metabolized and excreted in soluble form. It is thus difficult to systematically compare the fate of assimilated material measured in different forms. In addition, since growth of individual organisms is not usually measurable in short-term experiments, studies of population growth and estimates of production must usually be extrapolated to a reduced time scale. Conversely, short-term measurements of respiration and excretion must be extended to annual rates to be consistent with production studies.

Although these problems limit the interpretation of many budget calculations, available data does permit some conclusions to be drawn. Growth and assimilation vary greatly with food supply, although some organisms may maximize both measures when given a preferred food (Carefoot, 1967). There is no clear relation between production and assimilation efficiencies as suggested by Welch (1968) but often measures of assimilation are underestimated because loss of soluble excretory material is not measured. Also, the amount of assimilated material stored for growth varies seasonally and with age and it is clearly dependent on reproductive state. As organisms grow in size, the proportion of assimilated energy and material utilized for metabolic activity (respiration and excretion) increases. In mature slow-growing individuals these pathways account for almost all assimilation (Kleiber, 1961).

The interrelated nature of respiratory and excretory processes was demonstrated by Satomi and Pomeroy (1965) who observed a linear relation

$$y = -3 \cdot 1 + 0 \cdot 015x \tag{149}$$

between inorganic (molybdate reactive) phosphate release (y as μg-at PO_4/hr) and respiration (x as μg-at O/hr) per g dry weight in net zooplankton, oysters, and sea anenomes. Soluble organic phosphorus was released in a constant proportion to phosphate

TABLE 38. BUDGETS FOR ENERGY AND ELEMENTS, AS PERCENTAGE OF TOTAL ASSIMILATED PER UNIT TIME, FOR VARIOUS BENTHIC ORGANISMS

Organism	Basis	P	G	T	U	Reference
Modiolus	calories	25	5	70	—	Kuenzler, 1961a
demissus	P	6	3	—	90	Kuenzler, 1961b
(mollusc)	(annual)					
Aplysia						
punctata	calories					
(mollusc)	(annual)	21	15	51	13[a]	Carefoot, 1967
Scrobicularia						
plana	calories					
(mollusc)	(annual)	13	12	90	—	Hughes, 1970
Asellopsis						
intermedia						
(harpacticoid	C					
copepod)	(annual)	5	11	82	2[b]	Lasker *et al.*, 1970
Idothea baltica	calories					Tsikhon-Lukanina and
	(daily)	4	—	94	2[b]	Lukasheva, 1970
Sphaeroma pul-	calories					
chellum (isopod)	(daily)	15	—	49	36[c]	Hargrave, 1971
Hyalella azteca	calories					
(amphipod)	(annual)	16	2	75	7[b]	Mathias, 1971
Crangonyx	calories					
richmondensis	(annual)	17	0·5	79	4[b]	Mathias, 1971
Tellina tenuis	C	14	1	84	—	
(mollusc)	(annual	5	8	87	—	Trevallion, 1971
	3 yr)	12	16	71	—	
Nerita (3 spp.)	calories					
(mollusc)	(annual)	12	2	86	—	Hughes, 1971
Fissurella						
barbadensis	calories					
(mollusc)	(annual)	24	3	73		Hughes, 1971a
Tegula funebralis	calories					
(mollusc)	(annual)	13	1	77	10[a]	Paine, 1971b
Hippoglossoides						
platessoides	calories					
(teleost)	(annual)	31	20	72	—	MacKinnon, 1972
Strongylocentrotus						
droebachiensis	calories					
(echinoderm)	(annual)	6	1	25	68[a]	Miller and Mann, 1973
Tigriopus brevi-						
cornis (harpacti-	N					
coid copepod)	(annual)	4	23	73	0·4[a]	Harris, 1973
Neanthes	calories					
virens	(monthly	71	—	26	3	Kay and Brafield, 1973
(polychaete)	totals)					
Hydrobia ventrosa	C					
(mollusc)	(daily)	29	—	39	26	Kofoed, 1975

[a] Assumed value for mucus production, calculated by difference or assumed as a constant proportion of assimilation.

[b] Loss through moulted exoskeletons.

[c] Calculated from measured dissolved organic carbon released.

(P—PO$_4$:total P = 0·6). Webb and Johannes (1967) also observed that planktonic and benthic invertebrates released ammonia and dissolved free amino acids in a ratio of approximately 4:1. Complete oxidation of plankton with a C:N:P ratio of 106:16:1 would require 276 atoms of oxygen per atom of phosphorus and 16 atoms of oxygen per atom of nitrogen (Redfield, 1958). The O:P ratio of 267 observed in oysters and values of 54 to 72 in zooplankton (Satomi and Pomeroy, 1965) and O:N ratios as low as 1 in some invertebrates during winter (Conover, 1977) suggest variable and in-complete oxidation of food. Complete protein oxidation should yield an O:N ratio of 8 and much lower values could reflect processes such as gluconeogenesis which would alter expected relations between respiration and excretion during periods of low food supply (Mayzaud, 1973).

Respiratory and excretory processes respond similarly to changes in temperature and body size. Miller and Mann (1973) calculated an average Q_{10} value of 2·05 for respiration by fourteen species of poikilotherms. Q_{10} coefficients between 1·4 and 3·0 were observed for phosphorus and nitrogen released by zooplankton (Hargrave, 1966; Christiansen, 1968), although dissolved free amino acids released by zooplankton increased linearly with temperature above 6°C (Webb and Johannes, 1967). Similarly, the standardized size–metabolism relationship, with exponents which vary slightly with time of year and experimental temperature (Newell and Roy, 1973), permits comparison of respiration and excretion per unit weight in different organisms. Conover and Corner (1968) found oxygen uptake and nitrogen release to be correlated with body nitrogen in several planktonic crustaceans and carnivorous–omnivorous species had a higher weight-specific metabolism than herbivores. Johannes (1964a), on the other hand, attempted to evaluate the relative importance of dissolved phosphorus excretion by different sizes of marine invertebrates by calculating the time required for an animal to excrete an amount equal to its total phosphorus content (TT_B, the body-equivalent excretion time). A power relation existed between this measure and body dry weight (W) with the equation

$$TT_B = 1584W^{0·67} \tag{150}$$

derived for animals greater than 1 mg in weight and the equation

$$TT_B = 398W^{0·33} \tag{151}$$

describing the relation for animals between $10^{-7}\,\mu$g and 1 mg dry weight. When Johannes compared weight-specific phosphorus-excretion rates calculated from these equations with estimates of oxygen uptake based on body size (Zeuthen, 1953), O:P ratios were observed to decrease markedly with decreasing animal size. Thus, oxygen consumption was much less affected than apparent phosphorus release with decreasing size. Problems arise in interpreting the data, however, since excretion by animals weigh-ing less than 1 mg was determined with ^{32}P while colorimetric measurements were used for larger animals and also the phosphorus content of food ingested by different sizes of animal was probably different. In addition, the fraction of total soluble phos-phorus released as organic phosphorus could have been different in different sizes of animals. Analyses did not permit separation of the two fractions.

Johannes and Webb (1970) suggested that egestion, leakage across the body wall and true excretion are all involved in the release of dissolved organic compounds by aquatic organisms. There may also be an uptake, at least by some soft-bodied species, which may equal but seldom exceeds loss; however, few budget studies have quantified net flux due to both processes. Miller and Mann (1973) suggested that herbivorous sea urchins, like many animals which feed on a largely carbohydrate diet, may secrete soluble organic carbon to retain sufficient nitrogen. The identification of nitrogen-fixing bacteria and measurement of nitrogen fixation in sea urchin guts supports this idea and provides an additional mechanism to enhance nitrogen supply (Guerinot et al., 1977). Future studies of elemental or energy supply to organisms will have to consider the importance of these sources and sinks in budget calculations.

An additional aspect of biological transformation of energy and materials of importance in benthic communities concerns the biogeochemical effects of feeding and egestion. Deposit-feeding invertebrates which burrow, selectively feed, and have a high egestion rate can physically and chemically alter the sedimentary environment. These secondary effects may equal or exceed those directly associated with their own metabolism.
For example, the conversion of fine-grained deposits into fecal pellets affects sediment
stability and nutrient recycling (Rhoads and Young, 1970; Rhoads, 1973) and fecal
material provides sites for enhanced microbial activity (Hargrave, 1976). From a community point of view, these effects might be considerably greater than the metabolic

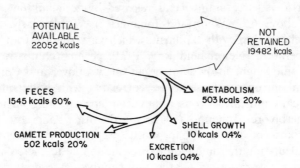

Fig. 81. Percentage distribution of annual calories available and utilized by a 1 m² oyster (*Crassostrea gigas*) reef (redrawn from Bernard, 1974).

activity of an organism itself. In addition, benthic suspension feeders are an important
link between suspended and sedimented organic particles since the production of biode-
posits may equal or exceed gravitational sedimentation (Bernard, 1974). Thus, in the
calculation of an annual gross energy budget of a mature Pacific oyster, only about
11% of the calories theoretically available were retained and 60% of these were deposited
as feces (Fig. 81). The high organic content of such material (28·5% in Bernard's study)
provides an enriched substrate for deposit feeding species. Tenore *et al* (1974) developed
a polyspecies aquaculture system using this principle.

6.4.4 POPULATION GROWTH AND PRODUCTION

A direct measure of growth rates in natural populations is possible if individual
organisms can be marked and recaptured at a later time (Ricker, 1971). The method
is widely used to study fish populations and it provides an estimate of total population
size if homogeneous mixing of marked and unmarked animals is assumed and if marking
does not affect growth and mortality rates. The method may be applied to molluscs
with hard shells and Mann (1972b) used a variation of the technique to measure length
increments of seaweed blades. Large crustaceans can also be tagged such that moulting
frequency and growth are determined simultaneously (Hancock and Edwards, 1967;
Newmann and Pollock, 1974).

The direct observation of increments in size permits calculation of an individual's
specific or instantaneous growth rate

$$G_i = \frac{1}{t} \ln \frac{W_t}{W_0} \tag{152}$$

where W_0 and W_t are weight (or length) before and after time period t. Values of G_i usually decrease as organisms age. The data may also be used to consider weight changes with time as a function of maximum attainable weight. This reflects the sigmoid growth pattern of organisms which Beverton and Holt (1957) quantified, following Bertalanffy, by the equation

$$W_t = W_\infty [1 - e^{-K(t-t_0)}]^3 \tag{153}$$

where W_∞ represents weight at maximum size, W_t is weight at time t, with t_0 the time of no weight and K a growth constant. K can be determined by plotting $\sqrt[e]{W_t}$ after the time interval $t + 1$ against t and if a constant time interval is used, a line with the slope K results which intersects the origin at $W_\infty^{1/3}$ (Walford, 1946). Length–age curves, transformed to straight lines of slope e^{-K} by plotting l_{n+1} against l_n, are termed 'Ford–Walford plots' (Crisp, 1971). The method has been used primarily to study fish growth but it has also been applied to compare growth rates of various bivalve molluscs (Warwick and Price, 1975). Conover (1977) also cited studies by Ivanova which applied a modification of the technique to produce generalized growth curves for naupliar and copepodite stages of calanoid and cyclopoid zooplankton. Specific growth rates for each stage were calculated by comparing copepodite size as a fraction of maximum adult female size and development time for each stage as a fraction of that for a female.

If the instantaneous growth rate (G_i) of an individual of weight (W_i) is known, then total population production (P) over time (t) is given by

$$P = \sum_{t=0}^{t=1} \sum_{0}^{N} G_i W_i \, \Delta t. \tag{154}$$

This calculation is based on survivors which are enumerated and it ignores the biomass produced in a population which is eliminated by predators. When yield from a population is of interest, a measure of elimination, either through natural processes or that due to harvesting, is necessary. Thus, if B_0 is the population standing crop biomass at the start and B_t that remaining after time t

$$P = (B_t - B_0) + E = \Delta B + E \tag{155}$$

where E is loss through mortality. E, the instantaneous mortality rate of an individual, may be calculated by determining the slope of the survivorship curve of a single age class of organisms. Plots of abundance against time may be fitted by linear or exponential equations over various portions of the curve. Usually a curvilinear plot best describes changes in density, particularly during early life-history stages; however, a linear decrease may occur during certain seasons or in larger weight classes.

Mann (1969) differentiated between these two measures of production by reference to early studies by Boysen Jensen (1919) who measured numbers and weight of macrofauna species annually in the Limfjord, Denmark. Numbers of specific age classes of molluscs and polychaetes decreased with time, but survivors increased in their individual weight. Therefore, E ('elimination' as production consumed by predators or lost by natural death if no immigration or emigration occurred) was calculated as

$$E = (N_1 - N_2).\tfrac{1}{2}(W_1 + W_2) \tag{156}$$

and production by the remaining animals which had increased in weight was determined as

$$P = \tfrac{1}{2}(N_1 + N_2).(W_2 - W_1). \tag{157}$$

The overall biomass change, ΔB, between two sampling times is

$$\Delta B = N_1 W_1 - N_2 W_2 \tag{158}$$

and from eqn. (155)

$$\Delta B = P - E. \tag{159}$$

Crisp (1971) discussed use of these and other calculations to estimate total population production. Weight–survivorship curves must be obtained for a single recruitment over a time interval equal to the whole or part of the lifetime of one generation. In populations with continuous or intermittent recruitment, similar methods may be used only when stocks (cohorts) belonging to different age classes can be recognized. Size frequency histograms of some character (length or width of a body part or growth marks) in individuals sampled from the population may allow age-class separation. When age classes cannot be recognized, instantaneous growth or mortality rates as a function of size must be determined and summed throughout the year for total numbers in each size class. Groups of organisms must be isolated and growth measured directly at various temperatures. If mortality can be attributed to harvesting or predation, some approximation to removal rates may be possible. Teal (1957), for example, isolated macrofauna populations in sediment cores from Root Spring and determined biomass increases over time. Mortality was estimated by comparing increased biomass with changes in the natural population over the same time period. In any population which maintains a constant biomass from year to year, that is $\Delta B = 0$, estimates of growth and mortality should be equivalent.

A third approach to measuring production of biomass which differs from the instantaneous growth rate and growth increment–survivorship curve methods is based on measures of respiration and growth efficiency. The 'physiological approach' (Winberg, 1968) is based on a calculation of growth efficiency (E_g), the total energy of production of body tissue and gonads ($\Delta B = P + G$) as a fraction of the energy ingested (R). Thus,

$$E_g = \frac{\Delta B}{R} = \frac{(P + G)}{R}. \tag{160}$$

This ratio should be calculated on an energy rather than weight basis since fats and carbohydrates have different weight-specific energy content and their proportion as storage products may change seasonally or with animal size. If caloric content of tissues is unknown, ΔB and R are expressed as weight and E_g becomes a 'conversion efficiency' (Crisp, 1971). In natural populations, R is seldom known and it is usually assumed to be approximated by the sum of $P + G + T$, with U neglected. While this is not justified (Section 6.4.3), many workers have followed the convention which then permits production to be calculated from estimates of respiration and growth efficiency by the equation

$$E_g = \frac{\Delta B}{\Delta B + T} \tag{161}$$

and

$$\Delta B = E_g(\Delta B + T). \tag{162}$$

Estimates of respiration and growth efficiency at different temperatures must be independently established. This introduces additional errors due to artificial experimental conditions but the method has the advantage that ΔB may be estimated from size-frequency histograms and no age-class separation is necessary.

The physiological approach provides no estimate of elimination from a population. A conservative estimate of this removal may be derived, however, when organisms consumed as food by specific predators can be measured. Since predation rates must be known or predicted for comparison with loss from a spatially defined prey population, the technique may be most useful when experimental enclosures are used to isolate populations. An example of the approach was used by Trevallion et al. (1970) in the case of plaice feeding on siphons of the mollusc Tellina. Only extended siphon tips were cropped and regeneration was possible. Tellina and plaice were held in experimental tanks in varying densities and ratios and energy requirements by plaice were compared to calculated energy requirements for siphon regeneration. Energy utilized to regenerate siphons (elimination) was equivalent to estimated energy requirements by the plaice populations.

Generation time (also termed 'doubling time' or 'turnover rate') has been used to estimate production from standing crop biomass determinations. When life-cycle or laboratory culture studies permit estimates of generation time, the figure can be multiplied by natural population standing crop to provide a measure of production potential. McIntyre (1964), Fenchel (1968), and Gerlach (1971) used the approach to estimate annual production which might be achieved by benthic meiofauna. The technique is most applicable to small organisms with a rapid generation time but growth and development rates in the laboratory are often more rapid than would occur in nature.

The concept or reproductive or replacement rate is implicit in this approach to measures of population growth. Smith (1954) suggested that this could be related to generation time (t) by the equation

$$R_0 = e^{r_m t} \tag{163}$$

where R_0 is the sum of the offspring of each female per unit time (reproductive rate) and r_m is the growth rate in the absence of any limiting conditions (intrinsic rate of natural increase). If $R_0 = 1$, a population replaces itself each generation. Although factors such as food availability and temperature greatly affect measures of t, particularly in small organisms in culture, Bonner (1965) and Sheldon et al. (1973) observed that doubling time was related to body size by a power function. The equation

$$\log_{10} t = 4 \cdot 35 + 0 \cdot 80 \log_{10} L \tag{164}$$

can be derived from their data which compares doubling time (t) in days with body length (L) in millimetres. Fenchel (1974) also demonstrated that values of r_m were related to average body weight (W) in various organisms by the equation

$$\log_{10} r_m = a - 0 \cdot 275 \log_{10} W \tag{165}$$

where the constant a differed for unicellular ($-1 \cdot 94$), heterotherm metazoan ($-1 \cdot 64$), and homoiotherm metazoan ($-1 \cdot 4$) organisms. The relationship is similar to that which describes metabolic rate per unit body weight and body size (exponent of $-0 \cdot 249$) (Zeuthen, 1953; Hemmingsen, 1960). Fenchel concluded that r_m is a measure of the maximum productivity (production potential) of an exponentially growing population which must be correlated with metabolic rate.

These equations may be useful for predicting the relative importance of production by different sizes of organisms in a community (Gerlach, 1971). They may also prove useful for observing regularities in size composition of some ecosystems. Kerr (1974), for example, by assuming that predator and prey sizes were simply related and that growth and metabolism were both power functions of body size, was able to calculate that the standing stock of prey organisms should typically be in the order of 1·2 times that of their predators. Sheldon *et al.* (1972) provided empirical evidence for this in pelagic marine ecosystems. Concentrations of particles (1–10^6 μm diameter) were relatively uniform, with a slight decrease (approximately two-fold with increasing size), when data were grouped in logarithmic size intervals.

The implied relationship between population production and metabolism (respiration) has also been observed. McNeill and Lawton (1970) extended earlier observations and compared annual calorific equivalents (kcal/m^2/yr) of production (*P*) and respiration (*T*) in populations of terrestrial and aquatic organisms. The equation

$$P = aT^b \tag{166}$$

was derived with the exponent $b = 1·0$ and different values of a for poikilotherms and homoiotherms. The P/T ratio for sea urchin (*Stronglylocentrotus*) populations approximated values predicted by this equation, but the slope coefficient ($b = 0·85$) was slightly lower (Miller and Mann, 1973). One difficulty in interpreting correlations between P and T, however, is that population size, as numbers or biomass, is used to calculate both parameters. Values are thus not statistically independent. Independent calculation of P and T for the amphipod *Hyalella* yielded a ratio of 0·31, however, which was comparable to an estimate derived from population studies (Hargrave, 1971).

Yet another approach to estimate production makes use of the concept of 'turnover time', defined as the reciprocal of the finite death rate (Mann, 1969). A population with a daily instantaneous death rate of 0·69 has a finite death rate of $l_n(0·69) = 0·5$ and its turnover time is 2 days. At average biomass $\bar{B}$, production is $0·5\,\bar{B}$ per day. The finite death rate is also equivalent to the ratio ($P/\bar{B}$), the 'turnover ratio' which, as weight-specific production, allows production by populations of different biomass to be compared on a common basis. Attempts to make general statements about turnover ratios in aquatic organisms are confounded by effects of temperature, organism size, and the duration of production and biomass estimates relative to the total life cycle. Regularities have appeared, however. Waters (1969) calculated that within the life span of an age class (i.e. per generation), populations with different mortality and growth patterns have turnover ratios between 2·5 and 5, with a modal value of 3·5. Annual turnover ratios are equivalent to this figure multiplied by the number of generations per year.

Sufficient data has accumulated to allow a test of Waters' idea of an approximately constant turnover ratio per generation. Studies* ($n = 13$) in which annual production and average annual biomass have been estimated for various macrobenthic species ($n = 55$) were compared on the basis of generation time and the equation

$$\log_{10} TR = 0·69 - 0·14t \tag{167}$$

---

* Boysen Jensen, 1919; Sanders, 1956; Richards and Riley, 1967; Peer, 1970; Zaika, 1970; Hughes, 1970; Miller and Mann, 1973; Burke and Mann, 1974; Buchanan and Warwick, 1974; Warwick and Price, 1975; Chambers and Milne, 1975; Klein *et al.*, 1975; Nichols, 1975.

was derived where t, life span in years, accounted for 60% of the variance in TR. Substitution of $t = 1$ into the equation yields a turnover ratio of 3·55, precisely the modal value suggested by Waters. The semi-logarithmic nature of the relation, also suggested by Zaika (1970), may arise from the lack of inclusion of data from organisms with short (less than 1 year) generation times. It is often difficult to relate published $P/\bar{B}$ estimates to specific cohort generation times for short-lived organisms and thus the conservative nature of lifetime turnover ratios cannot be assessed when $t < 1$. Gerlach (1971), however, estimated life-cycle duration in two species of nematodes and calculated cohort turnover ratios between 2 and 3. With three to twelve generations per year, meiofauna populations could have annual turnover ratios between 10 and 36, values comparable to those for most zooplankton. Allen (1971) cautioned against such calculations of production and pointed out that the relation between P and $\bar{B}$ over time depends on the shape of growth and mortality curves. Mean age and mean life span both equal the reciprocal of the $P/\bar{B}$ with constant exponential mortality. For other mortality functions, if growth in weight is linear, the $P/\bar{B}$ ratio equals the reciprocal of mean age. There is no simple relation if growth follows a non-linear pattern.

High annual turnover ratios for small-bodied organisms are predictable from the relationship between body size and generation time [eqn. (164)]. Turnover ratios for small organisms may also be more variable than those for larger organisms since their rapid generation time provides more opportunity for environmental factors (food availability, temperature) to affect growth and development rates. High-energy flux through populations of small organisms (plankton, micro-meiobenthos) is also usually accompanied by high spatial and temporal variability thought to indicate relatively weak self-regulatory mechanisms (Margalef, 1968; Johnson, 1972). In contrast, fish and macrobenthos, with their large body size and longer generation time, are less affected by short-term environmental variability once critical early growth periods are past.

The importance of temperature on turnover rates, largely independent of the effects of body size and generation time, was demonstrated by Johnson and Brinkhurst (1971a). Estimates of production by major taxonomic groups of macrobenthos (all with 1- to 2-year life cycles) in four different areas of a bay in Lake Ontario were derived from calculations of instantaneous growth. Although turnover ratios differed between groups, values were related to annual mean temperature by the equation

$$TR = \frac{T^2}{10} \tag{168}$$

where TR is the annual turnover ratio and T is the annual mean temperature (°C). Annual production was thus described by the equation

$$P = \bar{B}\left[\frac{T^2}{10}\right]. \tag{169}$$

The relation implies that if annual mean temperature is known, production may be calculated from measures of standing stock. Johnson and Brinkhurst suggested that biomass of organisms may be adjusted to utilize energy supplied consistent with a turnover ratio set by temperature. The importance of temperature in shortening generation time of pelagic organisms of different sizes was emphasized by Sheldon et al. (1973) and has been discussed above (Section 4.2.1).

Numerous estimates of production by benthic macrofauna populations exist but few studies have applied different methods of calculation in a single study. Kuenzler (1961a) compared estimates of instantaneous growth and mortality to calculate annual production by a population of the mussel *Modiolus* but poor agreement was achieved. Production by several macrobenthic species in Lake Ontario calculated by the growth efficiency-respiration technique was similar to estimates derived using instantaneous growth rates although some data was common to both sets of calculations (Johnson, 1974). Despite the difficulties of comparing estimates derived by different methods, there have been attempts to compare levels of production by predator and prey populations and to synthesize such information into a description of energy flow through an entire food chain or ecosystem (Section 4.2.3). The approach cannot indicate the precision of specific

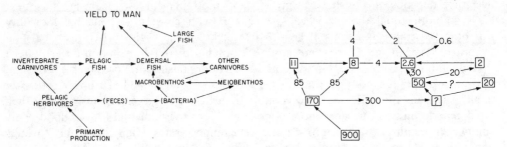

FIG. 82. A diagrammatic food web of major groups of organisms and their estimated annual production (kcal/m²) in the North Sea (redrawn from Steele, 1974).

production estimates and since most organisms derive food from several sources, ecological transfer efficiencies for particular predator–prey associations cannot be calculated. However, in some locations sufficient data has been collected to encourage comparison of measured primary production with production by major groups of organisms (herbivorous copepods, carnivorous zooplankton, benthos, and fish). Production by major groups of organisms can then be compared in view of expected ecological transfer efficiencies without specifying the actual food supply for each group.

Steele (1965, 1974) used this approach to summarize existing information on production by various organisms in the North Sea. When tentative estimates of production by major groups of organisms were compared (Fig. 82), the yield of commercial pelagic and demersal fish exceeded that expected on the basis of the structure and production of the underlying food web. The role of bacteria, both in the water column and sediments, which convert dissolved and particulate matter into biomass available to consumer organisms, was unquantified but such transfers were identified as being critical in determining the amount of deposited organic matter available for benthic production. In addition, invertebrate carnivores, both planktonic and benthic, consume unmeasured amounts of energy which further reduces the amount available to larger predators. Richards and Riley (1967) suggested that the abundance of epibenthic invertebrates in Long Island Sound might reduce food availability to demersal fish and Mills (1975) discussed the importance of these pathways as control mechanisms which regulate the amount of benthic production available as fish food. He suggested that benthic communities dominated by epibenthos are functionally distinct from those where infauna (bivalves, polychaetes) predominate. The effects of such structural differences on pathways and efficiencies of energy flow within and out of the benthos is unknown. However, if Steele's observations of demersal and pelagic fisheries yield are to be reconciled with observed production by benthos and plankton, transfer efficiencies as high as 20% must occur. Observations of predation effects on macrobenthos in St. Margaret's Bay substantiate the idea that predator–prey associations in food webs are tightly linked. Feeding by plaice eliminated approximately half of the annual macrofauna production (Peer, 1970; MacKinnon, 1973) and also dramatically altered community structure (Levings, 1975).

6.4.5 COMMUNITY METABOLISM

Energy exchange in biological systems can be measured by observing either anabolic (synthetic) or catabolic (degradative) processes. The diverse nature of methods of energy uptake makes measurements of production by all species in a community difficult. Energy release, however, as aerobic (oxidative) or anaerobic (sulfate and nitrate reduction, denitrification, and fermentation) metabolism involves exothermic biochemical reactions (Lehninger, 1965). Thus measures of heat release provide a common basis for comparing metabolic activity in all organisms.

Direct calorimetry has been widely used for measuring metabolism in homoiotherms (Kleiber, 1962) but measurements with small organisms like aquatic invertebrates are difficult because of their low rate of heat production. Also, endothermic reactions involved in chemical precipitation, dissolution, and ionization in water and sediments can result in heat absorption which greatly exceeds low rates of heat release by micro-organisms and invertebrates present in samples (Pamatmat, 1975). The use of dense cultures of micro-organisms or double-chambered calorimeters with metabolic activity confined to one chamber may avoid these problems (Brown, 1969). Pamatmat and Bhagwat (1973) were able to measure metabolic rates of 3 to 17 mcal/g/hr in sediment from Lake Washington placed in a gradient-type calorimeter; however, long periods (over 12 hr) were required for temperature stabilization and thermal equilibrium was not established. Also, when intact sediment cores were incubated, thermochemical processes resulted in net heat absorption which prevented detection of metabolic heat production (Pamatmat, 1975). These and other studies have suggested that chemical methods which measure metabolism by indirect means are more convenient for routine use.

Chemical measures of metabolic activity are based on the fact that all metabolism involves the transfer of H^+ through successive steps of dehydrogenation. High-energy adenosine tri-phosphate (ATP) is formed when electrons are passed from dehydrogenase enzymes to various electron acceptors. In aerobic respiration, the final transfer is to oxygen and thus oxygen uptake (or CO_2 release) is used as an indirect measure of hydrogen production. In anaerobic metabolism, however, electron acceptors other than oxygen are used (H_2S, CH_4, NH_4^+, and Fe^{++}, NO_3^- or reduced organic compounds in fermentation). Thus, whereas oxygen consumption only reflects aerobic metabolic activity in organisms, total (aerobic + anaerobic) metabolism may be measured as the rate of hydrogen or electron transfer, the activity of dehydrogenase enzymes which participate in the transfer, or ATP activity.

Pamatmat (1975) reviewed past studies which have used artificial electron acceptors and carriers to measure rates of electron transfer. Reduction of TTC (2,3,5-tri-phenyltetrazolium chloride) to formazan has been widely used as a technique for measuring total dehydrogenase activity (DHA) in plankton (Packard, 1971) and sediments (Pamatmat and Bhagwat, 1973). Refinement of the technique (Wieser and Zech, 1976) avoids prolonged incubation of samples by extraction and solubilization of enzymes and determination of initial and final activity by standardized kinetic analysis with added substrate. Incubation of extracted enzyme with high concentrations of substrates thus may represent the metabolic potential of organisms rather than *in situ* metabolic rates. Electron transport activity in anoxic Lake Washington sediment was positively related to metabolic heat release (Pamatmat and Bhagwat, 1973) and to ATP content in anaerobic Norwegian fjord sediments (Pamatmat and Skjoldal, 1974). Difficulties exist in interpreting these results in metabolic terms, however, since neither TTC reduction nor ATP content can readily be converted to a measure of metabolic rate. Studies with freshwater sediments indicated that chemical reduction of TTC to formazan occurred at similar rates before and after poisoning with formaldehyde (Pamatmat, 1975). Also, aside from difficulties in extracting ATP from sediments (Karl and LaRock, 1975), ATP content relative to other low-energy containing phosphate compounds and absolute dehydrogenase activity in organisms depends on whether phosphorylation involves an organic substrate (fermentation) or an electron transport system (oxidative phosphorylation). Thus, decreased ATP content and increased DHA with increased sediment depth (Pamatmat and Skjoldal, 1974) could reflect changes in species composition, metabolic type, or concentrations of reduced substances. The techniques may provide relative measures of metabolism, however. For example, Hobbie *et al.* (1972) compared respiratory rates in large water samples from several depths at two ocean stations. Estimates based on oxygen uptake, electron transport activity, and ATP were similar in surface water but different at deeper depths where oxygen consumption exceeded other measures of respiration. It was not possible to attribute high metabolic rates to the presence of particular organisms. As another example, Wieser and Zech (1976) estimated that 80% to 90% of total electron transport activity in carbonate beach sand was associated with particle surfaces. Thus, interstitial organisms were assumed to be of little importance to total energy release by the benthic community.

Measurements of metabolism in organisms calculated by conversion of oxygen uptake or carbon dioxide release to energy equivalents can be in error for two reasons. If anaerobic processes occur, reduced substances are formed which are not immediately oxidized. Secondly, if energy released is not converted into heat but stored as high-energy compounds, actual heat production is less than conversion of oxygen uptake by usual oxy-calorific equivalents would indicate. Crisp (1971) discussed these and other problems associated with interpretation of laboratory measurements of respiration. In view of the many other errors involved, the assumption that 1 ml of oxygen consumed at n.t.p. is equivalent to 4·8 calories (1 mg O_2 = 3·34 cal) as a mean value for oxidation of different substrates, introduces a small error.

A more serious problem of conversion and interpretation may exist when community oxygen uptake is measured with undisturbed sediment cores or stirred samples. If reduced inorganic metals (HS^-, Fe^{+++}) or reduced organic compounds are present, oxygen may be consumed by non-biological chemical oxidation which is measured as residual oxygen uptake after poisoning of biological respiration with buffered formalin or $HgCl_2$ (Teal and Kanwisher, 1961; Hargrave, 1972a). Since carbon dioxide is not evolved in these oxidation reactions, respiratory quotients are not comparable to those observed with individual organisms. Also, where aerobic and anaerobic heterotrophic and autotrophic organisms coexist, as in sediments, free carbon dioxide may be used as a carbon source (by chemoautotrophic bacteria) or it may be produced without oxygen uptake (by fermentation). Assumed RQ values for conversion of oxygen uptake to carbon dioxide release may thus be in error (Pamatmat, 1975).

Despite problems of interpretation, measures of oxygen and carbon dioxide exchange have been widely used as an index of biological activity in studies of metabolism in soil, water, and sediments. Oxygen depletion from bottom layers of stratified lakes was used as a basis for lake classification (Hutchinson, 1938) and Ohle (1956) used carbon dioxide accumulation in a similar way as a bioactivity index for whole lake metabolism. In contrast to these studies, the enclosure of water samples in dark bottles, with and without concentration of particulate matter, has been used extensively to measure respiration in planktonic communities (Riley, 1956b; Pomeroy and Johannes, 1966). Metabolic activity (aerobic respiration) is predominately due to μ-flagellates (2–10 μm) and bacteria which are associated with flocculated organic aggregates. Maximum respiration rates (up to 14 mg-at O/m^3/day) occurred in surface waters above the continental slope in the western North Atlantic and off the coast of Peru (Pomeroy and Johannes, 1968). Respiration below 500 m was estimated to be several orders of magnitude less than rates in surface water. Comparison of respiration rates with estimates of primary production determined with ^{14}C in other studies indicated that large spatial and temporal differences may exist between production and consumption of organic matter. The imbalance could arise from advective transport of particulate matter to or from other areas and loss through sinking.

Under steady-state conditions, with no accumulation, input of energy or elements to any system will be balanced by loss. This idea was used to calculate organic matter turnover (decomposition) rates in near-shore sediments from the ratio of supply/steady-state concentrations of carbon, nitrogen, and phosphorus (Seki et al., 1968; Chapter 5). Also, in ecosystems where no allochthonous supply of organic matter occurs, the ratio of photosynthesis:respiration (P/R) characterizes the balance between energy input and release (Odum, 1971). Physical removal (burial or transport) or accumulation of incompletely oxidized compounds resulting from anaerobic metabolism cause an underestimate of metabolic loss. When community metabolism has been determined by following changes in dissolved oxygen and carbon dioxide in a salt marsh over 24 hr ('diel curve method' described by Odum and Hoskin [1958]), however, RQ values fell between those expected for aerobic oxidation of fats and carbohydrates (0·7–1) (Nixon et al., 1976). Simultaneous measurements of community photosynthesis yielded PQ values of 1·1 and 1·2. Thus a diel balance appeared to exist where material produced during the day was respired at night. Production and respiration rates in laboratory microcosms also show a tendency to converge ($P/R = 1$) (Beyers, 1963) which suggests that tight coupling between metabolism and production exists in spatially limited ecosystems. Pomeroy (1970) considered this to be characteristic of systems where nutrients available for production are largely determined by recycling mechanisms. This would also explain Copeland's (1965) observation that gross photosynthesis and respiration in a laboratory microcosm reached similar levels when light intensity was changed despite completely altered species composition. Total nutrient content remained unchanged but photosynthesis could only be re-established to initial levels at a rate dependent on nutrient regeneration.

Teal and Kanwisher (1961) suggested that total oxygen uptake by undisturbed sediments represented an integrated measure of aerobic and anaerobic metabolism by an entire benthic community. Reduced metabolic by-products of anaerobic metabolism ac-

cumulate in many deposits, however, and chemical oxidation of some of these compounds may comprise a variable proportion of total oxygen availability (Pamatmat,
1968; 1971). Also, chemical oxidation cannot be assumed to be proportional to anaerobic
metabolism. Pamatmat and Bhagwat (1973) calculated that rates of metabolism indexed
by electron transport activity were up to 39 times greater than rates estimated from
chemical oxidation. On the other hand, Jørgensen and Fenchel (1974) observed that
in detritus-enriched sand in laboratory microcosms H_2S and FeS accumulation approximated the rate of sulfate reduction. Part of the sulfide that diffused to the sediment
surface was oxidized by CO_2 produced by photoautotrophic sulfur bacteria which would
not be detected by measures of chemical oxidation. Total net oxygen uptake (corrected
for oxygen produced by algal photosynthesis) over 6 months (470 μM O_2/cm^2), however,
could have oxidized 1·4 mg of carbohydrate/cm^2 to CO_2 which was approximately equal
to the total loss in organic matter observed. Thus, although total oxygen consumption
may under-estimate benthic community metabolism, the measure still appears to be
useful as an indicator of the amount of organic matter oxidized in sediments (Pamatmat,
1975).

Gradients in dissolved oxygen which form across sediment and particle surfaces suggest that uptake may be limited by rates of diffusion. Several studies have utilized
the basic equation,

$$\frac{dD}{dt} = D \frac{d^2C}{dx^2} \qquad (170)$$

where C is the concentration of diffusing material, t is time, D is the diffusion coefficient,
and x is distance, to describe changes in dissolved oxygen concentration with depth
in sediment and soil. For example, Hutchinson (1957) discussed ideas developed by
Grote which expressed the depth of oxygen penetration into sediment as a function
of oxygen concentration at the surface, molecular diffusion and metabolic consumption.
Thus, with a linear concentration gradient in the sediment, the depth at which oxygen
disappears (Z_0 in cm) is given by

$$Z_0 = \frac{k(O_2)}{R} \qquad (171)$$

where k is the molecular diffusion coefficient (cm^2/s) for oxygen in the sediment and
(O_2) is the oxygen concentration at the sediment surface (mg/l). R is the rate of oxygen
consumption per unit volume of sediment with unlimited oxygen supply. Greenwood
(1968) used a variation of the same equation to define the volume of a sphere where
anaerobic processes occur. If particles are spherical with a radius r, a volume V_2, with
an internal diffusion coefficient k, an oxygen-free volume at their centre of V_1 and
a rate of oxygen uptake per unit volume of R, then when the sphere is surrounded
by an oxygen concentration (O_2), the volume of the anaerobic sphere is given by

$$6k(O_2) = Rr^2 (1 + 2V_1/V_2 - 3[V_1/V_2]^{2/3}). \qquad (172)$$

Substitution of experimentally derived values for R (10^{-5} ml/ml/s) and k (10^{-5} cm^2/sec)
for moist soil aggregates with values for air saturated water (6 ml O_2/l) and no anaerobic
zone ($V_1 = 0$), gives an r value of 0·2 cm. Zones further than this distance from the
gas/water interface would become anaerobic.

Hutchinson (1957) made similar calculations, without empirical data, to estimate that
a diffusive flux of oxygen of 9·2 mg $O_2/cm^2/day$ would be required to deplete oxygen

to 2 cm in mud sediments. This is two orders of magnitude greater than values usually observed (Pamatmat, 1968) and it indicates that oxygen penetration by diffusion alone must only occur to shallow depths. Neame (1975) concluded that mass transport by respiratory and burrowing activity of benthic macrofauna was more important than diffusion in moving oxygen into sediments. Also, vertical profiles of dissolved oxygen in sediments (Hargrave, 1972a; Neame, 1975) do not show a linear decrease with depth indicating that processes other than diffusion must control transport. Despite this, various diffusion models have been used to relate oxygen uptake to the formation of an oxidized microzone in aerated sediment cores held in the laboratory. Bertru (1971) observed that the formation of an oxidized surface layer (to a maximum thickness of 4 cm in 10 days) was proportional to the square root of time. Oxygen demand was considered to represent a steady-state biological consumption and a variable uptake dependent on oxygen concentration was due to oxidation of diffusable material. A biological demand of 46 μg O_2/cm^2/day (50% of the total) was estimated—a value similar to that measured in surface sediments in Castle Lake (Neame, 1975). Bouldin (1968) also estimated oxygen flux from diffusion models which compared well with measures of uptake by salt marsh and lake sediment.

An alternative approach to quantify processes which control sediment oxygen uptake has been the use of measurements in a comparative way. Although attempts to attribute variance to different factors through empirical correlations do not demonstrate causal mechanisms, the relative importance of different factors can be evaluated. For example, the effect of limited oxygen diffusion into sediments was demonstrated by Baity (1938) who noted that oxygen uptake from water over sludge was not proportional to deposit depth but was described by the equation

$$R' = 2700D^{0.49} \tag{173}$$

where R' is oxygen consumption (mg/m^2/hr) and D is depth (cm). However, when cores deeper than 4 cm were used for determinations, depth had no significant effect on measured rates (Edwards and Rolley, 1965) suggesting that most oxygen consumption occurs in near-surface layers. These authors also observed that stirring only increased rates of uptake when scouring of surface sediment caused resuspension. Pamatmat (1971) and Davies (1975) made similar observations and attributed the effect to increased oxidation of reduced compounds exposed on stirring. However, flow rates up to 50 cm/min did not increase oxygen uptake when sediment was undisturbed (Pamatmat, 1965). Water movement as turbulence or advection across the sediment surface might be expected to increase uptake, particularly if an oxygen-depleted boundary layer exists. For example, Carey (1967) found that gentle stirring increased oxygen consumption by sediment cores from Long Island Sound. Berner (1976) and Suess (1976) discussed physical and chemical gradients which exist a few metres above deep-water marine sediments. Little is known, however, concerning conditions of water motion immediately above the bottom on a scale (cm to mm) which would affect the exchange of dissolved nutrients and gases such as oxygen.

The importance of diffusion-related processes for oxygen uptake by aquatic sediments has also been inferred from significant correlations between measured rates and dissolved oxygen concentration in overlying water and temperature. Baity (1938) observed that shallow layers (0·5 cm) of sludge consumed oxygen at rates which were independent of oxygen concentration between 2 and 5 mg/l. However, with deeper layers of sediment,

Edwards and Rolley (1965) and Hargrave (1969a) found that the equation

$$R' = a(O_2)^b \qquad (174)$$

described oxygen uptake R' (ml $O_2/m^2/hr$) where a (4·5) and b (0·66) were constants with an oxygen concentration in overlying water (O_2) between 2 and 8 mg/l. A relation of this form suggests that diffusion and active consumption occur simultaneously. Values of b calculated in the separate studies were 0·45 and 0·66 but the relationship was not adequate to describe data obtained from sediments which contained high densities of macrofauna. In contrast to these experimentally based studies, Pamatmat and Banse (1969) reported that seasonal changes in sediment oxygen uptake in Puget Sound were unrelated to changes in dissolved oxygen.

Seasonal variation in benthic oxygen consumption has also been related to temperature differences. Hargrave (1969a) compared measures of oxygen uptake (R') (ml $O_2/m^2/hr$) by sediments from different locations with temperature (T) (°C) and derived the equation

$$R' = 0·27T^{1·74}. \qquad (175)$$

Different slope coefficients can be derived from other studies (Pamatmat, 1968, 1971) with a minimal value (0·36) calculated by Smith (1973) for a sublittoral community off Georgia. Since seasonal fluctuations in factors other than temperature also affect oxygen uptake (for example, changes in the size of populations of respiring organisms and rates of chemical oxidation), estimates of Q_{10} cannot be interpreted on a physiological basis. However, highest values (4–11) occur in studies in northern latitudes (45–50°N) (Smith, 1973). This has been interpreted as an indication of a lack of temperature compensation and metabolic regulation at higher latitudes (Duff and Teal, 1965) but it could equally well reflect differences in seasonal patterns and variation of organic input to sediments. Davies (1975) demonstrated the effect of temperature alone ($Q_{10} = 1·9$) by incubating individual sediment cores from a Scottish sea loch at different temperatures.

Separate effects of season and temperature on sediment oxygen uptake were experimentally evaluated by Edwards and Rolley (1965) by incubating cores collected at different times to three different temperatures. A temperature coefficient

$$t = \frac{1}{T_2 - T_1} \log_e \frac{R'_2}{R'_1} \qquad (176)$$

was calculated where R'_1 and R'_2 were oxygen consumption rates (g $O_2/m^2/hr$) measured at temperatures T_1 and T_2 (°C). Average temperature coefficients of 0·065 and 0·077 (per degree Centigrade) between 10°C and 20°C were similar for sediments from different locations. Although periods of acclimatization were insufficient for complete temperature adaptation, minimum and maximum oxygen consumption occurred during winter and spring at each experimental temperature. Increased uptake during spring was attributed to growth of epibenthic algae and other benthic populations rather than only to an effect of increased temperature. Pamatmat (1971) also observed seasonal changes in total and chemical oxygen consumption which were independent of temperature at two sites in Puget Sound. The degree of seasonal effect was greatest at a location where biological respiration accounted for 48% of total consumption. There was a reduced seasonal difference after correction for temperature at a site where chemical oxidation accounted for most of the total annual consumption. Pamatmat concluded that differences between locations (maximum values during summer of 45 ml $O_2/m^2/hr$ in shallow water to minimum value of 9 ml $O_2/m^2/hr$ in deep water) reflected a lower organic supply or the presence of less oxidizable organic matter in deep-water sediment.

The idea that benthic oxygen consumption is quantitatively related to the flux of oxidizable organic matter to the sediment surface was first implied by Hutchinson's (1938) observation that plankton standing crop was directly related to oxygen loss from the hypolimnion of two stratified lakes. Thomas (1955) provided additional evidence by collecting sedimenting material in three lakes and observing that laboratory estimates of oxygen uptake by the material equalled observed hypolimnetic oxygen depletion. Oxygen consumption by sedimenting and deposited material thus appeared to be proportional to the supply of material from the euphotic zone. Pamatmat and Banse (1969) and Hargrave (1972b) considered this indirectly by comparing sediment oxygen uptake with ^{14}C estimates of phytoplankton primary production, on the basis of oxygen and carbon respectively, in Puget Sound and Lake Esrom. On an annual basis, total benthic oxygen consumption at two stations in Puget Sound corresponded to 17% and 25% of phytoplankton production. Biological respiration by sediments in central Lake Esrom amounted to 32% of the carbon fixed by phytoplankton. A seasonal cycle in benthic chemical oxidation and respiration which corresponded to a summer maximum in primary production existed in Puget Sound, but anoxic conditions in Lake Esrom resulted in low rates of consumption during summer when phytoplankton production was high. Increased rates of uptake at fall overturn, however, exceeded possible supply from phytoplankton suggesting that material accumulated throughout summer stratification was being oxidized.

Hargrave (1973) extended these comparisons by considering annual estimates of total benthic oxygen uptake and primary production in different freshwater and marine locations. Other sources of organic matter were not thought to be important in the sites which were compared and estimates of primary production provided a measure of organic carbon supply. The expression

$$C_0 = 55 \left[\frac{C_s}{Z_m} \right]^{0.39} \tag{177}$$

was empirically derived from multiple linear regression analysis which demonstrated that benthic oxygen consumption (C_0) (1 $O_2/m^2/yr$) was correlated inversely with mixed-layer depth (Z_m) (depth of the thermocline [m] during stratification), and directly with annual primary production (C_s) (g $C/m^2/yr$). Measures of benthic oxygen uptake in the deep sea, one to two orders of magnitude below rates in coastal waters (Smith and Teal, 1973), are less than would be predicted by extrapolation of this equation using values for C_s and Z_m typical for the open ocean. Deep-sea sediments may not receive organic matter directly from surface primary production. Also, if mineralization is largely complete within the surface mixed-layer, a small proportion of total carbon production may sediment and no clear relation between this input and decomposition at the sediment surface would exist (Riley, 1970). When sediment oxygen consumption was converted to carbon release (by assuming an RQ value of 0·85), however, calculations demonstrated that increased rates of primary production resulted in a proportionately smaller fraction respired at the sediment surface. This could indicate the increasing importance of metabolic loss to plankton at higher levels of primary production, or increased rates of burial of unoxidized organic matter. Pamatmat (1975) suggested that the latter seems most probable since with increasing deposition anaerobic metabolism may be expected to become more important than aerobic respiration. These and other data (Pamatmat and Bhagwat, 1973) suggest, despite the underestimate of total metabolism, that measures of annual oxygen uptake by sediments may be regarded as a fair estimate of the fraction of annual organic supply which is respired during a given year.

The lack of a significant effect of total water column depth on benthic oxygen consumption, in comparison to other factors, has been observed over a limited and extended depth range (Pamatmat, 1971; Hargrave, 1973). Depth (and pressure) effects *per se* may be of less importance than the oxidizable nature of organic matter which reaches the sediment. Rates of microbial degradation of organic matter added to deep-sea sediments, however, are one to two orders of magnitude lower than rates observed in surface waters (Jannasch *et al.*, 1971) and elevated pressure and low temperature both retard the rate of microbial metabolism (Wirsen and Jannasch, 1975). These observations, and those discussed earlier (Section 6.4.1.2) which indicate that most mineralization of organic matter occurs in near surface water layers, could account for the significant effect of mixed-layer rather than total depth on benthic oxygen consumption.

The observation that carbon sedimentation and benthic oxygen consumption are both related to the ratio of carbon supply:mixed-layer depth [eqns. (134) and (177)], implies that the flux of organic matter (particulate sedimentation) to benthic communities may be quantitatively related to benthic oxygen uptake. Smith (1974) observed that consumption by sediments under enriched surface water in an upwelling area off California was significantly higher (3·5% of estimated annual primary production) than uptake at comparable depths in more oligotrophic areas. The degree of oxidation of material deposited at the sediment surface, however, may reflect its origin. For example, material derived from terrestrial debris is mineralized slowly and may be buried without extensive degradation (Uyeno, 1966; Seki *et al.*, 1968). Johnson and Brinkhurst (1971b) clearly demonstrated the effect by comparing sedimentation and benthic oxygen consumption at four locations where terrestrial debris contributed varying amounts to sediments. The energy equivalent of annual sediment oxygen uptake was 23% of that sedimented and macroinvertebrates assimilated 2% when allochthonous supply was important. At other locations where sedimented material was derived mostly from phytoplankton production, 90% of the organic matter deposited was respired and macrofauna assimilated 30% of the supply. Davies (1975) also observed that the annual carbon equivalent of benthic oxygen uptake was not significantly different from carbon sedimented in a semi-enclosed marine bay where phytoplankton were the main source of organic matter. Davies' data for annual sedimentation and benthic oxygen consumption were related to the ratio of primary production:mixed-layer depth as described by eqns. (134) and (177).

Other studies have also indicated that supply and oxidation of organic matter in sediments may be linked to various degrees. Comparisons of annual epibenthic algal production and community respiration have shown that consumption measured as oxygen uptake amounted to only half of that produced in salt marsh sediments (Pomeroy, 1959) while comparable values for both measures were obtained for sediments in a shallow lake (Hargrave, 1969b). A similar correspondence between carbon input and CO_2 release from littoral sediments was observed in Lawrence Lake (Wetzel *et al.*, 1972) and Hartwig (1976) described a sublittoral marine area where carbon deposited or produced by epibenthic algae was either utilized immediately or resuspended. Marshall (1970) and McIntyre *et al.* (1970) also made estimates of apparent organic supply which were insufficient to meet organic carbon removed calculated from oxygen demand measurements in subtidal and artificial sand columns in the laboratory. Imbalances in such budget calculations may arise from methodological problems (such as some unmeasured fraction of dissolved or particulate matter). The calculations imply, however, that in sediments with low organic content but high rates of oxygen consumption, organic supply must be consumed rapidly to prevent accumulation.

This view of a rapid flux and microbial utilization of organic matter in many benthic communities is consistent with the general observation that benthic oxygen consumption is unrelated to standing stock of macrofauna and organic content (Pamatmat and Banse, 1969). Usually less than 20% of total oxygen uptake can be attributed to macrofauna respiration (Banse et al., 1971; Smith et al., 1972). Increased numbers of benthic macro-invertebrates can increase rates of oxygen consumption in sediment cores but increases · are not necessarily proportional to increased density. Interactions between organisms themselves and their food supply when this is primarily as micro-organisms modify any single relation which might be expected (Edwards and Rolley, 1965; Hargrave, 1970a). Consequently, sediment microflora are assumed to account for the bulk of ben-thic metabolic activity (Carey, 1967). This supports Marshall's (1970) idea that it is the flux of oxidizable organic matter and not that remaining as refractory substrates which fuels benthic metabolism. Waksman and Hotchkiss (1938) reached similar conclu-sions by comparing oxygen uptake in stirred sediment samples. Hargrave (1972d) empha-sized the importance of particle surface area in such measurements as a controlling factor for oxygen consumption. A positive correlation between oxygen uptake, bacterial biomass, and organic content also indicates that microbial aerobic respiration may be limited by factors other than only available organic matter.

In situ measurements of microbial respiration in sediments have been attempted by using antibiotics as respiratory inhibitors (Hargrave, 1969b; Smith et al., 1972; Smith, 1974). Dilute solutions of penicillin and streptomycin were shown not to affect inverte-brate respiration or algal ^{14}C uptake during short-term exposure and thus differences in rates of total oxygen uptake by cores with and without antibiotic treatment were assumed to reflect microbial respiration. Between 5 and 40% of total oxygen uptake was susceptible to inhibition in these studies and a seasonal effect was apparent. Antibio-tic-sensitive respiration increased with increasing temperature, but the proportion of this to total benthic community respiration decreased to a minimal value of 30% during summer (Hargrave, 1969b). Yetka and Wiebe (1974), however, examined the effectiveness of different antibiotics for respiratory inhibition in six strains of marine bacteria. Percent inhibition varied with growth phase, antibiotic and bacterial type and it was concluded that these antibiotics could not be used to delineate bacterial respiration in mixed mi-crobial communities. The method may thus have little quantitative value as a measure of bacterial respiration, but the susceptibility of early log-phase cells to inhibition sug-gests that some qualitative index of metabolic activity attributable to actively dividing cells may be provided.

A relationship between bacterial biomass and metabolic rate, measured as oxygen uptake, was first quantified in lake sediments by Hayes and MacAulay (1959) who observed that the equation

$$R = -100 + 12 \cdot 5B \qquad (178)$$

described the relation between oxygen consumed by intact sediment surfaces (R) (mg $O_2/m^2/hr$) and bacterial plate counts (B) ($\times 10^3/ml$ of wet sediment). Measures of oxygen uptake by resettled surface sediment were comparable to measurements with undisturbed cores. Although Hargrave (1975b) did not obtain equivalent rates of oxygen uptake in artificially restratified and freshly collected sediment cores from a lake and marine bay, it was apparent, as suggested by Hayes and MacAulay (1959), that the restabilized sediment–water interface consumed oxygen at rates which might be used to characterize different sediments. Hayes and MacAulay attempted to express this as a linear function of a lake quality index derived from estimates of fish population size, the length of the food chain and dimensions of various lakes. Extension of this type of analysis to marine systems would be profitable particularly if fish abundance, expressed on an area basis with demersal and pelagic species treated separately, could be compared with oxygen consumed by suspended particulate matter and bottom sediments.

Two recent studies suggest the existence of general quantitative relations between benthic community oxygen uptake and production by benthic fauna which follow from Hayes and MacAulay's (1959) observations. Sokolova (1972) correlated the biomass of meiobenthos (0·5 mm–0·5 cm) in eutrophic and oligotrophic regions of the Pacific on a semi-logarithmic basis with heterotrophic activity of microflora and oxygen consumption by stirred samples of surface sediment. The relationships were thought to reflect the supply of organic matter to sediments in different areas and its transformation by micro-organisms to produce levels of food supply which would support different amounts of meiobenthic biomass. Similar quantitative differences in abundance of macrobenthos between areas could be attributed to the level of oxidizable organic matter in surface deposits.

Observations by Johnson and Brinkhurst (1971b) also linked the production of benthic macrofauna in a Lake Ontario bay to the supply of oxidizable organic matter. Macrofauna production (P) was described by the equation

$$P = c(a \cdot b \cdot I_m) \tag{179}$$

where c is the proportion of growth to respired and exported energy by macrofauna (i.e. growth efficiency), a is the proportion of imported energy (I_m) (sedimentation) used by the total benthic community, and b is the proportion of this used by macroinvertebrates. All three proportions must remain constant if production is proportional to sedimentation. However, Johnson and Brinkhurst observed that the proportion of sedimented energy used by macrofauna decreased with increasing input. This could have been due to the importance of allochthonous debris at the site of highest deposition which was utilized with a low efficiency by the macrofauna. It could also reflect added respiratory costs associated with conditions of high sedimentation. In either case, the value of a (which approaches 1 when the total supply is fully oxidized) may be expected to vary directly with the proportion of material derived from autochthonous production. b may also decrease as the importance of refractory organic matter deposited increases and the lower its value the greater is the fraction of total community respiration attributable to non-macrofauna consumption. The relationships suggest that all three proportions may decrease as I_m increases, particularly if this is associated with the input of allochthonous material which is largely refractory to utilization by either micro-organisms or macroinvertebrates. With a nonrefractory organic supply, macroinvertebrate production should increase with increasing organic sedimentation up to a maximum level where the product $a.b.$ (the proportion of sedimentation used by macroinvertebrates) is maximized. If these relations have general application, it seems possible that macrofauna production could be predicted from measures of organic input and oxygen uptake by benthic communities.

CHAPTER 7

SOME PRACTICAL PROBLEMS IN
BIOLOGICAL OCEANOGRAPHY

BEFORE the turn of the last century the earliest publications in biological oceanography were primarily concerned with the taxonomy and distribution of marine planktonic organisms. With the introduction of extensive nutrient analyses in the 1930s, biological oceanography entered a more quantitative era and some efforts were soon made to use biological oceanographic information for the solution of practical problems. Early attempts to relate nutrient distributions to fisheries were not successful but the exercise served to demonstrate where more knowledge was required. Riley's (1946) work on modelling the plankton community led the way to a much greater understanding of integrated processes. The impediment to further development of models then lay in the inability to collect sufficient data over an area and time scale applicable to the models. This problem is being solved with the use of automated equipment such as autoanalysers coupled to shipboard computers (e.g. Walsh and Dugdale, 1971). Throughout this brief history of development, however, it has been continually necessary to summarize relationships between plankton and the environment in a variety of empirical expressions, many of which have been discussed in earlier sections of this text. These expressions (and future modifications) are the building blocks on which practical solutions to biological oceanographic problems must be attempted. While it is recognized that more knowledge would be helpful, solutions to many of today's problems are required today, and where answers are required they should be given to the best of the state of present-day knowledge.

In the following section some attempt has been made to identify a number of problems for which solutions may be attempted from a knowledge of biological oceanographic processes. The actual application of biological oceanographic information to the solution of a problem will depend heavily on the local environment and for this reason the following subjects are only considered by example, or from a point of view of general relationships.

7.1 EXAMPLE PROBLEMS IN POLLUTION AND
WATER MASS IDENTIFICATION

The survival of terrestrial bacteria in the marine environment has received considerable attention because of man's tendency to regard the sea as a septic tank. Thus one of the commonest forms of coastal pollution is the contamination of beaches and near-shore fisheries with pathogenic organisms from human excrement. Jones (1971) has estimated that 4.4×10^7 lb of feces are released per day in the United States and that an appreciable proportion of this is disposed of in the near-shore environment. While the distribution of this material on entering the sea is a problem in physical oceanography, the survival of pathogenic bacteria in sea water is a biological problem. The subject has recently been reviewed by Mitchell (1968) and Jones (1971). Of the many bacteria entering the sea, the presence of *Escherichia coli* is usually taken as an index of sewage pollution. The organism is not itself a pathogen but its presence in large numbers may indicate the presence of pathogenic bacteria and viruses. The actual 'acceptable' level of coliform contamination may vary depending on national standards and use of surrounding beach areas; generally a level of less than 100 coliforms/100 ml in 90% of the samples averaged over a 30-day period is considered an acceptable limit while a level of 1000 coliforms/100 ml would be considered contaminated.

Several mechanisms are considered to cause the destruction of terrestrial bacteria in the sea. The production of antibacterial substances by marine plankton (e.g. Duff *et al.*, 1966; Burkholder *et al.*, 1960) has been shown to include substances active against *Staphylococcus* although only about 40% of the algal species tested by Burkholder *et al.* (1960) displayed activity against this pathogen. A specific antibacterial activity of the phytoplankter, *Phaeocystis*, was identified by Sieburth (1960) as acrylic acid. In addition to the antibacterial activity of marine phytoplankton it has also been suggested that sunlight, temperature, and specific bacteriophages may cause mortality of terrestrial bacteria in seawater. However, most of these mechanisms might be considered to be equally effective against the indigenous species of marine bacteria and not specific agents in causing the high mortality of terrestrial species. A mechanism which is more specifically related to a difference in properties between freshwater and seawater is the salt content of the latter; salt (NaCl) in itself, however, is not considered to be the lethal agent since several authors have shown that some terrestrial bacteria can tolerate (but not necessarily grow) in NaCl concentrations higher than are found in the sea (Korinek, 1927; Burke and Baird, 1931). On the other hand, Jones (1963) found that *E. coli* was killed by natural seawater; a process which could be prevented by the addition of a chelating agent. Jones (1963, 1964 and 1967) concluded that it was the heavy metal content of seawater together with the low concentration of natural organic chelators which was lethal to *E. coli*, and other terrestrial bacteria. Under these circumstances it is apparent that near-shore environments which contain a large amount of organic material (from pollutants or natural sources) will allow for a greater *survival* of fecal bacteria in the marine environment. In addition to heavy metal toxicity, however, it is also apparent that substrate concentrations are generally much lower in sea-water than in terrigenous environments; thus the *growth* of terrigenous bacteria would be inhibited in seawater compared with the natural marine bacteria which have adapted to very low substrate concentrations.

In summary it appears that a function of the salt content of seawater (i.e. heavy metals) together with the presence or absence of chelating agents and organic substrates are the most likely factors determining the growth and survival of terrestrial bacteria in the sea. On the other hand the division of estuarine bacteria in terms of their tolerance to sodium chloride may be a useful ecological classification with which to differentiate between autochothonous and allochthonous populations. Larsen (1962) classified micro-organisms as halophobic or halopholic and described the former as organisms which grow best on a medium containing less than 2% NaCl. From an estuarine study, Seki *et al.* (1969) determined a regression line between the ratio of freshwater:saltwater colonies (*y*) and salinity (*x*). From this equation

$$y = -2 \cdot 2x + 5 \cdot 2 \tag{180}$$

it is apparent that bacteria in the estuarine environment studied by Seki *et al.* (1969) grew best in a freshwater medium (*y* > 1) when the salinity was less than 1·9% or approximately 19‰.

The assessment of microbial activity in polluted waters has recently been reviewed by Jannasch (1972). The author emphasizes the need to make a number of routine microbiological measurements in order to determine the biochemical activity of polluted waters, rather than simply obtaining an index of pollution, such as a coliform count. Five kinds of determinations are suggested; these are biomass determinations, respiratory activity, special metabolic activities (such as nitrification), growth-rate determinations, and specificity of substrate utilization.

Estuarine areas are of particular interest to biological oceanographers, both from the point of view of their unique biological activity and because they tend to be centres of human activity which includes the development of ports. The river above the estuary may be considered as a corridor for fish migrations, ship transport, and disposal of natural and man-made products which are carried from the land to the sea. In this situation a conflict often exists between the natural ecology of the estuary and its modification due to man's activities which have generally resulted in pollution. Obvious forms of pollution, such as organic and inorganic industrial poisons, cause direct and catastrophic effects on the natural estuarine environment and remedial action is generally best handled through chemical evidence and legal procedures. Slightly more subtle, however, are the effects of non-toxic substances which may decrease light penetration in the water column, take up oxygen, or cause toxic substance to accumulate through precipitation in the sediment near the river mouth. Substances which decrease the penetration of light into the water (e.g. silt, coal dust, pulp mill effluent) can cause a decrease in the primary productivity of the water column; this in turn reduces the amount of oxygen in the water and decreases the supply of phytoplankton at the primary trophic level. Since the effect of light-attenuating substances can be expressed in terms of changes in the extinction coefficient (*k*) of the water, a diagnosis of the overall effect of this form of pollution might be modelled through some of the expressions discussed in this text. Oxygen depletion over a period of time and distance can also be modelled using experimental data on the Biochemical Oxygen Demand (BOD) of samples of the polluting water (for BOD methodology see Amer. Pub. Health Assoc., 1965).

A property of the ecology of estuarine waters is their high productivity. In the immediate area of an estuary, where sea water and freshwater have mixed (often referred to as the river's 'plume') there is a zone of high primary production which in tropical environments, may persist throughout the year, or vary seasonally in temperate latitudes. This zone generally supports a high secondary productivity and often serves either as an area in which there is an extensive fishery, or perhaps more important, as a nursery ground for the young stages of many commercially important fish. There are a number of reasons why estuaries may be biologically productive. Firstly, there may be a natural enrichment of nutrient-poor surface sea waters by nutrients from the river; this was true for the Nile estuary, where the principal mechanism causing an enrichment of the nutrient-impoverished Mediterranean waters, prior to the construction of the Aswan dam, was the nitrogen and phosphorus content of the Nile silt (Halim, 1960). Secondly, the flow of river water into the sea may cause an upwelling of nutrient-rich deep salt water which continually fertilizes the river's plume; this occurs, for example, off the Columbia and Fraser rivers. Thirdly, there may be a process of eutrophication caused by nutrients contained in the river; these are generally derived from agricultural lands and domestic sewage. Classical examples of eutrophied estuarine areas are often found in quite local estuaries (e.g. Jeffries, 1962; Barlow et al., 1963) but eutrophication and upwelling may both be operative enrichment mechanisms in some of the larger estuarine areas (e.g. the New York Bight, Ketchum, 1967; the Columbia river estuary, Haertel et al., 1969). In addition in some limited areas a nutrient 'trap' may develop where the organisms grown at the surface, sediment to a layer where they are carried back into the inlet by the counter current to the surface flow. This situation was first described by Redfield (1955) and examples of self-enriching processes can generally be found in fjords with shallow outer sills; in some cases the waters in the inner basins become anoxic due to the accumulation of organic materials.

The identification of nutrient-polluted estuaries has been discussed by Ketchum (1967). In the simplest case it is apparent that if the nitrate or phosphate content of estuarine waters is higher than the maximum nutrient content of surrounding deep saline water, then a terrestrial source of nutrients may be expected. In the case of only two water types (i.e. freshwater and saltwater) the fraction of freshwater (F) can be obtained from the expression.

$$F = \left(1 - \frac{S}{S_0}\right) \tag{181}$$

where S is the salinity of the sample and S_0 is the salinity of the source sea water. Thus the concentration of a nutrient (or some other potential pollutant) at any point in the estuary, can be determined from the salinity of a sample and a knowledge of the nutrient content of the source waters. In a system in which there are three water types Ketchum (1967) assumed that *total* phosphorus could be used as a conservative property (since loss would only be by a small amount of sedimentation of particulate phosphorus) and together with the salinity, the fraction of any three water types could be found from the salinity and total phosphorus content of a sample. Thus in Ketchum's example three water types were identified in the New York Bight; these were

   A. Brackish river water: 30‰ S_a and total P_a 2·9 μg at/l
   B. Surface coastal water: 30·95‰ S_b and total P_b 0·5 μg at/l
   C. Deep ocean water: 34‰ S_c and total P_c 1·25 μg at/l

Then the equations used to determine the fraction of the three water types are

$$S_x = AS_a + BS_b + CS_c, \tag{182}$$

$$P_x = AP_a + BP_b + CP_c, \tag{183}$$

$$A + B + C = 1. \tag{184}$$

Where A, B, and C are the volume fractions of each water type and S and P are the salinities and total phosphorus contents of the three water types and the unknown sample, x.

This rather simplified diagnosis of water types can be used in conjunction with other environmental data to analyse changes in the estuarine environment. For example the maximum level of primary productivity is often found at some distance from a river mouth and is not necessarily associated with the maximum availability of nutrients. This is caused by a number of factors including the increased availability of light due to sedimentation of silt, decreased mixing processes with distance, and a time factor which allows the seed population of phytoplankton to increase exponentially as the waters move away from the river mouth; this situation has been documented as occurring under natural conditions in the Fraser river plume (Parsons et al., 1967) and is predictable from a simulated model of a sewage outfall entering the sea (Walsh, 1972) and from actual observations (Caperon et al., 1971). In the latter reference the authors also use a model (similar to Walsh's—1972) to predict the course of future and past eutrophication of a tropical estuary. The model employs a 4 component food chain in which each component is related to the rest by a hyperbolic function [eqn. (81)]; feedback is also introduced through nutrient regeneration. The model shows that in an oligotrophic environment, small changes in the input of nutrients are taken up rapidly by the population acting over the maximum response region of the hyperbolic function. However, the effect of a large perturbation, in terms of a massive injection of nutrients (e.g. sewage) could not be dampened out once the rate-compensating capacity of any trophic level had been exceeded. This resulted in a build-up of nutrients and the establishment of a large biomass of a new population at the primary trophic level; in other words the original ecosystem had become destabilized.

The effect of low nutrient additions on an oligotrophic environment has been considered experimentally (Parsons et al., 1972; Le Brasseur and Kennedy, 1972; Barraclough and Robinson, 1972) and theoretically (McAllister et al., 1972) in the fertilization of a 12,000 acre lake with 100 tons of nutrients. This experiment was set up to meet the requirements of hyperbolic growth responses discussed by Caperon et al. (1971). As a result it was found possible to increase the primary productivity of the lake's ecosystem without causing instability. Further there was good evidence that the increase in primary productivity was transferred up the food chain to cause an increased production at the secondary and tertiary levels.

The transfer of pollutants, such as heavy metals and chlorinated hydrocarbons, in the marine biota involves a study of the food web of organisms in benthic and pelagic environments and as such it is an area of practical interest to the biological oceanographer. Several general theories exist on the way in which pollutants are transferred between different trophic levels. These include a concentration theory which may operate in a case where a small amount of pollutant is selectively absorbed by certain tissues (e.g. DDT in fatty tissues); this then becomes more and more concentrated since the pollutant is not readily destroyed but the amount of tissue it is dissolved in becomes a smaller and smaller fraction of the total biomass as it moves towards the top of the food pyramid. This theory depends on the presence of a structured food chain in which there is a direct flow of food from the lowest trophic levels to the highest (e.g. the Bermuda petrel, Wurster and Wingate, 1968). However, other animals are very diverse feeders and as predators of many different types of food (e.g. sea gulls) they have unstructured food chains and are less subject to the accumulative effect of a pollutant. This has given rise to another theory on the bio-accumulation of pollutants which indicates that it may be more associated with the surface area of organisms and the adsorptive properties of these surfaces. For example, metals appear to be adsorbed by the plankton community and much larger amounts are adsorbed by the phytoplankton than the zooplankton (Mayzaud and Martin, 1975). On the other hand, Kerr and Vass (1973) conclude that the uptake of another group of pollutants (chlorinated hydrocarbons) is a function of metabolic rate and diffusion across respiratory surfaces. It is probable that the biological transfer of pollutants is in fact a function of any one or more of the three theories put forth above, and may in addition involve other processes as yet undefined.

Substances which have been shown to have caused some form of pollution in the marine hydrosphere include three major groups of elements or compounds. These are heavy metals, hydrocarbons (including both petroleum and chlorinated) and radionucliodes. Many papers and reviews of these subjects have been written and it is not our intention to give any exhaustive treatment to the subject of contaminants in the hydrosphere. However, in the following discussion we hope to show, by example, where biological oceanographers might be involved in some practical aspects of pollution research.

It has been shown that the concentrations of heavy metals have increased in some marine environments, especially in coastal areas. Since heavy metals such as lead, mercury, copper, and cadmium can be shown to be toxic to biological organisms, their presence is generally considered undesirable. Examples of intensive local pollution in nearshore areas are well known (e.g. the Minamata disease). However, it is quite another question whether or not traces of these elements in the oceanic environment of the sea are having any large-scale effect on the pelagic ecosystem. At present the lack of catastrophic changes in the marine biota of oceanic areas can generally be considered as indicating that the presence of a pollutant at very low concentrations (i.e. parts per billion) may not be harmful. However, the more subtle effects of such low concentrations really have not been investigated and the following brief review of current findings with respect to mercury serve as an illustration of where more research is needed by biologists.

The distribution of dissolved mercury in the eastern neritic zone of the Irish Sea shows values in excess of 200 ng/l (Gardner and Riley, 1972). These particularly high values have been in part attributed to the presence of sewage and sludge. Fish caught in the same area had an average mercury content of 0·53 ppm dry weight, or more than double the value for fish caught in the rest of the sea (average 0·21 ppm). Mercury may also enter the surface waters of much larger marine areas by being airborne. For example, Gardner (1975) found that the mean mercury content of North Atlantic waters under the northern jet stream was 26·5 ng/l compared with an average southern hemisphere value of 15·9 ng/l. However, the source of this mercury is not clearly established and may be a natural phenomenon caused either by degassing of the earth's crust in the northern land mass or by volcanic activity, such as in the region of Iceland (Siegel et al., 1973). Alternatively the high surface mercury values may be due to atmospheric transport of mercury from the urban industrial complex of North America. However, analysis of fish samples from the northeast Atlantic (Leatherland et al., 1973) failed to show any specific mercury contamination. In fact the highest mercury values in fish have been reported for the Black Marlin (7·3 ppm wet wt.; Mackay et al., 1975) caught off the coast of northeast Australia where man is unlikely to have caused extensive pollution of the ocean habitat. Thus from chemical analyses there are a number of unexplained differences in the distribution of mercury in the marine environment. Since the transfer of mercury into the food chain of the sea is in part a biological phenomenon, involving the formation of organic mercury compounds by bacteria (Jensen and Jernelöv, 1967; Wood et al., 1968; Fagerström and Jernelöv, 1972), it appears that biological oceanographers should be involved with the chemists in this type of problem.

The distribution of radioactive pollutants in the sea is a special case of marine pollution associated with nuclear power stations, dumping and earlier atmospheric testing of nuclear bombs. The distribution of isotopes, as with heavy metals and other pollutants, is a function of both biological and geochemical processes. The destructive effects of highly active isotopes on biological materials are well known and need not be discussed here further. However, it is of considerable interest that small traces of active isotopes, both artificial and natural, can be used to follow the transport of materials in the food web of the sea. For example, Williams et al. (1970) have shown that ^{14}C from weapons testing in 1961–2 has in some cases been rapidly transported into the tissue of bathypelagic animals, while other animals in the same habitat contained very little ^{14}C. Since ^{14}C has labelled the surface bicarbonate and has been taken up by phytoplankton in the euphotic zone it appears that in some cases there is a rapid vertical transport of this material to bathypelagic organisms (e.g. via overlapping vertical migrations, Vinogradov, 1955) while in other cases, bathypelagic organisms must rely for food on an organic carbon source which is 'old' relative to the age of surface bicarbonate.

Chemical evidence has indicated the presence of man-made synthetic pollutants in the marine food chain and the question to the biological oceanographer is how do these materials become so widely distributed and what is the consequence of their presence? At present the best documented example of a widely distributed pollutant is DDT and other chlorinated hydrocarbons. Reports on the distribution of these pesticides include measurements of their presence at great distances from the source of pollution (e.g. Antarctica—George and Frear, 1966) and in animals at the top of the food chain (e.g. sea birds—Risebrough et al., 1967). While pollutants can become slowly dispersed by ocean currents, or more rapidly by animal migrations, it is probable that the widespread dispersion of DDT is carried out in the atmosphere. On the other hand the accumulation of chlorinated hydrocarbons in tissues of higher animals must follow a structured marine food chain. In some cases the level of chlorinated hydrocarbons may be sufficient to interfere with the metabolism of higher animals, while the metabolism of organisms lower in the food chain may appear to be unaffected. The effect of DDT and other pesticides on lower organisms can be demonstrated, however, if these substances are used in rather high concentrations (e.g. 10 to 100 ppb). Wurster (1968) and Menzel et al. (1970) demonstrated photosynthetic inhibition of phytoplankton by chlorinated hydrocarbons. However, while there is no disagreement that qualitatively chlorinated hydrocarbons are toxic to the marine environment, the quantitative aspects of the problems indicate that at present DDT levels are generally well below any dangerous concentrations with possible exceptions to this to be found in some local nearshore areas and in particular among some marine birds; in the latter example it now appears well documented that egg shell thinning and reduced breeding success was associated with the presence of DDT residues in these animals (e.g. Hickey and Anderson, 1968; Wurster and Wingate, 1968).

Under some circumstances it is of interest to know the past history of a water mass. This may be traced in part from data on temperature and salinity but it is also possible to obtain information on the biological history of water from its oxygen and nutrient content (e.g. Park, 1967). The amount of oxygen which will dissolve in sea water at the sea surface is a function of sea water temperature and salinity. Any difference between the measured oxygen content of sea water and that expected from the known solubility at the temperature and salinity of the sample is called the 'apparent oxygen utilization' or AOU (Redfield, 1942). AOU may be negative (due to photosynthesis) or positive (due to mineralization). For example, if surface water is caused to sink at a convergence, oxidation of organic material will proceed at aphotic depths and the oxygen content of the water will be lowered (positive AOU). Assuming an elemental atomic ratio in plankton of $O:C:N:P \equiv 276:106:16:1$, the AOU can be expressed in terms of the oxidized state of these elements in a sea-water sample. Thus the *measured* nitrate and phosphate ($P_{meas.}$ and $N_{meas.}$) in a sea-water sample can be expressed as

$$P_{meas.} = P_p + P_{ox} \tag{185}$$

and
$$N_{meas.} = N_p + N_{ox}, \tag{186}$$

where P_p and N_p are the levels of preformed phosphate and nitrate, and P_{ox} and N_{ox} are the levels of phosphate and nitrate derived from the apparent oxygen utilization (AOU), respectively. Equations (109) and (110) can therefore be rewritten in terms of AOU, as

$$P_{meas.} = P_p + 0.0036\,AOU \tag{187}$$

and
$$N_{meas.} = N_p + 0.058\,AOU \tag{188}$$

where all units are in numbers of atoms (e.g. μg at/l). Sugiura (1965) has suggested that the preformed nutrients are a conservative property of sea water; as such, they can be used to characterize water masses by solving eqns. (187) and (188). The use of AOU values is not without difficulties, however, since it is assumed that no oxygen is lost to the atmosphere from the surface (Stefánsson and Richards, 1964). Loss of oxygen occurs when the surface concentration is greater than saturation. Since AOU is calculated from the difference between saturation and measured concentration, difficulties arise when the water leaves the surface with an oxygen content different from the equilibrium solubility.

The extent of ocean pollution from dissolved petroleum hydrocarbons is still a difficult question to answer. Since phytoplankton produce hydrocarbons it is not possible to differentiate between 'natural' hydrocarbons and petroleum hydrocarbons on the basis of the total hydrocarbon content of seawater. Barbier *et al.* (1973) suggest that the total amount of hydrocarbon in sea water is 1 to 5% of the total dissolved organic matter. If the lower figure represents an approximate background average for the open ocean, then it amounts to about 10 μg/l or 1.3×10^{10} tons in the whole ocean. This appears to be a fairly large number compared with Hunt and Blumer's (1971) estimate of 5×10^6 tons being the amount of petroleum hydrocarbons reaching the ocean annually. From the relatively small amount of aromatic hydrocarbons in sea water, Brown and Huffman (1976) suggest that the total hydrocarbon does not come predominately from petroleum but from natural hydrocarbon sources (i.e. the marine food chain). This can be argued on two bases; firstly, the aromatic hydrocarbons are more abundant in petroleum products than in hydrocarbons from phytoplankton, and secondly, aromatic hydrocarbons are preferentially more soluble in sea water (e.g. Anderson *et al.*, 1974). Both these factors would favour a high aromatic to total hydrocarbon ratio if petroleum was a major source of dissolved hydrocarbons in the marine environment.

In coastal waters (e.g. Barbier *et al.*, 1973) and along tanker routes (e.g. Brown and Huffman, 1973; Wong *et al.*, 1974) there appear to be clear examples of petroleum hydrocarbon contamination. On the other hand, a large oil spill, while producing most dramatic local damage, does not appear to cause any long-lasting effects (Nelson-Smith, 1970) and even the local concentration of dissolved hydrocarbons in the area of a major spill appears to diminish to ocean background levels after 1 or 2 years (e.g. Gordon and Michalik, 1971) in exposed waters.

In spite of these findings, however, it is reported from experiments that low levels of hydrocarbons (in the ppb range) may influence the growth of certain phytoplankton (Dunstan *et al.*, 1975; Parsons *et al.*, 1976). In particular it has been shown that small flagellate production may be enhanced over the production of large diatoms. One ecological consequence of this effect is indicated by Ryther's (1969) hypothesis which showed that small phytoplankton resulted in long food chains with less yield of harvestable resources than short food chains based on large-sized primary producers. Thus one may speculate that a possible effect of petroleum hydrocarbon pollution could be a loss in ocean fish production, particularly if there was a marked increase in the background of aromatics.

7.2 EXAMPLE PROBLEMS IN FISHERIES

A great deal of scientific effort has been expended on attempting to account for fluctuations in fish populations. Ahlstrom (1961) has emphasized that the two properties of a fishery which are most desirable to know are firstly the *abundance* of the species, including temporal fluctuations, and secondly the *location*, including factors which influence the aggregation or dispersion of a species. Attempts have been made to establish causal relationships between these two properties, and biotic and abiotic variables, such as winds, currents, temperature, plankton productivity; and predators. In some cases there are clearly defined relationships; for example, Uda (1961) has given the optimum temperature range for twenty-one different species of commercially important fish; Rae (1957) has shown a correlation between average wind strength and brood strength of haddock. In detailed studies on oceanography and the ecology of tuna, Blackburn (1965 and 1969) has concluded that there is a very general relationship between zooplankton abundance and tuna catch, but that efforts to show a definite monthly correlation between tuna and zooplankton over one degree rectangles usually failed. However, from a study of tuna distributions with temperature it was apparent that the tuna would not generally enter water colder than 20°C. Since high concentrations of food organisms often occurred in areas where the temperature was below 20°C, a direct relationship between tuna and zooplankton abundance was not apparent when data were collected from purely geographical coordinates. When temperature was taken into account it was concluded that the 20° isotherm determined the overall range of the tuna, while zooplankton concentration determined the distribution of tuna within the animal's temperature range.

In general, however, predictions of abundance or location of fish based on single or multiple component correlations have not been markedly successful. This conclusion has been reached independently by a number of authors studying very different associations in the oceans. Thus Ahlstrom (1961) concluded that there was no relationship between year class strength of sardines and high plankton productivity. Lillelund (1965) concluded from a review of the effects of abiotic factors on young stages of marine fish that abiotic factors had only an indirect effect on survival, and that the overall effects were complex and were more probably associated with biotic factors. Steele (1964) concluded that the prediction of fishereis could not be based on levels of primary productivity and assumptions regarding ecological efficiencies, but that more detailed knowledge of food-web structure was required.

While relationships between adult stocks and biological oceanographic data are tenuous, several authors have recognized the importance of survival rates during the early stages of the life history of a fish (e.g. Dementjeva, 1965; Gulland 1965; Shelbourne, 1957). Thus while most marine teleosts produce large numbers of eggs, only a few animals survive to become adults. In the case of cod, the numbers of eggs laid in the lifetime of an adult amounts to several millions, and yet on the average only two cod survive to become adults. Gulland (1965) has estimated that the annual mortality of commercially exploited adult cod does not exceed 30 to 75% but that in the first few months of a cod's life, mortality must be as high as 99·9999%. The important point in these approximations is not in fact the mortality of the early life stages, but the 0·0001% survival; obviously conditions which might lead to a change in the survival rate from one-tenthousandths of a percent to two or three-tenthousandths of a percent are more easily conceived as factors in determining a large year class, than other factors, which would have to affect the survival of the more mature fish stages by several hundred percent in order to account for the wide fluctuations in fish abundance which are observed in nature. Inasmuch as the early life stages of many fish are planktonic (e.g. herring, plaice, cod, tuna), their survival in the plankton community should be an integral part of biological oceanographic studies.

From detailed observations on the survival of plaice larvae in the North Sea, Shelbourne (1957) has described larvae in a deteriorating condition during the month of January when suitable planktonic food items were scarce. Deterioration of larvae started as the yolk reserves became exhausted; similar observations conducted during March showed the presence of healthy larvae which were feeding on plant and animal plankton. The importance of specific food items in the diet of plaice larvae was also observed; during the earliest feeding stage the larvae fed off some large diatoms (e.g. *Biddulphia* and *Coscinodiscus*) but an abrupt change to a zooplankton diet (primarily *Oikopleura*) occurred very soon after feeding started.

The feeding and survival of herring larvae has been studied extensively both in the field (e.g. Blaxter, 1963, Lisivnenko, 1961) and under laboratory conditions (e.g. Rosenthal and Hempel, 1970). Blaxter's (1963) detailed observations have shown that there is some obvious selectivity for prey items depending on size, while visual sighting and attacks on individual prey items depended to some extent on the amount of light available. Nutritional differences between prey items consumed were also demonstrated to have an effect on the growth and survival of the larvae. From field studies conducted by a number of authors, Blaxter (1963) concluded that larvae were most abundant when food items were present at a concentration of *ca.* 30 organisms per litre. This figure is similar to the concentration of prey items determined experimentally by Rosenthal and Hempel (1970), if it is assumed that only a certain fraction of the prey items seen by a larvae are effectively captured. Lisivnenko (1961) studied the abundance of herring larvae on the concentration of food organisms over a period of 5 years in the Gulf of Riga; her results show a strong correlation between an approximate 5-fold increase in larval abundance and an increase in food items from *ca.* 5 to 20 organisms/litre.

Sysoeva and Degtereva (1965) showed that the main food item of the Arctic–Norwegian cod larvae and fry was the copepod, *Calanus finmarchicus.* Both larvae and fry fed off *C. finmarchicus* during different stages of the copepod's life cycle and this is illustrated in Fig. 83 together with the secondary importance of *Oithona* nauplii and some other food organisms. The absolute concentration of food organisms required by cod fry for successful feeding was expressed as the number of organisms per m^2 in a 50-m water column. Concentrations of $>18,000$ *C. finmarchicus* $/m^2$ provided a sufficient food supply while the intensity of feeding decreased over the range from 18,000 down to 5000 organisms/m^2. The food of juvenile and adult cod has been further studied by Dann (1973) in the North Sea. Starting with animals of *ca.* 10 cm, he showed that food preferences continued to change with age and also with region and season. For example, young cod fed mostly on euphausiids or shrimps, depending on region, while older cod fed predominately on small fish. Thus the trophic position of a fish in the food web of the sea may vary considerably over its life cycle.

From the examples of larval fish survival and the concentration of food organisms given above, it is apparent that there is an interdependent relationship which may be generally expressed as follows: above a certain concentration of organisms, survival of larvae will be high and largely independent of further increases in the concentration of food items. Over a range of concentrations below this maximum threshold for survival, larval feeding success will depend on the concentration of food items; however, at some lower threshold concentration of food items, the absorption of yolk sacs will terminate the survival of most larvae since they must have a certain minimum amount of food in order to exist as effective predators on the plankton. At this point in the predator/prey relationship, the energy that is being expended in obtaining food is equal or more than the energy obtained from the food; this effectively reduces the benefit from the hunt to zero and the population of larvae dies. Ivlev (1944) expressed these observations as an equation which may be of some value in examining larval survival and food item concentration. This relationship is represented in Fig. 84 as the time spent in hunting by a larval fish predator (t) plotted against the corresponding prey density (p). In the equation for the curve, α and β are constants and (p_1) and (t_1) are arbitrarily chosen points on the curve. The curve takes the form of a hyperbola and the asymptotes on the axes have real biological significance. The asymptote parallel to the prey axis will define the predator's limiting rate of food consumption (i.e. rate of food consumption which is independent of the concentration of food items); the asymptote parallel to the t-axis represents the minimum concentration of prey at which the predator can obtain a ration and survive.

Food conditions leading to larval survival as summarized above were first suggested by Hjort (1914). While the general form of Ivlev's curve (Fig. 84) may apply to larvae of different ages, Hjort (1914) considered that the 'critical period' in which year class strength of an adult fish population was determined was during the earliest larval stages. This concept has been reviewed by May (1975) and Hempel (1975) who found the idea to be difficult to support *in toto* while agreeing that from circumstantial evidence the concept may be valid for some fish species under some conditions. The most significant departure from endorsing Hjort's idea comes in the lack of field data demonstrating that high larval mortality has resulted in a poor year class for any particular species

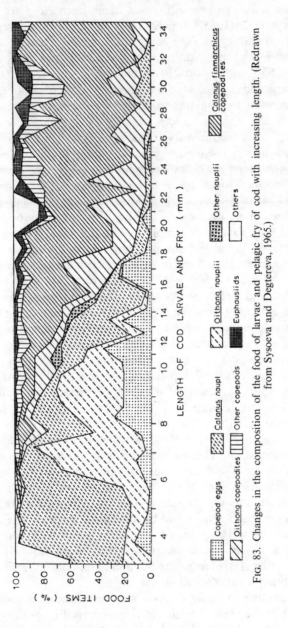

FIG. 83. Changes in the composition of the food of larvae and pelagic fry of cod with increasing length. (Redrawn from Sysoeva and Degtereva, 1965).

of fish. Since such data are very difficult to obtain it is probable that larval fish studies will continue to be highly respected as contributing to our knowledge of year class strength, but in the absence of direct evidence, the question of other mechanisms (e.g. pelagic egg survival, predation, temperature) continue to be offered as alternatives, some of which are not necessarily mutually exclusive as causative agents. The biggest problem appears to be one of experimental design, either in field sampling or laboratory experiments. Lasker (1975) in a unique series of experiments has combined both laboratory and field techniques to derive a specific relationship supporting Hjort's original 'critical

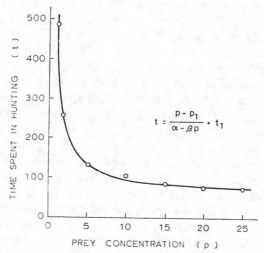

$$t = \frac{p - p_1}{\alpha - \beta p} + t_1$$

FIG. 84. Dependence of the time spent in hunting by a larval fish on the concentration of food organisms (redrawn from Ivlev, 1944).

period' content. In Lasker's (1975) experiments it was shown that first-feeding anchovy larvae require phytoplankton aggregations of >20 cells/ml within $2\frac{1}{2}$ days after the larvae are ready to feed and that individual cells had to be about $40\,\mu m$ diameter. The authors found that such feeding conditions were quite transient but usually associated with subsurface chlorophyll maxima.

The type of relationship given above appears to be important as an indication of the processes involved in the dependence of fish larvae on their food supply. Among some species of fish, however, there is essentially no pelagic larval stage and the animals emerge into the pelagic environment as juvenile fish (*ca.* 35 mm length) with a much greater ability to search for prey than is found among pelagic larvae. Perhaps the most important example of this group of fish are the salmonids. Processes governing the early pelagic life of young salmonids with respect to their food supply in the plankton community are different in some respects from those governing larval fish survival.

The ability of juvenile fish to catch prey will depend on individual species and environments, but an indication of some of the mechanisms involved have been given in studies by a number of authors (e.g. Paloheimo and Dickie, 1965 and 1966a; LeBrasseur, 1969; Leong and O'Connell, 1969; Parsons and LeBrasseur, 1970; Parker, 1971).

From extensive studies of fish-feeding data reported by various authors Paloheimo and Dickie (1966b) concluded that growth efficiency (K_1) was related to the amount of food eaten by an animal (R) such that,

$$\ln K_1 = -a - bR. \tag{189}$$

When $\ln K_1$ is plotted against R for different rations, different values of a and b are obtained. The relationship may not apply to hand-fed laboratory fish cultures but there is considerable justification for its use under field conditions. The importance of eqn. (189) from the point of view of the production of food items is that constants 'a' and 'b' must be related in some way to the metabolic cost of behaviour patterns associated with particular types of prey. As an example of this effect the authors suggested that the size of prey items might be one factor in determining differences between growth efficiency and ration (i.e. in the differences in 'a' and 'b'). Differences in the ability of young salmonids to graze off different-sized prey has been demonstrated in short-term experiments, such as those reported by Parsons and LeBrasseur (1970). In the latter experiments the same biomass of three different sized zooplankton (two species of copepods and a euphausiid) were fed to young salmon over a concentration range from 0·5 to 100 g/m³ wet weight of zooplankton. The results showed that the salmon were able to eat more of the medium-size copepods per unit time than of the large euphausiids or the very small copepods. From this observation it may be concluded that not only is the absolute abundance of prey items important to the growth of juvenile fish, but also there are marked differences in the efficiency of fish depending on the type of prey as indicated by eqn. (189). With some fish species the method of capturing prey change and this will also alter the rate of feeding at similar prey densities. This is illustrated in Fig. 85 from Leong and O'Connell (1969). The experimental data show that the Northern Anchovy obtained small *Artemia* nauplii by filter feeding at a slower rate than they obtain *Artemia* adults by raptorial feeding. Furthermore the shapes of the feeding curves are different and show that the greatest difference occurs during the shortest feeding interval. In a later experiment O'Connel (1972) was able to show that raptorial feeding was the exclusive form of feeding down to an adult *Artemia* concentration of 7% of the total biomass of mixed adult and nauplii *Artemia* prey. Only when the large prey items were reduced to less than 2% of the total biomass did the form of feeding become greater than 50% in favour of filtering. In similar experiments using the much larger Pacific mackerel (*Scomber japonicus*), O'Connel and Zweifel (1972) showed that *Artemia* nauplii were ignored as food items but that *Artemia* adults were fed on by raptorial (biting) feeding at low prey densities (1 or 2/litre) and by filter feeding at high prey densities (22 to 112/litre).

From studies carried out by Parker (1971) it is also apparent that feeding and growth rate of fish are an integral part of its chances for survival. Thus in the case of young pink salmon, Parker (1971) was able to show that the true growth rate of a pink salmon population was 1·4%/day and that this growth rate was greater than that of the pink salmon's predator (2-year-old coho salmon) which grew at 0·7% per day. Thus the survival of the juvenile pink salmon was assured as they 'outgrew' being suitable prey items for coho salmon. This type of mechanism is illustrated in Fig. 86 where it is seen that depending on the amount of zooplankton present young pink salmon may grow at different rates. Those that grow faster than the growth rate of the predator will eventually escape attacks as suitable prey items.

The associations described in this section between the plankton community and fisheries serve as illustrations of how biological oceanographic data may be utilized in understanding mechanisms contributing to the survival of fish. More complex treatments

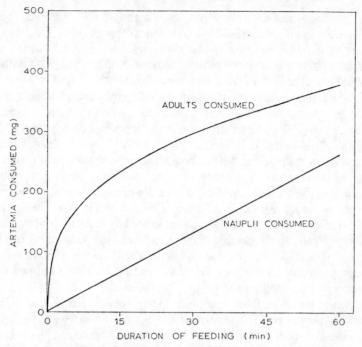

FIG. 85. Comparison of particulate and filter feeding rates for a 4-g anchovy. Curve showing 'adults consumed' was determined for *Artemia* concentrations of 2- to 50-mg adults/litre; curve showing 'nauplii consumed' was determined for *Artemia* concentrations of 4·4 mg nauplii/litre. (Figure redrawn from Leong and O'Connell, 1969).

of this subject, particularly those which include the physiological requirements of a fish while hunting for its prey, have been discussed by a number of authors. In particular a recent presentation by Kerr (1971) gives an indication of how growth efficiency must include a consideration of the animal's metabolic requirements for hunting, basal metabolism, and the internal cost of food utilization. These factors must then be related to the size and concentration of prey available, and the swimming speed of the predator.

While it is possible to conduct intensive studies on local species under local conditions, the difficulty of extrapolating these findings to huge areas of ocean in order to benefit

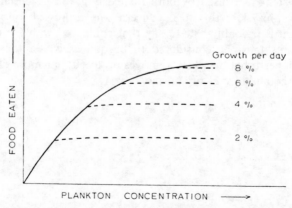

FIG. 86. Schematic relationship between plankton concentration, food eaten and the daily growth rate of juvenile fish.

fisheries is another problem facing the biological oceanographer, especially under conditions of limited research ship facilities. One solution to this problem has been the Hardy
plankton recorder program (Glover *et al.*, 1974 and Section 1.3.2) which is operated
from commercial vessels crossing the North Sea and North Atlantic. Another possible
solution is the initiation of studies from aircraft and satellites. Aerial and remote sensing
of biological parameters in the ocean, such as chlorophyll and antarctic krill, have
been suggested by several authors (e.g. Yentsch, 1973; El-Sayed and Green, 1974 and
references cited therein). The implementation of such studies on a routine basis should
provide the kind of temporal and spatial data bank necessary for testing the effect
of the environment on fisheries, as well as providing information on unexploited stocks.
New attempts to understand large-scale oceanic variability should also be made in
association with physical oceanographers so that the causitive agent in multiple correlation approaches, involving fish catch data and climatic effects can be fully understood.
For example, Sutcliffe (1972; 1973) has demonstrated correlations between fish catch
for several different fisheries (including halibut, lobster, and clam) and discharge from
the St. Lawrence river. Subsequent discussion of these results (Sutcliffe *et al.*, 1976;
Sutcliffe *et al.*, in press) have been pursued in association with an attempt to understand
the physical processes controlling the biology of the sea.

Among the world fisheries there is now a clear trend towards the exploitation of
new stocks which have not previously formed the basis of an extensive fishery. Examples
of these new fisheries are the landing of 150,000 tons of sand eel from the northeast
Atlantic, 250,000 tons of capelin from the northwest Atlantic, the Japanese and Soviet
expeditions to the Antarctic for krill and the landings of squid in Japan which currently
amount to about 10% of that country's total catch. In addition, mixed species fisheries
for processing as Fish Protein Concentrate (FPC) are replacing the traditional concept
of single-species fisheries on which fisheries management strategies are based. In the
future Gulland (1970) has estimated that total quantities of fish coming from new fisheries in millions of tons per year could amount to: zooplankton (e.g. krill and galatheids,
etc.) >50, small fishes (e.g. lantern fish, saury, sand eel) >100, squid 10–100 and new
mollusc fisheries 10–25. Compared with existing traditional world fisheries of *ca.* 60
million tons, it is clear that these new fisheries may come to predominate the total
harvest of protein from the sea. The practical mission for the biological oceanographer
in assisting in fisheries management is to understand the effect of these perturbations
on the marine ecosystem, as caused by both the existing and potentially huge fisheries
of the future. In Section 7.1 of this text, concern was expressed for various forms of
ocean pollution; yet it is probably true to say that compared with the effects of fishing,
the effect of pollutants must be considered to be quite secondary on a global scale
when one considers a perturbation which removes up to 50% or more of the top trophic
levels in the marine community of organisms.

REFERENCES

ABDULLAH, M. I., L. G. ROYLE and A. W. MORRIS. 1972. Heavy metal concentration in coastal waters. *Nature*, **235**, 158–160.

ACKMAN, R. G. 1965. Occurrence of odd-numbered fatty acids in the mullet *Mugil cephalus*. *Nature*, **208**, 1213–1214.

ACKMAN, R. G. and M. G. CORMIER. 1967. α-Trocopherol in some Atlantic fish and shellfish with particular reference to live-holding without food. *J. Fish. Res. Bd. Canada*, **24**, 357–373.

ACKMAN, R. G., C. A. EATON, J. C. SIPOS, S. N. HOOPER and J. D. CASTELL. 1970. Lipids and fatty acids of two species of North Atlantic krill *Meganyctiphanes norvegica* and *Thysanoessa inermis* and their role in the aquatic food web. *J. Fish. Res. Bd. Canada*, **27**, 513–533.

ACKMAN, R. G., C. S. TOCHER and J. McLACHLAN. 1968. Marine phytoplankter fatty acids. *J. Fish. Res. Bd. Canada*, **25**, 1603–1620.

ADAMS, J. A. and J. H. STEELE. 1966. Shipboard experiments on the feeding of *Calanus finmarchicus* (Gunnerus). In *Some Contemporary Studies in Marine Science*, Ed. H. BARNES, Allen & Unwin Ltd., London, pp. 19–35.

ADAMS, S. M. and J. W. ANGELOVIC. 1970. Assimilation of detritus and its associated bacteria by three species of estuarine animals. *Chesapeake Sci.* **11**, 249–254.

AHLSTROM, E. H. 1961. Fisheries oceanography. *Calif. Coop. Oceanic Fish Invest.* **8**, 71–72.

ALI, R. M. 1970. The influence of suspension density and temperature on filtration rate of *Hiatella arctica*. *Mar. Biol.* **6**, 291–302.

ALLAN, G. G., J. LEWIN and P. G. JOHNSON. 1972. Marine polymers. IV. Diatom polysaccharides. *Botanica Marina*, **15**, 102–108.

ALLEN, K. R. 1971. Relation between production and biomass. *J. Fish. Res. Bd. Canada*, **28**, 1573–1589.

ALLEN, M. B. 1956. Excretion of organic compounds by *Chlamydomonas*. *Arch. Mikrobiol.* **24**, 163–168.

ALLEN, M. B. 1963. Nitrogen fixing organisms in the sea. In *Symposium on Marine Microbiology*, Ed. C. H. OPPENHEIMER, C. C. Thomas, Springfield, Illinois, pp. 85–105.

ALLEN, M. B., T. W. GOODWIN and S. PHAGPOLNGARM. 1960. Carotenoid distribution in certain naturally occurring algae and in some artificially induced mutants of *Chlorella pyrenoidosa*. *J. Gen. Microbiol.* **23**, 93–103.

ALLEN, M. B., L. FRIES, T. W. GOODWIN and D. M. THOMAS. 1964. The carotenoids of algae. Pigments from some cryptomonads, a heterokont and some Rhodophyceae. *J. Gen. Microbiol.* **34**, 259–267.

ALVERSON, D. L., A. R. LONGHURST and J. A. GULLAND. 1970. How much food from the sea? *Science*, **168**, 503–505.

Amer. Publ. Health Assn., Amer. Water Works Assn., Water Pollution Contr. Fed. 1965. *Standard Methods for the Examination of Water and Waste Water including Bottom Sediments and Sludges*, 12th ed. Amer. Publ. Health Assn. Inc., New York. 769 pp.

ANDERSON, G. C. 1964. The seasonal and geographic distribution of primary productivity off the Washington and Oregon Coasts. *Limnol. Oceanogr.* **9**, 284–302.

ANDERSON, G. C. 1965. Fractionation of phytoplankton communities off the Washington and Oregon Coasts. *Limnol. Oceanogr.* **10**, 477–480.

ANDERSON, G. C. 1969. Subsurface chlorophyll maximum in the northeast Pacific Ocean. *Linmol. Oceanogr.* **14**, 386–391.

ANDERSON, G. C. and R. P. ZEUTSCHEL. 1970. Release of dissolved organic matter by marine phytoplankton in coastal and offshore areas of the northeast Pacific Ocean. *Limnol. Oceanogr.* **15**, 402–407.

ANDERSON, J. W., J. M. NEFF, B. A. COX, H. E. TATEM and G. M. HIGHTOWER. 1974. Characteristics of dispersions and water soluble extracts of crude and refined oils and their toxicity to estuarine crustaceans and fish. *Mar. Biol.* **27**, 75–88.

ANDREWS, P. and P. J. LeB. WILLIAMS. 1971. Heterotrophic utilization of dissolved organic compounds in the sea. III. Measurement of the oxidation rates and concentrations of glucose and amino acids in sea water. *J. Mar. Biol. Ass. U.K.* **51**, 111–125.

ANGEL, H. H. and M. V. ANGEL. 1967. Distribution pattern analysis in a marine benthic community. *Helgoländer wiss. Meeresunters*, **15**, 445–454.

ANGEL, M. V. and M. J. R. FASHAM. 1973. SOND Cruise 1965: factor and cluster analysis of the plankton results, a general summary. *J. Mar. Biol. Ass. U.K.* **53**, 185–231.

ANKAR, S. and R. ELMGREN. 1975. A survey of the benthic macro and meiofauna of the Askö–Landsort area. *Merentutkimuslait. Julk./Havsforskningsinst.*, Skr. **239**, 257–264.

ANRAKU, M. and M. OMORI. 1963. Preliminary survey of the relationship between the feeding habit and the structure of the mouthparts of marine copepods. *Limnol. Oceanogr.* **8**, 116–126.

ANSELL, A. D. 1974. Sedimentation of organic detritus in Lochs Etive and Creran, Argyll, Scotland. *Mar. Biol.* **27**, 263–273.

ANSELL, A. D. 1975. Seasonal changes in biochemical composition of the bivalve *Astarte montagu*: in the Clyde Sea area. *Mar. Biol.* **29**, 235–243.

ANSELL, A. D. and A. TREVALLION. 1967. Studies on *Tellina tenuis* da Costa I. Seasonal growth and biochemical cycle. *J. Exp. Mar. Biol. Ecol.* **1**, 220–235.

ANSELL, A. D. and P. SIVADAS. 1973. Some effects of temperature and starvation on the bivalve *Donax vittatus* da Costa in experimental laboratory populations. *J. Exp. Mar. Biol. Ecol.* **13**, 229–262.

ANTIA, N. J. and E. BILINSKI. 1967. A bacterial toxin type of phospholipase (Lecethinase C) in a marine phytoplanktonic chrysomonad. *J. Fish. Res. Bd. Canada*, **24**, 201–204.

ANTIA, N. J., C. D. MCALLISTER, T. R. PARSONS, K. STEPHENS and J. D. H. STRICKLAND. 1963. Further measurements of primary production using a large-volume plastic sphere. *Limnol. Oceanogr.* **8**, 166–183.

ANTIA, N. J., J. Y. CHENG and F. J. R. TAYLOR. 1969. The heterotrophic growth of a marine photosynthetic cryptomonad (*Chroomonas salina*). *Proc. Intl. Seaweed Symp.* **6**, 17–29.

ANTIA, N. J., B. R. BERLAND, D. J. BONIN and S. Y. MAESTRINI. 1975. Comparative evaluation of certain organic and inorganic sources of nitrogen for phototrophic growth of marine microalgae. *J. Mar. Biol. Ass. U.K.* **55**, 519–539.

APSTEIN, C. 1910. Hat ein Organismus in der Tiefe gelebt, in der er gefischt ist? *Int. Rev. ges. Hydrobiol. Hydrogr.* **3**, 17–33.

ARLT, G. 1973. Vertical and horizontal micro-distribution of the meiofauna in the Greifswolder Bodden. *Oikos Suppl.* **15**, 105–111.

ARMSTRONG, F. A. J. and W. R. G. ATKINS. 1950. The suspended matter of sea water. *J. Mar. Biol. Ass. U.K.* **29**, 139–143.

ARMSTRONG, F. A. J. and E. C. LAFOND. 1966. Chemical nutrient concentrations and their relationship to internal waves and turbidity off southern California. *Limnol. Oceanogr.* **11**, 538–547.

ARMSTRONG, F. A. J., C. R. STEARNS and J. D. H. STRICKLAND. 1967. The measurement of upwelling and subsequent biological processes by means of the Technicon Autoanalyser® and associated equipment. *Deep-Sea Res.* **14**, 381–389.

ARUGA, Y. 1965a. Ecological studies of photosynthesis and matter production of phytoplankton. I. Seasonal changes in photosynthesis of natural phytoplankton. *Bot. Mag., Tokyo*, **78**, 280–288.

ARUGA, Y. 1965b. Ecological studies of photosynthesis and matter production of phytoplankton. II. Photosynthesis of algae in relation to light intensity and temperature. *Bot. Mag., Tokyo*, **78**, 360–365.

ARUGA, Y. 1966. Ecological studies of photosynthesis and matter production of phytoplankton. III. Relationship between chlorophyll amount in water and primary productivity. *Bot. Mag. Tokyo*, **79**, 20–27.

ARUGA, Y. and S. ICHIMURA, 1968. Characteristics of photosynthesis of phytoplankton and primary production in the Kuroshio. *Bull. Misaki Marine Biol. Inst. Kyoto Univ.*, No. 12 (*Proceedings of the U.S.–Japan Seminar on Marine Microbiology*, August 1966 in Tokyo), 3–20.

ARUGA, Y., Y. YOKOHAMA and M. NAKANISHI. 1968. Primary productivity studies in February–March in the Northwestern Pacific off Japan. *J. Oceanogr. Soc. Japan*, **24**, 275–280.

BAAS BECKING, L. G. M. and E. J. F. WOOD. 1955. Biological processes in the estuarine environment. I and II. Ecology of the sulfur cycle. *Koninkl. Ned. Akad. Wetenschap. Proc.*, **B58**, 160–181.

BAAS BECKING, L. G. M., I. R. KAPLAN and D. MOORE. 1960. Limits of the natural environment in terms of pH and oxidation–reduction potentials. *J. Geol.* **68**, 243–284.

BAASSHAM, J. A. and M. KIRK. 1962. The effect of oxygen on reduction of CO_2 to glycolic acid and other products during photosynthesis by *Chlorella*. *Biochem. Biophys. Res. Commun.* **9**, 376–380.

BADER, H. 1970. The hyperbolic distribution of particle sizes. *J. Geophys. Res.* **75**, 2822–2830.

BADER, R. G. 1954. The role of organic matter in determining the distribution of pelecypods in marine sediments. *J. Mar. Res.* **13**(1), 32–47.

BADER, R. G., B. W. HOOD and J. B. SMITH. 1960. Recovery of dissolved organic matter in seawater and organic sorption by particulate matter. *Geochim. Cosmochim. Acta*, **19**, 236–243.

BAGGE, P. 1969. Effects of pollution on estuarine ecosystems. I. Effects of effluents from wood-processing industries on the hydrography, bottom and fauna of Saltkällefjord (W. Sweden). *Merentutkimuslait. Julk./Havsforskningsinst.*, Skr. **228**, 3–118.

BAINBRIDGE, R. 1957. The size, shape and density of marine phytoplankton concentrations. *Biol. Rev.* **32**, 91–115.

BAITY, H. G. 1938. Some factors affecting the aerobic decomposition of sewage sludge deposits. *Sewage Wks. J.* **10**, 539–568.

BALY, E. C. C. 1935. The kinetics of photosynthesis. *Proc. R. Soc. Cond.* Ser. B, **117**, 218–239.

BANNISTER, T. T. 1974. Production equations in terms of chlorophyll concentration, quantum yield, and upper limit to production. *Limnol. Oceanogr.* **19**, 1–12.

BANSE, K. 1964. On the vertical distribution of zooplankton in the sea. *Prog. Oceanogr.* **2**, 56–125.

BANSE, K. 1974a. On the interpretation of data for the carbon-to-nitrogen ratio of phytoplankton. *Limnol. Oceanogr.* **19**, 695–699.

BANSE, K. 1974b. The nitrogen-to phosphorus ratio in the photic zone of the sea and the elemental composition of the plankton. *Deep-Sea Res.* **21**, 767–771.

BANSE, K. 1975. Pleuston and Neuston: On the categories of organisms in the uppermost pelagial. *Int. Rev. ges. Hydrobiol.* **60**, 439–447.

BANSE, K. (in press) Determining the carbon-to-chlorophyll ratio of natural phytoplankton. *Mar Biol.*

BANSE, K. and G. C. ANDERSON. 1967. Computations of chlorophyll concentrations from spectrophotometric readings. *Limnol. Oceanogr.* **12**, 696–697.

BANSE, K., F. H. NICHOLS and D. R. MAY. 1971. Oxygen consumption by the seabed. III. On the role of the macrofauna at three stations. *Vie Milieu*, Suppl. **22**, 31–52.

BARBER, R. T. 1966. Interaction of bubbles and bacteria in the formation of organic aggregates in sea water. *Nature*, **211**, 257–258.

BARBER, R. T., R. C. DUGDALE, J. J. MACISAAC and R. L. SMITH. 1971. Variations in phytoplankton growth associated with the source and conditioning of upwelling water. *Inv. Pesq.* **35**, 171–193.

BARBIER, M., D. JOLY, A. SALIOT and D. TOURRES. 1973. Hydrocarbons from sea water. *Deep-Sea Res.* **20**, 305–314.

BARHAM, E. G., N. J. AYER, JR. and R. E. BOYCE. 1967. Macrobenthos of the San Diego Trough: photographic census and observations from the bathyscape, *Trieste*. *Deep-Sea Res.* **14**, 773–784.

BARLOW, J. P., C. J. LORENZEN and R. T. MYREN. 1963. Eutrophication of a tidal estuary. *Limnol. Oceanogr.* **8**, 251–262.

BARNES, H. 1949. On the volume measurement of water filtered by a plankton pump, with an observation on the distribution of plankton animals. *J. Mar. Biol. Ass. U.K.* **28**, 651–662.

BARNES, H. 1952. The use of transformations in marine biological statistics. *J. Cons. Int. Explor. Mer*, **18**, 61–71.

BARNES, H. 1956. *Balanus balanoides* (L.) in the Firth of Clyde: the development and annual variation of the larval population, and the causative factors. *J. Anim. Ecol.* **25**, 72–84.

BARNES, H. 1959. *Oceanography and Marine Biology: A Book of Techniques*, Hafner (New York), 218 pp.

BARNES, H. and S. M. MARSHALL. 1951. On the variability of replicate plankton samples and some application of 'contagious' series to the statistical distribution of catches over restricted periods. *J. Mar. Biol. Ass. U.K.* **30**, 233–263.

BARRACLOUGH, W. E., R. J. LEBRASSEUR and O. D. KENNEDY. 1969. Shallow scattering layer in the subarctic Pacific Ocean: detection by high-frequency echo sounder. *Science*, **166**, 611–613.

BARRACLOUGH, W. E. and D. ROBINSON. 1972. The fertilization of Great Central Lake. III. Effect on juvenile sockeye salmon. *Fish. Bull.* **70**, 37–48.

BARSDATE, R. J., R. T. PRENTKI and T. FENCHEL. 1974. Phosphorus cycle of model ecosystems: significance for decomposer food chains and effect of bacterial grazers. *Oikos*, **25**, 239–251.

BARY, B. McK. 1959. Species of zooplankton as a means of identifying different surface waters and demonstrating their movements and mixing. *Pac. Sci.* **13**, 14–34.

BARY, B. McK. 1963. Distributions of Atlantic pelagic organisms in relation to surface water bodies. In *Marine Distributions*, Ed. M. J. DUNBAR, Royal Soc. Canada, Sp. Publ., No. 5: 51–67.

BARY, B. McK. 1967. Diel vertical migrations of underwater scattering, mostly in Saanich Inlet, British Columbia. *Deep-Sea Res.* **14**, 35–50.

BAYLOR, E. R. and W. H. SUTCLIFFE. 1963. Dissolved organic matter in sea water as a source of particulate food. *Limnol. Oceanogr.* **8**, 369–371.

BEARDSLEY, G. F. JR., H. PAK and K. CARDER. 1970. Light scattering and suspended particles in the eastern equatorial Pacific Ocean. *J. Geophys. Res.* **75**, 2837–2845.

BEATTIE, A., E. L. HIRST and E. PERCIVAL. 1961. Studies on the metabolism of the Chrysophyceae. Comparative structural investigations on leucosin (Chrysolaminarin) separated from diatoms and laminarin from the brown algae. *Biochem. J.* **79**, 531–537.

BEERS, J. R. 1966. Studies on the chemical composition of the major zooplankton groups in the Sargasso Sea off Bermuda. *Limnol. Oceanogr.* **11**, 520–528.

BEERS, J. R., M. R. STEVENSON, R. W. EPPLEY and E. R. BROOKS. 1971. Plankton populations and upwelling off the coast of Peru, June 1969. *Fish. Bull.* **69**, 859–876.

BEKLEMISHEV, K. V. 1954. The feeding of some common plankton copepods in far eastern seas. *Zool. J., Inst. Oceanol. Acad. Sci. USSR*, **33**, 1210–1230.

BELL, R. K. and F. J. WARD. 1970. Incorporation of organic carbon by *Daphnia pulex*. *Limnol. Oceanogr.* **15**, 713–726.

BELL, W. and R. MITCHELL. 1972. Chemotactic and growth responses of marine bacteria to algal extracellular products. *Biol. Bull.* **143**, 265–277.

BELLAMY, D. J., D. M. JOHN and A. WHITTICK. 1968. The 'kelp forest ecosystem' as a 'phytometer' in the study of pollution of the inshore environment. *Underwater Ass. Rep.*, 1968, 79–82.

BENSON, A. A. and R. F. LEE. 1975. The role of wax in oceanic food chains. *Sci. Amer.* **232**, 76–83.

BEN-YAAKOV, S. 1973. pH buffering of pore water of recent anoxic marine sediments. *Limnol. Oceanogr.* **18**, 86–94.

BERNARD, F. 1963. Vitesses de chute en mer des amas palmelloides de *Cyclococcolithus*. Ses conséquences pour le cycle vital des mers chaudes. *Pelagos*, **1**, 5–34.

BERNARD. F. R. 1974. Annual biodeposition and gross energy budget of mature Pacific oysters, *Crassostrea gigas. J. Fish. Res. Bd. Canada*, **31**, 185–190.

BERNER, R. A. 1963. Electrode studies of hydrogen sulfide in marine sediments. *Geochim. Cosmochim. Acta*, **27**, 563–575.

BERNER, R. A. 1971. *Principles of Chemical Sedimentology*. McGraw Hill, New York. 240 pp.

BERNER, R. A. 1976. The benthic boundary layer from the viewpoint of a geochemist. In *The Benthic Boundary Layer*, Ed. I. N. MCCAVE, Plenum Press, New York, pp. 33–55.

BERTRU, G. 1971. La microzone oxydée et la consommation d'oxygène dans les sédiments des étangs oligotrophes. *Arch. Hydrobiol.* **68**, 277–287.

BEVERTON, R. J. H. and S. J. HOLT. 1957. On the dynamics of exploited fish populations. *Fish. Inv., Lond.*, Ser. II, **19**, 1–533.

BEYERS, R. J. 1963. The metabolism of twelve aquatic laboratory microecosystems. *Ecol. Monogr.* **33**, 281–306.

BLACKBURN, M. 1965. Oceanography and the ecology of tunas. *Oceanogr. Mar. Biol. Ann. Rev.* **3**, 299–322.

BLACKBURN, M. 1969. Conditions related to upwelling which determine distribution of tropical tunas off western Baja California. *Fish. Bull.* **68**, 147–176.

BLACKBURN, M., R. M. LAURS, R. W. OWEN and B. ZEITZSCHEL. 1970. Seasonal and areal changes in standing stocks of phytoplankton, zooplankton and micronekton in the eastern tropical Pacific. *Mar. Biol.* **7**, 14–31.

BLACKBURN, T. H., P. KLIEBER and T. FENCHEL. 1975. Photosynthetic sulfide oxidation in marine sediments. *Oikos*, **26**, 103–108.

BLACKMAN, F. F. 1905. Optima and limiting factors. *Ann. Bot.* **19**, 281–295.

BLAXTER, J. H. S. 1963. The feeding of herring larvae and their ecology in relation to feeding. *Calif. Coop. Oceanic Fish Investig. Rep.* **10**, 79–88.

BLOOM, S. A., J. L. SIMON and V. D. HUNTER. 1972. Animal–sediment relations and community analysis of a Florida estuary. *Mar. Biol.* **13**, 43–56.

BLUMER, M., R. R. L. GUILLARD and T. CHASE. 1971. Hydrocarbons of marine phytoplankton. *Mar. Biol.* **8**, 183–189.

BLUMER, M., M. M. MULLIN and R. R. L. GUILLARD. 1970. A polyunsaturated hydrocarbon (3, 6, 9, 12, 15, 18-heneicosahexaene) in the marine food web. *Mar. Biol.* **6**, 226–235.

BODIOU, J.-Y. and P. CHARDY. 1973. Analyse en composantes principales du cycle annuel d'un peuplement de copépods harpacticoides des sables fins infralittoraux de Banyuls-sur-Mer. *Mar. Biol.* **20**, 27–34.

BODUNGEN, B. VON, K. VON BRÖCKEL, V. SMETACEK and B. ZEITZSCHEL. 1975. Ecological studies on the plankton in Kiel Bight. 1. Phytoplankton. *Merentutkimuslait. Julk./Havsforskningsinst. Skr.* **239**, 179–186.

BOGDANOV, Y. A. 1965. Particulate organic matter in the waters of the Pacific Ocean. *Okeanologiia*, **5**, 286–297.

BONNER, J. T. 1965. *Size and Cycle*. Princeton University Press, Princeton, New Jersey. 219 pp.

BORDOVSKY, O. Y. 1965. Accumulation and transformation of organic substance in marine sediments. Parts 1–3. *Mar. Geol.* **3**, 3–114.

BOULDIN, D. R. 1968. Models for describing the diffusion of oxygen and other mobile constituents across the mud-water interface. *J. Ecol.* **56**, 77–87.

BOURGET, E. and G. LACROIX. 1973. Aspects saisonniers de la fixation de l'epifaune benthique de l'étage infralittoral de l'estuaire du Saint-Laurent. *J. Fish. Res. Bd. Canada*, **30**, 867–880.

BOYD, C. M. 1973. Small-scale spatial patterns of marine zooplankton examined by an electronic *in situ* zooplankton detecting device. *Netherlands J. Sea Res.* **7**, 103–111.

BOYD, C. M. 1976. Selection of particle sizes by filter-feeding copepods: a plea for reason. *Limnol. Oceanogr.* **21**, 175–180.

BOYSEN JENSEN, P. 1919. Valuation of the Limfjord 1. *Rep. Dan. biol. Stn.* **26**, 1–24.

BOYSEN JENSEN, P. 1932. Die Stoffproduktion der Pflanzen. *Publ. Gustav Fischer (Jena)*. 108 pp.

BRAARUD, T. 1963. Reproduction in the marine coccolithophorid *Coccolithus huxleyi* in culture. *Pubbl. staz. zool. Napoli*, **33**, 110–116.

BRAARUD, T. and B. FØYN. 1931. Beiträge zur Kenntnis des Stoffwechsels in Meer. *Avh. norske Vidensk Akad. Oslo*, **14**, 24 pp.

BREEN, P. A. and K. H. MANN. 1976. Destructive grazing by sea urchins in Eastern Canada. *J. Fish. Res. Bd. Canada*, **33**, 1278–1283.

BRINKHUIS, B. H., N. R. TEMPEL and R. F. JONES. 1976. Photosynthesis and respiration of exposed salt-marsh fucoids. *Mar. Biol.* **34**, 349–359.

BROCK, M. L. and T. D. BROCK. 1968. The application of microautoradiographic techniques to ecological studies. *Mitt. Int. Verein. Limnol.* **15**, 1–29.

BROCK, T. D. 1966. *Principals of Microbial Ecology*, Prentice-Hall, Inc., Englewood Cliffs, New Jersey. 306 pp.

BROCKSEN, R. W., G. E. DAVIS and C. E. WARREN. 1970. Analysis of trophic processes on the basis of density-dependent functions. In *Marine Food Chains*, Ed. J. H. STEELE. Oliver & Boyd, Edinburgh, pp. 468–498.

BROECKER, W. S. 1974. *Chemical Oceanography*, Harcourt Brace Javanovich Inc., N.Y. 214 pp.

BROOKS, J. L. and S. I. DODSON. 1965. Predation, body size, and composition of plankton. *Science*, **150**, 28–35.

BROWN, D. H., C. E. GIBBY and M. HICKMAN. 1972. Photosynthetic rhythms in epipelic algal populations. *Br. Phycol. J.* **7**, 37–44.

BROWN, D. L. and E. B. TREGUNNA. 1967. Inhibition of respiration during photosynthesis by some algae. *Can. J. Bot.* **45**, 1135–1143.

BROWN, H. R. 1969. *Biochemical Microcalorimetry*, Academic Press, New York, 338 pp.

BROWN, R. A. and H. L. HUFFMAN, JR. 1976. Hydrocarbons in open ocean waters. *Science*, **191**, 847–849.

BROWN, T. E. and F. L. RICHARDSON. 1968. The effect of growth environment on the physiology of algae: light intensity. *J. Phycol.* **4**, 38–54.

BUCHANAN, J. B. 1963. The bottom fauna communities and their sediment relationships off the coast of Northumberland. *Oikos*, **14**, 154–175.

BUCHANAN, J. B. and J. M. KAIN. 1971. Measurement of the physical and chemical environment. In *Methods for the Study of Marine Benthos*, IBP Handbook No. 16, Eds. N. A. HOLME and A. D. MACINTYRE, Blackwell Scientific Publ., Oxford.

BUCHANAN, J. B., P. F. KINGSTON and M. SHEADER. 1974. Long-term population trends of the benthic macrofauna in the offshore mud of the Northumberland coast. *J. Mar. Biol. Ass. U.K.* **54**, 785–795.

BUCHANAN, J. B. and R. M. WARWICK. 1974. An estimate of benthic macrofaunal production in the offshore mud of the Northumberland coast. *J. Mar. Biol. Ass. U.K.* **54**, 197–222.

BUGGELN, R. G. and J. S. CRAIGIE. 1973. The physiology and biochemistry of *Chondrus crispus* Stackhouse. In *Chondrus crispus*, Eds. M. J. HARVEY and J. MCLACHLAN, *Proceedings of the Nova Scotia Institute of Science*, Suppl. **27**, 81–102.

BUNKER, H. J. 1936. A review of the physiology and biochemistry of the sulfur bacteria. *Dept. Sci. Ind. Res., Chem. Res. Spec. Rpt.* 3.

BUNT, J. S. 1964a. Primary productivity under sea ice in Antarctic waters. 1. Concentrations and photosynthetic activities of microalgae in the waters of McMurdo Sound, Antarctica. *Antarctic Res. Ser.* **1**, 13–26.

BUNT, J. S. 1964b. Primary productivity under sea ice in Antarctic waters. 2. Influence of light and other factors on photosynthetic activities of Antarctic marine microalgae. *Antarctic Res. Ser.* **1**, 27–31.

BUNT, J. S. 1965. Measurements of photosynthesis and respiration in a marine diatom with the mass spectrometer and with carbon-14. *Nature*, **207**, 1373–1375.

BUNT, J. S. and C. C. LEE. 1970. Seasonal primary production in Antarctic sea ice at McMurdo Sound in 1967. *J. Mar. Res.* **28**, 304–320.

BURKE, V. and L. A. BAIRD. 1931. Fate of fresh water bacteria in the sea. *J. Bacteriol.* **21**, 287–298.

BURKE, M. V. and K. H. MANN. 1974. Productivity and production: biomass ratios of bivalve and gastropod populations in an eastern Canadian estuary. *J. Fish. Res. Bd. Canada.* **31**, 167–177.

BURKHOLDER, P. R. and L. M. BURKHOLDER. 1956. Vitamin B_{12} in suspended solids and marsh muds collected along the coast of Georgia. *Limnol. Oceanogr.* **1**, 202–208.

BURKHOLDER, P. R., L. M. BURKHOLDER and L. P. ALMODÓVAR. 1960. Antibiotic activity of some marine algae of Puerto Rico. *Botanica Mar.* **2**, 149–156.

BURKHOLDER, P. R. and E. F. MANDELLI. 1965. Productivity of microalgae in Antarctic sea ice. *Science*, **149**, 872–874.

BURKHOLDER, P. R., A. REPARK and J. SIEBERT. 1965. Studies on some Long Island Sound littoral communities of microorganisms and their primary productivity. *Bull. Torrey Bot. Club*, **92**, 378–402.

BURNS, N. M. and A. E. PASHLEY. 1974. *In situ* measurement of the settling velocity profile of particulate organic carbon in Lake Ontario. *J. Fish. Res. Bd. Canada*, **31**, 291–297.

BUTLER, E. I., E. D. S. CORNER and S. M. MARSHALL. 1970. On the nutrition and metabolism of zooplankton. VII. Seasonal survey on nitrogen and phosphorus excretion by *Calanus* in the Clyde sea area. *J. Mar. Biol. Ass. U.K.* **50**, 525–560.

CALOW, P. 1975. The feeding strategies of two freshwater gastropods, *Ancylus fluviatilis* Müll, and *Planorbis contortus* Linn. (Pulmonata), in terms of ingestion rates and absorption efficiencies. *Oecologia*, **20**, 33–49.

CALOW, P. and C. R. FLETCHER. 1972. A new radiotracer technique involving ^{14}C and ^{51}Cr for estimating the assimilation efficiencies of aquatic, primary consumers. *Oecologia*, **9**, 155–170.

CALVIN, M. and J. A. BAASHAM. 1962. *The Photosynthesis of Carbon Compounds*, Benjamin, New York. 127 pp.

CAPERON, J. 1967. Population growth in micro-organisms limited by food supply. *Ecology*, **48**, 715–722.

CAPERON, J. 1968. Population growth response of *Isochrysis galbana* to nitrate variation at limiting concentrations. *Ecology*, **49**, 866–872.

CAPERON, J., S. A. CATTELL and G. KRASNICK. 1971. Phytoplankton kinetics in a subtropical estuary: eutrophication. *Limnol. Oceanogr.* **16**, 599–607.

CAREFOOT, T. H. 1967. Growth and nutrition of *Aplysia punctata* feeding on a variety of marine algae. *J. Mar. Biol. Ass. U.K.* **47**, 565–589.

CAREY, A. G. JR. 1967. Energetics of the benthos of Long Island Sound. I. Oxygen utilization of sediment. *Bull. Bingham Oceanogr. Coll.* **19**, 136–144.

CAREY, A. G. Jr. 1972. Ecological observations on the benthic invertebrates from the central Oregon continental shelf. In *The Columbia River Estuary and Adjacent Ocean Waters: Bioenvironmental Studies*, Eds. A. T. PRUTER and D. L. ALVERSON, University of Washington Press, pp. 422–443.

CARLUCCI, A. F. and P. M. BOWES. 1970. Production of vitamin B_{12}, thiamine, and biotin by phytoplankton. *J. Phycol.* **6**, 351–357.

CARLUCCI, A. F. and P. M. WILLIAMS. 1965. Concentration of bacteria from seawater by bubble scavenging. *J. Cons. Int. Explor. Mer.* **30**, 28–33.

CARLUCCI, A. F. and P. M. MCNALLY. 1969. Nitrification by marine bacteria in low concentrations of substrate and oxygen. *Limnol. Oceanogr.* **14**, 736–739.

CARPENTER, E. J. and R. R. L. GUILLARD. 1971. Intraspecific differences in nitrate half-saturation constants for three species of marine phytoplankton. *Ecology*, **52**, 183–185.

CASSIE, R. M. 1959. Micro-distribution of plankton. *New Zealand J. Sci.* **2**, 398–409.

CASSIE, R. M. 1962a. Frequency distribution models in the ecology of plankton and other organisms. *J. Anim. Ecol.* **31**, 65–92.

CASSIE, R. M. 1962b. Microdistribution and other error components of C^{14} primary production estimates. *Limnol. Oceanogr.* **7**, 121–130.

CASSIE, R. M. 1963. Multivariate analysis in the interpretation of numerical plankton data. *New Zealand J. Sci.* **6**, 36–59.

CASSIE, R. M. 1968. Sample design. Zooplankton sampling. *Mon. Oceanogr. Method. Unesco (Paris)*, **2**, 105–121.

CASSIE, R. M. and A. D. MICHAEL. 1968. Fauna and sediments of an intertidal mud flat: a multivariate analysis. *J. Exp. Mar. Biol. Ecol.* **2**(1), 1–23.

CASTENHOLZ, R. W. 1961. The effect of grazing on marine littoral diatom populations. *Ecol.* **42**, 783–794.

CHAMBERS, M. R. and H. MILNE. 1975. The production of *Macoma balthica* (L.) in the Ythan estuary. *Estuar. Coast. Mar. Sci.* **3**, 133–144.

CHANG, B. D. and T. R. PARSONS. 1975. Metabolic studies on the amphipod *Anisogammarus pugettensis* in relation to its trophic position in the food web of young salmonids. *J. Fish. Res. Bd. Canada*, **32**, 243–247.

CHARDY, P., M. GLEMAREC and A. LAUREC. 1976. Application of inertia methods to benthic marine ecology: practical implications of the basic options. *Estuar. Coast. Mar. Sci.* **4**, 179–205.

CHAU, Y. K., L. CHUECAS and J. P. RILEY. 1967. The component combined amino acids of some marine phytoplankton species. *J. Mar. Biol. Ass. U.K.* **47**, 543–554.

CHAVE, K. E. 1965. Carbonates: association with organic matter in surface seawater. *Science*, **148**, 1723–1724.

CHAVE, K. E. 1970. Carbonate-organic interactions in sea water. *Symp. Organic Matter in Natural Waters*, Ed. D. W. HOOD, University of Alaska, pp. 373–385.

CHINDONOVA, Y. G. 1959. The nutrition of certain groups of abyssal macroplankton in the north-western area of the Pacific Ocean. *Trudy Institute Okeanologii*, **30**, 166–189. (Nat. Inst. Oceanogr. Translation No. 131, Wormley, U.K.)

CHISLENKO, L. L. 1968. Nomographs for determination of weights of aquatic organisms by size and body form (marine mesobenthos and plankton). Acad. Sci. USSR. Zool. Inst. 'Science' Publishing House, Leningrad. (Translation by J. Marliave, Univ. British Columbia.)

CHRISTIAN, R. R., K. BANCROFT and W. J. WIEBE. 1975. Distribution of microbial adenosine triphosphate in salt marsh sediments at Sapelo Island, Georgia. *Soil Sci.* **119**, 89–96.

CHRISTIANSEN, F. E. 1968. *Nitrogen Excretion by some Planktonic Copepods of Bras-d'Or Lake, Nova Scotia*, M.S. Thesis, Dalhousie University, Halifax, 107 pp.

CHUECAS, L. and J. P. RILEY. 1969. Component fatty acids of the total lipids of some marine phytoplankton. *J. Mar. Biol. Ass. U.K.* **49**, 97–116.

CLARKE, G. L. 1965. *Elements of Ecology*, John Wiley & Sons, New York. 560 pp.

CLARKE, G. L. and H. R. JAMES. 1939. Laboratory analysis of the selective absorption of light by sea water. *J. Opt. Soc. Amer.* **29**, 43–55.

CLARK, M. E., G. A. JACKSON and W. J. NORTH. 1972. Dissolved free amino acids in southern California coastal waters. *Limnol. Oceanogr.* **17**, 749–758.

CLARK, R. B. and A. MILNE. 1955. The sublittoral fauna of two sandy bays on the isle of Cumbrae, Firth of Clyde. *J. Mar. Biol. Ass. U.K.* **34**, 161–180.

CLARKE, M. R. and N. MERRETT. 1972. The significance of squid, whale and other remains from the stomachs of bottom-living deep sea fish. *J. Mar. Biol. Ass. U.K.* **52**, 599–603.

CLENDENNING, K. A. 1971. Organic productivity in kelp areas. In: *The Biology of Giant Kelp Beds (Macrocystis)*
in California, Ed. W. J. NORTH. *Nova Hedwigia*, **32** (Suppl.), 259–263.

CLUTTER, R. I. and G. H. THEILACKER. 1971. Ecological efficiency of a pelagic mysid shrimp. Estimates
from growth, energy budget, and mortality studies. *Fish. Bull.* **69**, 93–115.

COLE, L. C. 1954. The population consequences of life history phenomena. *Quart. Rev. Biol.* **29**, 103–137.

COLEBROOK, J. M., R. S. GLOVER and G. A. ROBINSON. 1961. Continuous plankton records: contributions
towards a plankton atlas of the north-eastern Atlantic and the North Sea. *Bull. Mar. Ecol.* **5**, 67–80.

COLES, S. L. 1969. Quantitative estimates of feeding and respiration for three scleractinian corals. *Limnol.
Oceanogr.* **14**, 949–953.

COMITA, G. W. 1968. Oxygen consumption in *Diaptomus*. *Limnol. Oceanogr.* **13**, 51–57.

CONNELL, J. 1961. The influence of interspecific competition and other factors on the distribution of the
barnacle *Chthamalus stellatus*. *Ecology*, **42**, 133–146.

CONOVER, J. T. 1968. The importance of natural diffusion gradients and transport of substances related to
benthic marine plant metabolism. *Bot. Mar.* **11**, 1–9.

CONOVER, R. J. 1966a. Assimilation of organic matter by zooplankton. *Limnol. Oceanogr.* **11**, 338–345.

CONOVER, R. J. 1966b. Factors affecting the assimilation of organic matter by zooplankton and the question
of superfluous feeding. *Limnol. Oceanogr.* **11**, 346–354.

CONOVER, R. J. 1966c. Feeding on large particles by *Calanus hyperboreus* (Kröyer). In *Some Contemporary
Studies in Marine Science*, Ed. H. BARNES, Allen & Unwin, London, pp. 187–194.

CONOVER, R. J. 1968. Zooplankton—life in a nutritionally dilute environment. *Amer Zoologist*, **8**, 107–118.

CONOVER, R. J. (In press). Transformation of organic matter. In *Marine Ecology. IV. Dynamics*, Ed. O. KINNE,
Wiley–Interscience, London.

CONOVER, R. J. and E. D. S. CORNER. 1968. Respiration and nitrogen excretion by some marine zooplankton
in relation to their life cycles. *J. Mar. Biol. Ass. U.K.* **48**, 49–75.

CONOVER, R. J. and V. FRANCIS. 1973. The use of radioactive isotopes to measure the transfer of materials
in aquatic food chains. *Mar. Biol.* **18**, 272–283.

CONOVER, R. J. and C. M. LALLI. 1974. Feeding and growth in *Clione limacina* (Phipps), a pteropod mollusc.
II. Assimilation, metabolism and growth efficiency. *J. Exp. Mar. Biol. Ecol.* **16**, 134–154.

COOMBS, J., W. M. DARLEY, O. HOLM-HANSEN and B. E. VOLCANI. 1967a. Studies on the biochemistry and
fine structure of silica shell formation in diatoms. Chemical composition of *Navicula pelliculsa* in silicon-
starvation synchrony. *Plant Physiol.* **42**, 1601–1606.

COOMBS, J., C. SPANIS and B. E. VOLCANI. 1967b. Studies on the biochemistry and fine structure of silica
shell formation in diatoms. Photosynthesis and respiration in silicon-starvation synchrony of *Navicula
pelliculosa*. *Plant Physiol.* **42**, 1607–1611.

COOMBS, J., P. J. HALICKI, O. HOLM-HANSEN and B. E. VOLCANI. 1967c. Studies on the biochemistry and
fine structure of silica shell formation in diatoms. II. Changes in the concentration of neucloside triphos-
phates in silicon starvation synchrony of *Navicula pelliculosa* (Bréb.). *Hilse, Exptl. Cell Res.* **47**, 315–328.

COOPER, L. H. N. 1937. On the ratio of nitrogen to phosphorus in the sea. *J. Mar. Biol. Ass. U.K.* **22**,
177–182.

COPELAND, B. J. 1965. Evidence for regulation of community metabolism in a marine ecosystem. *Ecology*,
46, 563–564.

CORNER, E. D. S. 1961. On the nutrition and metabolism of zooplankton. I. Preliminary observations on
the feeding of the marine copepod, *Calanus helgolandicus* (Claus). *J. Mar. Biol. Ass. U.K.* **41**, 5–16.

CORNER, E. D. S. and B. S. NEWELL. 1967. On the nutrition and metabolism of zooplankton. IV. The forms
of nitrogen excreted by *Calanus*. *J. Mar. Biol. Ass. U.K.* **47**, 113–120.

CORNER, E. D. S., C. B. COWEY and S. M. MARSHALL. 1967. On the nutrition and metabolism of zooplankton.
V. Feeding efficiency of *Calanus finmarchicus*. *J. Mar. Biol. Ass. U.K.* **47**, 259–270.

COSPER, T. C. and M. R. REEVE. 1975. Digestive efficiency of the chaetognath *Sagitta hispida* Conant. *J.
Exp. Mar. Biol. Ecol.* **17**, 33–38.

COULL, B. C. 1972. Species diversity and fauna affinities of meiobenthic Copepoda in the deep sea. *Mar.
Biol.* **14**, 48–51.

COWEY, C. B. and E. D. S. CORNER. 1963. On the nutrition and metabolism of zooplankton. II. The relationship
between the marine copepod *Calanus helgolandicus* and particulate material in Plymouth sea water, in
terms of amino acid composition. *J. Mar. Biol. Ass. U.K.* **43**, 495–511.

COWEY, C. B. and E. D. S. CORNER. 1966. The amino-acid composition of certain unicellular algae, and
of the faecal pellets produced by *Calanus finmarchicus* when feeding on them. In *Some Contemporary
Studies in Marine Science*, Ed. H. BARNES, George Allen & Unwin, London, pp. 225–231.

CRISP, D. J. 1971. Energy flow measurements. In *Methods for the Study of Marine Benthos*, Eds. N. A. HOLME
and A. D. McINTYRE, I.B.P. Handbook, **16**. Oxford and Edinburgh: Blackwell Scientific Publications,
pp. 197–279.

CRUZ, A. A. DE LA and W. E. POE. 1975. Amino acid content of marsh plants. *Estuar. Coast. Mar. Sci.*
3, 243–246.

CULKIN, F. and R. J. MORRIS. 1970. The fatty acid composition of two marine filter-feeders in relation to
a phytoplankton diet. *Deep-Sea Res.* **17**, 861–865.

CUPP, E. E. 1943. *Marine Plankton Diatoms of the West Coast of North America*, University of California Press, Berkeley. 237 pp.

CURL, H. 1962. Standing crops of carbon, nitrogen, and phosphorus and transfer between trophic levels, in continental shelf waters south of New York. *Rapp. Proc.-Verb. Cons. int. Explor. Mer.* **153**, 183–189.

CURRIE, R. I. 1962. Pigments in zooplankton faeces. *Nature*, **193**, 956–957.

CUSHING, D. H. 1955. Production of a pelagic fishery in the sea. *Fish. Invest. London*, Ser. 2, **18**, 103.

CUSHING, D. H. 1962. Patchiness. *Rapp. Proc.-Verb. Cons. int. Explor. Mer*, **153**, 152–164.

CUSHING, D. H. 1964. The work of grazing in the sea. In *Grazing in Terrestrial and Marine Environments*, Ed. D. J. CRISP, Blackwell, London, pp. 207–225.

CUSHING, D. H. 1971. Upwelling and fish production. *Adv. Mar. Biol.* **9**, 255–334.

CUSHING, D. H. 1973. Production in the Indian Ocean and transfer from primary to secondary level. In *The Biology of the Indian Ocean*, Ed. B. ZEITZSCHEL, Publ. Springer-Verlag (Berlin), pp. 475–486.

CUSHING, D. H. and H. F. NICHOLSON. 1966. Method of estimating algal production rates at sea. *Nature*, **212**, 310–311.

CUSHING, D. H. and T. VUCETIC. 1963. Studies on a *Calanus* patch. III. The quantity of food eaten by *Calanus finmarchicus*. *J. Mar. Biol. Ass. U.K.* **43**, 349–371.

CUSHMAN, J. A. 1931. The Foraminifera of the Atlantic Ocean. *Bull. U.S. Nat. Mus.* **104**, pp. 55.

DAAN, N. 1973. A quantitative analysis of the food intake of North Sea cod, *Gadus morhua*. *Neth. J. Sea Res.* **6**, 479–517.

DAL PONT, G. and B. NEWELL. 1963. Suspended organic matter in the Tasman Sea. *Aust. J. Mar. Freshw. Res.* **14**, 155–165.

DALE, N. G. 1974. Bacteria in intertidal sediments: factors related to their distribution. *Limnol. Oceanogr.* **19**(3), 509–518.

DALES, R. P. 1960. On the pigments of the Chrysophyceae. *J. Mar. Biol. Ass. U.K.* **39**, 693–699.

DAVIES, J. M. 1975. Energy flow through the benthos in a Scottish sea loch. *Mar. Biol.* **31**, 353–362.

DAY, J. H., J. G. FIELD and M. POTTS MONTGOMERY. 1971. The use of numerical methods to determine the distribution of the benthic fauna across the continental shelf at North Carolina. *J. Anim. Ecol.* **40**, 93–125.

DAYTON, P. K. 1975. Experimental evaluation of ecological dominance in a rocky intertidal algal community. *Ecol. Monogr.* **45**(2), 137–159.

DAYTON, P. K. and R. R. HESSLER. 1972. Role of biological disturbance in maintaining diversity in the deep sea. *Deep-Sea Res.* **19**, 199–208.

DEGENS, E. T. 1970. Molecular nature of nitrogenous compounds in sea water and recent marine sediments. In *Organic Matter in Natural Waters*, Ed. D. W. HOOD. University of Alaska, pp. 77–106.

DEMENTJEVA, T. F. 1965. Changes in recruitment to the stock in relation to the environment, with reference to mathematical modelling. *ICNAF Spec. Publ.* **6**, 381–385.

DEUSER, H. G. 1971. Organic-carbon budget of the Black Sea. *Deep-Sea Res.* **18**, 995–1004.

DICKIE, L. M. (1972). Food chains and fish production. *ICNAF Spec. Publ.* **8**, 201–221.

DOTY, M. S. 1959. Phytoplankton photosynthetic periodicity as a function of latitude. *J. Mar. Biol. Ass. India*, **1**, 66–68.

DOTY, M. S. and M. OGURI. 1957. Evidence for a photosynthetic daily periodicity. *Limnol. Oceanogr.* **2**, 37–40.

DOYLE, R. W. 1972. Genetic variation of *Ophiomusium lymani* (Echinodermata) populations in the deep sea. *Deep-Sea Res.* **19**(9), 661–664.

DRAKE, D. E. 1971. Suspended sediment and thermal stratification in Santa Barbara Channel, California. *Deep-Sea Res.* **18**, 763–769.

DRAL, A. D. G. 1967. The movements of the latero-frontal cilia and the mechanism of particle retention in the mussel (*Mytilus edulis*). *Neth. J. Sea Res.* **3**, 391–422.

DRISCOLL, E. G. and D. E. BRANDON. 1973. Mollusc–sediment relationships in northwestern Buzzards Bay, Massachusetts, U.S.A. *Malacologia*, **12**, 13–46.

DROOP, M. R. 1968. Vitamin B_{12} and marine ecology. IV. The kinetics of uptake, growth and inhibition in *Monochrysis lutheri*. *J. Mar. Biol. Ass. U.K.* **48**, 689–733.

DROOP, M. R. 1970. Vitamin B_{12} and marine ecology. V. Continuous culture as an approach to nutritional kinetics. *Helgoländer wiss. Meeresunters.* **20**, 629–636.

DROOP, M. R., J. J. A. McLAUGHLIN, I. J. PINTER and L. PROVASOLI. 1959. Specificity of some protophytes toward vitamin B-like compounds. *Preprints Int. Oceanogr. Congr.* (*New York*), pp. 916–918.

DUFF, D. C. B., D. L. BRUCE and N. J. ANTIA. 1966. The antibacterial activity of marine planktonic algae. *Can. J. Microbiol.* **12**, 877–884.

DUFF, S. and J. M. TEAL. 1965. Temperature change and gas exchange in Nova Scotia and Georgia salt-marsh muds. *Limnol. Oceanogr.* **10**, 67–73.

DUGDALE, R. C. 1967. Nutrient limitation in the sea: dynamics, identification and significance. *Limnol. Oceanogr.* **12**, 685–695.

DUGDALE, R. C., J. J. GOERING and J. H. RYTHER. 1964. High-nitrogen fixation rates in the Sargasso Sea and the Arabian Sea. *Limnol. Oceanogr.* **9**, 507–510.

DUGDALE, R. C. and J. J. GOERING. 1967. Uptake of new and regenerated forms of nitrogen in primary productivity. *Limnol. Oceanogr.* **12,** 196–206.

DUNBAR, M. J. 1960. The evolution of stability in marine environments. Natural selection at the level of the ecosystem. *Am. Nat.* **94,** 129–136.

DUNBAR, M. J. 1962. The life cycle of *Sagitta elegans* in arctic and subarctic seas, and the modifying effects of hydrographic differences in the environment. *J. Mar. Res.* **20,** 76–91.

DUNSTAN, W. M., L. P. ATKINSON and J. MATOLI. 1975. Simulation and inhibition of phytoplankton growth by low molecular weight hydrocarbons. *Mar. Biol.* **31,** 305–310.

DUSSART, B. M. 1965. Les différentes catégories de planction. *Hydrobiologia,* **26,** 72–74 (with Erratum).

DUURSMA, E. K. 1960. *Dissolved Organic Carbon, Nitrogen and Phosphorus in the Sea*, Ph.D. Theses, J. B. Wolters, Groningen. 147 pp.

DUURSMA, E. K. 1965. The dissolved organic constituents of sea water. In *Chemical Oceanography*, Ed. J. P. RILEY and G. SKIRROW, Academic Press, New York, pp. 433–475.

DUVAL, W. S. and G. H. GEEN. 1976. Diel feeding and respiration rhythms in zooplankton. *Limnol. Oceanogr.* (in press).

EAGLE, R. A. 1975. Natural fluctuations in a soft bottom benthic community. *J. Mar. Biol. Ass. U.K.* **55,** 865–878.

EDMONDSON, W. T. and G. G. WINBERG. 1971. (Eds) *A Manual on Methods for the Assessment of Secondary Productivity in Fresh Waters*, I.B.P. Handbook, **17.** Blackwell Scientific Publications, Oxford. 358 pp.

EDWARDS, R. R. C. 1967. Estimates of the respiratory rate of young plaice (*Pleuronectes platessa* L.) in natural conditions using zinc-65. *Nature,* **216,** 1335–1337.

EDWARDS, R. R. C. 1973. Production ecology of two Caribbean marine ecosystems. II. Metabolism and energy flow. *Estuar. Coast. Mar. Sci.* **1,** 319–333.

EDWARDS, R. R. C., D. M. FINLAYSON and J. H. STEELE. 1969. The ecology of 0-group plaice and common dabs in Loch Ewe. II. Experimental studies of metabolism. *J. Exp. Mar. Biol. Ecol.* **8,** 299–309.

EDWARDS, R. W. and H. L. ROLLEY. 1965. Oxygen consumption of river muds. *J. Ecol.* **53,** 1–19.

EGLINTON, G. and M. T. J. MURPHY. 1969. *Organic Geochemistry: Methods and Results*, Springer-Verlag, New York. 828 pp.

ELLIOTT, J. M. 1971. Some methods for the statistical analysis of samples of benthic invertebrates. *Freshwater Biol. Ass. Publ.* **25,** 144 pp.

ELMGREN, R. 1975. Benthic meiofauna as indicators of oxygen conditions in the northern Baltic proper. *Merentutkimuslait. Julk./Havsforskningsinst.* Skr. **239,** 265–271.

EL-SAYED, S. Z. and K. A. GREEN. 1974. Use of remote sensing in the study of Antarctic marine resources. In *Approaches to Earth Survey Problems through use of Space Techniques*, Ed. P. BOCK, Publ. Akademic-Verlag (Berlin), pp. 47–63.

ELTRINGHAM, S. K. 1971. *Life in Mud and Sand*, The English Universities Press Ltd., London. 218 pp.

EMERSON, R. and C. M. Lewis. 1942. The photosynthetic efficiency of phycocyanin *Chroococcus*, and the problem of carotenoid participation in photosynthesis. *J. Gen. Physiol.* **25,** 579–595.

EMLEN, J. M. 1966. The role of time and energy in food preference. *Amer. Nat.* **100,** 611–617.

EMLEN, J. M. 1973. *Ecology: an Evolutionary Approach*, Addison–Wesley, Massachusetts. 493 pp.

EPPLEY, R. W. 1972. Temperature and phytoplankton growth in the sea. *Fish. Bull.* **70,** 1063–1085.

EPPLEY, R. W. and P. R. SLOAN. 1965. Carbon balance experiments with marine phytoplankton. *J. Fish. Res. Bd. Canada,* **22,** 1083–1097.

EPPLEY, R. W., R. W. HOLMES and J. D. H. STRICKLAND. 1967. Sinking rates of marine phytoplankton measured with a fluorometer. *J. Exp. Mar. Biol. Ecol.* **1,** 191–208.

EPPLEY, R. W. and J. L. COATSWORTH. 1968. Uptake of nitrate and nitrite by *Ditylum brightwellii*—kinetics and mechanisms. *J. Phycol.* **4,** 151–156.

EPPLEY, R. W., O. HOLM-HANSEN and J. D. H. STRICKLAND. 1968. Some observations on the vertical migration of dinoflagellates. *J. Phycol.* **4,** 333–340.

EPPLEY, R. W. and J. D. H. STRICKLAND. 1968. Kinetics of marine phytoplankton growth. In *Advances in Microbiology of the Sea*, Eds. M. R. DROOP and E. J. F. WOOD, Academic Press, London, Vol. 1, pp. 23–62.

EPPLEY, R. W., J. L. COATSWORTH and L. SOLORZANO. 1969a. Studies of nitrate reductase in marine phytoplankton. *Limnol. Oceanogr.* **14,** 194–205.

EPPLEY, R. W., J. N. ROGERS and J. J. MCCARTHY. 1969b. Half saturation constants for uptake of nitrate and ammonium by marine phytoplankton. *Limnol. Oceanogr.* **14,** 912–920.

EPPLEY, R. W. and W. H. THOMAS. 1969. Comparison of half-saturation constants for growth and nitrate uptake of marine phytoplankton. *J. Phycol.* **5,** 375–379.

EPPLEY, R. W., A. F. CARLUCCI, O. HOLM-HANSEN, D. KIEFER, J. J. MCCARTHY and P. M. WILLIAMS. 1972. Evidence for eutrophication in the sea near southern California coastal sewage outfalls, July 1970. *Calcofi Rep.*

ESAIAS, W. E. and H. C. CURL, JR. 1972. Effect of dinoflagellate bioluminescence on copepod ingestion rates. *Limnol. Oceanogr.* **17,** 901–906.

ESTRADA, M., I. VALIELA and J. M. TEAL. 1974. Concentration and distribution of chlorophyll in fertilized plots in a Massachusetts salt marsh. *J. Exp. Mar. Biol. Ecol.* **14,** 47–56.

EWING, J. A. 1973. Wave induced bottom currents on the outer shelf. *Mar. Geol.* **15**, M31–M35.

FAGER, E. W. 1957. Determination and analysis of recurrent groups. *Ecology*, **38**(4), 586–595.

FAGER, E. W. 1964. Marine sediments: effects of a tube-building polychaete. *Science*, **143**, 356–359.

FAGER, E. W. 1972. Diversity: a sampling study. *Amer. Nat.* **106**, 293–309.

FAGER, E. W. and J. A. McGOWAN. 1963. Zooplankton species groups in the North Pacific. *Science*, **140**, 453–460.

FAGER, E. W. and A. R. LONGHURST. 1968. Recurrent group analysis of species assemblage of demersal fish in the Gulf of Guinea. *J. Fish. Res. Bd. Canada* **25**(7), 1405–1421.

FAGERSTRÖM, T. and A. JERNELOV. 1972. Some aspects of the quantitative ecology of mercury. *Water Res.* **6**, 1193–1202.

FALLER, A. J. 1971. Oceanic turbulence and the Langmuir circulation. *Ann. Rev. Ecol. Syst.* **2**, 201–236.

FANNING, F. A. and M. E. Q. PILSON. 1974. The diffusion of dissolved silica out of deep-sea sediments. *J. Geophys. Res.* **79**, 1293–1297.

FARRINGTON, J. W. and P. A. MEYER. 1975. Hydrocarbons in the marine environment. In *Environmental Chemistry: Air and Water Pollution*, Eds. H. S. STOKER and S. L. SEAGAR, Scott, Foresman, Glenview, Illinois, pp. 109–136.

FASHAM, M. J. R., M. V. ANGEL and H. S. J. ROE. 1974. An investigation of the spatial pattern of zooplankton using the Longhurst–Hardy plankton recorder. *J. Exp. Mar. Biol. Ecol.* **16**, 93–112.

FEE, E. J. 1969. A numerical model for the estimation of photosynthetic production, integrated over time and depth, in natural waters. *Limnol. Oceanogr.* **14**, 906–911.

FEE, E. J. 1971. *A Numerical Model for the Estimation of Integral Primary Production and its Application to Lake Michigan*, Ph.D. Thesis, University of Wisconsin. 169 pp.

FENCHEL, T. 1968. The ecology of marine microbenthos. III. The reproductive potential of ciliates. *Ophelia*, **5**, 123–136.

FENCHEL, T. 1969. The ecology of marine microbenthos. IV. Structure and function of the benthic ecosystem, its chemical and physical factors and the microfauna communities with special reference to the ciliated protozoa. *Ophelia*, **6**, 1–182.

FENCHEL, T. 1970. Studies on the decomposition of organic detritus derived from the turtle grass *Thalassia testudinum. Limnol. Oceanogr.* **15**, 14–20.

FENCHEL, T. 1971. The reduction-oxidation properties of marine sediments and the vertical distribution of the microfauna. *Vie Milieu*, **22**, 509–521.

FENCHEL, T. 1972. Aspects of decomposer food chains in marine benthos. *Verh. Dtsch. Zool. Ges.* **65**(5), 14–22.

FENCHEL, T. 1974. Intrinsic rate of natural increase: the relationship with body size. *Oecologia*, **14**, 317–326.

FENCHEL, T. 1975. Factors determining the distribution patterns of mud snails (Hydrobiidae). *Oecologia*, **20**, 1–17.

FENCHEL, T. and R. H. RIEDL. 1970. The sulphide system: a new biotic community underneath the oxidized layer of marine sand bottoms. *Mar. Biol.* **7**, 255–268.

FENCHEL, T. and B. J. STRAARUP. 1971. Vertical distribution of photosynthetic pigments and the penetration of light in marine sediments. *Oikos*, **22**, 172–182.

FENCHEL, T., L. H. KOFOED and A. LAPPALAINEN. 1975. Particle size–selection of two deposit feeders: the amphipod *Corophium volutator* and the prosobranch *Hydrobia ulvae. Mar. Biol.* **30**, 119–128.

FERGUSON, C. F. and J. K. B. RAYMONT. 1974. Biochemical studies on marine zooplankton. XII. Further investigations on *Euphausia superba* (Dana). *J. Mar. Biol. Ass. U.K.* **54**, 719–725.

FIELD, J. G. 1971. A numerical analysis of changes in the soft-bottom fauna along a transect across False Bay, South Africa. *J. Exp. Mar. Biol. Ecol.* **7**, 215–253.

FIELD, J. G. 1972. Some observations on the release of dissolved organic carbon by the sea urchin, *Strongylocentrotus droebachiensis. Limnol. Oceanogr.* **17**, 759–761.

FIELD, J. G. and G. McFARLANE. 1968. Numerical methods in marine ecology. I. A quantitative 'similarity' analysis of rocky shore samples in False Bay, South Africa. *Zool. Afr.* **3**, 119–137.

FITZGERALD, G. P. 1968. Detection of limiting or surplus nitrogen in algae and aquatic weeds. *J. Phycol.* **4**, 121–126.

FOERSTER, R. E. 1968. The sockeye salmon, *Oncorhynchus nerka. Bull. Fish. Res. Bd. Canada, Bull.* **162**, pp. 422.

FOERSTER, R. E. and W. E. RICKER. 1941. The effect of reduction of predaceous fish on survival of young sockeye salmon at Cultus Lake. *J. Fish. Res. Bd. Canada*, **5**, 315–336.

FOGG, G. E. 1952. The production of extracellular nitrogenous substances by a blue-green alga. *Proc. Roy. Soc. (London)*, B, **139**, 372–397.

FOGG, G. E. 1966. The extracellular products of algae. *Oceanogr. Mar. Biol. Ann. Rev.* **4**, 195–212.

FOGG, G. E., C. NALEWAJKO and W. D. WATT. 1965. Extracellular products of phytoplankton photosynthesis. *Proc. Royal Soc. (London)*, B, **162**, 517–534.

FORD, C. W. and E. PERCIVAL. 1965. The carbohydrates of *Phaedactylum tricornutum*. Part 1. Preliminary examination of the organism and characterization of low molecular weight material and of a glucan. *J. Chem. Soc.* (5), 7035–7042.

FOURNIER, R. O. 1966. North Atlantic deep-sea fertility. *Science,* **153,** 1250–1252.

FOURNIER, R. O. 1971. The transport of organic carbon to organisms living in the deep oceans. *Proc. Roy. Soc. Edinburgh,* **73,** 203–211.

FOURNIER, R. O. 1973. Studies of pigmented microorganisms from aphotic marine environments. III. Evidence of apparent utilization by benthic and pelagic tunicata. *Limnol. Oceanogr.* **18,** 38–43.

FOX, D. L., D. M. UPDEGRAFF and G. D. NOVELLI. 1944. Carotenoid pigments in the ocean floor. *Arch. Biochem.* **5,** 1–23.

FOX, D. L., S. C. CRANE and B. M. McCONNAUGHEY. 1948. A biochemical study of the marine annelid worm, *Thoracophelia mucronata,* its food, biochromes and carotenoid metabolism. *J. Mar. Res.* **7,** 567–585.

FRANKENBERG, D. and R. J. Menzies. 1968. Some quantitative analyses of deep-sea benthos off Peru. *Deep-Sea Res.* **15,** 623–626.

FRAZER, H. J. 1935. Experimental study of porosity and permeability of clastic sediments. *J. Geol.* **43,** 910–1010.

FRITSCH, F. E. 1956. *The Structure and Reproduction of the Algae,* Vol. 1. Cambridge University Press, London. 791 pp.

FRITZ, S. 1957. Solar energy on clear and cloudy days. *Sci. Month.* **84,** 55–65.

FRONTIER, S. 1973. Étude statistique de la dispersion du zooplankton. *J. Exp. Mar. Biol. Ecol.* **12,** 229–262.

FROST, B. W. 1972. Effects of size and concentration of food particles on the feeding behaviour of the marine planktonic copepod *Calanus pacificus. Limnol. Oceanogr.* **17,** 805–815.

FROST, B. W. 1974. Feeding processes at lower trophic levels in pelagic communities. In *The Biology of the Oceanic Pacific,* Publ. Oregon State Univ. Press, pp. 59–77.

FROST, B. W. 1975. A threshold feeding behaviour in *Calanus pacificus. Limnol. Oceanogr.* **20,** 263–266.

FUJITA, Y. 1970. Photosynthesis and plant pigments. *Bull. Plankton Soc. Japan,* **17,** 20–31.

FULTON, J. 1972. Trials with an automated plankton counter. *J. Fish. Res. Bd. Canada,* **29,** 1075–1078.

FULTON, J. 1973. Some aspects of the life history of *Calanus plumchrus* in the Strait of Georgia. *J. Fish. Res. Bd. Canada,* **30,** 811–815.

GAGE, J. 1972. Community structure of the benthos in Scottish sea-lochs. I. Introduction and species diversity. *Mar. Biol.* **14,** 281–297.

GAGE, J. and A. D. GEEKIE. 1973. Community structure of the benthos in Scottish sea-lochs. II. Spatial pattern. *Mar. Biol.* **19,** 41–53.

GAGE, J. and P. B. TETT. 1973. The use of log normal statistics to describe the benthos of Lochs Etine and Creran. *J. Anim. Ecol.* **42,** 373–382.

GALLAGHER, J. L. and F. C. DAIBER. 1973. Diel rhythms in edaphic community metabolism in a Delaware salt marsh. *Ecol.* **54,** 1160–1163.

GÄRDEFORS, D. and L. Orrhage. 1968. Patchiness of some marine bottom animals. A methodological study. *Oikos,* **19**(2), 311–322.

GARDINER, A. C. 1933. Vertical distribution in *Calanus finmarchicus. J. Mar. Biol. Ass. U.K.* **18,** 575–610.

GARDNER, D. 1975. Observations on the distribution of dissolved mercury in the ocean. *Mar. Poll. Bull.* **6,** 43–46.

GARDNER, D. and J. P. RILEY. 1973. Distribution of dissolved mercury in the Irish Sea. *Nature,* **241,** 526–527.

GARGAS, E. 1970. Measurements of primary production, dark fixation and vertical distribution of the microbenthic algae in the Øresund. *Ophelia,* **8,** 251–253.

GARGAS, E. 1971. 'Sun-shade' adaptation in microbenthic algae from the Øresund. *Ophelia,* **9,** 107–112.

GARRETT, W. D. 1964. *The Organic Chemical Composition of the Ocean Surface.* NRL Report, 6201, 1–12.

GARRETT, W. D. 1965. Collection of slick-forming materials from the sea surface. *Limnol. Oceanogr.* **10,** 602–605.

GARRETT, W. D. 1967. The organic chemical composition of the ocean surface. *Deep-Sea Res.* **14,** 221–227.

GASITH, A. 1975. Tripton sedimentation in eutrophic lakes—simple correction for the resuspended matter. *Verh. Int. Verein. Limnol.* **19,** 116–122.

GATTEN, R. R. and J. R. SARGENT. 1973. Wax ester biosynthesis in calanoid copepods in relation to vertical migration. *Neth. J. Sea Res.* **7,** 150–158.

GAUDY, R. 1974. Feeding four species of pelagic copepods under experimental conditions. *Mar. Biol.* **25,** 125–141.

GAULD, D. T. 1951. The grazing rate of planktonic copepods. *J. Mar. Biol. Ass. U.K.* **29,** 695–706.

GAULD, D. T. 1966. The swimming and feeding of planktonic copepods. In *Some Contemporary Studies in Marine Science,* Ed. H. BARNES, Allen & Unwin, London, pp. 313–334.

GEORGE, J. L. and D. E. H. FREAR. 1966. Pesticides in the Antarctic. *J. appl. Ecol.* **3**(suppl.), 155–167.

GERLACH, S. A. 1971. On the importance of marine meiofauna for benthos communities. *Oecologia,* **6,** 176–190.

GIBBS, M. and J. A. SCHIFF. 1960. Chemosynthesis: the energy relations of chemoautotrophic organisms. In *Plant Physiology,* vol. 1B. *Photosynthesis and Chemosynthesis,* Ed. F. C. STEWARD, Academic Press, London, pp. 279–319.

GILMARTIN, M. and N. REVELANTE. 1974. The 'Island Mass' effect on the phytoplankton and primary production of the Hawaiian Islands. *J. Exp. Mar. Biol. Ecol.* **16,** 181–204.

GILMER, R. W. 1974. Some aspects of feeding in thecosomatous pteropod molluscs. *J. Exp. Mar. Biol. Ecol.* **15,** 127–144.

GLOVER, R. S., G. A. ROBINSON and J. M. COLEBROOK. 1974. Marine biological surveillance. *Environ. Change*, **2**, 395–402.

GNAIGER, E. and G. GLUTH. (In press) pH and temperature fluctuations in an inter-tidal beach in Bermuda. *Limnol. Oceanogr.*

GOERING, J. J., R. C. DUGDALE and D. W. MENZEL. 1966. Estimates of *in situ* rates of nitrogen uptake by *Trichodesmium* sp. in the tropical Atlantic Ocean. *Limnol. Oceanogr.* **11**, 614–620.

GOERING, J. J., D. D. WALLEN and R. M. NAUMAN. 1970. Nitrogen uptake by phytoplankton in the disconti-nuity layer of the eastern subtropical Pacific Ocean. *Limnol. Oceanogr.* **15**, 789–796.

GOERING, J. J. and M. PAMATMAT. 1971. Denitrification in sediments off Peru. *Inv. Pesq.* **35**, 233–242.

GOERING, J. J., D. M. NELSON and J. A. CARTER. 1973. Silicic acid uptake by natural populations of marine phytoplankton. *Deep-Sea Res.* **20**, 777–789.

GOLDBERG, E. D. 1971. River-ocean interactions. In *Fertility of the Sea*, Ed. J. D. COSTLOW, Gordon & Breach, New York, Vol. 1, pp. 143–156.

GOODWIN, T. W. 1955. Carotenoids. In *Modern Methods of Plant Analysis*, Eds. K. PAECH and M. V. TRACEAY, Springer-Verlag (Berlin), Vol. 3, pp. 272–311.

GOODWIN, T. W. 1957. The nature and distribution of carotenoids in some blue-green algae. *J. Gen. Microbiol.* **17**, 467–473.

GORDON, D. C. 1970a. A microscopic study of organic particles in the North Atlantic Ocean. *Deep-Sea Res.* **17**, 175–185.

GORDON, D. C. 1970b. Some studies on the distribution and composition of particulate organic carbon in the North Atlantic Ocean. *Deep-Sea Res.* **17**, 233–243.

GORDON, D. C. Jr. 1971. Distribution of particulate organic carbon and nitrogen at an oceanic station in the central Pacific. *Deep-Sea Res.* **8**, 1127–1134.

GORDON, D. C. Jr. Variability of particulate organic carbon and nitrogen along the Halifax–Bermuda section. *Deep-Sea Res.* **24** (in press).

GORDON, D. C. and P. A. MICHALIK. 1971. Concentration of bunker C fuel oil in the waters of Chedabucto Bay, April 1971. *J. Fish. Res. Bd. Canada*, **28**, 1912–1914.

GORDON, L. I., P. K. PARK, S. W. HAGER and T. R. PARSONS. 1971. Carbon dioxide partial pressures in north Pacific surface waters—time variations. *J. Oceanogr. Soc. Japan*, **27**, 81–90.

GORDON, W. G. and E. O. WHITTIER. 1966. In *Fundamentals of Dairy Chemistry*, Eds. B. H. WEBB and A. H. JOHNSON, Avi Publishing Co., Westport, Conn., pp. 60.

GRAN, H. H. and T. BRAARUD. 1935. A quantitative study of the phytoplankton in the Bay of Fundy and the Gulf of Maine including observations on hydrography, chemistry and turbidity. *J. Biol. Bd. Canada*, **1**, 219–467.

GRANT, B. R. 1967. The action of light on nitrate and nitrite assimilation by the marine chlorophyte, *Dunaliella tertiolecta* (Butcher). *J. Gen. Microbiol.* **48**, 379–389.

GRASSLE, J. F. and J. P. GRASSLE. 1974. Opportunistic life histories and genetic systems in marine benthic polychaetes. *J. Mar. Res.* **32**, 253–284.

GRASSLE, J. F., H. L. SANDERS, R. R. HESSLER, G. T. ROWE and T. McLELLAN. 1975. Pattern and zonation: a study of the bathyal megafauna using the research submersible *Alvin. Deep-Sea Res.* **22**, 457–481.

GRAY, J. S. 1966. The attractive factor of intertidal sands to *Protodrilus symbiaticus. J. Mar. Biol. Ass. U.K.* **46**, 627–645.

GRAY, J. S. 1967. Substrate selection by the archiannelid *Protodrilus hypoleucus* Armenante. *J. Exp. Mar. Biol. Ecol.* 1, 47–54.

GRAY, J. S. 1974. Animal–sediment relationships. *Oceanogr. Mar. Biol. Ann. Rev.* **12**, 223–261.

GRAY, J. S. and R. M. RIEGER. 1971. A quantitative study of the meiofauna of an exposed sandy beach at Robin Hood's Bay, Yorkshire. *J. Mar. Biol. Ass. U.K.* **51**, 1–19.

GREENSLATE, J. 1974. Micro-organisms participate in the construction of manganese nodules. *Nature*, **249**, 181–183.

GREENWOOD, D. J. 1968. Measurements of microbial metabolism in soil. In *The Ecology of Soil Bacteria*, Eds. T. R. G. GRAY and D. PARKINSON, Univ. Toronto Press, pp. 138–157.

GREGORY, R. R. F. 1971. *Biochemistry of Photosynthesis*, Wiley–Interscience, London, pp. 202.

GREVE, W. 1971. Ökologische Untersuchungen an *Pleurobrachia pileus* 1. Freilanduntersuchungen. *Helgoländer Wiss. Meeresunters.* **22**, 303–325.

GREVE, W. 1972. Ökologische Untersuchungen an *Pleurobrachia pileus* 2. Laboratoriums-untersuchungen. *Helgoländer Wiss. Meeresunters.* **23**, 141–164.

GREZE, V. N. and E. P. Baldina. 1964. *Trudy Sevastopol biol. Sta. 18*, Fish, Res. Bd. Canada, Translation 893 (Population dynamics and the annual production of *Acartia clausii* Giesbr. and *Centropages kroyeri* in the neritic zone of the Black Sea).

GRICE, G. D. and V. R. GIBSON. 1975. Occurrence, viability and significance of resting eggs of the calanoid copepod *Labidocera aestiva. Mar. Biol.* **31**, 335–337.

GRIFFIN, J. J., H. WINDOM and E. D. GOLDBERG. 1968. The distribution of clay minerals in the world ocean. *Deep-Sea Res.* **15**, 433–459.

GRIFFITHS, D. 1975. Prey availability and the food of predators. *Ecology*, **56**, 1209–1214.

GRØNTVED, J. 1960. On the productivity of microbenthos and phytoplankton in some Danish fjords. *Meddr. Danm. Fisk. – og Havunders.*, N.S. **3**, 55–92.

GROSS, F. and E. ZEUTHEN. 1948. The buoyancy of plankton diatoms: a problem of cell physiology. *Proc. Roy. Soc. (London)*, **135**, 382–389.

GUERINOT, M. L., W. FONG and D. G. PATREQUIN. 1977. Nitrogen fixation (acetylene reduction) associated with sea urchins (*Strongylocentrotus droebachiensis*) feeding on seaweeds and eelgrass. *J. Fish. Res. Bd. Canada* **34**, (in press).

GUILLARD, R. R. L. 1963. Organic sources of nitrogen for marine centric diatoms. In *Symposium on Marine Microbiology*, Ed. C. H. OPPENHEIMER, C. C. Thomas, Springfield, Illinois, pp. 93–104.

GUILLARD, R. R. L., P. KILHAM and T. A. JACKSON. 1973. Kinetics of silicon-limited growth in the marine diatom *Thalassiosira pseudonana* Hasle and Heimdal (= *Cyclotella nana* Hustedt). *J. Phycol.* **9**, 233–237.

GULLAND, J. A. 1965. Survival of the youngest stages of fish, and its relation to year-class strength. *ICNAF Spec. Publ.* **6**, 363–371.

GULLAND, J. A. 1970. The fish resources of the oceans. *FAO Fish. Tech. Pap.* No. 97, p. 423.

GUNDERSEN, K. 1968. The formation and utilization of reducing power in aerobic chemo-autotrophic bacteria. *Zeitschrift. Allg. Mikrobiol.* **8**, 445–457.

HAERTEL, L., C. OSTERBERG, H. CURL, JR. and P. K. PARK. 1969. Nutrient and plankton ecology of the Columbia River estuary. *Ecology*, **50**, 962–978.

HÅKANSON, L. 1976. A bottom sediment trap for recent sedimentary deposits. *Limnol. Oceanogr.* **21**, 170–174.

HALIM, Y. 1960. Observations on the Nile bloom of phytoplankton in the Mediterranean. *J. Cons. Int. Explor. Mer*, **26**, 57–67.

HALL, K. J., W. C. WEIMER and G. F. LEE. 1970. Amino acids in an estuarine environment. *Limnol. Oceanogr.* **15**, 162–164.

HALL, K. J., P. M. KLEIBER and I. YESAKI. 1972. Heterotrophic uptake of organic solutes by microorganisms in the sediment. *Mem. Ist. Ital. Idiobiol.* **29** (Suppl), 441–447.

HALLBERG, R. O. 1968. Some factors of significance in the formation of sedimentary metal sulfides. *Stockholm Contr. Geol.* **15**, 39–66.

HALLBERG, R. O. 1973. The microbiological C–N–S cycles in sediments and their effect on the ecology of the sediment–water interface. *Oikos Suppl.* **15**, 51–62.

HALLBERG, R. O. 1974. Metal distribution along a profile of an intertidal area. *Estuar. Coast. Mar. Sci.* **2**, 153–170.

HALLDAL, P. 1958. Pigment formation and growth in blue-green algae in crossed gradients of light intensity and temperature. *Physiol. Plant.* **11**, 401–420.

HALTINER, G. J. and F. L. MARTIN. 1957. *Dynamical and Physical Meteorology*, McGraw Hill, New York. 470 pp.

HAMILTON, R. D. and O. HOLM-HANSEN. 1967. Adenosine triphosphate content of marine bacteria. *Limnol. Oceanogr.* **12**, 319–324.

HAMILTON, R. D., O. HOLM-HANSEN and J. D. H. STRICKLAND. 1968. Notes on the occurrence of living microscopic organisms in deep water. *Deep-Sea Res.* **15**, 651–656.

HAMILTON, R. D. and J. E. PRESLAN. 1970. Observations on heterotrophic activity in the eastern tropical Pacific. *Limnol. Oceanogr.* **15**, 395–401.

HANCOCK, D. A. and E. EDWARDS. 1967. Estimation of annual growth in the edible crab (*Cancer pagurus* L.). *J. Cons. int. Explor. Mer*, **31**, 246–264.

HANDA, N. 1970. Dissolved and particulate carbohydrates. *Symp. Organic Matter in Natural Waters*, Ed. D. W. HOOD, University of Alaska, pp. 129–152.

HANDA, N. 1969. Carbohydrate metabolism in the marine diatom *Skeletonema costatum*. *Mar. Biol.* **4**, 208–214.

HANDA, N. and H. TOMINAGA. 1969. A detailed analysis of carbohydrates in marine particulate matter. *Mar. Biol.* **2**, 228–235.

HANDA, N. and K. YANAGI. 1969. Studies on water extractable carbohydrates of the particulate matter from the northwest Pacific Ocean. *Mar. Biol.* **4**, 197–207.

HARDER, W. 1968. Reactions of plankton organisms to water stratification. *Limnol. Oceanogr.* **13**, 156–168.

HARDING, G. C. H. 1974. The food of deep-sea copepods. *J. Mar. Biol. Ass. U.K.* **54**, 141–155.

HARDY, A. C. 1936. The continuous plankton recorder. *Discovery Rep.* **11**, 457–510.

HARGRAVE, B. T. 1966. Some aspects of phosphorus excretion by planktonic and benthic crustaceans of Bras d'Or Lake, Nova Scotia, MS Thesis, Dalhousie Univ., Halifax. 87 pp.

HARGRAVE, B. T. 1969a. Similarity of oxygen uptake by benthic communities. *Limnol. Oceanogr.* **14**, 801–805.

HARGRAVE, B. T. 1969b. Epibenthic algal production and community respiration in the sediments of Marion Lake. *J. Fish. Res. Bd. Canada*, **26**, 2003–2026.

HARGRAVE, B. T. 1970a. The effect of a deposit-feeding amphipod on the metabolism of benthic microflora. *Limnol. Oceanogr.* **15**, 21–30.

HARGRAVE, B. T. 1970b. The utilization of benthic microflora by *Hyalella azteca* (Amphipoda). *J. Anim. Ecol.* **39**, 427–437.

HARGRAVE, B. T. 1971. An energy budget for a deposit-feeding amphipod. *Limnol. Oceanogr.* **16**, 99–103.

HARGRAVE, B. T. 1972a. Oxidation-reduction potentials, oxygen concentration and oxygen uptake of profundal sediments in a eutrophic lake. *Oikos*, **23**, 167–177.

HARGRAVE, B. T. 1972b. A comparison of sediment oxygen uptake, hypolimnetic oxygen deficit and primary production in Lake Esrom, Denmark. *Verh. Int. Verein. Limnol.* **18**, 134–139.

HARGRAVE, B. T. 1972c. Prediction of egestion by the deposit-feeding amphipod *Hyalella azteca. Oikos*, **23**, 116–124.

HARGRAVE, B. T. 1972d. Aerobic decomposition of sediment and detritus as a function of particle surface area and organic content. *Limnol. Oceanogr.* **17**(4), 583–596.

HARGRAVE, B. T. 1973. Coupling carbon flow through some pelagic and benthic communities. *J. Fish. Res. Bd. Canada*, **30**, 1317–1326.

HARGRAVE, B. T. 1975a. The importance of total and mixed-layer depth in the supply of organic material to bottom communities. *Symp. Biol. Hung.* **15**, 157–165.

HARGRAVE, B. T. 1975b. Stability in structure and function of the mud–water interface. *Verh. Int. Verein. Limnol.* **19**, 1073–1079.

HARGRAVE, B. T. 1976. *The Central Role of Invertebrate Feces in Sediment Decomposition*, Brit. Ecol. Soc. Decomp. Symp., Eds. A. MACFADYEN and J. M. ANDERSON, Blackwell Scientific Publ., Oxford, pp. 301–321.

HARGRAVE, B. T. and G. H. GEEN. 1970. Effects of copepods grazing on two natural phytoplankton populations (*A. tonsa*). *J. Fish. Res. Bd. Canada*, **27**, 1395–1403.

HARGRAVE, B. T. and D. L. PEER. 1973. Comparison of benthic biomass with depth and primary production in some Canadian east coast inshore waters. ICES Manuscript. 1973/E.

HARGRAVE, B. T., G. A. PHILLIPS and S. TAGUCHI. 1976. Sedimentation measurements in Bedford Basin, 1973–1974. *Fish. Mar. Ser. Rept.* **608**, 129 pp.

HARGRAVE, B. T. and G. A. PHILLIPS. 1977. Oxygen uptake by microbial communities on solid surfaces. In *Aquatic Microbial Communities*, Ed. J. CAIRNS, Jr., Garland Publ. Inc., New York (in press).

HARRIS, E. 1959. The nitrogen cycle in Long Island Sound. *Bull. Bingham Oceanogr.* **17**(1), 31–65.

HARRIS, R. P. 1973. Feeding, growth, reproduction and nitrogen utilization by the harpacticoid copepod, *Trigriopus brevicornis. J. Mar. Biol. Ass. U.K.* **53**, 785–800.

HARRISON, M. J., R. T. WRIGHT and R. Y. MORITA. 1971. Method for measuring mineralization in lake sediments. *Appl. Microbiol.* **21**, 698–702.

HARRISON, P. G. and K. H. MANN. 1975. Chemical changes during the seasonal cycle of growth and decay in eelgrass (*Zostera marina*) on the Atlantic coast of Canada. *J. Fish Res. Bd. Canada*, **32**, 615–621.

HARTWIG, E. O. 1976. Nutrient cycling between the water column and a marine sediment. I. Organic carbon. *Mar. Biol.* **34**, 285–295.

HARVEY, G. W. 1966. Microlayer collection from the sea surface: A new method and initial results. *Limnol. Oceanogr.* **11**, 608–613.

HARVEY, H. W. 1957. *The Chemistry and Fertility of Sea Water*, 2nd edn. Cambridge University Press. 234 pp.

HARVEY, P. H., J. S. RYLAND and P. J. HAYWARD. 1976. Pattern analysis in bryozoan and spirorbid communities. II. Distance sampling methods. *J. Exp. Mar. Biol. Ecol.* **21**, 99–108.

HATCH, M. D. and C. R. SLACK. 1970. Photosynthetic CO_2-fixation pathways. *Ann. Rev. Plant Physiol.* **21**, 141–162.

HATTORI, A. 1962a. Light-induced reduction of nitrate, nitrite and hydroxylamine in a blue-green alga, *Anabaena cylindrica. Plant and Cell Physiol.* **3**, 355–369.

HATTORI, A. 1962b. Adaptive formation of nitrate reducing system in *Anabaena cylindrica. Plant and Cell Physiol.* **3**, 371–377.

HATTORI, A. and J. MYERS. 1966. Reduction of nitrate and nitrite by subcellular preparations of *Anabaena cylindrica*. I. Reduction of nitrite to ammonia. *Plant Physiol.* **41**, 1031–1036.

HATTORI, A. and E. WADA. 1971. Nitrite distribution and its regulating processes in the equatorial Pacific Ocean. *Deep-Sea Res.* **18**, 557–568.

HAUG, A. and S. MYKLESTAD. 1973. Studies on the phytoplankton ecology of the Trondheimsfjord. I. The chemical composition of phytoplankton populations. *J. Exp. Mar. Biol. Ecol.* **11**, 15–26.

HAUG, A., S. MYKLESTAD and E. SAKSHAUG. 1973. Studies on the phytoplankton ecology of Trondheimsfjord. I. The chemical composition of phytoplankton populations. *J. Exp. Mar. Biol. Ecol.* **11**, 15–26.

HAXO, F. T. 1960. The wavelength dependence of photosynthesis and the role of accessory pigments. In *Comparative Biochemistry of Photoreactive Systems*, Ed. M. B. ALLEN, Academic Press, New York, pp. 339–360.

HAXO, F. T. and D. C. FORK. 1959. Photosynthetically active accessory pigments of cryptomonads. *Nature*, **184**, 1051–1052.

HAYES, F. R. 1964. The mud-water interface. In: *Oceanogr. Mar. Biol. Ann. Rev.* 2, Ed. H. Barnes. Publ. George Allen and Unwin Ltd., London, pp. 121–145.

HAYES, F. R. and M. A. MACAULAY. 1959. Lake water and sediment. V. Oxygen consumed in water over sediment cores. *Limnol. Oceanogr.* **4**, 291–298.

HECKY, R. E. and P. KILHAM. 1974. Environmental control of phytoplankton cell size. *Limnol. Oceanogr.***19**, 361–366.

HEDGPETH, J. W. 1957. Classification of marine environments. *Mem. Geol. Soc. Amer.* **67**, 17–28.

HEEZEN, B. C., M. EWING and R. J. MENZIES. 1955. The influence of submarine turbidity currents on abyssal productivity. *Oikos*, **6**, 170–182.

HEEZEN, B. C., M. TARP and M. EWING. 1959. The floors of the ocean. I. North Atlantic. *Geol. Soc. Amer. Spec. Paper* **65**, 122 pp.

HEINRICH, A. K. 1962. The life histories of plankton animals and seasonal cycles of plankton communities in the oceans. *J. Cons. Int. Explor. Mer*, **27**, 15–24.

HEIP, C. and P. ENGELS. 1974. Comparing species diversity and evenness indices. *J. Mar. Biol. Ass. U.K.* **54**, 559–563.

HELLEBUST, J. A. 1965. Excretion of some organic compounds by marine phytoplankton. *Limnol. Oceanogr.* **10**, 192–206.

HELLEBUST, J. A. 1970. The uptake and utilisation of organic substances by marine phytoplankters. In *Organic Matter in Natural Waters*, Ed. D. W. HOOD, University of Alaska, pp. 225–256.

HEMMINGSEN, A. M. 1960. Energy metabolism as related to body size and respiratory surfaces and its evolution. *Rep. Steno. Hosp., Copenh.* **9**, 1–110.

HEMPEL, G. 1975. Summing up the symposium on the early life history of fish. In *The Early Life History of Fish*, Ed. J. H. S. BLAXTER, Publ. Springer-Verlag (New York), pp. 755–759.

HEMPEL, G. and H. WEIKERT. 1972. The neuston of the subtropical and boreal North-eastern Atlantic Ocean. A review. *Mar. Biol.* **13**, 70–88.

HERBERT, R. A. 1975. Heterotrophic nitrogen fixation in shallow estuarine sediments. *J. Exp. Mar. Biol. Ecol.* **18**, 215–225.

HERON, A. C. 1973. A specialized predator–prey relationship between the copepod *Sapphirina angusta* and the pelagic tunicate *Thalia democratica*. *J. Mar. Biol. Ass. U.K.* **53**, 429–435.

HERSEY, J. B. and R. H. BACKUS. 1962. Sound scattering by marine organisms. In *The Sea*, Ed. M. N. Hill, Interscience New York, vol. 1, pp. 498–539.

HEWITT, E. J. 1957. Some aspects of micronutrient element metabolism in plants. *Nature*, **180**, 1020–1022.

HEYRAUD, M., S. W. FOWLER, T. M. BEASLEY and R. D. CHEERY. 1976. Polonium-210 in euphausiids: a detailed study. *Mar. Biol.* **34**, 127–136.

HICKEY, J. J. and D. W. ANDERSON. 1968. Chlorinated hydrocarbons and eggshell changes in raptorial and fish-eating birds. *Science*, **162**, 271–273.

HILL, R. and C. P. WHITTINGHAM. 1955. *Photosynthesis*, Methuen's Monographs. John Wiley, New York. 165 pp.

HIMMELMAN, J. H. and T. H. CAREFOOT. 1975. Seasonal changes in calorific value of three Pacific coast seaweeds, and their significance to some marine invertebrate herbivores. *J. Exp. Mar. Biol. Ecol.* **18**, 139–151.

HINCHCLIFFE, P. R. and J. P. RILEY. 1972. The effect of diet and the component fatty acid composition of *Artemia salma*. *J. Mar. Biol. Ass. U.K.* **52**, 203–211.

HJORT, J. 1914. Fluctuations in the great fisheries of northern Europe viewed in the light of biological research. *Rapp. Proc. Verbaux*, **20**, 1–228.

HOAR, W. S. 1966. *General and Comparative Physiology*, Prentice-Hall, Englewood Cliffs, New Jersey. 815 pp.

HOBBIE, J. E., O. HOLM-HANSEN, T. T. PACKARD, L. R. POMEROY, R. W. SHELDON, J. P. THOMAS and W. J. WIEBE. 1972. A study of the distribution and activity of microorganisms in ocean water. *Limnol. Oceanogr.* **17**(4), 544–555.

HOBSON, L. A. 1966. Some influences of the Columbia River effluent on marine phytoplankton during January, 1961. *Limnol. Oceanogr.* **11**, 223–234.

HOBSON, L. A. 1971. Relationships between particulate organic carbon and micro-organisms in upwelling areas off Southwest Africa. *Invest. Pesq.* **35**, 195–208.

HOBSON, L. A. and C. J. LORENZEN. 1972. Relationship of chlorophyll maxima to density structure in the Atlantic Ocean and Gulf of Mexico. *Deep-Sea Res.* **19**, 297–306.

HOCK, C. W. 1940. Decomposition of chitin by marine bacteria. *Biol. Bull.* **79**, 199–206.

HODSON, R. E., O. HOLM-HANSEN and F. AZAM. 1976. Improved methodology for ATP determination in marine environments. *Mar. Biol.*, **34**, 143–149.

HOFMANN, T. and H. LEES. 1952. The biochemistry of the nitrifying organisms. 2. The free-energy efficiency of *Nitrosomonas*. *Biochem. J.* **52**, 140–142.

HOGETSU, K., M. SAKAMOTO and H. SUMIKAWA. 1959. On the high photosynthetic activity of *Skeletonema costatum* under the strong light intensity. *Bot. Mag., Tokyo*, **72**, 421–422.

HOLME, N. A. 1950. Population dispersion in *Tellina tennis* da Costa. *J. Mar. Biol. Ass. U.K.* **29**, 267–280.

HOLME, N. A. and A. D. MCINTYRE. 1971. (Eds.) *Methods for the Study of Marine Benthos*, I.B.P. Handbook, **16**. Blackwell Scientific Publications, Oxford. 334 pp.

HOLMES, R. W. and T. M. WIDRIG. 1956. The enumeration and collection of marine phytoplankton. *J. Cons. Int. Explor. Mer*, **22**, 21–32.

HOLM-HANSEN, O. 1968. Ecology, physiology, and biochemistry of blue-green algae. *Ann. Rev. Microbiol.* **22**, 47–70.

HOLM-HANSEN, O. 1969a. Determination of microbial biomass in ocean profiles. *Limnol. Oceanogr.* **14**, 740–747.

HOLM-HANSEN, O. 1969b. Algae: amounts of DNA and organic carbon in single cells. *Science,* **163,** 87–88.

HOLM-HANSEN, O. 1970a. ATP levels in algal cells as influenced by environmental conditions. *Plant. Cell Physiol.* **11,** 689–700.

HOLM-HANSEN, O. 1970b. Determination of microbial biomass in deep ocean water. In *Organic Matter in Natural Waters,* Ed. D. W. HOOD, Inst. Mar. Sci. Publ. 1, Univ. of Alaska, pp. 287–300.

HOLM-HANSEN, O. and C. R. BOOTH. 1966. The measurement of adenosine triphosphate in the ocean and its ecological significance. *Limnol. Oceanogr.* **11,** 510–519.

HOLM-HANSEN, O., J. D. H. STRICKLAND and P. M. WILLIAMS. 1966. A detailed analysis of biologically important substances in a profile off southern California. *Limnol. Oceanogr.* **11,** 548–561.

HONEYMAN, D. 1874. The ocean (account of a public lecture). *Acadian Recorder,* Sat., Feb. 14.

HUGHES, R. N. 1969. A study of feeding in *Scrobicularia plana. J. Mar. Biol. Ass. U.K.* **49,** 805–823.

HUGHES, R. N. 1970. An energy budget for a tidal flat population of the bivalve *Scrobicularia plaua* Da Costa. *J. Anim. Ecol.* **39,** 357–381.

HUGHES, R. N. 1971a. Ecological energetics of the keyhole limpet *Fissurella barbadensis* Gmelin. *J. Exp. Mar. Biol. Ecol.* **6,** 167–178.

HUGHES, R. N. 1971b. Ecological energetics of *Nerita* (Archaeogastropoda, Neritacea) populations on Barbados, West Indies. *Mar. Biol.* **11,** 12–22.

HUGHES, R. N., D. L. PEER and K. H. MANN. 1972. Use of multivariate analysis to identify functional components of the benthos in St. Margaret's Bay, Nova Scotia. *Limnol. Oceanogr.* **17**(1), 111–121.

HULBERT, E. M. 1970. Competition for nutrients by marine phytoplankton in oceanic, coastal, and estuarine regions. *Ecology,* **51,** 475–484.

HULBERT, E. M., J. H. RYTHER and R. R. L. GUILLARD. 1960. The phytoplankton of the Sargasso Sea off Bermuda. *J. Cons. int. explor. Mer,* **25,** 115–128.

HUMPHREY, G. F. 1975. The photosynthesis: Respiration ratio of some unicellular marine algae. *J. Exp. Mar. Biol. Ecol.* **18,** 111–119.

HUMPHREY, G. F. and D. V. SUBBA RAO. 1967. Photosynthetic rate of the marine diatom *Cylindrotheca closterium. Austral. J. Mar. Freshw. Res.* **18,** 123–127.

HUNDING, C. 1971. Production of benthic microalgae in the littoral zone of a eutrophic lake. *Oikos,* **22,** 389–397.

HUNDING, C. 1973. Diel variation in oxygen production and uptake in a microbenthic littoral community of a nutrient-poor lake. *Oikos,* **24,** 352–360.

HUNDING, C. and B. T. HARGRAVE. 1973. A comparison of benthic microalgal production measured by C^{14} and oxygen methods. *J. Fish. Res. Bd. Canada,* **30,** 309–312.

HURD, L. E., M. V. MELLINGER, L. L. WOLFE and S. J. McNAUGHTON. 1971. Stability and diversity of three trophic levels in terrestrial successional ecosystems. *Science,* **173,** 1134–1136.

HURLBERT, S. H. 1971. The nonconcept of species diversity: a critique and alternative parameters. *Ecology,* **52**(4), 577–586.

HUTCHINSON, G. E. 1938. On the relation between the oxygen deficit and the productivity and typology of lakes. *Int. Rev. ges. Hydrobiol.* **36,** 336–355.

HUTCHINSON, G. E. 1957. *A Treatise on Limnology,* Vol. I, John Wiley & Sons, Inc., New York. 1015 pp.

HUTCHINSON, G. E. 1969. Eutrophication, past and present. In *Eutrophication: Causes, Consequences, Correctives,* Nat. Acad. Sci., Washington, D.C., pp. 17–26.

HUTTON, W. E. and C. E. ZOBELL. 1949. The occurrence and characteristics of methane oxidizing bacteria in marine sediments. *J. Bacteriol.* **58,** 463–473.

HUTTON, W. E. and C. E. ZOBELL. 1953. Production of nitrite from ammonia by methane-oxidizing bacteria. *J. Bact.* **65,** 216–219.

HYLLEBERG, J. 1972. Carbohydrases of some marine invertebrates with notes on their food and on the natural occurrence of the carbohydrates studied. *Mar. Biol.* **14,** 130–142.

HYLLEBERG, J. 1975. Selective feeding by *Abarenicola pacifica* with notes on *Abarenicola vagabunda* and a concept of gardening in lugworms. *Ophelia,* **14,** 113–137.

HYLLEBERG, J. 1976. Resource partitioning on the basis of hydrolytic enzymes in deposit-feeding mud snails (Hydrobiidae). II. Studies on niche overlap. *Oecologia,* **23,** 115–125.

HYLLEBERG, J. and V. GALLUCCI. 1975. Selectivity in feeding by the deposit-feeding bivalve *Macoma nasuta. Mar. Biol.* **32,** 167–178.

HYMAN, L. H. 1951. *The Invertebrates. III. Acanthocephala, Aschelminthes, and Entoprocta,* McGraw-Hill, New York. 572 pp.

HYMAN, L. H. 1959. *The Invertebrates: Smaller Coelomate Groups,* McGraw-Hill, New York. 783 pp.

ICHIMURA, S. 1956a. On the ecological meaning of transparency for the production of matter in phytoplankton community of lake. *Bot. Mag., Tokyo,* **69,** 219–226.

ICHIMURA, S. 1956b. On the standing crop and productive structure of phytoplankton community in some lakes of central Japan. *Bot. Mag., Tokyo,* **69,** 7–16.

ICHIMURA, S. 1960. Diurnal fluctuation of chlorophyll content in lake water. *Bot. Mag., Tokyo,* **73,** 217–224.

ICHIMURA, S. 1967. Environmental gradient and its relation to primary productivity in Tokyo Bay. *Records Oceanogr. Works, Japan,* **9,** 115–128.

ICHIMURA, S. and Y. SAIJO. 1958. On the application of ^{14}C-method to measuring organic matter production in the lake. *Bot. Mag., Tokyo,* **17,** 174–180.

ICHIMURA, S., Y. SAIJO and Y. ARUGA. 1962. Photosynthetic characteristics of marine phytoplankton and their ecological meaning in the chlorophyll method. *Bot. Mag., Tokyo,* **75,** 212–220.

ICHIMURA, S. and Y. ARUGA. 1964. Photosynthetic natures of natural algal communities in Japanese waters. In *Recent Researches in the Fields of Hydrosphere, Atmosphere and Nuclear Geochemistry,* Eds. Y. MIYAKE and T. KOYOMA, Maruzen, Tokyo, pp. 13–37.

ICHIMURA, S., S. NAGASAWA and T. TANAKA. 1968. On the oxygen and chlorophyll maxima found in the metalimnion of a mesotrophic lake. *Bot. Mag., Tokyo,* **81,** 1–10.

IDSO, S. B. and R. G. GILBERT. 1974. On the universality of the Poole and Atkins Secchi disk-light extinction equation. *J. Appl. Ecol.* **11,** 399–401.

IKEDA, T. 1970. Relationship between respiration rate and body size in marine plankton animals as a function of the temperature of habitat. *Bull. Fac. Fish. Hokkaido Univ.* **21,** 91–112.

IKUSHIMA, I. 1967. Ecological studies on the productivity of aquatic plant communities. III. Effect of depth on daily photosynthesis in submerged macrophytes. *Bot. Mag., Tokyo,* **80,** 57–67.

ITOH, K. 1970. A consideration on feeding habits of planktonic copepods in relation to the structure of their oral parts. *Bull. Plankton Soc. Japan,* **17,** 1–10.

IVLEV, V. S. 1944. The time of hunting and the path followed by the predator in relation to the density of the prey population. *Zool. Zh.* **23**(4), 139–145 (translation by L. Birkett, Lowestoft).

IVLEV, V. S. 1945. The biological productivity of waters. *Usp. Sovrem. Biol.* **19,** 98–120.

IVLEV, V. S. 1961. *Experimental Ecology of the Feeding of Fishes.* Translated by D. SCOTT. Yale Univ. Press, New Haven, 302 pp.

IVLEVA, I. V. 1970. The influence of temperature on the transformation of matter in marine invertebrates. In *Marine Food Chains,* Ed. J. H. STEELE, Publ. Oliver & Boyd (Edinburgh), pp 96–112.

JANNASCH, H. W. 1958. Studies on planktonic bacteria by means of a direct membrane filter method. *J. Gen. Microbiol.* **18,** 609–620.

JANNASCH, H. W. 1960. Versuche über denitrifikation und die Verfügbarkeit des Sauerstoffes in Wassen und Schlamm. *Arch. F. Hydrobiol.* **56,** 355–369.

JANNASCH, H. W. 1969. Currrent concepts in aquatic microbiology. *Verh. Int. Verein. Limnol.* **17,** 25–39.

JANNASCH, H. W. 1970. Threshold concentrations of carbon sources limiting bacterial growth in sea water. In *Organic Matter in Natural Waters,* Ed. D. W. HOOD, University of Alaska, pp. 321–328.

JANNASCH, H. W. 1972. New approaches to assessment of microbial activity in polluted waters. In *Water Pollution Microbiology,* Ed. R. MITCHELL, J. Wiley, New York, pp. 291–303.

JANNASCH, H. W., K. EIMHJELLEN, C. O. WIRSEN and A. FARMANFARMAIAN. 1971. Microbial degradation of organic matter in the deep sea. *Science,* **171,** 672–675.

JANNASCH, H. W. and P. H. PRITCHARD. 1972. The role of inert particulate matter in the activity of aquatic microorganisms. *Mem. Ist. Ital. Idiobiol.* **29**(Suppl.), 289–308.

JANNASCH, H. W. and C. O. WIRSEN. 1973. Deep-sea micro-organisms: *in situ* response to nutrient enrichment. *Science,* **180,** 641–643.

JANSSON, B. O. 1967. The availability of oxygen for the interstitial fauna of sandy beaches. *J. Exp. Mar. Biol. Ecol.* **1,** 123–143.

JANSSON, B. O. 1968. Quantitative and experimental studies of the interstitial fauna in four Swedish sandy beaches. *Ophelia,* **5,** 1–71.

JASSBY, A. D. and C. R. GOLDMAN. 1974. Loss rates from a lake phytoplankton community. *Limnol. Oceanogr.* **19,** 618–627.

JASSBY, A. D. and T. PLATT. 1976. Mathematical formulation of the relationship between photosynthesis and light for phytoplankton. *Limnol. Oceanogr.* **21,** 540–547.

JEFFREY, S. W. 1961. Paper chromatographic separation of chlorophylls and carotenoids from marine algae. *Biochem. J.* **80,** 336–342.

JEFFREY, S. W. 1969. Properties of two spectrally different components in chlorophyll *c* preparations. *Biochim. Biophys. Acta,* **177,** 456–467.

JEFFREY, S. W. 1974. Profiles of photosynthetic pigments in the ocean using thin-layer chromatography. *Mar. Biol.* **26,** 101–110.

JEFFREY, S. W. and M. B. ALLEN. 1964. Pigments, growth and photosynthesis in cultures of two chrysomonads, *Coccolithus huxleyi* and a *Hymenomonas* sp. *J. Gen. Microbiol.* **36,** 277–288.

JEFFRIES, H. P. 1962. Environmental characteristics of Raritan Bay, a polluted estuary. *Limnol. Oceanogr.* **7,** 21–31.

JEFFRIES, H. P. 1969. Seasonal composition of temperate plankton communities: free amino acids. *Limnol. Oceanogr.* **14,** 41–52.

JEFFRIES, H. P. 1970. Seasonal composition of temperate plankton communities: fatty acids. *Limnol. Oceanogr.* **15,** 419–426.

JEFFRIES, H. P. 1972. Fatty-acid ecology of a tidal marsh. *Limnol. Oceanogr.* **17,** 433–440.

JENSEN, S. and A. JERNELOV. 1967. Biosyntes av metylkvicksilver. *Biocid information,* **10,** 3–4.

JERLOV, N. G. 1957. Optical studies of ocean waters. *Rep. Sweden Deep Sea Exped.* **3,** 1–59.

JERLOV, N. G. 1968. *Optical Oceanography*, Elsevier, New York. 194 pp.

JITTS, H. R. 1963. The simulated *in situ* measurement of oceanic primary production. *Aust. J. Mar. Freshw. Res.* **14**, 139–147.

JOHANNES, R. E. 1964a. Phosphorus excretion as related to body size in marine animals: microzooplankton and nutrient regeneration. *Science*, **146**, 923–924.

JOHANNES, R. E. 1964b. Uptake and release of phosphorus by a benthic marine amphipod. *Limnol. Oceanogr.* **9**, 235–242.

JOHANNES, R. E. 1965. Influence of marine protozoa on nutrient regeneration. *Limnol. Oceanogr.* **10**, 434–442.

JOHANNES, R. E. 1968. Nutrient regeneration in lakes and oceans. In *Advances in Microbiology of the Sea*. Eds. M. R. DROOP and E. J. FERGUSON WOOD, Academic Press, New York, pp. 203–213.

JOHANNES, R. E. and K. L. WEBB. 1965. Release of dissolved amino acids by marine zooplankton. *Science*, **150**, 76–77.

JOHANNES, R. E., S. L. COLES and N. T. KUENZEL. 1970. The role of zooplankton in the nutrition of some scleractinian corals. *Limnol. Oceanogr.* **15**, 579–586.

JOHANNES, R. E. and M. SATOMI. 1967. Measuring organic matter retained by aquatic invertebrates. *J. Fish. Res. Bd. Canada* **24**, 2467–2471.

JOHANNES, R. E. and K. L. WEBB. 1970. Release of dissolved organic compounds by marine and fresh water invertebrates. In *Organic Matter in Natural Waters*, Ed. D. W. HOOD, University of Alaska, pp. 257–273.

JOHNSON, M. G. 1974. Production and productivity. In *The Benthos of Lakes*, Ed. R. O. BRINKHURST, Macmillan Press, London, pp. 46–64.

JOHNSON, M. G. and R. O. BRINKHURST. 1971a. Production of benthic macroinvertebrates of Bay of Quinte and Lake Ontario. *J. Fish. Res. Bd. Canada*, **28**, 1699–1714.

JOHNSON, M. G. and R. O. BRINKHURST. 1971b. Benthic community metabolism in Bay of Quinte and Lake Ontario. *J. Fish. Res. Bd. Canada*, **28**, 1715–1725.

JOHNSTON, R. 1962. An equation for the depth distribution of deep-sea zooplankton and fishes. *Rapp. Proc.-Verb. Cons. int. Explor. Mer*, **153**, 217–219.

JOHNSTON, R. 1963. Seawater, the natural medium of phytoplankton. I. General features. *J. Mar. Biol. Ass. U.K.* **43**, 427–456.

JOHNSON, R. G. 1972. Conceptual models of benthic communities. In *Models in Paleobiology*. Ed. T. J. M. SCHOPF, Freeman & Cooper, San Francisco, pp. 149–150.

JOHNSON, R. G. 1974. Particulate matter at the sediment–water interface in coastal environments. *J. Mar. Res.* **32**, 313–330.

JONES, D. and M. S. WILLS. 1956. The attenuation of light in sea and estuarine waters in relation to the concentration of suspended solid matter. *J. Mar. Biol. Ass. U.K.* **35**, 431–444.

JONES, G. E. 1963. Suppression of bacterial growth by sea water. In *Symposium on Marine Microbiology*, Ed. C. H. OPPENHEIMER, C. C. Thomas, Springfield, Illinois, pp. 572–579.

JONES, G. E. 1964. Effect of chelating agents on the growth of *Escherichia coli* in sea water. *J. Bacteriol.* **87**, 483–499.

JONES, G. E. 1967. Precipitates from autoclaved sea water. *Limnol. Oceanogr.* **13**, 165–167.

JONES, G. E. 1971. The fate of freshwater bacteria in the sea. *Developments in Indust. Microbiol.* **12**, 141–151.

JONES, J. G. and B. M. SIMON. 1975. Some observations on the fluorometric determination of glucose in freshwater. *Limnol. Oceanogr.* **20**, 882–887.

JONES, K. 1974. Nitrogen fixation in a salt marsh. *J. Ecol.* **62**, 553–565.

JONES, N. S. 1950. Bottom fauna communities. *Biol. Rev.* **25**, 283–313.

JONES, N. S. 1956. The fauna and biomass of a muddy sand deposit off Port Erin, I.O.M. *J. Anim. Ecol.* **25**, 217–252.

JØRGENSEN, B. B. and T. FENCHEL. 1974. The sulfur cycle of a marine sediment model system. *Mar. Biol.* **24**, 189–201.

JØRGENSEN, C. B. 1966. *Biology of Suspension Feeding*, Publ. Pergamon Press, London. 357 pp.

JØRGENSEN, E. G. 1964. Adaptation to different light intensities in the diatom *Cyclostella memeghiniana* Küts. *Physiol. Plant.* **17**, 136–145.

JØRGENSEN, E. G. 1966. Photosynthetic activity during the life cycle of synchronous *Skeletonema* cells. *Physiol. Plant.* **19**, 789–799.

JUMARS, P. A. 1975a. Methods for measurement of community structure in deep-sea macrobenthos. *Mar. Biol.* **30**, 245–252.

JUMARS, P. A. 1975b. Environmental grain and polychaete species' diversity in a bathyal benthic community. *Mar. Biol.* **30**, 253–266.

JUMARS, P. A. 1976. Deep-sea species diversity: does it have a characteristic scale? *J. Mar. Res.* **34**(2), 217–246.

KAJIHARA, M. 1971. Settling velocity and porosity of large suspended particles. *J. Oceanogr. Soc. Japan*, **27**, 158–162.

KAMYKOVSKI, D. 1974. Possible interactions between phytoplankton and semidiurnal internal tides. *J. Mar. Res.* **32**, 67–89.

KANE, J. E. 1967. Organic aggregates in the surface waters of the Ligurian Sea. *Limnol. Oceanogr.* **12**, 287–294.

KANNEWORFF, E. 1965. Life cycle, food, and growth of the amphipod *Ampelisca macrocephala* Liljeborg from the Øresund. *Ophelia*, **2**, 305–318.

KANWISHER, J. W. 1966. Photosynthesis and respiration in some seaweeds. In *Some Contemporary Studies in Marine Science*, Ed. H. BARNES, Allen & Unwin Ltd., London, pp. 407–420.

KAPLAN, I. R., K. O. EMERY and S. C. RITTENBERG. 1963. The distribution and isotopic abundance of sulfur in recent marine sediments of Southern California. *Geochim. Cosmochim. Acta*, **27**, 297–331.

KARL, D. M., J. MORSE, R. SCOTT and L. BUTLER. 1974. ATP and DOC measurements in the mid-Atlantic. *Trans. Amer. Geophys. Union*, **55**, 294.

KARL, D. M. and P. A. LaROCK. 1975. Adenosine triphosphate measurements in soil and marine sediments. *J. Fish. Res. Bd. Canada*, **32**, 599–607.

KARL, D. M., P. A. LaROCK, J. M. MORSE and W. STURGES. 1976. Adenosine triphosphate in the North Atlantic ocean and its relationship to the oxygen minimum. *Deep-Sea Res.* **23**, 81–88.

KATO, K. 1956. Chemical investigations on marine humus in bottom sediments. *Mem. Fac. Fisheries, Hokkaido Univ.* **4**, 91–209.

KAY, D. C. and A. E. BRAFIELD. 1973. The energy relations of the polychaete *Neanthes* (=*Nereis*) *virens* Sars. *J. Anim. Ecol.* **42**, 673–692.

KAYAMA, M. 1964. Fatty acid metabolism of fishes. *Bull. Japanese Soc. Sci. Fish.* **30**, 647–659 (In Japanese).

KERFOOT, W. B. 1970. Bioenergetics of vertical migration. *Amer. Nat.* **104**, 529–546.

KERR, S. R. 1971. A simulation model of lake trout growth. *J. Fish. Res. Bd. Canada*, **28**, 815–819.

KERR, S. R. 1974. Theory of size distribution in ecological communities. *J. Fish. Res. Bd. Canada*, **31**, 1859–1862.

KERR, S. R. and N. V. MARTIN. 1970. Trophic-dynamics of lake trout production systems. In *Marine Food Chains*, Ed. J. H. STEELE, Oliver & Boyd, Edinburgh, pp. 365–376.

KETCHUM, B. H. 1939a. The absorption of phosphate and nitrate by illuminated cultures of *Nitzschia closterium*. *Am. J. Bot.* **26**, 399–407.

KETCHUM, B. H. 1939b. The development and restoration of deficiencies in the phosphorus and nitrogen composition of unicellular plants. *J. Cell. Comp. Physiol.* **13**, 373–381.

KETCHUM, B. H. 1967. Phytoplankton nutrients in estuaries. In *Estuaries*, Ed. G. H. LAUFF, AAAS, Washington, pp. 329–335.

KEVERN, N. R. 1966. Feeding rate of carp estimated by a radio-isotopic method. *Trans. Amer. Fish. Soc.* **95**, 363–371.

KHAYLOR, K. M. and V. YE. YERKOHIN. 1971. Utilization of dissolved organic matter by the crustaceans *Tigriopus brevicornis* and *Calanus finmarchicus*. *Oceanology*, **11**, 95–103.

KHMELEVA, N. N. 1972. Intensity of generative growth in crustaceans. *Dokl. Akad. Nauk SSSR*, **207**, 707–710.

KHMELEVA, N. N. 1967. Role of radiolarians in the estimation of the primary production in the Red Sea and the Gulf of Aden. *Dokl. Akad. Nauk SSSR*, **172**, 1430–1433 (In Russian). *Dokl. (Proc.) Acad. Sci. USSR*, **172**, 70–72 (English Transl.).

KIBBY, H. V. 1971. Effect of temperature on the feeding behaviour of *Daphnia rosea*. *Limnol. Oceanogr.* **16**, 580–581.

KIEFER, D. and J. D. H. STRICKLAND. 1970. A comparative study of photosynthesis in seawater samples incubated under two types of light attenuator. *Limnol. Oceanogr.* **15**, 408–412.

KIMATA, M., H. KADOTA, Y. HATA and T. TAJIMA. 1955. Studies on the marine sulfite-reducing bacteria. I. Distribution of marine sulfate-reducing bacteria in the coastal waters receiving a considerable amount of pulp-mill drainage (Japanese, English summary). *Bull. Japan Soc. Sci. Fisheries*, **21**, 102–108.

KIRCHNER, H. B. 1975. An evaluation of sediment trap methodology. *Limnol. Oceanogr.* **20**, 657–661.

KLEIBER, M. 1961. *The Fire of Life, an Introduction to Animal Energetics*, John Wiley & Sons, Inc., New York, N.Y. 454 pp.

KLEIN, G., E. RACHOR and S. A. GERLACH. 1975. Dynamics and productivity of two populations of the benthic tube-dwelling amphipod *Ampelisca brevicornis* Costa in Helgoland Bight. *Ophelia*, **14**, 139–159.

KNIGHT-JONES, E. W. and J. MOYSE. 1961. Intraspecific competition in sedentary marine animals. *Symp. Soc. Exp. Biol.* **15**, 72–95.

KNOECHEL, R. and J. KALFF. 1976. The applicability of grain density auto-radiography to the quantitative determination of algal species production: a critique. *Limnol. Oceanogr.* **21**, 583–589.

KOBLENTZ-MISHKE, O. J., V. V. VOLKOVISNKY and J. G. KABANOVA. 1970. Plankton primary production of the world ocean. In *Scientific Exploration of the South Pacific*, Standard Book No. 309-01755-6, Nat. Acad. Sci., Wash., pp. 183–193.

KOFOED, L. H. 1975. The feeding biology of *Hydrobia ventrosa* Montagu. II. Allocation of the components of the carbon-budget and the significance of the secretion of dissolved organic material. *J. Exp. Mar. Biol. Ecol.* **19**, 243–256.

KOK, B. 1960. Efficiency of photosynthesis. In *Handbuch der Pflanzenphsiologie*, Ed. W. RUHLAND, Springer Verlag, New York, Vol. 5, Part 1, pp. 563–633.

KOK, B. and G. HOCH. 1961. Spectral changes in photosynthesis. In *A Symposium on Light and Life*, Eds. W. D. McELROY and B. GLASS, The John Hopkins Press, Baltimore, pp. 397–423.

KORINEK, J. 1927. Ein Beitrag zur Mikrobiologie des Meeres. *Zentralbl. Bakteriol.* **71**, 73–79.

KRANCK, K. 1973. Flocculation of suspended sediment in the sea. *Nature,* **246,** 348–350.

KRANCK, K. 1975. Sediment deposition from flocculated suspensions. *Sedimentology,* **22,** 111–123.

KRAUSE, H. R. 1961. Einige Bemerkungen über den postmortalen Abbau von Süsswasser—Zooplankton unter laboratoriums—und Freiland bedingungen. *Arch. Hydrobiol.* **57,** 539–543.

KRAUSE, H. R. 1962. Investigation of the decomposition of organic matter in natural waters. *FAO Fish. Biol. Rep.* No. **34,** 19 pp.

KRAUSE, H. R., L. MOCHEL and M. STEGMANN. 1961. Organische Sauren als geloste Intermediarprodukte des postmortalen Abbaues von Süsswasser. *Zooplankton Naturwissenschaften,* **48,** 434–435.

KRISS, A. E. 1963. *Marine Microbiology (Deep Sea),* translated by J. M. SHEWAN and Z. KABATA, Oliver & Boyd, London. 536 pp.

KRISS, A. E., M. N. LEBEDEVA and I. N. MITZKEVICH. 1960. Microorganisms as indicators of hydrological phenomena in seas and oceans. II. Investigation of the deep circulation of the Indian Ocean using microbiological methods. *Deep-Sea Res.* **6,** 173–183.

KUENEN, P. H. 1950. *Marine Geology,* John Wiley & Sons, New York, 568 pp.

KUENZLER, E. J. 1961a. Structure and energy flow of a mussel population in a Georgia salt marsh. *Limnol. Oceanogr.* **6,** 191–204.

KUENZLER, E. J. 1961b. Phosphorus budget of a mussel population. *Limnol. Oceanogr.* **6,** 400–415.

KUENZLER, E. J. 1965. Glucose-6-phosphate utilization by marine algae. *J. Phycol.* **1,** 156–164.

KUENZLER, E. J. 1970. Dissolved organic phosphorus excretion by marine phytoplankton. *J. Phycol.* **6,** 7–13.

KUENZLER, E. J. and B. H. KETCHUM. 1962. Rate of phosphorus uptake by *Phaeodactylum tricornutum. Biol. Bull.* **123,** 134–145.

KUZNETSOV, S. I. 1958. A study of the size of bacterial populations and of organic matter formation due to the photo- and chemo-synthesis in water bodies of different types. *Verh. Int. Verein. Limnol.* **13,** 156–159.

KUZNETSOV, S. I. 1959. *Die Rolle der Mikroorganismen im Stoffkreislauf der Seen,* VEB Deutscher Verlag der Wissenschafften, Berlin. 301 pp.

KUZNETSOV, S. I. 1968. Recent studies on the role of microorganisms in the cycling of substances in lakes. *Limnol. Oceanogr.* **13,** 211–224.

LACKEY, J. B. 1961. Bottom sampling and environmental niches. *Limnol. Oceanogr.* **6,** 271–279.

LAFOND, E. C. and K. G. LAFOND. 1971. Oceanography and its relation to marine organic production. In *Fertility of the Sea,* Ed. J. D. COSTLOW, Publ. Gordon & Breach (New York), Vol. 1, pp. 241–265.

LALLI, C. M. 1970. Structure and function of the buccal apparatus of *Clione limacina* (Phipps) with a review of feeding in gymnosomatous pteropods. *J. Exp. Mar. Biol. Ecol.* **4,** 101–118.

LALLI, C. M. and R. J. CONOVER. 1973. Reproduction and development of *Paedoclione doliiformis,* and a comparison with *Clione limacina* (Opisthobranchia: Gymnosomata). *Mar. Biol.* **19,** 13–22.

LAM, R. K. and B. W. FROST. 1976. Model of copepod filtering response to changes in size and concentration of food. *Limnol. Oceanogr.* **21,** 490–500.

LAMPORT, W. 1974. A method for determining food selection by zooplankton. *Limnol. Oceanogr.* **19,** 995–997.

LANCE, J. 1962. Effects of water of reduced salinity on the vertical migration of zooplankton. *J. Mar. Biol. Ass. U.K.* **42,** 131–154.

LARMAN, V. N. and P. A. GABBOTT. 1975. Settlement of cyprid larvae of *Balanus balanoides* and *Elminius modestus* induced by extracts of adult barnacles and other marine animals. *J. Mar. Biol. Ass. U.K.* **55,** 183–190.

LARSEN, H. 1962. Halophilism. In *The Bacteria. A Treatise on Structure and Function. IV. The Physiology of Growth,* Eds. I. C. GUNSALUS and R. Y. STANIER, Academic Press, New York, pp. 297–342.

LASENBY, D. C. and R. R. LANGFORD. 1973. Feeding and assimilation of *Mysis relicta. Limnol. Oceanogr.* **18,** 280–285.

LASKER, R. 1960. Utilization of organic carbon by a marine crustacean. Analysis with Carbon-14. *Science,* **131,** 1098–1100.

LASKER, R. 1966. Feeding, growth, respiration and carbon utilization of a euphausiid crustacean. *J. Fish. Res. Bd. Canada,* **23,** 1291–1317.

LASKER, R. 1975. Field criteria for survival of anchovy larvae: the relation between inshore chlorophyll maximum layers and successful first feeding. *Fish. Bull.* **73,** 453–462.

LASKER, R., J. B. J. WELLS and A. D. McINTYRE. 1970. Growth, reproduction, respiration and carbon utilization of the sand-dwelling harpacticoid copepod, *Asellopsis intermedia. J. Mar. Biol. Ass. U.K.* **50,** 147–160.

LAWS, E. A. 1975. The importance of respiration losses in controlling the size distribution of marine phytoplankton. *Ecology,* **56,** 419–426.

LEACH, J. H. 1970. Epibenthic algal production in an intertidal mudflat. *Limnol. Oceanogr.* **15,** 514–521.

LEATHERLAND, T. M., J. D. BURTON, F. CULKIN, M. J. McCARTNEY and R. J. MORRIS. 1973. Concentrations of some trace metals in pelagic organisms and of mercury in Northeast Atlantic Ocean water. *Deep-Sea Res.* **20,** 679–685.

LEBRASSEUR, R. J. 1969. Growth of juvenile chum salmon (*Oncorhynchus keta*) under different feeding regimes. *J. Fish. Res. Bd. Canada,* **26,** 1631–1645.

LeBrasseur, R. J. and J. Fulton. 1967. A guide to zooplankton of the north western Pacific Ocean. *Fish. Res. Bd. Canada, Circular No. 84.* 34 pp.

LeBrasseur, R. J. and O. D. Kennedy. 1972. The fertilization of Great Central Lake. II. Zooplankton standing stock. *Fish. Bull.* **70**, 25–36.

Lee, R. F., J. C. Nevenzel and G. A. Paffenhöfer. 1970. Wax esters in marine copepods. *Science,* **167**, 1510–1511.

Lee, R. F., J. Hirota and A. M. Barnett. 1971. Distribution and importance of wax esters in marine copepods and other zooplankton. *Deep-Sea Res.* **18**, 1147–1165.

Lee, R. F., J. C. Nevenzel and G. A. Paffenhöfer. 1971. Importance of wax esters and other lipids in the marine food chain: phytoplankton and copepods. *Mar. Biol.* **9**, 99–108.

Lehman, J. T. 1976. The filter-feeder as an optimal forager, and the predicted shapes of feeding curves. *Limnol. Oceanogr.* **21**, 501–516.

Lehninger, A. L. 1965. *Bioenergetics. The Molecular Basis of Biological Energy Transformation,* Benjamin, New York. 258 pp.

Leong, R. J. H. and C. P. O'Connell. 1969. A laboratory study of particulate and filter feeding of northern anchovy (*Engraulis mordas*). *J. Fish. Res. Bd. Canada,* **26**, 557–582.

Leppäkoski, E. 1973. Benthic recolonization of the Bornholm Basin (Southern Baltic) in 1969–71. *Thalassia Jugosl.* **7**(1), 171–179.

Levanidov, V. I. 1949. The significance of allochthonous material as food resource in a water-basin and its consumption by *Asellus aquaticus. Trudy uses. gidrobiol. Obslch.* **1**, (in Russian).

Levi, D. and T. Wyatt. 1971. On the dependence of phaeopigment abundance on grazing by herbivores. *Thal. Jugoslavica,* **7**, 181–184.

Levings, C. D. 1973. Dominance and movements of winter flounder (*Pseudopleuronectes americanus*) at St. Margaret's Bay, Nova Scotia. *Naturaliste Can.* **100**, 337–345.

Levings, C. D. 1975. Analyses of temporal variations in the structure of a shallow-water benthic community in Nova Scotia. *Int. Rev. ges. Hydrobiol.* **60**, 449–470.

Levinton, J. 1972a. Spatial distribution of *Nucula proxima* Say (Protobranchia): an experimental approach. *Biol. Bull.* **143**, 175–183.

Levinton, J. 1972b. Stability and trophic structure in deposit-feeding and suspension-feeding communities. *Amer. Nat.* **106**, 472–486.

Levinton, J. S. 1975. Levels of genetic polymorphism at two enzyme encoding loci in eight species of the genus *Macoma* (Mollusca: Bivalvia). *Mar. Biol.* **33**, 41–47.

Lewin, J. C. 1957. Silicon metabolism in diatoms. IV. Growth and frustule formation in *Navicula pelliculosa. Can. J. Microbiol.* **3**, 427–433.

Lewin, J. C., R. A. Lewin and D. E. Philpott. 1958. Observations on *Phaeodactylum tricornutum. J. Gen. Microbiol.* **18**, 418–426.

Lewin, J. C. and R. A. Lewin. 1960. Auxotrophy and heterotrophy in marine littoral diatoms. *Can. J. Microbiol.* **6**, 127–134.

Lewis, A. G. 1967. An enrichment solution for culturing the early developmental stages of the planktonic marine copepod *Euchaeta japonica.* Marukawa. *Limnol. Oceanogr.* **12**, 147–148.

Lewis, A. G., R. Ramnarine and M. S. Evans. 1971. Natural chelators—an indication of activity with the calanoid copepod *Euchaeta japonica. Mar. Biol.* **11**, 1–4.

Lewis, R. W. 1969. The fatty acid composition of Arctic marine phytoplankton and zooplankton with special reference to minor acids. *Limnol. Oceanogr.* **14**, 35–40.

Lie, U. 1968. A quantitative study of benthic infauna in Puget Sound, Washington, U.S.A. in 1963–1964. *Fiskerdirektorat. Skrifter. Ser. Havunders,* **14**, 229–556.

Lie, U. 1969. Standing crop of benthic infauna in Puget Sound and off the coast of Washington. *J. Fish. Res. Bd. Canada,* **26**, 55–62.

Lie, U. and R. A. Evans. 1973. Long-term variability in the structure of subtidal benthic communities in Puget Sound, Washington, U.S.A. *Mar. Biol.* **21**, 122–126.

Liebig, J. 1840. *Chemistry in its Application to Agriculture and Physiology,* Taylor & Walton, London (4th ed., 1847), pp. 352.

Lillelund, K. 1965. Effect of abiotic factors in young stages of marine fish. *ICNAF Spec. Publ.* **6**, 674–686.

Lisivnenko, L. N. 1961. Plankton and the food of larval Baltic herring, in the Gulf of Riga. *Trudy N.–I. Instituta Rybnogo Khoziaistva Soveta Narodnogo Khoziaistva Latviiskoi SSR,* **3**, 105–138 (Fish. Res. Bd. Canada, Trans. No. 444. pp. 36, 1963).

List, R. J. 1958. *Smithsonian Meteorological Tables,* 6th revised edition, Vol. 114, Publication 4014, Smithsonian Institution, Washington, D.C. 527 pp.

Litchfield, C. D. and G. D. Floodgate. 1975. Biochemistry and microbiology of some Irish Sea sediments: II. Bacteriological analyses. *Mar. Biol.* **30**, 97–103.

Lloyd, M. and R. J. Ghellardi. 1964. A table for calculating the 'equitability' component of species diversity. *J. Anim. Ecol.* **33**, 217–225.

Lloyd, M., J. H. Zar and J. R. Karr. 1968. On the calculation of information—theoretical measures of diversity. *Amer. Midland Nat.* **79**, 257–272.

LOEBLICH, A. R., III. 1966. Aspects of the physiology and biochemistry of the Pyrrhophyta. *Phykos, Prof. Iyengar Memorial Volume,* **5,** 216–255.

LONGBOTTOM, M. R. 1970. The distribution of *Arenicola marina* (L) with particular reference to the effects of particle size and organic matter of the sediments. *J. Exp. Mar. Biol. Ecol.* **5,** 138–157.

LONGHURST, A. R., A. D. REITH, R. E. BOWER and D. L. R. SEIBERT. 1966. A new system for the collection of multiple serial plankton samples. *Deep-Sea Res.* **13,** 213–222.

LORD, J. M., G. A. CODD and M. J. MERRITT. 1970. The effect of light quality on glycolate formation and excretion in algae. *Plant Physiol.* **46,** 885–856.

LORENZEN, C. J. 1963. Diurnal variation in photosynthetic activity of natural phytoplankton populations. *Limnol. Oceanogr.* **8,** 56–62.

LORENZEN, C. J. 1965. A note on the chlorophyll and phaeophytin content of the chlorophyll maximum. *Limnol. Oceanogr.* **10,** 482–483.

LORENZEN, C. J. 1972. Extinction of light in the ocean by phytoplankton. *J. Cons. Int. Explor. Mer.* **34,** 262–267.

LORENZEN, C. J. 1976. Primary production in the sea. In *Ecology of the Seas,* Eds. D. H. CUSHING and J. J. WALSH, Blackwell Scientific Publications, Oxford. 467 pp.

LUND, J. W. G. 1959. Buoyancy in relation to the ecology of the freshwater phytoplankton. *Brit. Phycol. Bull.* **7,** 1–17.

LYMAN, J. and R. H. FLEMING. 1940. Composition of sea water. *J. Mar. Res.* **3,** 134–146.

McALLISTER, C. D. 1970. Zooplankton rations, phytoplankton mortality and the estimation of marine production. In *Marine Food Chains,* Ed. J. H. STEELE, Oliver & Boyd, Edinburgh, pp. 419–457.

McALLISTER, C. D., N. SHAH and J. D. H. STRICKLAND. 1964. Marine phytoplankton photosynthesis as a function of light intensity: a comparison of methods. *J. Fish. Res. Bd. Canada,* **21,** 159–181.

McALLISTER, C. D., R. J. LeBRASSEUR and T. R. PARSONS. 1972. Stability of enriched aquatic ecosystems. *Science,* **175,** 562–564.

McCARTHY, J. J. 1970. A urease method for urea in seawater. *Limnol. Oceanogr.* **15,** 309–313.

McCARTHY, J. J. 1970a. The role of urea in marine phytoplankton ecology. Ph.D. thesis. Scripps Institution of Oceanography, La Jolla, Calif. 165 pp.

McCARTHY, J. J. 1972. The uptake of urea by natural populations of marine phytoplankton. *Limnol. Oceanogr.* **17,** 738–748.

McCARTHY, J. J. and R. W. EPPLEY. 1972. A comparison of chemical, isotopic, and enzymatic methods for measuring nitrogen assimilation of marine phytoplankton. *Limnol. Oceanogr.* **17,** 371–382.

McCAVE, I. N. 1975. Vertical flux of particles in the ocean. *Deep-Sea Res.* **22,** 491–502.

McEACHRAN, J. D., D. F. BOESCH and J. A. MUSICK. 1976. Food division within two sympatric species—pairs of skates (Pisces: Rajidae). *Mar. Biol.* **35,** 301–317.

McGOWAN, J. A. 1971. Oceanic biogeography of the Pacific. In *The Micropaleontology of Oceans,* Eds. B. M. FUNNELL and W. R. RIEDEL, Cambridge University Press, pp. 3–74.

McINTYRE, A. and A. W. H. BÉ. 1967. Modern Coccolithophoridae of the Atlantic Ocean. 1. Placoliths and cyrtoliths. *Deep-Sea Res.* **14,** 561–597.

McINTYRE, A. D. 1961. Quantitative differences in the fauna of boreal mud associations. *J. Mar. Biol. Ass. U.K.* **41,** 599–616.

McINTYRE, A. D. 1964. Meiobenthos of sub-littoral muds. *J. Mar. Biol. Ass. U.K.* **44,** 665–674.

McINTYRE, A. D. 1969. Ecology of marine meiobenthos. *Biol. Rev.* **44,** 245–290.

McINTYRE, A. D., A. L. S. MUNRO and J. H. STEELE. 1970. Energy flow in a sand ecosystem. In *Marine Food Chains,* Ed. J. H. STEELE, Oliver & Boyd, Edinburgh, pp. 19–31.

McLAREN, I. A. 1963. Effects of temperature on growth of zooplankton and the adaptive value of vertical migration. *J. Fish. Res. Bd. Canada,* **20,** 685–727.

McLAREN, I. A. 1965. Some relationships between temperature and egg size, body size, development rate and fecundity of the copepod *Pseudocalanus. Limnol. Oceanogr.* **10,** 528–538.

McLAREN, I. A. 1966a. Predicting development rate of copepod eggs. *Biol. Bull.* **131,** 457–469.

McLAREN, I. A. 1966b. Adaptive significance of large size and long life of the chaetognath, *Sagitta elegans* in the Arctic. *Ecology,* **47,** 852–855.

McLAREN, I. A. 1974. Demographic strategy of vertical migration by a marine copepod. *Amer. Nat.* **108,** 91–102.

McLAUGHLIN, J. J. A. and P. A. ZAHL. 1966. Endozoic algae. In *Symbiosis,* Ed. S. M. HENRY, Academic Press, New York, pp. 257–297.

McNAUGHT, D. C. and A. D. HASLER. 1964. Rate of movement of populations of *Daphnia* in relation to changes in light intensity. *J. Fish. Res. Bd. Canada,* **21,** 291–318.

McNEILL, S. and J. H. LAWTON. 1970. Annual production and respiration in animal populations. *Nature,* **225,** 472–474.

McROY, C. P. 1974. Seagrass productivity: carbon uptake experiments in eelgrass, *Zostera marina. Aquaculture,* **4,** 131–137.

MacARTHUR, R. H. 1955. Fluctuations of animal populations and a measure of community stability. *Ecology,* **36,** 533–536.

MacArthur, R. H. and E. O. Wilson. 1967. *The Theory of Island Biogeography*, Princeton Univ. Press, Princeton, N.J. 203 pp.

MacIsaac, J. J. and R. C. Dugdale. 1969. The kinetics of nitrate and ammonia uptake by natural populations of marine phytoplankton. *Deep-Sea Res.* **16**, 45–57.

MacIsaac, J. J. and R. C. Dugdale. 1972. Interactions of light and inorganic nitrogen in controlling nitrogen uptake in the sea. *Deep-Sea Res.* **19**, 209–232.

MacKay, N. J., M. N. Kazacos, R. H. Williams and M. I. Leedow. 1975. Selenium and heavy metals in black marlin. *Mar. Poll. Bull.* **6**, 57–61.

MacKinnon, D. L. and R. S. J. Hawes. 1961. *An Introduction to the Study of Protozoa.* Clarendon Press, Oxford. 506 pp.

MacKinnon, J. C. 1972. Summer storage of energy and its use for winter metabolism and gonad maturation in American plaice (*Hippoglossoides platessoides*). *J. Fish. Res. Bd. Canada*, **29**, 1749–1759.

MacKinnon, J. C. 1973. Analysis of energy flow and production in an unexploited marine flatfish population. *J. Fish. Res. Bd. Canada*, **30**, 1717–1728.

MacLeod, R. A. 1965. The question of the existence of specific marine bacteria. *Bact. Rev.* **29**, 9–23.

Machan, R. and K. Fedra. 1975. A new towed underwater camera system for wide-ranged benthic surveys. *Mar. Biol.* **33**, 75–84.

Makarova, N. P. and V. Ye Zaika. 1971. Relationship between animal growth and quantity of assimilated food. *Hydrobiol. J.* **7**, 1–8.

Malone, T. C. 1971. Diurnal rhythms in netplankton and nannoplankton assimilation ratios. *Mar. Biol.* **10**, 285–289.

Malone, T. C. 1971a. The relative importance of nannoplankton and netplankton as primary producers in the California current system. *Mar. Biol.* **10**, 285–289.

Malone, T. C. 1971b. The relative importance of nannoplankton and netplankton as primary producers in tropical oceanic and neritic phytoplankton communities. *Limnol. Oceanogr.* **16**, 633–639.

Malone, T. C. 1975. Environmental control of phytoplankton cell size. *Limnol. Oceanogr.* **20**, 490.

Maly, E. J. 1969. A laboratory study of the interaction between the predatory rotifer *Aspalanchina* and *Paramecium. Ecology*, **50**, 59–73.

Maly, E. J. 1970. The influence of predation on the adult sex ratios of two copepod species. *Limnol. Oceanogr.* **15**, 566–573.

Mann, K. H. 1965. Energy transformations by a population of fish in the River Thames, *J. Anim. Ecol.* **34**, 253–275.

Mann, K. H. 1969. The dynamics of aquatic ecosystems. *Adv. Ecol. Res.* **6**, 1–81.

Mann, K. H. 1972a. Macrophyte production and detritus food chains in coastal waters. *Mem. 1st. Ital. Idrobiol.* **29**, Suppl., 353–383.

Mann, K. H. 1972b. Ecological energetics of the sea-weed zone in a marine bay on the Atlantic coast of Canada. II. Productivity of the seaweeds. *Mar. Biol.* **14**, 199–209.

Mann, K. H. 1973. Seaweeds: their productivity and strategy for growth. *Science*, **182**, 975–981.

Mann, K. H. and P. A. Breen. 1972. Relations between lobsters, sea urchins and kelp beds. *J. Fish. Res. Bd. Canada*, **29**, 603–609.

Marcotte, B. M. 1974. Two new harpacticoid copepods from the North Adriatic and a revision of the genus *Paramphiacella. Zool. J. Lin. Soc.* **55**(1), 65–82.

Mare, M. F. 1942. A study of a marine benthic community with special references to the micro-organisms. *J. Mar. Biol. Ass. U.K.* **25**. 517–554.

Margalef, D. R. 1951. Diversidad de especies en les communidades naturales. *Publ. Inst. Biol. apl., Barcelona*, **9**, 5–27.

Margalef, D. R. 1957. La teoria de la informacion en ecologia. *Mem. Real. Acad. Ciencias y Artes de Barcelona*, **23**, 373–449. (Translation by Wendell Hall, General Systems, Yearbook, **3**, 36–71, 1957.)

Margalef, D. R. 1958. Temporal succession and spatial heterogeneity in phytoplankton. *Perceptives inMarine Biology*, Ed. A. A. Buzzati-Traverso, University California Press, Berkeley, pp. 323–349.

Margalef, D. R. 1961. Correlations entre certains caractères synthétiques des populations de phytoplankton. *Hydrobiologia*, **18**, 155–164.

Margalef, D. R. 1965. Ecological correlations and the relationship between primary productivity and community structure. In *Primary Productivity in Aquatic Environments*, Ed. C. H. Goldman, University of California Press, Berkeley. *Mem. 1st. Ital. Idrobiol.* **18** Suppl., 355–364.

Margalef, R. 1968. *Perspectives in Ecological Theory*, University of Chicago Press, Chicago, Ill. 111 pp.

Margalef, R. 1969. Diversity and stability: A practical proposal and a model of interdependence. In *Diversity and Stability in Ecological Systems*, Eds. G. M. Woodwell and H. H. Smith, *Brookhaven Symp. Biol.* **22**, 25–37.

Marr, J. W. S. 1962. The natural history and geography of the Antarctic krill (*Euphausia superba* Dana). *Discovery Rep.* **32**, 33–464.

Marshall, N. 1970. Food transfer through the lower trophic levels of the benthic environment. In *Marine Food Chains*, Ed. J. H. Steele, Oliver & Boyd, Edinburgh, pp. 52–66.

Marshall, P. T. 1958. Primary production in the Arctic. *J. Cons. Int. Explor. Mer*, **23**, 173–177.

MARSHALL, S. M. and A. P. ORR. 1948. Further experiments on the fertilization of a sea loch (Loch Criglin). The effect of different plant nutrients on the phytoplankton. *J. Mar. Biol. Ass. U.K.* **27**, 360–379.

MARSHALL, S. M. and A. P. ORR. 1955. *The Biology of a Marine Copepod, Calanus finmarchicus* (Gunnerus), Oliver & Boyd, Edinburgh. 188 pp.

MARTENS, C. S. 1974. A method for measuring dissolved gases in pore water. *Limnol. Oceanogr.* **19**, 525–530.

MARTENS, C. S. and R. A. BERNER. 1974. Methane production in the interstitial waters of sulfate-depleted marine sediments. *Science*, **185**, 1167–1169.

MARTIN, J. H. 1968. Phytoplankton—zooplankton relationships in Narragansett Bay. III. Seasonal changes in zooplankton excretion rates in relation to phytoplankton abundance. *Limnol. Oceanogr.* **13**, 63–71.

MARUMO, R., N. TAGA and T. NAKAI. 1971. Neustonic bacteria and phytoplankton in surface microlayers of the equatorial waters. *Bull. Plankton Soc. Japan*, **18**, 36–41.

MARUMO, R. and O. ASAOKA. 1974. Trichodesmium in the East China Sea. I. Distribution of *Trichodesmium thiebautii* Gomont during 1961–1967. *J. Oceanogr. Soc. Japan*, **30**, 298–303.

MASSÉ, H. 1972. Quantitative investigations of sand bottom macrofauna along the Mediterranean north-west coast. *Mar. Biol.* **15**, 209–220.

MATHIAS, J. A. 1971. Energy flow and secondary production of the amphipods *Hyalella azteca* and *Crangonyx richmondensis occidentalis* in Marion Lake, British Columbia. *J. Fish. Res. Bd. Canada*, **28**, 711–726.

MATHIESON, A. C. and J. S. PRINCE. 1973. Ecology of *Chondrus crispus* Stackhouse. In *Chondrus crispus*, Eds. M. J. HARVEY and J. McLACHLAN, *Proc. N.S. Inst. of Sci.* **27**, 53–80.

MATSUYAMA, M. 1973. Organic substances in sediment and settling matter during spring in a meromictic Lake Suigestau. *J. Oceanogr. Soc. Japan*, **29**, 53–60.

MAUCHLINE, J. 1972. The biology of bathypelagic organisms, especially Crustacea. *Deep-Sea Res.* **19**, 753–780.

MAUCHLINE, J. 1973. The broods of British Mysidacea (Crustacea). *J. Mar. Biol. Ass. U.K.* **53**, 801–817.

MAY, R. C. 1975. Larval mortality in marine fishes and the critical period concept. In *The Early Life History of Fish*, Ed. J. H. S. BLAXTER, Publ. Springer-Verlag (New York), pp. 3–19.

MAYZAUD, P. 1973. Respiration and nitrogen excretion of zooplankton. II. Studies of the metabolic characteristics of starved animals. *Mar. Biol.* **21**, 19–28.

MAYZAUD, P. and J.-L. M. MARTIN. 1975. Some aspects of the biochemical and mineral composition of marine plankton. *J. Exp. Mar. Biol. Ecol.* **17**, 297–310.

MEADOWS, P. S. and J. G. ANDERSON. 1968. Micro-organisms attached to marine sand grains. *J. Mar. Biol. Ass. U.K.* **48**, 161–175.

MEADOWS, P. S. and J. I. CAMPBELL. 1972. Habitat selection by aquatic invertebrates. In *Advances in Marine Biology*, Eds. F. S. RUSSELL and M. YONGE, Academic Press, London and New York, 271–382.

MELAND, S. M. 1962. Marine alginate-decomposing bacteria from north Norway. *Nytt Mag. Bot.* **10**, 53–80.

MENGE, B. A. 1972. Foraging strategy of a starfish in relation to actual prey availability and environmental predictability. *Ecol. Monogr.* **42**, 25–50.

MENSHUTKIN, V. V., M. E. VINOGRADOV and E. A. SHUSHKINA. 1974. Mathematical model of the pelagic ecosystem in the Sea of Japan. *Oceanology*, **14**, 717–723 (English).

MENZEL, D. W. 1964. The distribution of dissolved organic carbon in the Western Indian Ocean. *Deep-Sea Res.* **11**, 757–765.

MENZEL, D. W. 1966. Bubbling of sea water and the production of organic particles: a re-evaluation. *Deep-Sea Res.* **13**, 963–966.

MENZEL, D. W. 1967. Particulate organic carbon in the deep sea. *Deep-Sea Res.* **14**, 229–238.

MENZEL, D. W. 1974. Primary productivity, dissolved and particulate organic matter and the sites of oxidation of organic matter. In *The Sea*, Ed. E. GOLDBERG, Interscience, New York, Vol. 5, pp. 659–678.

MENZEL, D. W. and J. H. RYTHER. 1960. The annual cycle of primary production in the Sargasso Sea off Bermuda. *Deep-Sea Res.* **6**, 351–367.

MENZEL, D. W. and J. H. RYTHER. 1961. Nutrients limiting the production of phytoplankton in the Sargasso Sea, with special reference to iron. *Deep-Sea Res.* **7**, 276–281.

MENZEL, D. W., E. M. HULBERT and J. H. RYTHER. 1963. The effects of enriching Sargasso Sea water on the production and species composition of the phytoplankton. *Deep-Sea Res.* **10**, 209–219.

MENZEL, D. W. and J. H. RYTHER. 1964. The composition of particulate organic matter in the western North Atlantic. *Limnol. Oceanogr.* **9**, 179–186.

MENZEL, D. W. and J. J. GOERING. 1966. The distribution of organic detritus in the ocean. *Limnol. Oceanogr.* **11**, 333–337.

MENZEL, D. W., J. ANDERSON and A. RADTKE. 1970. Marine phytoplankton vary in their response to chlorinated hydrocarbons. *Science*, **167**, 1724–1726.

MENZIES, R. J. and G. T. ROWE. 1968. The distribution and significance of turtle grass, *Thalassia testudinum*, on the deep-sea floor off North Carolina. *Int. Rev. gest. Hydrobiol.* **5**, 217–222.

MENZIES, R. J., R. Y. GEORGE and G. T. ROWE. 1973. *Abyssal Environment and Ecology of the World Oceans*, Wiley-Interscience, New York. 488 pp.

MERRIMAN, D. 1965. Edward Forbes-Manxman. *Prog. Oceanogr.* **3**, 191–206.

MILLER, R. J. and K. H. MANN. 1973. Ecological energetics of the seaweed zone in a marine bay on the Atlantic coast of Canada. III. Energy transformations by sea urchins. *Mar. Biol.* **18**, 99–114.

MILLS, E. L. 1967. The biology of an ampeliscid amphipod crustacean sibling species pair. *J. Fish. Res. Bd. Canada*, **24**, 305–355.

MILLS, E. L. 1969. The community concept in marine zoology, with comments on continua and instability in some marine communities: a review. *J. Fish. Res. Bd. Canada*, **26**, 1415–1428.

MILLS, E. L. 1973. H.M.S. *Challenger, Halifax*, and the Reverend Dr. Honeyman. *Dal. Rev.* **53**, 529–545.

MILLS, E. L. 1975. Benthic organisms and the structure of marine ecosystems. *J. Fish. Res. Bd. Canada*, **32**, 1657–1663.

MINDERMAN, G. 1968. Addition, decomposition and accumulation of organic matter in forests. *J. Ecology*, **56**, 335–362.

MISHIMA, J. and E. P. ODUM. 1963. Excretion rate of Zn^{65} by *Littorina irrorata* in relation to temperature and body size. *Limnol. Oceanogr.* **8**, 39–44.

MITCHELL, R. 1968. Factors affecting the decline of non-marine microorganisms in seawater. *Water. Res.* **2**, 535–543.

MONOD, J. 1949. The growth of bacterial cultures. *Ann. Rev. Microbiol.* **3**, 371–394.

MOORE, H. B. 1938. Algal production and the food requirements of a limpet. *Proc. Malacol. Soc.* **23**, 117–118.

MOORE, J. W. 1976. The proximate and fatty acid composition of some estuarine crustaceans. *Estuar. Coast. Mar. Sci.* **4**, 215–224.

MOORE, L. R. 1969. Geomicrobiology and geomicrobiological attack on sedimented organic matter. In *Organic Geochemistry*, Eds. G. EGLINTON and M. T. J. MURPHY, Springer-Verlag, New York, pp. 265–303.

MOOTZ, C. A. and C. E. EPIFANIO. 1974. An energy budget for *Menippe mercenaria* larvae fed *Artemia* nauplii. *Biol. Bull.* **146**, 44–55.

MORCOS, S. A. 1973. A table for the ionic composition of sea water based on 1967 atomic weights. *J. Cons. Int. Explor. Mer*, **35**, 94–95.

MOREL, A. and R. C. SMITH. 1974. Relation between total quanta and total energy for aquatic photosynthesis. *Limnol Oceanogr.* **19**, 591–600.

MORGAN, E. 1970. The effect of environmental factors on the distribution of the amphipod *Pectenogammarus planicrurus*, with particular reference to grain size. *J. Mar. Biol. Ass. U.K.* **50**, 769–785.

MORIARTY, D. J. W. 1975. A method for estimating the biomass of bacteria in aquatic sediments and its application to trophic studies. *Oecologia*, **20**, 219–229.

MORRIS, I. and P. SYRETT. 1963. The development of nitrate reductase in *Chlorella* and its repression by ammonium. *Arch. Mikrobiol.* **47**, 32–41.

MORRIS, R. J. 1972. The occurrence of wax esters in crustaceans from the Northeast Atlantic Ocean. *Mar. Biol.* **16**, 102–107.

MORSE, J. W. 1974. Calculation of diffusive fluxes across the sediment–water interface. *J. Geophys. Res.* **79**, 5045–5048.

MORTIMER, C. H. 1971. Chemical exchanges between sediments and water in the Great Lakes—speculations on probable regulatory mechanisms. *Limnol. Oceanogr.* **16**, 387–404.

MOSHKINA, L. V. 1961. Photosynthesis by dinoflagellatae of the Black Sea. *Fiziologiya Rastenii*, **8**, 172–177 (in Russian). *Plant. Physiol.* **8**, 129–132. (English translation.)

MOSS, B. 1968. The chlorophyll *a* content of some benthic algal communities. *Arch. Hydrobiol.* **65**, 51–62.

MULLIN, M. M. 1963. Some factors affecting the feeding of marine copepods of the genus *Calanus*. *Limnol. Oceanogr.* **8**, 239–250.

MULLIN, M. M. 1965. Size fractionation of particulate organic carbon in the surface waters of the western Indian Ocean. *Limnol. Oceanogr.* **10**, 459–462.

MULLIN, M. M. and E. R. BROOKS. 1967. Laboratory culture, growth rate, and feeding behavior of a planktonic marine copepod. *Limnol. Oceanogr.* **12**, 657–666.

MULLIN, M. M. and E. R. BROOKS. 1970. Growth and metabolism of two planktonic marine copepods as influenced by temperature and type of food. In *Marine Food Chains.*, Ed. J. H. STEELE, Oliver & Boyd, Edinburgh, pp. 74–95.

MULLIN, M. M., E. F. STEWART and F. J. FUGLISTER. 1975. Ingestion by planktonic grazers as a function of concentration of food. *Limnol. Oceanogr.* **20**, 259–262.

MUNK, W. H. and G. A. RILEY. 1952. Absorbtion of nutrients by aquatic plants. *J. Mar. Res.* **11**, 215–240.

MURRAY, J. 1895. General observations on the distribution of marine organisms. *Rept. Sci. Res. Voy. H.M.S. 'Challenger', Summary Sci. Res.* **2**, 1431–1462.

MURRAY, S. P. 1970. Settling velocities and vertical diffusion of particles in turbulent water. *J. Geophys. Res.* **75**, 1647–1654.

MUUS, B. 1967. The fauna of Danish estuaries and lagoons. *Meddr. Danm. Fisk. –og. Havunders., N.S.*, **5**, 1–316.

MYERS, J. 1953. In *The Metabolism of Algae*, by G. E. FOGG, Methuen, London. 149 pp.

MYKLESTAD, S. 1974. Production of carbohydrates by marine planktonic diatoms. I. Comparison of nine different species in culture. *J. Exp. Mar. Biol. Ecol.* **15**, 261–274.

MYKLESTAD, S., A. HAUG and B. LARSEN. 1972. Production of carbohydrates by the marine diatom *Chaetoceros affinis* var. *willei* (Gran) Hustedt. II. Preliminary investigation of the extracellular polysaccharide. *J. Exp. Mar. Biol. Ecol.* **9**, 137–144.

NAKAJIMA, K. and S. NISHIZAWA. 1968. Seasonal cycles of chlorophyll and seston in the surface water of the Tsugaru Strait area. *Records Oceanogr. Works, Japan,* **9,** 219–246.

NAKAJIMA, K. and S. NISHIZAWA. 1972. Exponential decrease in particulate carbon concentration in a limited depth interval in the surface layer of the Bering Sea. In *Biological Oceanography of the Northern North Pacific Ocean,* Ed. A. Y. TAKENOUTI *et al.,* Idemitsu Shoten, Tokyo, Japan, pp. 495–505.

NAKANISHI, M. and M. MONJI. 1965. Effect of variation in salinity on photosynthesis of phytoplankton growing in estuaries. *J. Fac. Sci., Univ. Tokyo, Sec. III. Botany,* **9,** 19–42.

NASSOGNE, A. 1970. Influence of food organisms on the development and culture of pelagic copepods. *Helgoländer wiss. Meeresunters.* **20,** 333–345.

NEAME, P. A. 1975. Oxygen uptake of sediments in Castle Lake, California. *Verh. Int. Verein. Limnol.* **19,** 792–799.

NEES, J. C. 1949. *A Contribution to Aquatic Population Dynamics,* Ph.D. Thesis, University of Wisconsin, Madison.

NELSON-SMITH, A. 1970. The problem of oil pollution of the sea. *Adv. Mar. Biol.* **8,** 215–306.

NEMOTO, T. 1967. Feeding pattern of euphausiids and differentiations in their body characteristics. *Inf. Bull. Plankton. Japan* (Comm. No. Dr. Y. Matsue), pp. 157–171.

NEMOTO, T. and K. ISHIKAWA. 1969. Organic particulate and aggregate matters stained by histological reagents in the East China Sea. *J. Oceanogr. Soc. Japan,* **25,** 281–290.

NEMOTO, T., K. KAMADA and K. HARA. 1972. Fecundity of a euphausiid crustacean, *Nematoscelis difficilis,* in the North Pacific Ocean. *Mar. Biol.* **14,** 41–47.

NESIS, K. N. 1965. Biocoenoses and biomass of benthos of the Newfoundland–Labrador region. *Trudy VNIRO,* **57,** 453–489 (*Fish. Res. Bd. Canada Trans.* No. 2951).

NEWELL, B. S. 1967. The determination of ammonia in seawater. *J. Mar. Biol. Ass. U.K.* **47,** 271–280.

NEWELL, G. E. and R. C. NEWELL. 1963. *Marine Plankton,* Hutchinson Educational Ltd., London (Revised edition), 1966, pp. 221.

NEWELL, R. 1965. The role of detritus in the nutrition of two marine deposit feeders, the prosobranch *Hydrobia ulvae* and the bivalve *Macoma balthica. Proc. Zool. Soc. London,* **144,** 25–45.

NEWELL, R. C. 1970. *Biology of Intertidal Animals,* Logos Press, London. 555 pp.

NEWELL, R. C. and A. ROY. 1973. A statistical model relating the oxygen consumption of a mollusk (*Littorina littorea*) to activity, body size, and environmental conditions. *Physiol. Zool.* **46,** 253–275.

NEWHOUSE, J., M. S. DOTY and R. T. TSUDA. 1967. Some diurnal features of a neritic surface plankton population. *Limnol. Oceanogr.* **12,** 207–212.

NEWMANN, G. G. and D. E. POLLOCK. 1974. Growth of the rock lobster *Jasus lalandii* and its relationship to benthos. *Mar. Biol.* **24,** 339–346.

NICHOL, J. A. C. 1960. *The Biology of Marine Animals,* Pitman, London, 707 pp.

NICHOLS, F. H. 1975. Dynamics and energetics of three deposit-feeding benthic invertebrate populations in Puget Sound, Washington. *Ecol. Monogr.* **45,** 57–82.

NICOLAISEN, W. and E. KANNEWORFF. 1969. On the burrowing and feeding habits of the amphipods *Bathyporeia pilosa* Lindström and *Bathyporeia sarsi* Watkin. *Ophelia,* **6,** 231–250.

NIHEI, T., T. SASA, S. MIYACHI, K. SUZUKI and H. TAMIYA. 1954. Change of photosynthetic activity of *Chlorella* cells during the course of their normal life cycle. *Archiv. Mikrobiol.* **21,** 156–166.

NISHIZAWA, S. 1969. Suspended material in the sea. II. Re-evaluation of the hypotheses. *Bull. Plankton Soc. Japan,* **16,** 1–42.

NISHIZAWA, S. 1971. Concentration of organic and inorganic material in the surface skin at the equator, 155°W. *Bull. Plankton Soc. Japan,* **18,** 42–44.

NISSENBAUM, A., M. J. BAEDECKER and I. R. KAPLAN. 1972. Studies on dissolved organic matter from interstitial water of a reducing marine fjord. In *Advances in Organic Geochemistry,* 1971, Eds. H. R. V. GAERTNER and H. WEHNER, Pergamon Press, Oxford, pp. 427–440.

NIVAL, P. and S. NIVAL. 1973. Efficacité de filtration des copépodes planctoniques. *Ann. Inst. Oceanogr.* **49,** 135–144.

NIVAL, P. and S. NIVAL. 1976. Particle retention efficiencies of an herbivorous copepod, *Acartia clausi* (adult and copepodite states); effects on grazing. *Limnol. Oceanogr.* **21,** 24–38.

NIXON, S. W., C. A. OVIATT, J. GARBER and V. LEE. 1976. Diel metabolism and nutrient dynamics in a salt marsh embayment. *Ecology,* **57,** 740–750.

NORTH, B. B. 1975. Primary amines in California coastal waters: Utilization by phytoplankton. *Limnol. Oceanogr.* **20,** 20–27.

NORTH, B. B. and G. C. STEPHENS. 1971. Uptake and assimilation of amino acids by *Platymonas.* II. Increased uptake in nitrogen-deficient cells. *Biol. Bull.* **140,** 242–254.

NORTH, B. B. and G. C. STEPHENS. 1972. Amino acid transport in *Nitzschia ovalis* Arnott. *J. Phycol.* **8,** 64–68.

O'CONNELL, C. P. 1972. The interrelation of biting and filtering in the feeding activity of the northern anchovy (*Engraulis mordax*). *J. Fish. Res. Bd. Canada,* **29,** 285–293.

O'CONNELL, C. P. and J. R. ZWEIFEL. 1972. A laboratory study of particulate and filter feeding of the Pacific mackerel *Scomber japonicus. Fish. Bull.* **70,** 973–981.

ODUM, E. P. 1971. *Fundamentals of Ecology*, W. B. Saunders, Philadelphia. 202 pp.

ODUM, E. P. and A. A. DE LA CRUZ. 1963. Detritus as a major component of ecosystems. *A.I.B.S. Bull.* **13**, 39–40.

ODUM, E. P. and A. A. DE LA CRUZ. 1967. Particulate organic detritus in a Georgia salt marsh–estuarine ecosystem. In *Estuaries*, Ed. G. H. LAUFF, Amer. Ass. Adv. Sci. Publ. **83**, Washington, pp. 383–388.

ODUM, H. T. and C. M. HOSKIN. 1958. Comparative studies on the metabolism of marine waters. *Publ. Inst. Mar. Sci. Univ. Tex.* **5**, 16–46.

ODUM, W. E. 1970. Utilization of the direct grazing and plant detritus food chains by the striped mullet *Mugil cephalus*. In *Marine Food Chains*, Ed. J. H. STEELE, Oliver & Boyd, Edinburgh, pp. 222–240.

OGURA, N. 1972a. Decomposition of dissolved organic matter derived from dead phytoplankton. In *Biological Oceanography of the Northern North Pacific Ocean*, Ed. A. Y. TAKENOUTI, Publ. Idemitsu Shoten (Tokyo), pp. 508–515.

OGURA, N. 1972b. Rate and extent of decomposition of dissolved organic matter in surface seawater. *Mar. Biol.* **13**, 89–93.

OHEOCHA, C. and M. RAFTERY. 1959. Phycoerythrins and phycocyanins of cryptomonads. *Nature*, **184**, 1049–1051.

OHLE, W. 1956. Bioactivity, production, and energy utilization of lakes. *Limnol. Oceanogr.* **1**, 139–149.

OLSON, J. S. 1963. Energy storage and the balance of producers and decomposers in ecological systems. *Ecology*, **44**, 322–331.

OLSSON, I. and B. ERIKSSON. 1974. Horizontal distribution of meio-fauna within a small area, with special reference to Foraminifera. *Zoon*, **2**, 67–84.

OMORI, M. 1965. The distribution of zooplankton in the Bering Sea and northern North Pacific, as observed by high-speed sampling of the surface waters, with special reference to the copepods. *J. Oceanogr. Soc. Japan*, **21**, 18–27.

OMORI, M. 1969. Weight and chemical composition of some important oceanic zooplankton in the North Pacific Ocean, *Mar. Biol.* **3**, 4–10.

ORR, W. L., K. O. EMERY and J. R. GRADY. 1958. Preservation of chlorophyll derivatives in sediments off Southern California. *Bull. Amer. Ass. Petrol. Geol.* **42**, 925–962.

OSTERBERG, C., A. G. CAREY and H. CURL. 1963. Acceleration of sinking rates of radionuclides in the ocean. *Nature*, **200**, 1276–1277.

OTSUKI, A. and T. HANYA. 1968. On the production of dissolved nitrogen-rich organic matter. *Limnol. Oceanogr.* **13**, 183–185.

OVIATT, C. A. and S. W. NIXON. 1975. Sediment resuspension and deposition in Narragansett Bay. *Estuar. Coast Mar. Sci.* **3**, 201–217.

OWEN, R. W. and B. ZEIZSCHEL. 1970. Phytoplankton production: seasonal changes in the oceanic eastern tropical Pacific. *Mar. Biol.* **7**, 32–36.

PAASCHE, E. 1962. Coccolith formation. *Nature*, **193**, 1094–1095.

PAASCHE, E. 1966. Action spectrum of coccolith formation. *Physiol. Plant*, **19**, 770–779.

PAASCHE, E. 1973. Silicon and the ecology of marine plankton diatoms. II. Silicate-uptake kinetics in five diatom species. *Mar. Biol.* **19**, 262–269.

PACKARD, T. T. 1971. The measurement of respiratory electron-transport activity in marine phytoplankton. *J. Mar. Res.* **29**, 235–244.

PAFFENHÖFER, G. A. 1970. Cultivation of *Calanus helgolandicus* under controlled conditions. *Helgoländer wiss. Meeresunters*, **20**, 346–359.

PAFFENHÖFER, G. A. and J. D. H. STRICKLAND. 1970. A note on the feeding of *Calanus helgolandicus* on detritus. *Mar. Biol.* **5**, 97–99.

PAINE, R. T. 1965. Natural history, limiting factors and energetics of the opisthobranch *Navanax inermis*. *Ecol.* **46**, 603–619.

PAINE, R. T. 1966. Food web complexity and species diversity. *Amer. Nat.* **100**(910), 65–75.

PAINE, R. T. 1969. A note on trophic complexity and community stability. *Amer. Nat.* **103**(929), 91–93.

PAINE, R. T. 1971a. A short term experimental investigation of resource partitioning in a New Zealand rocky intertidal habitat. *Ecology*, **52**(6), 1096–1106.

PAINE, R. T. 1971b. Energy flow in a natural population of the herbivorous gastropod *Tegula funebralis*. *Limnol. Oceanogr.* **16**, 86–98.

PAINE, R. T. and R. L. VADAS. 1969. The effects of grazing by sea urchins, *Strongylocentrotus spp.*, on benthic algal populations. *Limnol. Oceanogr.* **14**, 710–719.

PAINE, R. T. and R. L. VADAS. 1969b. Calorific values of benthic marine algae and their postulated relation to invertebrate food preference. *Mar. Biol.* **4**, 79–86.

PAINTER, H. A. 1970. A review of literature on inorganic nitrogen metabolism in microorganisms. *Water Res.* **4**, 393–450.

PALOHEIMO, J. E. and L. M. DICKIE. 1965. Food and growth of fishes. I. A growth curve derived from experimental data. *J. Fish. Res. Bd. Canada*, **22**, 521–542.

PALOHEIMO, J. E. and L. M. DICKIE. 1966a. Food and growth of fishes. II. Effects of food and temperature on the relation between metabolism and body weight. *J. Fish. Res. Bd. Canada*, **23**, 869–908.

PALOHEIMO, J. E. and L. M. DICKIE. 1966b. Food and growth of fishes. III. Relation among food, body size and growth efficiency. *J. Fish. Res. Bd. Canada*, **23**, 1209–1248.

PAMATMAT, M. M. 1965. A continuous-flow apparatus for measuring metabolism of benthic communities. *Limnol. Oceanogr.* **10**, 486–489.

PAMATMAT, M. M. 1968. Ecology and metabolism of a benthic community on an intertidal sand flat. *Int. Rev. ges. Hydrobiol.* **53**, 211–298.

PAMATMAT, M. M. 1971. Oxygen consumption by the sea bed. VI. Seasonal cycle of chemical oxidation and respiration in Puget Sound. *Int. Rev. ges. Hydrobiol.* **56**, 769–793.

PAMATMAT, M. M. 1975. *In situ* metabolism of benthic communities. *Cah. Biol. Mar.* **16**, 613–633.

PAMATMAT, M. M. and K. BANSE. 1969. Oxygen consumption by the seabed. II. *In situ* measurements to a depth of 180 m. *Limnol. Oceanogr.* **14**, 250–259.

PAMATMAT, M. M. and A. BHAGWAT. 1973. Anaerobic metabolism in Lake Washington sediments. *Limnol. Oceanogr.* **18**, 611–627.

PAMATMAT, M. M. and H. R. SKJOLDAL. 1974. Dehydrogenase activity and adenosine triphosphate concentration of marine sediments in Lindaspollene, Norway. *Sarsia*, **56**, 1–12.

PANDIAN, T. J. 1975. Mechanisms of heterotrophy. In *Marine Ecology*, II, Ed. O. KINNE, Wiley–Interscience, London, New York, pp. 61–249.

PARK, K. 1967. Nutrient regeneration of preformed nutrients off Oregon. *Limnol. Oceanogr.* **12**, 353–357.

PARKE, M. and P. S. DIXON. 1968. Check-list of British Marine algae–second revision. *J. Mar. Biol. Ass. U.K.* **48**, 783–832.

PARKER, P. L., C. VAN BAALEN and L. MAURER. 1967. Fatty acids in eleven species of blue-green algae: geochemical significance. *Science*, **155**, 707–708.

PARKER, R. A. 1974. Empirical functions relating metabolic processes in aquatic systems to environmental variables. *J. Fish. Res. Bd. Canada*, **31**, 1550–1552.

PARKER, R. H. 1975. *The Study of Benthic Communities: a Model and a Review*, Elsevier Scientif. Publ. Com., Amsterdam. 279 pp.

PARKER, R. R. 1971. Size selective predation among juvenile salmonid fishes in a British Columbia inlet. *J. Fish. Res. Bd. Canada*, **28**, 1503–1510.

PARSONS, T. R. 1961. On the pigment composition of eleven species of marine phytoplankters. *J. Fish. Res. Bd. Canada*, **18**, 1017–1025.

PARSONS, T. R. 1969. The use of particle size spectra in determining the structure of a plankton community. *J. Oceanogr. Soc. Japan*, **25**, 172–181.

PARSONS, T. R. 1976. The structure of life in the sea. In *The Ecology of the Seas*, Eds. D. H. CUSHING and J. J. WALSH, Blackwell Scientific Publications, Oxford, pp. 81–97.

PARSONS, T. R., K. STEPHENS and J. D. H. STRICKLAND. 1961. On the chemical composition of eleven species of marine phytoplankters. *J. Fish. Res. Bd. Canada*, **18**, 1001–1016.

PARSONS, T. R. and J. D. H. STRICKLAND. 1962a. Oceanic detritus. *Science*, **136**, 313–314.

PARSONS, T. R. and J. D. H. STRICKLAND. 1962b. On the production of particulate organic carbon by heterotrophic processes in sea water. *Deep-Sea Res.* **8**, 211–222.

PARSONS, T. R., R. J. LeBRASSEUR and J. D. FULTON. 1967. Some observations on the dependence of zooplankton grazing on the cell size and concentration of phytoplankton blooms. *J. Oceanogr. Soc. Japan*, **23**, 10–17.

PARSONS, T. R., R. J. LeBRASSEUR, J. D. FULTON and O. D. KENNEDY. 1969. Production studies in the Strait of Georgia. Part II. Secondary production under the Fraser River plume, February to May, 1967. *J. Exp. Mar. Biol. Ecol.* **3**, 39–50.

PARSONS, T. R. and R. J. LeBRASSEUR. 1970. The availability of food to different trophic levels in the marine food chain. In *Marine Food Chains*, Ed. J. H. STEELE, Oliver & Boyd, Edinburgh, pp. 325–343.

PARSONS, T. R. and H. SEKI. 1970. Importance and general implications of organic matter in aquatic environments. In *Organic Matter in Natural Waters*, Ed. D. W. HOOD, University of Alaska, pp. 1–27.

PARSONS, T. R., K. STEPHENS and M. TAKAHASHI. 1972. The fertilization of Great Central Lake. I. Effect of primary production. *Fish. Bull.* **70**, 13–23.

PARSONS, T. R. and M. TAKAHASHI. 1973. Environmental control of phytoplankton cell size. *Limnol. Oceanogr.* **18**, 511–515.

PARSONS, T. R., W. K. W. LI and R. WATERS. 1976. Some preliminary observations on the enhancement of phytoplankton growth by low levels of mineral hydrocarbons. *Hydrobiol.* (in press).

PATTEN, B. C. 1959. An introduction to the cybernetics of the ecosystem: the trophic-dynamic aspect. *Ecology*, **40**, 221–231.

PATTEN, B. C. 1961. Preliminary method for estimating stability in plankton. *Science*, **134**, 1010–1011.

PATTEN, B. C. 1962a. Species diversity in net phytoplankton of Raritan Bay. *J. Mar. Res.* **20**, 57–75.

PATTEN, B. C. 1962b. Improved method for estimating stability in plankton. *Limnol. Oceanogr.* **7**, 266–268.

PATTEN, B. C. 1968. Mathematical models of plankton production. *Int. Rev. ges. Hydrobiol.* **53**, 357–408.

PATTON, S., P. T. CHANDLER, E. B. KALAN, A. R. LOEBLICH III, G. FULLER and A. A. BENSON. 1967. Food value of red tide (*Gonyaulax polyedra*). *Science*, **158**, 789–790.

PEARSON, T. H. 1975. The benthic ecology of Loch Linnhe and Loch Eil, a sea-loch system on the west coast of Scotland. IV. Changes in the benthic fauna attributable to organic enrichment. *J. Exp. Mar. biol. Ecol.* **20**, 1–41.

PEASE, A. K. 1976. Studies of the relationship of RNA/DNA ratios and the rate of protein synthesis to growth in the oyster, *Crassostrea virginica*. *Fish. Mar. Ser. Tech. Rpt.* **622**, 78 pp.

PEER, D. L. 1970. Relation between biomass, productivity and loss to predators in a population of a marine benthic polychaete *Pectinaria hypoborea*. *J. Fish. Res. Bd. Canada*, **27**, 2143–2153.

PERKINS, E. J. 1963. Penetration of light into littoral soils. *J. Ecol.* **51**, 687–692.

PETERS, R. H. 1975. Phosphorus excretions and the measurement of feeding and assimilation by zooplankton. *Limnol. Oceanogr.* **20**, 858–859.

PETERSEN, C. G. J. 1913. Valuation of the sea. II. The animal communities of the sea bottom and their importance for marine zoogeography. *Rep. Dan. biol. Stn.* **21**, 1–44.

PETERSEN, R. 1975. The paradox of the plankton: an equilibrium hypothesis. *Amer. Nat.* **109**, 35–49.

PETERSON, C. H. 1975. Stability of species and of community for the benthos of two lagoons. *Ecology*, **56**, 958–965.

PETIPA, T. S. 1966. Relationship between growth, energy metabolism and ration in *Acartia clausi* Giesbr. Physiology of marine animals. Akademiya Nauk SSSR. *Oceanogr. Comm.* 82–91 (translated by M. A. PARANJAPE, Univ. Wash.).

PETIPA, T. S. and N. P. MAKAROVA. 1969. Dependence of phytoplankton production on rhythm and rate of elimination. *Mar. Biol.* **3**, 191–195.

PETIPA, T. S., E. V. PAVLOVA and G. N. MIDONOV. 1970. The food web structure, utilization and transport of energy by trophic levels in the planktonic communities. In *Marine Food Chains*, Ed. J. H. STEELE, Oliver & Boyd, Edinburgh, pp. 142–167.

PICKARD, G. L. 1964. *Descriptive Physical Oceanography*. Pergamon Press, Oxford. 199 pp.

PIELOU, E. C. 1966. The measurement of diversity in different types of biological collections. *J. Theoret. Biol.* **13**, 131–144.

PIELOU, E. C. 1975. *Population and Community Ecology*, Gordon & Breach, New York. 424 pp.

PLATT, T. 1969. The concept of energy efficiency in primary production. *Limnol. Oceanogr.* **14**, 653–659.

PLATT, T. 1971. The annual production by phytoplankton in St. Margaret's Bay, Nova Scotia. *J. Cons. Int. Explor. Mer*, **33**, 324–334.

PLATT, T., V. M. BRAWN and B. IRWIN. 1969. Caloric and carbon equivalents of zooplankton biomass. *J. Fish. Res. Bd. Canada*, **26**, 2345–2349.

PLATT, T., L. M. DICKIE and R. W. TRITES. 1970. Spatial heterogeneity of phytoplankton in a near-shore environment. *J. Fish. Res. Bd. Canada*, **27**, 1453–1473.

PLATT, T., A. PRAKASH and B. IRWIN. 1972. Phytoplankton nutrients and flushing of inlets on the coast of Nova Scotia. *Naturaliste Can.* **99**, 253–261.

PLATT, T. and C. FILION. 1973. Spatial variability of the productivity: biomass ratio for phytoplankton in a small marine basin. *Limnol. Oceanogr.* **18**, 743–749.

PLATT, T. and B. IRWIN. 1973. Caloric content of phytoplankton. *Limnol. Oceanogr.* **18**, 306–310.

PLATT, T., K. L. DENMAN and A. D. JASSBY. 1975. *The Mathematical Representation and Prediction of Phytoplankton Productivity*, Tech. Rep. 523, Fish. Mar. Serv., Environ. Canada. 110 pp.

PLATT, T. and K. L. DENMAN. Spectral analysis in ecology. *Ann. Rev. Ecol. System*, **6**, 37 pp. (in press).

POCKLINGTON, R. 1976. Terrigenous organic matter in surface sediments from the Gulf of St. Lawrence. *J. Fish. Res. Bd. Canada*, **33**, 93–97.

POMERAT, C. M. and C. M. WEISS. 1946. The influence of texture and composition of surface on the attachment of sedentary marine organisms. *Biol. Bull.* **91**, 57–65.

POMEROY, L. R. 1959. Algal productivity in salt marshes of Georgia. *Limnol. Oceanogr.* **4**, 386–397.

POMEROY, L. R. 1970. The strategy of mineral cycling. *Ann. Rev. Ecol. Syst.* **1**, 171–190.

POMEROY, L. R. 1974. The ocean's food web, a changing paradigm. *Bioscience*, **24**, 499–504.

POMEROY, L. R., H. M. MATHEWS and H. S. MIN. 1963. Excretion of phosphate and soluble organic phosphorus compounds by zooplankton. *Limnol. Oceanogr.* **8**, 50–55.

POMEROY, L. R. and R. E. JOHANNES. 1966. Total plankton respiration. *Deep-Sea Res.* **13**, 971–973.

POMEROY, L. R. and R. E. JOHANNES. 1968. Occurrence and respiration of ultraplankton in the upper 500 metres of the ocean. *Deep-Sea Res.* **15**, 381–391.

PORTER, J. R. 1946. *Bacterial Chemistry and Physiology*, Chapman & Hall, London, pp. 1073.

PORTER, K. G. 1973. Selective grazing and differential digestion of algae by zooplankton. *Nature*, **244**, 179–180.

POSTMA, H. 1967. Sediment transport and sedimentation in the estuarine environment. In *Estuaries*, Ed. G. H. LAUFF, Amer. Ass. Adv. Sci. Publ. **83**, 158–179.

POULET, S. A. 1974. Seasonal grazing of *Pseudocalanus minutus* on particles. *Mar. Biol.* **25**, 109–123.

PRAKASH, A. 1971. Terrigenous organic matter and coastal phytoplankton fertility. In *Fertility of the Sea*, Ed. J. D. COSTLOW, Gordon & Breach, New York, Vol. 2, pp. 351–368.

PRAKASH, A. and M. A. RASHID. 1968. Influence of humic substances on the growth of marine phytoplankton: dinoflagellates. *Linmol. Oceanogr.* **13**, 598–606.

PRATT, D. W. 1966. Competition between *Skeletonema costatum* and *Olithodiscus luteus* in Narragansett Bay and in culture. *Limnol. Oceanogr.* **11**, 447–455.

PROVASOLI, L. 1958. Nutrition and ecology of protozoa and algae. *Ann. Rev. Microbiol.* **12**, 279–308.

PROVASOLI, L., J. J. A. McLAUGHLIN and M. R. DROOP. 1957. The development of artificial media for marine algae. *Archiv. Mikrobiol.* **25**, 392–428.

PROVASOLI, L., K. SHIRAISHI and J. R. LANCE. 1959. Nutritional idiosyncrasies of *Artemia* and *Tigriopus* in monoxenic culture. *Ann. New York Acad. Sci.* **77**, 250–261.

PROVASOLI, L. and J. J. A. McLAUGHLIN. 1963. Limited heterotrophy of some photosynthetic dinoflagellates. In *Symposium on Marine Microbiology*, Ed. C. H. OPPENHEIMER, C. C. Thomas, Springfield, Illinois, pp. 105–113.

PSHENIN, L. N. 1963. Distribution and ecology of *Azotobacter* in the Black Sea. In *Symposium on Marine Microbiology*, Ed. C. H. OPPENHEIMER, C. C. Thomas, Springfield, Illinois, pp. 383–391.

PULLIAM, H. R. 1974. On the theory of optimal diets. *Amer. Nat.* **108**, 59–74.

PÜTTER, A. 1909. *Die Ernährung der Wassertiere und der Stoffhaushalt der Gewässer*, Fischer, Jena. 168 pp.

RABINOWITCH, E. I. 1951. *Photosynthesis and Related Processes*, Vol. II, Part I, Interscience, New York, pp. 1211–2088.

RACHOR, E. and S. A. GERLACH. 1975. Variations in macrobenthos in the German Bight. Symp. on The Changes in the North Sea Fish Stocks and their Causes. *ICES*, No. 11. 16 pp.

RAE, K. M. 1957. A relationship between wind, plankton distribution and haddock brood strength. *Bull. Mar. Ecol.* **4**, 347–369.

RAYMONT, J. E. G. and R. J. CONOVER. 1961. Further investigations on the carbohydrate content of marine zooplankton. *Limnol. Oceanogr.* **6**, 154–164.

RAYMONT, J. E. G., R. T. SRINIVASAGAM and J. K. B. RAYMONT. 1969a. Biochemical studies on marine zooplankton. VII. Observations on certain deep sea zooplankton. *Int. Rev. ges. Hydrobiol.* **54**, 357–365.

RAYMONT, J. E. G., R. T. SRINIVASAGAM and J. K. B. RAYMONT. 1969b. Biochemical studies on marine zooplankton. VI. Investigations on *Meganyctiphanes norvegica* (M. Sars). *Deep-Sea Res.* **16**, 141–156.

RAYMONT, J. E. G., C. F. FERGUSON and J. K. B. RAYMONT. 1973. Biochemical studies on marine zooplankton. XI. The amino acid composition of some local species. *Spl. Publ. Mar. Biol. Ass. India*, pp. 91–99.

REDFIELD, A. C. 1934. On the proportions of organic derivatives in sea water and their relation to the composition of plankton. *James Johnstone Memorial Volume (Liverpool)*, pp. 176.

REDFIELD, A. C. 1942. The processes determining the concentration of oxygen, phosphate and other organic derivatives within the depths of the Atlantic Ocean. *Pap. Phys. Oceanogr. Met.* **9**, 1–22.

REDFIELD, A. C. 1955. The hydrography of the Gulf of Venezuela. Papers Marine Biol. Oceanogry. *Deep-Sea Res. Suppl.* **3**, 115–133.

REDFIELD, A. C. 1958. The biological control of chemical factors in the environment. *Amer. Sci.* **46**, 205–221.

REEBURGH, W. S. 1969. Observations of gases in Chesapeake Bay sediments. *Limnol. Oceanogr.* **14**, 368–375.

REEVE, M. R. 1963. The filter-feeding of *Artemia*. I. In pure cultures of plant cells. *J. Exptl. Biol.* **40**, 195–205.

REEVE, M. R. 1964. Feeding of zooplankton, with special reference to some experiments with *Sagitta*. *Nature*, **201**, 211–213.

REEVE, M. R. and M. A. WALTER. 1972. Conditions of culture, food-size selection, and the effects of temperature and salinity on growth rate and generation time in *Sagitta hispida* Conant. *J. Exp. Mar. Biol. Ecol.* **9**, 191–200.

REEVE, M. R. and T. C. COSPER. 1975. Chaetognatha. In *Reproduction of Marine Invertebrates*, Ed. A. C. GIESE and J. S. PEARSE, Publ. Academic Press (New York), pp. 157–184.

REINECK, H.-E. 1968. Die Sturmflutlagen. In *Sedimentologie, Faunenzonierung und Faziesabfolge vor der Ostküste der inneren Deutschen Bucht*, Ed. H.-E. REINECK, Senckenberg. leth. **49**, 270–272.

REMSEN, C. C. 1971. The distribution of urea in coastal and oceanic waters. *Limnol. Oceanogr.* **16**, 732–740.

Report of the Committee on Terms and Equivalents. 1958. *Rapp. Proces-Verbaux. Reunions*, **144**, 15–16.

RHOADS, D. C. 1973. The influence of deposit-feeding benthos on water turbidity and nutrient recycling. *Amer. J. Sci.* **273**, 1–22.

RHOADS, D. C. 1974. Organism–sediment relations on the muddy sea floor. *Oceanogr. Mar. Biol. Ann. Rev.* **12**, 263–300.

RHOADS, D. C. and D. K. YOUNG. 1970. The influence of deposit-feeding organisms on sediment stability and community trophic structure. *J. Mar. Res.* **28**(2), 150–178.

RHOADS, D. C. and D. K. YOUNG. 1971. Animal–sediment relations in Cape Cod Bay, Massachusetts. II. Reworking by *Malpedia oolitica* (Holothuroidea). *Mar. Biol.* **11**, 255–261.

RICHARDS, F. A. and A. C. REDFIELD. 1954. A correlation between the oxygen content of sea water and the organic content of marine sediments. *Deep-Sea Res.* **1**, 279–281.

RICHARDS, S. W. and G. A. RILEY. 1967. The benthic epifauna of Long Island Sound. *Bull. Bingham Oceanogr. Coll.* **19**, 89–135.

RICHERSON, P., R. ARMSTRONG and C. R. GOLDMAN. 1970. Contemporaneous disequilibrium, a new hypothesis to explain the 'Paradox of the Plankton'. *Proc. Nat. Acad. Sci.* **67**, 1710–1714.

RICHMAN, S. 1958. The transformation of energy by *Daphnia pulex*. *Ecol. Monogr.* **28**, 273–291.

RICHMAN, S. and J. N. ROGERS. 1969. The feeding of *Calanus helgolandicus* on synchronously growing populations of the marine diatom *Ditylium brightwelli*. *Limnol. Oceanogr.* **14**, 701–709.

RICKER, W. E. 1937. Statistical treatment of sampling processes useful in the enumeration of plankton organisms. *Arch. Hydrobiol.* (*Plankt.*), **31**, 68–84.

RICKER, W. E. 1958. Handbook of computations for biological statistics of fish populations. *Fish. Res. Bd. Bull.* **119**, pp. 300.

RICKER, W. E. 1968. Food from the sea. In *Resources and Man*, Publ. Nat. Aca. Sci.-Nat. Res. Council, W. H. Freeman Co., San Francisco, pp. 87–108.

RICKER, W. E. 1971. Ed. *Methods for Assessment of Fish Production in Fresh Waters*, I.B.P. Handbook, **3**. Oxford and Edinburgh: Blackwell Scientific Publications. 326 pp.

RICKETTS, T. R. 1966a. On the chemical composition of some unicellular algae. *Phytochem.* **5**, 67–76.

RICKETTS, T. R. 1966b. The carotenoids of the phytoflagellate, *Micromonas pusilla*. *Phytochem.* **5**, 571–580.

RICKETTS, T. R. 1967a. The pigment composition of some flagellates possessing scaly flagella. *Phytochem.* **6**, 669–676.

RICKETTS, T. R. 1967b. Further investigations into the pigment composition of green flagellates possessing scaly flagella. *Phytochem.* **6**, 1375–1386.

RICKETTS, T. R. 1970. The pigments of the Prasinophyceae and related organisms. *Phytochem.* **9**, 1835–1842.

RIEDL, R. J., N. HUANG and R. MACHAN. 1972. The subtidal pump, a mechanism of intertidal water exchange by wave action. *Mar. Biol.* **13**(3), 210–221.

RIEDL, R. J. and R. MACHAN. 1972. Hydrodynamic patterns in lotic intertidal sands and their bioclimatological implications. *Mar. Biol.* **13**(3), 179–209.

RIGLER, F. H. 1971. Feeding rates. Zooplankton. In *A Manual on Methods for the Assessment of Secondary Productivity in Fresh Waters*, Eds. W. T. EDMONDSON and G. G. WINBERG, Blackwell Scientific, Edinburgh, pp. 228–256.

RILEY, G. A. 1946. Factors controlling phytoplankton populations on Georges Bank. *J. Mar. Res.* **6**, 54–73.

RILEY, G. A. 1947. A theoretical analysis of the zooplankton population of Georges Bank. *J. Mar. Res.* **6**, 104–113.

RILEY, G. A. 1951. Oxygen, phosphate and nitrate in the Atlantic Ocean. *Bull. Bingham Oceanogr. Coll.* **13**, 1–126.

RILEY, G. A. 1956a. Oceanography of Long Island Sound, 1952–54. II. Physical Oceanography. *Bull. Bingham Oceanogr. Coll.* **15**, 15–46.

RILEY, G. A. 1956b. Oceanography of Long Island Sound, 1952–54. IX. Production and utilization of organic matter. *Bull. Bingham Oceanogr. Coll.* **15**, 324–343.

RILEY, G. A. 1963. Organic aggregates in sea water and the dynamics of their formation and utilization. *Limnol. Oceanogr.* **8**, 372–381.

RILEY, G. A. 1970. Particulate and organic matter in sea water. *Adv. Mar. Biol.* **8**, 1–118.

RILEY, G. A. 1975. Transparency–chlorophyll relations. *Limnol. Oceanogr.* **20**, 150–152.

RILEY, G. A., H. STOMMEL and D. A. BUMPUS. 1949. Quantitative ecology of the plankton of the western North Atlantic. *Bull. Bingham Oceanogr. Coll.* **12**, 1–169.

RILEY, J. P. and T. R. S. WILSON. 1967. The pigments of some marine phytoplankton species. *J. Mar. Biol. Ass. U.K.* **47**, 351–362.

RILEY, J. P. and D. A. SEGAR. 1969. The pigments of some further marine phytoplankton species. *J. Mar. Biol. Ass. U.K.* **49**, 1047–1056.

RILEY, J. P. and D. A. SEGAR. 1970. The seasonal variation of the free and combined dissolved amino acids in the Irish Sea. *J. Mar. Biol. Ass. U.K.* **50**, 713–720.

RILEY, J. P. and I. ROTH. 1971. The distribution of trace elements in some species of phytoplankton grown in culture. *J. Mar. Biol. Ass. U.K.* **51**, 63–72.

RILEY, J. P. and R. CHESTER. 1971. *Introduction to Marine Chemistry*, Academic Press, London. 465 pp.

RILEY, J. P. and G. SKIRROW. 1975. *Chemical Oceanography*, Vol. 1–4, Academic Press, New York.

RISEBROUGH, R. W., D. W. MENZEL, D. J. MARTIN and H. S. OLCOTT. 1967. DDT residues in Pacific sea birds: a persistent insecticide in marine food chains. *Nature*, **216**, 589–591.

RITTENBERG, S. C., K. O. EMERY and W. L. ORR. 1955. Regeneration of nutrients in sediments of marine basins. *Deep-Sea Res.* **3**, 23–45.

RIZNYK, R. Z. and H. K. PHINNEY. 1972. Manometric assessment of intertidal microalgal production in two estuarine sediments. *Oecologia*, **10**, 193–203.

ROKOP, F. J. 1974. Reproductive patterns in the deep-sea benthos. *Science*, **186**, 743–745.

ROMANENKO, W. I. 1964a. Potential capacity of the microflora of sludge sediments for heterotrophic assimilation of carbon dioxide and for chemosynthesis. *Mikrobiologiya*, **33**, 134–139 (in Russian).

ROMANENKO, W. I. 1964b. Heterotrophic assimilation of CO_2 by the aquatic microflora. *Mikrobiologiya*, **33**, 679–683 (in Russian).

ROMANENKO, W. I. 1964c. The relationship between the amounts of the O_2 and CO_2 required by heterotrophic bacteria. *Dokl. Akad. Nauk. Sci. SSSR*, **157**, 178–179 (in Russian). *Dokl.* (*Proc.*) *Acad. Sci. USSR*, **157**, 562–563 (English transl.).

ROSENBERG, R. 1973. Succession in benthic macrofauna in a Swedish fjord subsequent to the closure of a sulphite pulp mill. *Oikos*, **24**, 1–16.

ROSENBERG, R. 1974. Spatial dispersion of an estuarine benthic faunal community. *J. Exp. Mar. Biol. Ecol.* **15**, 69–80.

ROSENTHAL, H. and G. HEMPEL. 1970. Experimental studies in feeding and food requirements of herring larvae (*Clupea harengus* L.). In *Marine Food Chains*, Ed. J. H. STEELE, Oliver & Boyd, Edinburgh, pp. 344–364.

ROSENZWEIG, M. L. and R. H. MACARTHUR. 1963. Graphical representation and stability conditions of predator-prey interactions. *Amer. Nat.* **97**, 209–223.

ROWE, G. T. 1971a. Observations on bottom currents and epibenthic populations in Hatteras submarine canyon. *Deep-Sea Res.* **18**, 569–581.

ROWE, G. T. 1971b. Benthic biomass and surface productivity. In *Fertility of the Sea*, Ed. J. D. COSTLOW, Gordon & Breach, New York, pp. 441–454.

ROWE, G. T. and R. J. MENZIES. 1969. Zonation of large benthic invertebrates in the deep sea off the Carolinas. *Deep-Sea Res.* **16**, 531–537.

ROWE, G. T., G. KELLER, H. EDGERTON, N. STARESINIC and J. MACILVAINE. 1974a. Time-lapse photography of the biological reworking of sediments in Hudson submarine canyon. *J. Sed. Petrology*, **44**(2), 549–552.

ROWE, G. T., P. T. PALLONI and S. G. HORNER. 1974b. Benthic biomass estimates from the northwestern Atlantic Ocean and the Northern Gulf of Mexico. *Deep-Sea Res.* **21**, 641–650.

ROWE, G. T., C. H. CLIFFORD, K. L. SMITH, Jr. and P. C. HAMILTON. 1975a. Benthic nutrient regeneration and its coupling to primary productivity in coastal waters. *Nature*, **255**, 215–217.

ROWE, G. T., P. T. POLLONI and R. L. HAEDRICH. 1975b. Quantitative biological assessment of the benthic fauna in deep basins of the Gulf of Maine. *J. Fish. Res. Bd. Canada*, **32**, 1805–1812.

RUSSEL, F. S., A. J. SOUTHWARD, G. T. BOALCH and E. I. BUTLER. 1971. Changes in biological conditions in the English Channel off Plymouth during the last half century. *Nature*, **234**, 468–470.

RYTHER, J. H. 1954. The ratio of photosynthesis to respiration in marine plankton algae and its effect upon the measurement of productivity. *Deep-Sea Res.* **21**, 134–139.

RYTHER, J. H. 1956. Photosynthesis in the ocean as a function of light intensity. *Limnol. Oceanogr.* **1**, 61–70.

RYTHER, J. H. 1963. IV. Biological Oceanography. 17. Geographic variations in productivity. In *The Sea*, Ed. M. N. HILL, Interscience Publishers, New York, Vol. 2, pp. 347–380.

RYTHER, J. H. 1965. The measurement of primary production. *Limnol. Oceanogr.* **1**, 72–84.

RYTHER, J. H. 1969. Photosynthesis and fish production in the sea. The production of organic matter and its conversion to higher forms of life vary throughout the world ocean. *Science*, **166**, 72–76.

RYTHER, J. H. and D. D. KRAMER. 1961. Relative iron requirement of some coastal and off-shore plankton algae. *Ecology*, **42**, 444–446.

RYTHER, J. H. and R. R. L. GUILLARD. 1962. Studies of marine planktonic diatoms. III. Some effects of temperature on respiration of five species. *Can. J. Microbiol.* **8**, 447–453.

RYTHER, J. H. and D. W. MENZEL. 1965. On the production, composition, and distribution of organic matter in the West Arabian Sea. *Deep-Sea.* **12**, 199–209.

SAIJO, Y. 1969. Chlorophyll pigments in the deep sea. *Bull. Japanese Soc. Fish. Oceanogr.* Special No. (Prof. Uda's Commemorative Papers), pp. 179–182.

SAIJO, Y. and S. ICHIMURA. 1962. Some considerations on photosynthesis of phytoplankton from the point of view of productivity measurement. *J. Oceanogr. Soc. Japan, 20th Anniv. Vol.*, pp. 687–693.

SAIJO, Y. and K. TAKESUE. 1965. Further studies on the size distribution of photosynthesizing phytoplankton in the Indian Ocean. *J. Oceanogr. Soc. Japan*, **20**, 10–17.

SAILA, S. B., R. A. PIBANOWSKI and D. S. VAUGHAN. 1976. Optimum allocation strategies for sampling benthos in the New York Bight. *Estuar. Coast. Mar. Sci.* **4**, 119–128.

SAKAMOTO, M. 1966. The chlorophyll amount in the euphotic zone in some Japanese lakes and its significance in the photosynthetic production of phytoplankton community. *Bot. Mag., Tokyo*, **79**, 77–88.

SANDERS, H. L. 1956. Oceanography of Long Island Sound, 1952–1954. X. The biology of marine bottom communities. *Bull. Bingham Ocean. Coll.* **15**, 345–414.

SANDERS, H. L. 1958. Benthic studies in Buzzards Bay. I. Animal–sediment relationships. *Limnol. Oceanogr.* **3**(3), 245–258.

SANDERS, H. L. 1960. Benthic studies in Buzzards Bay. III. The structure of the soft-bottom community. *Limnol. Oceanogr.* **5**, 138–153.

SANDERS, H. L. 1968. Marine benthic diversity: a comparative study. *Amer. Nat.* **102**, 243–282.

SANDERS, H. L., R. R. HESSLER and G. R. HAMPSON. 1965. An introduction to the study of the deep-sea benthic faunal assemblages along the Gay–Head–Bermuda transect. *Deep-Sea Res.* **12**, 845–867.

SANDERS, H. L. and R. R. HESSLER. 1969. Ecology of the deep-sea benthos. *Science*, **163**, 1419–1424.

SATOMI, M. and L. R. POMEROY. 1965. Respiration and phosphorus excretion in some marine populations. *Ecology*, **46**, 877–881.

SCHAEFER, M. B. 1965. The potential harvest of the sea. *Trans. Amer. Fish. Soc.* **94**, 123–128.

SCHAFER, C. T. 1971. Sampling and spatial distribution of benthonic foraminifera. *Limnol. Oceanogr.* **16**, 944–951.

SCHELTEMA, R. S. 1971. Larval dispersal as a means of genetic exchange between geographically separated populations of shallow water benthic marine gastropods. *Biol. Bull.* **140**, 284–322.

SCHOENER, A. and G. T. ROWE. 1970. Pelagic *Sargassum* and its presence among the deep-sea benthos. *Deep-Sea Res.* **17**, 923–925.

SCHOENER, T. W. 1971. Theory of feeding strategies. *Ann. Rev. Ecol. Syst.* **2**, 369–404.

SEIWELL, H. R. and G. E. SEIWELL. 1938. The sinking of decomposing plankton in sea water and its relationship to oxygen consumption and phosphorus liberation. *Proc. Amer. Phil. Soc.* **78**, 465–481.

SEKI, H. 1965a. Microbial studies on the decomposition of chitin in the marine environment. IX. Rough estimation of chitin decomposition in the ocean. *J. Oceanogr. Soc. Japan*, **21**, 253–260.

SEKI, H. 1965b. Decomposition of chitin in marine sediments. *J. Oceanogr. Soc. Japan*, **21**, 261–268.

SEKI, H. 1968. Relation between production and mineralization of organic matter in Aburatsubo Inlet, Japan. *J. Fish. Res. Bd. Canada*, **25**, 625–637.

SEKI, H. 1972. Formation of Anoxic Zones in Seawater. In *Biological Oceanography of the Northern North Pacific Ocean*, Eds. Y. TAKENOUCHI *et al.*, Idemitsu Shoten, Tokyo, pp. 487–493.

SEKI, H., J. SKELDING and T. R. PARSONS. 1968. Observations on the decomposition of a marine sediment. *Limnol. Oceanogr.* **13**, 440–447.

SEKI, H., K. V. STEPHENS and T. R. PARSONS. 1969. The contribution of allochthonous bacteria and organic materials from a small river into a semi-enclosed sea. *Arch. Hydrobiol.* **66**, 37–47.

SEKI, H., T. TSUJI and A. HATTORI. 1974. Effect of zooplankton grazing on the formation of the anoxic layer in Tokyo Bay. *Estuar. Coast. Mar. Sci.* **2**, 145–151.

SEMINA, H. J. 1972. The size of phytoplankton cells in the Pacific Ocean. *Int. Rev. Ges. Hydrobiol.* **57**, 177–205.

SHANNON, C. E. and W. WEAVER. 1963. *The Mathematical Theory of Communication*, University of Illinois Press, Urbana. 125 pp.

SHARP, J. H. 1973. Size classes of organic carbon in seawater. *Limnol. Oceanogr.* **18**, 441–447.

SHELBOURNE, J. E. 1957. The feeding and condition of plaice larvae in good and bad plankton patches. *J. Mar. Biol. Ass. U.K.* **36**, 539–552.

SHELDON, R. W. 1969. A universal grade scale for particulate materials. *Proc. Geol. Soc. Lond.* **1659**, 293–295.

SHELDON, R. W. and P. J. WARREN. 1966. Transport of sediments by crustaceans and fish. *Nature*, **210**, 1171–1172.

SHELDON, R. W. and T. R. PARSONS. 1967a. *A Practical Manual on the use of the Coulter Counter in Marine Science*, Coulter Electronics Sales Co., Canada, Toronto. 66 pp.

SHELDON, R. W. and T. R. PARSONS. 1967b. A continuous size spectrum for particulate matter in the sea. *J. Fish. Res. Bd. Canada*, **24**, 909–915.

SHELDON, R. W., T. P. T. EVELYN and T. R. PARSONS. 1967. On the occurrence and formation of small particles in sea water. *Limnol. Oceanogr.* **12**, 367–375.

SHELDON, R. W. and W. H. SUTCLIFFE, JR. 1969. Retention of marine particles by screens and filters. *Limnol. Oceanogr.* **14**, 441–444.

SHELDON, R. W., A. PRAKASH and W. H. SUTCLIFFE, JR. 1972. The size distribution of particles in the ocean. *Limnol. Oceanogr.* **17**(3), 327–340.

SHELDON, R. W., W. H. SUTCLIFFE, JR. and A. PRAKASH. 1973. The production of particles in the surface waters of the ocean with particular reference to the Sargasso Sea. *Limnol. Oceanogr.* **18**, 719–733.

SHIMADA, B. M. 1958. Diurnal fluctuations in photosynthetic rate and chlorophyll *a* content of phytoplankton from eastern Pacific waters. *Limnol. Oceanogr.* **3**, 336–339.

SHOLKOVITZ, E. and A. SOUTAR. 1975. Changes in the composition of the bottom water of the Santa Barbara Basin: effect of turbidity currents. *Deep-Sea Res.* **22**, 13–21.

SIEBURTH, J. McN. 1960. Acrylic acid, an 'antibiotic' principle in *Phaeocystis* blooms in Antarctic waters. *Science*, **132**, 676–677.

SIEBURTH, J. McN. 1961. Antibiotic properties of acrylic acid, a factor in the gastrointestinal antibiosis of polar marine animals. *J. Bacteriol.* **82**, 72–79.

SIEBURTH, J. McN. 1964. Antibacterial substances produced by marine algae. In *Developments in Industrial Microbiology*, Soc. Industr. Microbiol., Washington, D.C., pp. 124–134.

SIEBURTH, J. McN. 1968. Observations on bacteria planktonic in Narragansett Bay, Rhode Island; A résumé. Proceedings of the U.S.–Japan seminar on marine microbiology. *Bull. Misaki Marine Biol. Inst., Kyoto Univ.* **12**, 49–64.

SIEBURTH, J. McN. 1969. Studies on algal substances in the sea. III. The production of extracellular organic matter by littoral marine algae. *J. Exp. Mar. Biol. Ecol.* **3**, 290–309.

SIEBURTH, J. McN. 1971. Distribution and activity of oceanic bacteria. *Deep-Sea Res.* **18**, 1111–1121.

SIEBURTH, J. McN. and A. JENSEN. 1969. Studies on algal substances in the sea. II. The formation of Gelbstoff (humic material) by exudates of Phaeophyta. *J. Exp. Mar. Biol. Ecol.* **3**, 275–289.

SIEBURTH, J. McN., R. D. BROOKS, R. V. GESSNER, C. D. THOMAS and J. L. TOOTLE. 1974. Microbial colonization of marine plant surfaces as observed by scanning electron microscopy. In *Effect of the Ocean Environment on Microbial Activities*, Ed. R. R. COLWELL and R. Y. MORITA, University Park Press, Baltimore, pp. 418–432.

SIEGEL, B. Z., S. M. SIEGEL and F. THORARINSSON. 1973. Icelandic Geothermal activity and the mercury of the Greenland Icecap. *Nature*, **241**, 526.

SKOPINTSEV, B. A. 1966. Some aspects of the distribution and composition of organic matter in the waters of the ocean. *Oceanology*, **6**, 441–450. (Fish. Res. Bd. Canada, Transl. No. 930.)

SKOPINTSEV, B. A. 1971. Recent advances in the study of organic matter in oceans. *Oceanology*, **11**, 775–789.

SLOAN, P. R. and J. D. H. STRICKLAND. 1966. Heterotrophy of four marine phytoplankters at low substrate concentrations. *J. Phycol.* **2**, 29–32.

SLOBODKIN, L. B. 1961. *Growth and Regulation of Animal Populations* (Ch. 12), Holt, Rinehart, & Winston, New York. 184 pp.

SMAYDA, T. J. 1958. Biogeographical studies of marine phytoplankton. *Oikos*, **9**, 158–191.

SMAYDA, T. J. 1963. Succession of phytoplankton and the ocean as an holocoenotic environment. In *Symposium on Marine Microbiology*, Ed. C. H. OPPENHEIMER, C. C. Thomas, Springfield, Illinois, pp. 260–274.

SMAYDA, T. J. 1964. Enrichment experiments using the marine centric diatom *Cyclotella nana* (Clone 13-1) as an assay organism. In *Proceedings of Symposium on Experimental Marine Ecology*, Occasional Publication No. 2, Graduate School of Oceanography, University of Rhode Island, pp. 25–32.

SMAYDA, T. J. 1969. Some measurements of the sinking rate of fecal pellets. *Limnol. Oceanogr.* **14**, 621–625.

SMAYDA, T. J. 1970a. The suspension and sinking of phytoplankton in the sea. *Oceanogr. Mar. Biol. Ann. Rev.* **8**, 353–414.

SMAYDA, T. J. 1970b. Growth potential bioassay of water masses using diatom cultures: Phosphorescent Bay (Puerto Rico) and Caribbean Waters. *Helgoländer wiss. Meeresunters.* **20**, 172–194.

SMAYDA, T. J. 1971. Normal and accelerated sinking of phytoplankton in the sea. *Mar. Geol.* **11**, 105–122.

SMAYDA, T. J. and B. J. BOLEYN. 1966. Experimental observations on the flotation of marine diatoms. II. *Skeletonema costatum* and *Rhizosolenia setigera. Limnol. Oceanogr.* **11**, 18–34.

SMETACEK, V., B. VON BODUNGEN, K. VON BROCKEL and B. ZEITSCHEL. 1976. The plankton tower. II. Release of nutrients from sediments due to changes in the density of bottom water. *Mar. Biol.* **34**, 373–378.

SMITH, E. L. 1936. Photosynthesis in relation to light and carbon dioxide. *Proc. Nat. Acad. Science, Wash.* **22**, 504–511.

SMITH, F. E. 1954. Quantitative aspects of population growth. In *Dynamics of Growth Processes*, Ed. E. BOELL, Princeton University Press, Princeton, New Jersey, pp. 274–294.

SMITH, F. E. and E. R. BAYLOR. 1953. Color responses in the Cladocera and their ecological significance. *Amer. Nat.* **87**, 49–55.

SMITH, K. L., JR. 1973. Respiration of a sublittoral community. *Ecology*, **54**, 1065–1075.

SMITH, K. L., JR. 1974. Oxygen demands of San Diego trough sediments: an *in situ* study. *Limnol. Oceanogr.* **19**, 939–944.

SMITH, K. L., JR., H. A. BURNS and J. M. TEAL. 1972. *In situ* respiration of benthic communities in Castle Harbour, Bermuda. *Mar. Biol.* **12**, 196–199.

SMITH, K. L., JR. and J. M. TEAL. 1973. Deep-sea benthic community respiration: an *in situ* study at 1850 meters. *Science*, **179**, 282–283.

SMITH, R. I. 1964. *Keys to Marine Invertebrates of the Woods Hole Region*, Contrib. No. 11, Systematics–Ecology Program. Mar. Biol. Lab., Woods Hole, Mass. 208 pp.

SOKOLOVA, G. A. and G. I. KARAVAIKO. 1964. *Physiology and geochemical activity of thiobacilli*. Akademiya Nauk SSSR Institut Mikrobiologii, Moskva [translated from Russian, Israel Program for Scientific Translation (Jerusalem), 1968]. 283 pp.

SOKOLOVA, M. N. 1972. Trophic structure of deep-sea macrobenthos. *Mar. Biol.* **16**, 1–12.

SOLÓRZANO, L. and J. D. H. STRICKLAND. 1968. Polyphosphate in sea water. *Limnol. Oceanogr.* **13**, 515–518.

SOROKIN, YU. I. 1961. Heterotrophic carbon dioxide assimilation by micro-organisms. *Zhurnal Obshchei Biologii*, **22**, 265–272 (in Russian).

SOROKIN, YU. I. 1964a. On the trophic role of chemosynthesis in water bodies. *Int. Rev. ges. Hydrobiol.* **49**, 307–324.

SOROKIN, YU. I. 1964b. A quantitative study of the microflora in the central Pacific Ocean. *J. Cons. Int. Explor. Mer*, **29**, 25–40.

SOROKIN, YU. I. 1964c. On the primary production and bacterial activities in the Black Sea. *J. Cons. Int. Explor. Mer*, **29**, 41–60.

SOROKIN, YU. I. 1965. On the trophic role of chemosynthesis and bacterial biosynthesis in water bodies. *Mem. Ist. Ital. Idiobiol.* **18**(Suppl.), 187–205.

SOROKIN, YU. I. 1966. On the carbon dioxide uptake during the cell synthesis by microorganisms. *Zeitschrift Allg. Mikrobiol.* **6**, 69–73.

SOROKIN, YU, I. 1968. The use of ^{14}C in the study of nutrition of aquatic animals. *Mitt. Int. Verein. Limnol.* **16**, 1–41.

SOROKIN, YU. I. 1969. On the trophic role of chemosynthesis and bacterial biosynthesis in water bodies. In *Primary Productivity in Aquatic Environments*, Ed. C. R. GOLDMAN, Mem. Ist. Ital. Idrobiol., 18 Suppl., University of California Press, Berkeley, pp. 187–205.

SOROKIN, YU. I. 1970. Determination of the activity of heterotrophic microflora in the ocean using C^{14} containing organic matter. *Mikrobiologiya*, **39**, 149–156 (in Russian).

SOROKIN, YU. I. 1971. On the role of bacteria in the productivity of tropical oceanic waters. *Int. Rev. ges. Hydrobiol.* **56**, 1–48.

SOROKIN, YU. I., T. S. PETIPA and YE. V. PAVLOVA. 1970. Quantitative estimate of marine bacterioplankton as a source of food. *Oceanology*, **10**, 253–260. (Trans. Scripta Technia inc. for Amer. Geophys. Union.)

SOURNIA, M. A. 1967. Rythme nychéméral du rapport 'intensité photosynthétique chlorophylle' dans le plancton marin. *C. R. Acad. Sci. Paris*, **265**, 1000–1003.

SOURNIA, M. A. 1969. Cycle annuel du phytoplancton et de la production primaire dans les mers tropicales. *Mar. Biol.* **3**, 287–303.

SOURNIA, A. 1974. Circadian periodicities in natural populations of marine phytoplankton. *Adv. Mar. Biol.* **12**, 325–386.

SOUTHWARD, A. J. 1974. Changes in the plankton community of the Western English Channel. *Nature*, **249**, 180–181.

SOUTHWARD, A. J. and E. C. SOUTHWARD. 1972. Observations on the role of dissolved organic compounds in the nutrition of benthic invertebrates II. Uptake by other animals living in the same habitat as Pogonophores, and by some littoral Polychaeta. *Sarsia*, **48**, 61–70.

SOUTHWARD, A. J. and E. C. SOUTHWARD. 1974. Observations on the role of dissolved organic compounds in the nutrition of benthic invertebrates. III. Uptake in relation to organic content of the habitat. *Sarsia*, **50**, 29–46.

SPENCER, C. P. 1954. Studies on the culture of a marine diatom. *J. Mar. Biol. Ass. U.K.* **33**, 265–290.

SPOEHR, H. A. and H. W. MILNER. 1949. The chemical composition of *Chlorella*, effect of environmental conditions. *Plant Physiol.* **24**, 120–149.

SPRATT, T. A. B. and E. FORBES. 1847. *Travels in Lycia, Milyas, the Cibyratis*, London, 2 vols.

STARIKOVA, N. D. 1970. Vertical distribution patterns of dissolved organic carbon in sea water and interstitial solutions. *Oceanology*, **10**, 796–807.

STARR, T. J. 1956. Relative amounts of vitamin B_{12} in detritus from oceanic and estuarine environments near Sapelo Island, Georgia. *Ecology*, **37**, 658–664.

STAVN, R. H. 1971. The horizontal–vertical distribution hypothesis: Langmuir circulation and *Daphnia* distributions. *Limnol. Oceanogr.* **16**, 453–466.

STEELE, J. H. 1962. Environmental control of photosynthesis in the sea. *Limnol. Oceanogr.* **7**, 137–150.

STEELE, J. H. 1964. Some problems in the study of marine resources. *ICNAF Environ. Symp. Rome*, 1964. Contrib. No. C-4. 11 pp.

STEELE, J. H. 1965. Some problems in the study of marine resources. *Int. Comm. M. W. Atlantic Fish.*, spec. publ. **6**, 463–476.

STEELE, J. H. 1974. *The Structure of Marine Ecosystems*, Harvard Univ. Press, Cambridge, Mass. 128 pp.

STEELE, J. H. 1974b. Spatial heterogeneity and population stability. *Nature*, **248**, 83.

STEELE, J. H. and C. S. YENTSCH 1960. The vertical distribution of chlorophyll. *J. Mar. Biol. Ass. U.K.* **39**, 217–226.

STEELE, J. H. and I. E. BAIRD. 1968. Production ecology of a sandy beach. *Limnol. Oceanogr.* **13**, 14–25.

STEELE, J. H., A. L. S. MUNRO and G. S. GIESE. 1970. Environmental factors controlling the epipsammic flora on beach and sublittoral sands. *J. Mar. Biol. Ass. U.K.* **50**, 907–918.

STEELE, J. H. and I. E. BAIRD. 1972. Sedimentation of organic matter in a Scottish sea loch. *Mem. Ist. Ital. Idrobiol.* **29**(Suppl.), 73–88.

STEEMANN NIELSEN, E. 1952. The use of radioactive carbon (C^{14}) for measuring organic production in the sea. *J. Cons. Int. Explor. Mer*, **18**, 117–140.

STEEMAN NIELSEN, E. 1958. Experimental methods for measuring organic production in the sea. *Rapp. Proc.-Verb. Cons. Int. Explor. Mer*, **144**, 38–46.

STEEMANN NIELSEN, E. 1961. Chlorophyll concentration and rate of photosynthesis in *Chlorella vulgaris*. *Physiol. Plant.* **14**, 868–876.

STEEMANN NIELSEN, E. 1962. On the maximum quantity of plankton chlorophyll per surface unit of a lake or the sea. *Int. Rev. ges. Hydrobiol.* **47**, 333–338.

STEEMANN NIELSEN, E. 1965. On the determination of the activity in ^{14}C ampoules for measuring primary production. *Limnol. Oceanogr.*, Suppl. 10: R247–252.

STEEMANN NIELSEN, E. and V. KR. HANSEN. 1959a. Light adaptation in marine phytoplankton populations and its interrelation with temperature. *Physiol. Plant.* **12**, 353–370.

STEEMANN NIELSEN, E. and V. KR. HANSEN. 1959b. Measurements with the carbon-14 technique of the respiration rates in natural populations of phytoplankton. *Deep-Sea Res.* **5**, 222–233.

STEEMANN NIELSEN, E. and V. KR. HANSEN. 1961. Influence of surface illumination on plankton photosynthesis in Danish waters (56°N) throughout the year. *Physiol. Plant.* **14**, 595–613.

STEEMANN NIELSEN, E. and E. G. JØRGENSEN. 1962. The physiological background for using chlorophyll measurements in hydrobiology and a theory explaining daily variations in chlorophyll concentration. *Arch. Hydrobiol.* **58**, 349–357.

STEEMANN NIELSEN, E., V. KR. HANSEN and E. G. JØRGENSEN. 1962. The adaptation to different light intensities in *Chlorella vulgaris* and the time dependence on transfer to a new light intensity. *Physiol. Plant.* **15**, 505–517.

STEEMANN NIELSEN, E. and T. S. PARK. 1964. On the time course in adapting to low light intensities in marine phytoplankton. *J. Cons. Int. Explor. Mer*, **29**, 19–24.

STEEMANN NIELSEN, E. and E. G. JØRGENSEN. 1968. The adaptation of algae. I. General part. *Physiol. Plant.* **21**, 401–413.

STEEMANN NIELSEN, E. and M. WILLEMOËS. 1971. How to measure the illumination rate when investigating the rate of photosynthesis of unicellular algae under various light conditions. *Int. Rev. ges. Hydrobiol.* **56**, 541–556.

STEFÁNSSON, U. and F. A. RICHARDS. 1964. Distributions of dissolved oxygen, density and nutrients off the Washington and Oregon coasts. *Deep-Sea Res.* **11**, 355–380.

STEPHEN, A. C. 1938. Production of large broods in certain marine lamellibranchs with a possible relation to weather conditions. *J. Anim. Ecol.* **1**, 130–143.

STEPHENS, G. C. 1967. Dissolved organic material as a nutritional source for marine and estuarine invertebrates. In *Estuaries*, Ed. G. H. LAUFF, Publ. Am. Ass. Adv. Sci. **83**, 367–373.

STEPHENS, K. 1970. Automated measurement of dissolved nutrients. *Deep-Sea Res.* **17**, 393–396.

STEPHENS, K., R. H. SHELDON and T. R. PARSONS. 1967. Seasonal variations in the availability of food for benthos in a coastal environment. *Ecology*, **48**(5), 852–855.

STEVEN, D. M. and R. GLOMBITZA. 1972. Oscillatory variation of a phytoplankton population in a tropical ocean. *Nature*, **237**, 105–107.

STEVENSON, L. H., C. E. MILLWOOD and B. H. HEBELER. 1974. Aerobic, heterotrophic bacterial populations in estuarine water and sediments. In *Effect of the Ocean Environment on Microbial Activities*, Eds. R. R. COWELL and R. Y. MORITA, Univ. Park Press, Baltimore, pp. 268–285.

STOMMEL, H. 1949. Trajectories of small bodies sinking slowly through convection cells. *J. Mar. Res.* **8**, 24–29.

STRAIN, H. H. 1951. The pigments of algae. In *Manual of Phycology*, Chronica Botanica, Waltham, Mass., pp. 243–262.

STRAIN, H. H. 1958. *Chloroplast Pigments and Chromatographic Analysis*, 32nd Priestley Lecture, Pennsylvania State University Press. 180 pp.

STRAIN, H. H. 1966. Fat-soluble chloroplast pigments: their identification and distribution in various Australian plants. In *Biochemistry of Chloroplasts*, Ed. T. W. GOODWIN, Academic Press, New York, Vol. 1, pp. 387–406.

STRATHMANN, R. R. 1967. Estimating the organic carbon content of phytoplankton from cell volume or plasma volume. *Limnol. Oceanogr.* **12**, 411–418.

STRICKLAND, J. D. H. 1958. Solar radiation penetrating the ocean. A review of requirements, data and methods of measurement, with particular reference to photosynthetic productivity. *J. Fish. Res. Bd. Canada*, **15**, 453–493.

STRICKLAND, J. D. H. 1960. Measuring the production of marine phytoplankton. *Fish. Res. Bd. Canada Bull.* **122**, 172.

STRICKLAND, J. D. H. 1965. Production of organic matter in the primary stages of the marine food chain. In *Chemical Oceanography*, Eds. J. P. RILEY and G. SKIRROW, Academic Press, London, Vol. 1, pp. 477–610.

STRICKLAND, J. D. H. 1968. A comparison of profiles of nutrient and chlorophyll concentrations taken from discrete depths and by continuous recording. *Limnol. Oceanogr.* **13**, 388–391.

STRICKLAND, J. D. H. 1970. Introduction. In *Marine Food Chains*, Ed. J. H. STEELE, Oliver & Boyd, Edinburgh, pp. 3–5.

STRICKLAND, J. D. H. and K. H. AUSTIN. 1960. On the forms, balance and cycle of phosphorus observed in the coastal and oceanic waters of the Northeastern Pacific. *J. Fish. Res. Bd. Canada*, **17**, 337–345.

STRICKLAND, J. D. H. and T. R. PARSONS. 1968. A practical handbook of seawater analysis. *Fish. Res. Bd. Canada, Bull.* **167**, 311.

STRICKLAND, J. D. H., R. W. EPPLEY and B. ROJAS DE MENDIOLA. 1969. Phytoplankton populations, nutrients and photosynthesis in Peruvian coastal waters. *Bol. Inst. del Mar del Peru*, **2**, 1–45.

STROSS, R. G., S. W. CHISHOLM and T. A. DOWNING. 1973. Causes of daily rhythms in photosynthetic rates of phytoplankton. *Biol. Bull.* **145**, 200–209.

STULL, E. A., E. DE AMEZAGA and C. R. GOLDMAN. 1973. The contribution of individual species of algae to primary productivity of Castle Lake, California. *Verh. Int. Verein. Limnol.* **18**, 1776–1783.

SUESS, E. 1976. Nutrients near the depositional interface. In *The Benthic Boundary Layer*, Ed. I. N. McCAVE, Plenum Press, New York, pp. 57–79.

SUGIURA, Y. 1965. On the reserved nutrient matters. *Bull. Soc. Franco–japonaise d'Oceanographie*, **2**, 7–11.

SUSHCHENYA, L. M. 1962. Quantitative data on nutrition and energy balance of *Artemia salina* (L.). *Dokl. Akad. Nauk SSSR*, **143**, 1205–1207 (in Russian). English transl. *Dokl. (Proc.) Acad. Sci. USSR*, **143**, 329–330.

SUSHCHENYA, L. M. 1970. Food rations, metabolism and growth of crustaceans. In *Marine Food Chains*, Ed. J. H. STEELE, Oliver & Boyd, Edinburgh, pp. 127–141.

SUTCLIFFE, W. H. JR. 1969. Relationship between growth rate and ribonucleic acid concentration in some invertebrates. *J. Fish. Res. Bd. Canada*, **27**, 606–609.

SUTCLIFFE, W. H. JR. 1972. Some relations of land drainage, nutrients, particulate material and fish catch in two eastern Canadian bays. *J. Fish. Res. Bd. Canada*, **29**, 357–362.

SUTCLIFFE, W. H. JR. 1973. Correlations between seasonal river discharge and local landings of American lobster (*Homarus americanus*) and Atlantic halibut (*Hippoglossus hippoglossus*) in the Gulf of St. Lawrence. *J. Fish. Res. Bd. Canada.* **30**, 856–859.

SUTCLIFFE, W. H. JR., E. R. BAYLOR and D. W. MENZEL. 1963. Sea surface chemistry and Langmuir circulation. *Deep-Sea Res.* **10**, 233–243.

SUTCLIFFE, W. H. JR., R. H. LOUCKS and K. F. DRINKWATER. 1976. Coastal circulation and physical oceanography of the Scotian Shelf and the Gulf of Maine. *J. Fish. Res. Bd.* **33**, 98–115.

SUTCLIFFE, W. H. JR., K. DRINKWATER and B. S. MUIR. Correlations of fish catch and environmental factors in the Gulf of Maine. *J. Fish. Res. Bd. Canada* (in press).

SUYAMA, M., K. NAKAJIMA and J. NONAKA. 1965. Studies on the protein and non-protein nitrogenous constituents of *Euphausia. Bull. Japanese Soc. Sci. Fish.* **31**, 302–306 (Japanese with English summary).

SUZUKI, N. and K. KATO. 1953. Studies on suspended materials 'marine snow' in the sea. Part I. Sources of marine snow. *Bull. Fac. Fish., Hokkaido University*, **4**, 132–137.

SVERDRUP, H. U. 1953. On conditions for the vernal blooming of phytoplankton. *J. Cons. Explor. Mer*, **18**, 287–295.

SVERDRUP, H. U., M. W. JOHNSON and R. H. FLEMING. 1946. *The Oceans, their Physics, Chemistry and General Biology*, Prentice-Hall, New York. 1087 pp.

SWEDMARK, B. 1964. The interstitial fauna of marine sand. *Biol. Rev.* **39**, 1–42.

SYSOEVA, T. K. and A. A. DEGTEREVA. 1965. The relation between the feeding of cod larvae and pelagic fry and the distribution and abundance of their principal food organisms. *ICNAF Spec. Publ.* **6**, 411–416.

SZEICZ, G. 1966. Field measurements of energy in the 0·4–0·7 micron range. In *Light as an Ecological Factor*, Eds. R. BAINBRIDGE, G. C. EVANS and O. RACTHAM, *Symp. Brist. Ecol. Soc.* **6**, 41–51. Blackwell Scientific Publications, Oxford.

SZEKIELDA, K. 1967. Some remarks on the influence of hydrographic conditions on the concentration of particulate carbon in sea water. *IBP Symposium (Amsterdam)*, Eds. H. L. GOLTERMAN and R. S. CLYMO, Noord-Hollandsche Vitgevers Maatschappi, Amsterdam, pp. 314–322.

TAGUCHI, S. and K. NAKAJIMA. 1971. Plankton and seston in the sea surface of three inlets of Japan. *Bull. Plankton Soc. Japan*, **18**, 20–36.

TAKAHASHI, M., S. SHIMURA, Y. YAMAGUCHI and Y. FUJITA. 1971. Photoinhibition of phytoplankton photosynthesis as a function of exposure time. *J. Oceanogr. Soc. Japan*, **27**, 43–50.

TAKAHASHI, M., K. SATAKE and N. NAKAMOTO. 1972. Chlorophyll profile and photosynthetic activity in the north and equatorial Pacific Ocean. *J. Oceanogr. Soc. Japan.* **28**, 27–36.

TAKAHASHI, M. and T. R. PARSONS. 1972. The maximization of the standing stock and primary productivity of marine phytoplankton under natural conditions. *India J. Mar. Sci.* **1**.

TAKAHASHI, M. and F. NASH. 1973. The effect of nutrient enrichment on algal photosynthesis in Great Central Lake, British Columbia, Canada. *Arch. Hydrobiol.* **71**, 166–182.

TAKAHASHI, M. and T. IKEDA. 1975. Excretion of ammonia and inorganic phosphorus by *Euphausia pacifica* and *Metridia pacifica* at different concentrations of phytoplankton. *J. Fish. Res. Bd. Canada*, **32**, 2189–2195.

TAKAHASHI, M., W. H. THOMAS, D. L. R. SEIBERT, J. BEERS, P. KOELLER and T. R. PARSONS. 1975. The replication of biological events in enclosed water columns. *Arch. Hydrobiol.* **76**, 5–23.

TALLING, J. F. 1957a. The phytoplankton population as a compound photosynthetic system. *New Phytol.* **56**, 133–149.

TALLING, J. F. 1957b. Photosynthetic characteristics of some freshwater plankton diatoms in relation to underwater radiation. *New Phytol.* **56**, 29–50.

TALLING, J. F. 1960. Comparative laboratory and field studies of photosynthesis by a marine planktonic diatom. *Linmol. Oceanogr.* **5**, 62–77.

TAMIYA, H., E. HASE, K. SHIBATA, A. MITSUYA, T. IWAMURA, T. NIHEI and T. SASA. 1953. Kinetics of growth of *Chlorella* with special reference to its dependence on quantity of available light and on temperature. In *Algal Culture: from Laboratory to Pilot Plant*, Ed. J. S. BURLOW, Carnegie Inst. Publ. No. 600, pp. 204–232.

TANADA, T. 1951. The photosynthetic efficiency of carotenoid pigments in *Navicula minima. Amer. J. Bot.* **38**, 276–283.

TANIGUCHI, A. 1973. Phytoplankton–zooplankton relationships in the Western Pacific Ocean and adjacent seas. *Mar. Biol.* **21**, 115–121.

TAYLOR, W. R. 1964. Light and photosynthesis in intertidal benthic diatoms. *Helgölander wiss. Meeresunters.* **10**, 29–37.

TAYLOR, W. R. and C. D. GEBELEIN. 1966. Plant pigments and light penetration in intertidal sediments. *Helgoländer wiss. Meeresunters*, **13**, 229–237.

TEAL, J. M. 1957. Community metabolism in a temperate cold spring. *Ecol. Monogr.* **27**, 283–302.

TEAL, J. M. 1962. Energy flow in the salt marsh ecosystem of Georgia. *Ecology*, **43**, 614–624.

TEAL, J. M. and J. KANWISHER. 1961. Gas exchange in a Georgia salt marsh. *Limnol. Oceanogr.* **6**, 388–399.

TEIXEIRA, C., J. TUNDISI and J. SANTORO. 1967. Plankton studies in a mangrove environment. IV. Size fraction-
    ation of the phytoplankton. *Bolm Inst. Oceanogr. S. Paulo*, **16**, 39–42.
TENORE, K. R., M. G. BROWNE and E. J. CHESNEY, JR. 1974. Polyspecies aquaculture systems: the detrital
    trophic level. *J. Mar. Res.* **32**, 425–432.
TETT, P. B. 1973. The use of log-normal statistics to describe phytoplankton population from the Firth
    of Lorne area. *J. Exp. Mar. Biol. Ecol.* **11**, 121–136.
THEEDE, H., A. PONAT, K. HIROKI and C. SCHLIEPER. 1969. Studies on the resistance of marine bottom inverte-
    brates to oxygen deficiency and hydrogen sulphide. *Mar. Biol.* **2**, 325–337.
THIEL, H. 1975. The size structure of the deep-sea benthos. *Int. Rev. ges. Hydrobiol.* **60**, 575–606.
THOMAS, E. 1955. Stoffhaushalt und Sedimentation im oligotrophen Aegerisee und im eutrophen Pfäffiker
    und Greifensee. *Mem Ist Ital. Idrobiol.*, Suppl. **8**, 357–465.
THOMAS, W. H. 1968. Nutrient requirements and utilization: Algae. In *Metabolism*, Eds. P. L. ALTMAN and
    D. S. DITTMER, Fed. Am. Soc. Exptl. Biol., Bethesda, Md., pp. 210–228.
THOMAS, W. H. 1970a. On nitrogen deficiency in tropical Pacific oceanic phytoplankton: photosynthetic par-
    ameters in poor and rich water. *Limnol. Oceanogr.* **15**, 380–385.
THOMAS, W. H. 1970b. Effect of ammonium and nitrate concentration on chlorophyll increases in natural
    tropical Pacific phytoplankton populations. *Limnol. Oceanogr.* **15**, 386–394.
THOMAS, W. H. and R. W. OWEN, JR. 1971. Estimating phytoplankton production from ammonium and
    chlorophyll concentrations in nutrient-poor water of the eastern tropical Pacific Ocean. *Fish. Bull.* **69**,
    87–92.
THOMAS, W. H. and A. N. DODSON. 1974. Effect of interactions between temperature and nitrate supply
    on the cell-division rates of two marine phytoflagellates. *Mar. Biol.* **24**, 213–217.
THORSON, G. 1950. Reproductive and larval ecology of marine bottom invertebrates. *Biol. Rev.* **25**, 1–45.
THORSON, G. 1957. Bottom communities. *Mem. Geol. Soc. Amer.* **67**(1), 461–534.
THORSON, G. 1958. Parallel level bottom communities, their temperature adaptation, and their 'balance'
    between predators and food animals. In *Perspectives in Marine Biology*, Ed. A. A. BUZZATI-TRAVERSO,
    Univ. Calif. Press, Berkeley, Calif. 621 pp.
THORSON, G. 1966. Some factors influencing the recruitment and establishment of marine benthic communities.
    Netherlands J. Sea. Res. **3**, 267–293.
THRONDSEN, J. 1973. Motility in some marine nanoplankton flagellates. *Norw. J. Zool.* **21**, 193–200.
TIETJEN, J. H. 1971. Ecology and distribution of deep-sea meiobenthos off North Carolina. *Deep-Sea Res.*
    18, 941–957.
TILTON, R. C. 1968. The distribution and characterization of marine sulfur bacteria. *Rev. Int. Oceanogr. Med.*
    9, 237–253.
TILTON, R. C., A. B. COBER and G. E. JONES. 1967. Marine thiobacilli. I. Isolation and distribution. *Can.
    J. Microbiol.* **13**, 1521–1528.
TOLBERT, N. E. 1974. Photorespiration. In *Algal Physiology and Biochemistry*, Ed. W. D. STEWART, Univ.
    Calif., pp. 474–504.
TOMINAGA, H. and S. ICHIMURA. 1966. Ecological studies on the organic matter production in a mountain
    river ecosystem. *Bot. Mag., Tokyo*, **79**, 815–829.
TOOMING, H. 1970. Mathematical description of net photosynthesis and adaptation processes in the photosyn-
    thetic apparatus of plant communities. In *Prediction and Measurement of Photosynthetic Productivity*,
    Ed. I. MALEK *et al.*, Pudoc, Wageningen, pp. 103–113.
TRACEY, M. L., K. NELSON, D. HEDGECOCK, R. A. SHLESER and M. L. PRESSICK. 1975. Biochemical genetics
    of lobsters: genetic variation and the structure of american lobster (*Homarus americanus*) populations.
    J. Fish. Res. Bd. Canada, **32**, 2091–2101.
TRANTER, D. J. and B. S. NEWELL. 1963. Enrichment experiments in the Indian Ocean. *Deep-Sea Res.* **10**,
    1–9.
TRAVERS, M. 1971. Diversité du microplancton du Golfe de Marseille en 1964. *Mar. Biol.* **8**, 308–343.
TREVALLION, A. 1967. An investigation of detritus in Southampton Water. *J. Mar. Biol. Ass. U.K.* **47**, 523–532.
TREVALLION, A. 1971. Studies on *Tellina tenuis* da Costa. III. Aspects of general biology and energy flow.
    J. Exp. Mar. Biol. Ecol. **7**, 95–122.
TREVALLION, A., R. R. C. EDWARDS and J. H. STEELE. 1970. Dynamics of a benthic bivalve. In *Marine Food
    Chains*, Ed. J. H. STEELE, Oliver & Boyd, Edinburgh, pp. 285–295.
TSIKHON-LUKANINA, YE. A. and T. A. LUKASHEVA. 1970. Conversion of food energy in the young of some
    marine isopods. *Oceanology*, **10**, 553–556.
TSYBAN, A. V. 1971. Marine bacterioneuston. *J. Oceanogr. Soc. Japan*, **27**, 56–66.
TURNER, R. 1973. Wood boring bivalves, opportunistic species in the deep sea. *Science*, **180**, 1377–1379.
TYLER, A. V. 1971. Surges of winter flounder, *Pseudopleuronectes americanus*, into the intertidal zone. *J.
    Fish. Res. Bd. Canada*, **28**, 1727–1732.
UDA, M. 1961. Fisheries oceanography in Japan, especially on the principles of fish distribution, concentration,
    dispersal and fluctuation. *Calif. Coop. Oceanic. Fish. Invest.* **8**, 25–31.
UNESCO. 1973. A guide to the measurement of marine primary production under some special conditions.
    In *Monographs on Oceanographic Methodology*, **3**, Paris. 73 pp.

UYENO, F. 1966. Nutrient and energy cycles in an estuarine oyster area. *J. Fish. Res. Bd. Canada*, **23**, 1635–1652.

VACCARO, R. F. and H. RYTHER. 1959. Marine phytoplankton and the distribution of nitrite in the sea. *J. Cons. Int. Explor. Mer*, **25**, 260–271.

VACCARO, R. F. and H. W. JANNASCH. 1966. Studies on the heterotrophic activity in seawater based on glucose assimilation. *Limnol. Oceanogr.* **11**, 596–607.

VACCARO, R. F. and H. W. JANNASCH. 1967. Variations in uptake kinetics for glucose by natural populations in seawater. *Limnol. Oceanogr.* **12**, 540–542.

VACCARO, R. F., S. E. HICKS, H. W. JANNASCH and F. G. CAREY. 1968. The occurrence and role of glucose in seawater. *Limnol. Oceanogr.* **13**, 356–360.

VAHL, O. 1972. Efficiency of particle retention in *Mytilus edulis* L. *Ophelia*, **10**, 17–25.

VALIELA, I. and J. M. TEAL. 1974. Nutrient limitation in salt marsh vegetation. In *Ecology of Halophytes*, Eds. R. J. REIMOLD and W. H. QUEEN, Academic Press, New York and London, pp. 547–563.

VALIELA, I., J. M. TEAL and N. Y. PERSSON. 1976. Production dynamics of experimentally enriched salt marsh vegetation: below-ground biomass. *Limnol. Oceanogr.* **21**, 245–252.

VALLENTYNE, J. R. 1955. Sedimentary chlorophyll determination as a paleobotanical method. *Can. J. Bot.* **33**, 304–313.

VALLENTYNE, J. R. 1969. Sedimentary organic matter and paleolimnology. *Mitt. Int. Verein. Limnol.* **17**, 104–110.

VAN RAALTE, C. D., I. VALIELA and J. M. TEAL. 1976. Production of epibenthic salt marsh algae: light and nutrient limitation. *Limnol. Oceanogr.* **21**, 862–872.

VENRICK, E. L. 1972. Small-scale distributions of oceanic diatoms. *Fish. Bull.* **70**, 363–372.

VENRICK, E. L., J. A. McGOWAN and A. W. MANTYLA. 1972. Deep maxima of photosynthetic chlorophyll in the Pacific Ocean. *Fish. Bull.* **71**, 41–52.

VILKS, G. 1967. Quantitative analysis of foraminifera in Bras d'Or Lakes (Unpubl. manuscript). *Atl. Oceanogr. Lab. Rep.* **67**(1). 83 pp.

VILKS, G., E. H. ANTHONY and W. T. WILLIAMS. 1970. Application of association-analysis to distribution studies of recent Foraminifera. *Can. J. Earth Sci.* **7**(6), 1462–1469.

VINOGRADOV, A. P. 1953. The elementary chemical composition of marine organisms. Translation by EFRON and SELTON. *Memoir. Sears Found. Mar. Res.* **2**, 130–146.

VINOGRADOV, M. E. 1955. Vertical migrations of zooplankton and their importance for the nutrition of abyssal pelagic fauna. *Trudy Inst. Okean.* **13**, 71–76.

VINOGRADOV, M. E. 1962. Feeding of deep-sea zooplankton. *Rapp. Proc. Verb. Cons. Perm. Int. Explor. Mer*, **153**, 114–120.

VINOGRADOV, M. E. 1968. *Vertical Distribution of the Oceanic Zooplankton*, Nauka Publ. House, Moscow. (Transl. Israel. Prog. for Sci. Transl. Jerusalem, 1970.) 320 pp.

VINOGRADOV, M. E., I. I. GITELZON and YU. I. SOROKIN. 1970. The vertical structure of a pelagic community in the tropical ocean. *Mar. Biol.* **6**, 187–194.

VINOGRADOV, M. E., V. V. MENSHUTKIN and E. A. SHUSHKINA. 1972. On mathematical simulation of a pelagic ecosystem in tropical waters of the ocean. *Mar. Biol.* **16**, 261–268.

VINOGRADOV, M. E., V. F. KRAPIVIN, V. V. MENSHUTKIN, B. S. FLEYSHMAN and E. A. SHUSKINA. 1973. Mathematical model of the functions of the pelagical ecosystem in tropical regions (from the 50th voyage of the R/V *Vityaz*). *Oceanology*, **13**, 704–717 (English translation).

VINOGRADOVA, N. G. 1962. Some problems of the study of deep-sea bottom fauna. *J. Oceanogr. Soc. Jap., 20th Ann.*, pp. 724–741.

VINOGRADOVA, Z. A. and V. V. KOVAL'SKIY. 1962. Elemental composition of the Black Sea plankton. *Dokl. Akad. Nauk SSSR*, **147**, 1458–1460.

VLYMEN, W. J. 1970. Energy expenditure of swimming copepods. *Limnol. Oceanogr.* **15**, 348–356.

VOGEL, K. and B. J. D. MEEUSE. 1968. Characterization of the reserve granules from the dinoflagellate *Thecadinium inclinatum* Balech. *J. Phycol.* **4**, 317–318.

VOLKMANN, C. M. and C. H. OPPENHEIMER. 1962. The microbial decomposition of organic carbon in surface sediments of marine bays of the Central Texas Gulf Coast. *Publ. Inst. Mar. Sci.* **8**, 80–96.

VOLLENWEIDER, R. A. 1965. Calculation models of photosynthesis—depth curves and some implications regarding day rate estimates in primary production measurements. In *Primary Productivity in Aquatic Environments*, Ed. C. R. GOLDMAN, Mem. Ist. Ital., Idrobiol., *18* Suppl.: The University of California Press, Berkeley, pp. 425–457.

VOLLENWEIDER, R. A. (Ed.) 1969. A manual on methods for measuring primary production in aquatic environments including a chapter on bacteria. *IBP Handbook No. 12*, F. A. Davis Co., Philadelphia. 213 pp.

VOLLENWEIDER, R. A. and A. NAUWERCK. 1961. Some observations on the ^{14}C-method for measuring primary production. *Verh. int. Limnol.* **14**, 134–139.

VON ARX, W. S. 1962. *An Introduction to Physical Oceanography*, Addison–Wesley, London. 135 pp.

WADA, E. and A. HATTORI. 1971. Nitrite metabolism in the euphotic layer of the central North Pacific Ocean. *Limnol. Oceanogr.* **16**, 766–722.

WAILES, G. H. 1937. *Canadian Pacific Fauna*. 1. Protozoa. 1a. Lobosa, 1b. Reticulosa, 1c. Heliozoa, 1d. Radiolaria, Biological Board of Canada, Toronto. 14 pp.

WAILES, G. H. 1939. *Canadian Pacific Fauna.* 1. Protozoa. 1e. Mastigophora, Biological Board of Canada, Toronto, 45 pp.

WAILES, G. H. 1943. *Canadian Pacific Fauna.* 1. Protozoa. If. Ciliata, 1g. Suctoria, Biological Board of Canada, Toronto, pp. 46.

WAKSMAN, S. A. 1933. On the distribution of organic matter in the sea bottom and the chemical nature and origin of marine humus. *Soil. Sci.* **36**, 125–147.

WAKSMAN, S. A., C. L. CAREY and H. W. REUSZER. 1933. Marine bacteria and their role in the cycle of life in the sea. I. Decomposition of marine plant and animal residues by bacteria. *Biol. Bull.* **65**, 57–79.

WAKSMAN, S. A. and M. HOTCHKISS. 1938. On the oxidation of organic matter in marine sediments by bacteria. *J. Mar. Res.* **15**, 101–118.

WALFORD, L. A. 1946. A new graphic method of describing the growth of animals. *Biol. Bull.* **90**, 141–147.

WALLEN, D. G. and G. H. GEEN. 1971a. Light quality in relation to growth, photosynthetic rates and carbon metabolism in two species of marine plankton algae. *Mar. Biol.* **10**, 34–43.

WALLEN, D. G. and G. H. Geen. 1971b. Light quality and concentration of proteins, RNA, DNA and photosynthetic pigments in two species of marine plankton algae. *Mar. Biol.* **10**, 44–51.

WALLEN, D. G. and G. H. GEEN. 1971c. The nature of the photosynthate in natural phytoplankton populations in relation to light quality. *Mar. Biol.* **10**, 157–168.

WALNE, P. R. 1972. The influence of current speed, body size and water temperature on the filtration rate of five species of bivalve. *J. Mar. Biol. Ass. U.K.* **32**, 345–374.

WALSH, J. J. 1971. Relative importance of habitat variable in predicting the distribution of phytoplankton at the ecotone of the antarctic upwelling ecosystem. *Ecol. Monographs,* **41**, 291–309.

WALSH, J. J. 1972. Implications of a systems approach to oceanography. *Science,* **176**, 969–975.

WALSH, J. J. and R. C. DUGDALE. 1971. A simulation model of the nitrogen flow in the Peruvian upwelling system. *Inv. Resq.* **35**, 309–330.

WALTER, M. D. 1973. Fressverhalten und Darminhaltsunter-suchungen bei Sipunculiden. *Helgoländer wiss. Meeresunters,* **25**, 486–494.

WANGERSKY, P. J. 1965. The organic chemistry of sea water. *Amer. Scient.* **53**, 358–374.

WANGERSKY, P. J. 1974. Particulate organic carbon: sampling variability. *Limnol. Oceanogr.* **19**, 980–984.

WANGERSKY, P. J. 1975. *Heterogeneous Distribution of Organic Matter in the Ocean and its Ecological Significance.* Presented at the International symposium of interactions between water and living matter, October 6–10, 1975, Odessa, USSR.

WANGERSKY, P. J. 1976. Particulate organic carbon in the Atlantic and Pacific oceans. *Deep-Sea Res.* **23**, 457–465.

WANGERSKY, P. J. and D. C. GORDON. 1965. Particulate carbonate, organic carbon, and Mn^{++} in the open ocean. *Limnol. Oceanogr.* **10**, 544–550.

WARWICK, R. M. and J. B. BUCHANAN. 1970. The meiofauna off the coast of Northumberland. I. The structure of the nematode population. *J. Mar. Biol. Ass. U.K.* **50**, 129–146.

WARWICK, R. M. and R. PRICE. 1975. Macrofauna production in an estuarine mudflat. *J. Mar. Biol. Ass. U.K.* **55**, 1–18.

WATANABE, T. and R. G. ACKMAN. 1974. Lipids and fatty acids of the American *Crassostrea virginica* and European flat *Ostrea edulis* oysters from a common habitat, before and after feeding with *Dicrateria inoruata* or *Isochrysis galbana. J. Fish. Res. Bd. Canada,* **31**, 403–409.

WATERS, T. F. 1969. The turnover ratio in production ecology of freshwater invertebrates. *Amer. Nat.* **103**, 173–185.

WATLING, L. 1975. Analysis of structural variations in a shallow estuarine deposit-feeding community. *J. Exp. Mar. Biol. Ecol.* **19**, 275–313.

WATSON, S. W. 1965. Characteristics of a marine nitrifying bacterium, *Nitrosocystis oceanus* sp. nov. *Limnol. Oceanogr. Suppl.* **10** (Redfield 75th Anniv. Vol.), R274–R289.

WATSON, S. W. 1971. Taxonomic considerations of the family Nitrobacteracea Buchanan. *Int. J. System. Bacteriol.* **21**, 254–270.

WATT, W. D. 1966. Release of dissolved organic material from the cells of phytoplankton populations. *Proc. Roy. Soc. (London),* B, **164**, 521–551.

WATT, W. D. 1971. Measuring the primary production rates of individual phytoplankton species in natural mixed populations. *Deep-Sea Res.* **18**, 329–340.

WEBB, J. 1969. Biologically significant properties of submerged marine sands. *Proc. R. Soc.,* B, **174**, 355–402.

WEBB, K. L. and R. E. JOHANNES. 1967. Studies of the release of dissolved free amino acids by marine zooplankton. *Limnol. Oceanogr.* **12**, 376–382.

WEBB, K. L. and R. E. JOHANNES. 1969. Do marine crustaceans release dissolved amino acids? *Comp. Biochem. Physiol.* **29**, 875–878.

WEBSTER, T. J. M., M. PARANJAPE and K. H. MANN. 1975. Sedimentation of organic matter in St. Margaret's Bay, Nova Scotia. *J. Fish. Res. Bd. Canada,* **32**, 1399–1407.

WEISS, P. 1969. The living system: determinism stratified. In *Beyond Reductionism,* Eds. A. KOESTLER and J. SMYTHIES. Hutchinson, London, pp. 3–55.

WELCH, E. B. and G. W. ISAAC. 1967. Chlorophyll variation with tide and with plankton productivity in an estuary. *J. Water Poll. Cont. Fed.* **39,** 360–366.

WELCH, H. E. 1968. Relationships between assimilation efficiencies and growth efficiencies for aquatic consumers. *Ecology,* **49,** 755–759.

WESTLAKE, D. F. 1963. Comparisons of plant productivity. *Biol. Rev.* **38,** 385–425.

WESTLAKE, D. F. 1965. Some problems in the measurement of radiation under water: A review. *Photochemistry and Photobiology,* **4,** 849–868.

WETZEL, R. G., P. H. RICH, M. C. MILLER and H. L. ALLEN. 1972. Metabolism of dissolved and particulate detrital carbon in a temperate hard-water lake. *Mem. Ist. Ital. Idrobial.* **29,** 185–243.

WHEELER, E. H. 1967. Copepod detritus in the deep sea. *Limnol. Oceanogr.* **12,** 697–701.

WHITFIELD, M. 1969. *Eh* as an operational parameter in estuarine studies. *Limnol. Oceanogr.* **14,** 547–558.

WHITLATCH, R. B. 1974. Food-resource partitioning in the deposit feeding polychaeta *Pectinaria gouldii. Biol. Bull.* **147,** 227–235.

WICKETT, W. P. 1967. Ekman transport and zooplankton concentration in the North Pacific Ocean. *J. Fish. Res. Bd.* **24,** 581–594.

WICKSTEAD, J. H. 1962. Food and feeding in pelagic copepods. *Proc. Zool. Soc. London,* **139,** 545–555.

WIEBE, P. H. 1970. Small-scale spatial distribution in oceanic zooplankton. *Limnol. Oceanogr.* **15,** 205–217.

WIEBE, P. H. 1971. A field investigation of the relationship between length of tow, size of net and sampling error. *J. Cons. Int. Explor. Mer,* **34,** 110–117.

WIEBE, P. H. and W. R. HOLLAND. 1968. Plankton patchiness: effects on repeated net tows. *Limnol. Oceanogr.* **13,** 315–321.

WIESER, W. 1960. Benthic studies in Buzzards Bay. II. The meiofauna. *Limnol. Oceanogr.* **5,** 121–137.

WIESER, W. and J. KANWISHER. 1961. Ecological and physiological studies on marine nematodes from a small salt marsh near Woods Hole, Massachusetts. *Limnol. Oceanogr.* **6,** 262–270.

WIESER, W., J. OTT, F. SCHIEMER and E. GNAIGER. 1974. An ecophysiological study of some meiofauna species inhabiting a sandy beach at Bermuda. *Mar. Biol.* **26,** 235–248.

WIESER, W. and M. ZECH. 1976. Dehydrogenases as tools in the study of marine sediments. *Mar. Biol.* **36,** 113–122.

WIGLEY, R. L. and A. D. MACINTYRE. 1964. Some quantitative comparisons of offshore meiobenthos and macrobenthos south of Martha's Vineyard. *Limnol. Oceanogr.* **9,** 485–493.

WILDISH, D. S. and N. J. POOLE. 1970. Cellulase activity in *Orchestia gammarella* (Pallas). *Comp. Biochem. Physiol.* **33,** 713–716.

WILHM, J. L. 1968. Use of biomass units in Shannon's Formula. *Ecology,* **49,** 153–156.

WILLIAMS, P. H. LEB. 1970. Heterotrophic utilization of dissolved organic compounds in the sea. I. Size distribution of population and relationship between respiration and incorporation of growth substrates. *J. Mar. Biol. Ass. U.K.* **50,** 859–870.

WILLIAMS, P. J. LEB. and R. W. GRAY. 1970. Heterotrophic utilization of dissolved organic compounds in the sea. II. Observations on the responses of heterotrophic marine populations to abrupt increases in amino acid concentration. *J. Mar. Biol. Ass. U.K.* **50,** 871–881.

WILLIAMS, P. J. LEB. 1975. Biological and chemical aspects of dissolved organic materials in sea water. In *Chemical Oceanography,* Ed. J. P. RILEY and G. SKIRROW, Publ. Academic Press, pp. 301–363.

WILLIAMS, P. M. 1965. Fatty acids derived from lipids of marine origin. *J. Fish. Res. Bd. Canada,* **22,** 1107–1122.

WILLIAMS, P. M. 1968. Organic and inorganic constituents of the Amazon River. *Nature,* **218,** 937–938.

WILLIAMS, P. M. and K. S. CHAN. 1966. Distribution and speciation of iron in natural waters: Transition from river water to a marine environment, British Columbia, Canada. *J. Fish. Res. Bd. Canada,* **23,** 575–593.

WILLIAMS, P. M., H. OESCHGER and P. KINNEY. 1969. Natural radiocarbon activity of the dissolved organic carbon in the North-east Pacific Ocean. *Nature,* **224,** 256–258.

WILLIAMS, P. M. and L. I. GORDON. 1970. Carbon-13: carbon-12 ratios in dissolved and particulate organic matter in the sea. *Deep-Sea Res.* **17,** 19–27.

WILLIAMS, P. M., J. A. MCGOWAN and M. STUIVER. 1970. Bomb carbon-14 in deep-sea organisms. *Nature,* **227,** 375–376.

WILLIAMS, R. 1972. The abundance and biomass of the interstitial fauna of a graded series of shell-gravels in relation to the available space. *J. Amin. Ecol.* **41,** 623–646.

WILLIAMSON, M. H. 1961. A method for studying the relation of plankton variations to hydrography. *Bull. Mar. Ecol.* **5,** 224–229.

WILSON, D. P. 1951. A biological difference between natural sea waters. *J. Mar. Biol. Ass. U.K.* **30,** 1–20.

WIMBUSH, M. 1976. The physics of the benthic boundary layer. In *The Benthic Boundary Layer,* Ed. I. N. McCAVE, Plenum Press, New York, pp. 3–10.

WINBERG, G. C. 1968. *Methods for the Estimation of Production of Aquatic Animals* (trans. by A. DUNCAN, 1971), Adv. Ecol. Res. Academic Press, London–New York. 175 pp.

WINSOR, C. P. and G. L. CLARKE. 1940. A statistical study of variation in the catch of plankton nets. *J. Mar. Res.* **3,** 1–34.

WINTER, J. E. 1969. Über den Einfluss der Nahrungskonzentration und anderer Faktoren auf Filtrierleistung und Nahrungsausnutzung der Muscheln *Arctica islandica* and *Modiolus modiolus*. *Mar. Biol.* **4**, 87–135.

WIRSEN, C. O. and H. W. JANNASCH. 1975. Activity of marine psychrophilic bacteria at elevated hydrostatic pressures and low temperature. *Mar. Biol.* **31**, 201–208.

WONG, C. S., D. R. GREEN and W. J. CRETNEY. 1974. Quantitative tar and plastic waste distributions in the Pacific Ocean. *Nature*, **247**, 30–32.

WOOD, E. J. F. 1953. Heterotrophic bacteria in marine environments of eastern Australia. *Aust. J. Mar. Freshw. Res.* **4**, 160–200.

WOOD, E. J. F. 1955. Fluorescent microscopy in marine microbiology. *J. Cons. Int. Explor. Mer.* **21**, 6–7.

WOOD, E. J. F. 1958. The significance of marine microbiology. *Bact. Rev.* **22**, 1–19.

WOOD, E. J. F. 1965. *Marine Microbial Ecology*, Reinhold, New York, pp. 243.

WOOD, H. G. and C. H. WERKMANN. 1935. The utilization of CO_2 by the propionic acid bacteria in the dissimulation of glycerol. *J. Bacteriol.* **30**, 332.

WOOD, H. G. and C. H. WERKMAN. 1936. The utilization of CO_2 in the dissimulation of glycerol by the propionic acid bacteria. *Biochem. J.* **30**, 48–53.

WOOD, H. G. and C. H. WERKMAN. 1940. The relationship of the bacterial utilization of CO_2 to succinic acid formation. *Biochem. J.* **34**, 129–138.

WOOD, J. M., F. SCOTT KENNEDY and C. G. ROSEN. 1968. Synthesis of methyl-mercury compounds by extracts of a methnogenic bacterium. *Nature*, **220**, 173–174.

WOOD, L. H. and K. E. CHUA. 1973. Glucose flux at the sediment-water interface of Toronto Harbour, Lake Ontario, with reference to pollution stress. *Can. J. Microbiol.* **19**, 413–420.

WOODIN, S. A. 1976. Adult-larval interactions in dense infaunal assemblages: patterns of abundance. *J. Mar. Res.* **34**, 25–41.

WOOSTER, W. S. and J. L. REID, JR. 1963. Eastern boundary currents. In *The Seas*, Ed. M. N. HILL, Interscience Publishers, New York, Vol. 2, pp. 253–280.

WRIGHT, R. T. 1964. Dynamics of a phytoplankton community in an ice-covered lake. *Limnol. Oceanogr.* **9**, 163–178.

WRIGHT, R. T. and J. E. HOBBIE. 1965. The uptake of organic solutes in lake water. *Limnol. Oceanogr.* **10**, 22–28.

WRIGHT, R. T. and J. E. HOBBIE. 1966. Use of glucose and acetate by bacteria and algae in aquatic ecosystems. *Ecology*, **47**, 447–464.

WURSTER, C. F. 1968. DDT reduces photosynthesis by marine phytoplankton. *Science*, **159**, 1474–1475.

WURSTER, C. F. and D. B. WINGATE. 1968. DDT residues and declining reproduction in the Bermuda petrel. *Science*, **159**, 979–981.

YAMADA, M. and T. OTA. 1970. Studies on the lipid of plankton. IV. Unsaponifiable matter of lipid of *Calanus plumchrus*. *J. Japan Oil Chem. Soc.* **19**, 377–382. (Fish. Res. Bd. Canada Trans. No. 1590.)

YENTSCH, C. S. 1965. Distribution of chlorophyll and phaeophytin in the open ocean. *Deep-Sea Res.* **12**, 653–666.

YENTSCH, C. S. 1973. Remote sensing for productivity in pelagic fisheries. *Nature*, **244**, 307–308.

YENTSCH, C. S. and J. H. RYTHER. 1957. Short-term variations in phytoplankton chlorophyll and their significance. *Limnol. Oceanogr.* **2**, 140–142.

YENTSCH, C. S. and J. H. RYTHER. 1959. Absorption curves of acetone extracts of deep water particulate matter. *Deep-Sea Res.* **6**, 72–74.

YENTSCH, C. S. and R. W. LEE. 1966. A study of photosynthetic light reactions, and a new interpretation of sun and shade phytoplankton. *J. Mar. Res.* **24**, 319–337.

YETKA, J. E. and W. J. WIEBE. 1974. Ecological application of antibiotics as respiratory inhibitors of bacterial populations. *Appl. Microbiol.* **28**, 1033–1039.

YINGST, J. Y. 1976. The utilization of organic matter in shallow marine sediments by an epibenthic deposit-feeding holothurian. *J. Exp. Mar. Biol. Ecol.* **23**, 55–69.

YOUNG, D. K. and D. C. RHOADS. 1971. Animal-sediment relations in Cape Cod Bay, Massachusetts. I. A transect study. *Mar. Biol.* **11**, 242–254.

YOUNGBLOOD, W. W., M. BLUMER, R. L. GUILLARD and F. FIORE. 1971. Saturated and unsaturated hydrocarbons in marine benthic algae. *Mar. Biol.* **8**, 190–201.

YONGE, C. M. 1928. Feeding mechanisms in the invertebrates. *Biol. Rev.* **3**, 21–76.

ZAIKA, V. E. 1970. Relationship between the productivity of marine molluscs and their life span. *Oceanology* **10**, 547–552.

ZAITZEV, YU. P. 1961. Surface pelagic biocoenose of the Black Sea. *Zool. Zh.* **40**, 818–825.

ZEITZSCHEL, B. 1965. The sedimentation of seston: an investigation of the biological content of sinking material and sediments in the western and middle Baltic. *Kiel. Meeresforch.* **21**, 55–80.

ZEITZSCHEL, B. 1970. The quantity, composition and distribution of suspended particulate matter in the Gulf of California. *Mar. Biol.* **7**, 305–318.

ZENKEVITCH, L. A. 1963. *Biology of the Seas of the U.S.S.R.*, Allen & Unwin Ltd., London. 955 pp.

ZEUTHEN, E. 1953. Oxygen uptake as related to body size in organisms. *Quart. Rev. Biol.* **28**, 1–12.

ZEUTHEN, E. 1970. Rate of living as related to body size in organisms. *Polskie Archiwum Hydrobiologii*, **17**, 21–30.

ZIEGELMEIER, E. 1970. Über massenvorkommen verschiedener makrobenthaler Wirbelloser während der Wiederbesiedlungsphase nach Schädigungen durch 'katastrophale' Umwelteinflüsse. *Helgoländer wiss. Meeresunters*. **21**, 9–20.

ZOBELL, C. E. 1946. *Marine Microbiology*, Chronica Botanica, Waltham, Mass. 240 pp.

ZOBELL, C. E. 1962. Geochemical aspects of the microbial modification of carbon compounds. In *Advances in Organic Geochemistry*, Pergamon Press, London, pp. 1–18.

ZOBELL, C. E. 1968. Bacterial life in the deep sea. In *Proceedings of the U.S.–Japan Seminar on Marine Microbiology*, August 1966 in Tokyo. *Bull. Misaki Marine Biol. Kyoto Inst. Univ.*, No. 12, 77–96.

INDEX